Rolf-Peter Holzapfel

Ball and Surface Arithmetics

Aspects of Mathematics

Edited by Klas Diederich

*A Publication of the Max-Planck-Institut für Mathematik, Bonn

Rolf-Peter Holzapfel

Ball and Surface Arithmetics

Prof. Dr. Rolf-Peter Holzapfel
Humboldt-Universität Berlin
Mathematisch-Naturwissenschaftliche Fakultät II
Institut für Mathematik
Unter den Linden 6
D-10117 Berlin
Germany
holzapfl@mathematik.hu-berlin.de

Die Deutsche Bibliothek - CIP-Einheitsaufnahme

Holzapfel, Rolf-Peter:
Ball and surface arithmetics / Rolf-Peter Holzapfel. -
Braunschweig; Wiesbaden: Vieweg, 1998
(Aspects of mathematics: E; Vol. 29)

Mathematics Subject Classification:
14 02, 14 J xx

Vieweg is a subsidiary company of the Bertelsmann Professional Information.

http://www.vieweg.de

Cover design: Wolfgang Nieger, Wiesbaden

ISSN 0179-2156

ISBN 978-3-322-90171-2 ISBN 978-3-322-90169-9 (eBook)
DOI 10.1007/978-3-322-90169-9

Contents

Preface

This monograph is based on the work of the author on surface theory connected with ball uniformizations and arithmetic ball lattices during several years appearing in a lot of special articles. The first four chapters present the heart of this work in a self-contained manner (up to well-known basic facts) increased by the new functorial concept of orbital heights living on orbital surfaces. It is extended in chapter 6 to an explicit HURWITZ theory for CHERN numbers of complex algebraic surfaces with the mildest singularities, which are necessary for general application and proofs. The chapter 5 is dedicated to the application of results in earlier chapters to rough and fine classifications of PICARD modular surfaces. For this part we need additionally the arithmetic work of FEUSTEL whose final results are presented without proofs but with complete references. We had helpful connections with Russian mathematicians around VENKOV, VINBERG, MANIN, SHAFAREVICH and the nice guide line of investigations of HILBERT modular surfaces started by HIRZEBRUCH in Bonn. More recently, we can refer to the independent (until now) study of Zeta functions of PICARD modular surfaces in the book [L-R] edited by LANGLANDS and RAMAKRISHNAN. The basic idea of introducing arrangements on surfaces comes from the monograph [BHH], (BARTHEL, HÖFER, HIRZEBRUCH) where linear arrangements on the complex projective plane $\mathbb{P}^2$ play the main role. Our monograph can be understood as a broad generalization of results presented in [BHH] for the important special case of (locally) finite coverings of the basic surface $\mathbb{P}^2$ whose branch loci consist of linear arrangements. Together with [HO4] our book is part of a trilogy. Our first book [HO4] gives relations with special systems of algebraic differential equations of PICARD-FUCHS-GAUSS-MANIN type. The second book [HO8] studies mainly special values of certain PICARD modular forms related with class fields and transcendence theory. I have to thank Mrs. B. Wüst, Mrs. D. Protzek, Mr. U. Bellack for their successful fight with new typing technics, and also Ms. A. Hegewald for her skillful setting of approx. 300 diagrams and drawings.

Berlin, January 1998 R.-P. Holzapfel

Introduction

The construction of (compact real) RIEMANN surfaces C is the starting point of fruitful ideas in mathematics. On the one hand, each such surface appears as finite covering of the compactified complex GAUSS plane $\mathbb{P}^1$, on the other hand it is a compactified quotient $\mathbb{D}/\Gamma$ of the complex unit disc $\mathbb{D}$ by a suitable discrete subgroup Γ (non-euclidean lattice) of the analytic automorphism group of $\mathbb{D}$. For the classification of RIEMANN surfaces the genus or equivalently, the EULER number, is the most important invariant. It connects algebraic properties, analytic integrals, metrics and topology (number of holes) with each other. For the calculation of this invariant the HURWITZ genus formula is most important. This formula reduces the genus calculation of a finite $\mathbb{P}^1$-covering to the determination of points and indices of ramification. The genus of the algebraic compactification of $\mathbb{D}/\Gamma$ is closely connected with a non-euclidean integral on a fundamental domain of Γ on the disc $\mathbb{D}$.

In higher dimensions the RIEMANN-ROCH theory is the most important tool for connecting analytic, algebraic and topological properties of manifolds. The monograph is concentrated to the complex dimension 2, more precisely to complex algebraic surfaces. In the smooth compact case there are two basic invariants: the EULER number c_2 and the signature. They are connected with arithmetic genus and canonical selfintersection c_1^2 of a canonical divisor by elementary formulas (proved by non-elementary RIEMANN-ROCH theory). Both kinds of constructions of RIEMANN surfaces described above, work also in higher dimensions, especially in complex dimension 2. The basic surface for finite coverings is the complex projective plane $\mathbb{P}^2$. Via general projections it is clear that each projective surface is a finite cover of $\mathbb{P}^2$.

The most natural domain generalizing $\mathbb{D}$ for the quotient construction is the complex unit ball $\mathbb{B}$. As symmetric domain it has a nice invariant hermitian metric well-known to differential geometers as BERGMANN metric. It has negative constant sectional curvature. Quotient surfaces $\mathbb{B}/\Gamma', \Gamma'$ a neat ball lattice, inherit this nice EINSTEIN-KÄHLER metric immediately in obvious manner. These are very special quasi-projective surfaces. Admitting also ball lattices Γ, which are not neat, one gets a much greater class of quasi-projective surfaces. Together with their compactifications we call them ball quotient surfaces. It is a working hypothesis or philosophy of the author that, up to birational equivalence and compactifications, all complex algebraic surfaces are ball quotients. No contradicting argument is known, at least to the author. On the other hand it is a fact that

Theorem 0.0.1 *On each projective complex surface there exists a* ZARISKI-*open part supporting an* EINSTEIN-KÄHLER *metric with negative constant sectional curvature coming from the ball.*

Proof Namely, we know that the complex projective plane $\mathbb{P}^2$ is a ball quotient surface. The branch locus of the corresponding locally finite covering $\mathbb{B} \longrightarrow \mathbb{P}^2 - \{4 \text{ points}\}$ consists of 6 projective lines (minus the 4 points), see Proposition 5.1.3. The complement of the 6 lines inherits the ball metric. Omitting branch and ramification loci this metric can be lifted to a ZARISKI-open part of each finite cover of $\mathbb{P}^2$.

□

Our point of view emphasizes the important role of finite coverings in surface theory. It is also clear that the philosophy forces to a productive work with non-smooth algebraic surfaces. The main role play quotient and (ball) cusp singularities. The latter come from compactifications of non-compact ball quotient surfaces, by definition. For a finite covering $f : X \to Y$ of two quasi-smooth surfaces it is not possible in general to resolve simultaneously the (quotient) singularities of X and Y preserving the finite covering relation. But with the trick of GALOIS closure it is not difficult to see that we can shrink the class of singularities by suitable simultaneous modifications. On this way our numerical studies of finite coverings can be reduced to surfaces with at most HIRZEBRUCH-JUNG (or cyclic) singularities and to GALOIS coverings.

The main purpose of the monograph is the introduction of special rational invariants, called orbital heights, and to work with them producing several interesting results. These heights are postulated to be invariant with respect to finite coverings up to a well-determined factor. The idea does not work for surfaces in the usual sense. A finer understanding is necessary. For this purpose orbital surfaces are introduced. These are complex surfaces together with an arrangement. An arrangement is a potentially assumed branch locus whose components consisting of (singular) points and irreducible curves are endowed with natural weights and rational coefficients. The only use of cycles instead seems, unfortunately, not fine enough for our aims. In order to find such heights of orbital surfaces one has to prove a lot of functorial properties going inductively through the categories of abelian points, orbital curves, orbital surfaces and the corresponding categories of relative objects. The diagram (D.4.3) in chapter 6 connects the functorial properties of several local and global invariants needed as contributions for orbital heights. It connects GALOIS theory, a fine arithmetic of singularities, weight shifting, reductions, localizations, base changes, additive and multiplicative properties with each other, going through all six categories just mentioned. This fundamental diagram is delegated to the last chapter because for its understanding the reader should be familiar at least with the first three chapters of the book, where the basic notions, relations and invariants are introduced stepwise for the cases of local and global GALOIS coverings.

In the first chapter we define the category of abelian points. These are embedded cyclic surface singularities together with two weighted embedded curve germs intersecting at this point. Most important are finite morphisms and modifications in this category. We refer to part 1.1 - 1.18 of the contents in order to indicate the type of arithmetic games we have to play for finding the 0-local height contributions. Weighted (electronic) graphs of abelian points are introduced. They classify abelian points and store the contributions in a convenient manner. In the short chapter 2 we introduce orbital curves as surface germes along a weighted compact curve supporting abelian points. They form again a category with the important class of finite morphisms. Based on the results of the first chapter we are able to find two kinds of orbital heights: the EULER heights and the signature heights of orbital curves. The first generalizes the EULER number and the second the selfintersection index of a curve on a smooth surface. Essentially for the first deeper functorial understanding including a geometric Local-Global Principle for orbital curves is section 2.3. Using graphs of abelian points we define star-like weighted (atomic) graphs of orbital curves. The rational EULER and signature heights can be read off from them.

The functorial procedure is inductively extended from the smaller dimensions to orbital surfaces with natural arrangements using only natural weights and coefficients. These natural orbital surfaces are sufficient to understand the GALOIS part of functorial properties defining EULER and signature heights on this way. In 3.5 we introduce quotient points as surface germ around a quotient singularity together with three weighted curve germs through it. Cusp points are defined in a similar manner using four weighted curve germs instead of three intersecting at a cusp singularity. Both types of these new orbital points are classified by means of atomic graphs. Allowing also quotient and cusp points we introduce open orbital surfaces supported by the open complement of all cusp singularities of the underlying compact surface. The arithmetic functorial game with invariants extends to open orbital surfaces defining the two above heights for them. Graphs of orbital curves and points are connected to introduce weighted (molecular) graphs of orbital surfaces or, more precisely, of the corresponding arrangements. They store the 0- and 1-local contributions for the calculation of EULER and signature heights. Additionally, one needs the CHERN numbers of the supporting surfaces and of the normalized irreducible curves of the arrangements. In the last section we present a formula containing the heights of the orbital curve components of arrangements as (weighted) summands. Alltogether, we can say that the existence proof for the two above heights combines the arithmetic of singularities with RIEMANN-ROCH and, more generally, with equivariant K-theory in a new functorial manner using geometric Local-Global Principles.

In the next chapter we understand ball quotient surfaces $\mathbb{B}/\Gamma, \Gamma$ a ball lattice, as open orbital surfaces denoted by $\mathbf{B}/\Gamma$. The support of the corresponding (open) arrangement $A(\Gamma)$ is the branch locus of the infinite GALOIS covering $\mathbb{B} \longrightarrow \mathbb{B}/\Gamma$. The preimages of the 0- or 1-components are Γ-elliptic points or Γ-reflection discs, respectively. The weights come from ramifiation indices. They can be read off from the stationary subgroups of elliptic points, cusps or the discs, respectively. Up to a universal factor, the heights of each curve component of $A(\Gamma)$ turn out to be non-euclidean volumes of a fundamental domain in the covering disc with respect to its normalizing subgroup of Γ. Similarly, the heights of $\mathbf{B}/\Gamma$ appear, up to universal factors again, as volume of a Γ-fundamental domain with respect to the ball metric. On this way one gets via universal factors universal relations between EULER and signature heights of $\mathbf{B}/\Gamma$ and of the components of $A(\Gamma)$. This is an extension of HIRZEBRUCH-MUMFORD'S proportionality theory in the case of the ball including all ball lattices now. The proportionality relations yield a strong criterion for (open) orbital surfaces to be ball quotients. We hope that it is precise. The proofs connect HIRZEBRUCH-MUMFORD'S proportionality theory for neat lattices with a relative variant for subdiscs of the ball, due to the author, and the results of the previous chapters.

Because of the above identities of heights and volumes, the data of the arrangement $A(\Gamma)$ stored in its molecular graph allow to turn the CHERN-invariants of compact models of $\mathbb{B}/\Gamma$ into the volume of a Γ-fundamental domain and vice versa. The last three sections are dedicated to first applications. For given suitable point-curve configuration A on a fixed surface X we can endow the components with weights such that A becomes an arrangement $\mathbf{A}$ and X an orbital surface $\mathbf{X} = (X, \mathbf{A})$. There are infinitely many possibilities in general. In order to get an orbital ball quotient surface the above proportionality relations have to be satisfied. By means of the orbital decomposition of heights the relations are transformed to a system of diophantine equations for the unknown weights. We prove that this system can be solved effectively in any case. Moreover, it turns out that there are at most finitely many solutions. This is the effective Finiteness Theorem of 4.10. We apply it to some simple blowing up's of $\mathbb{P}^2$ and the preimage A of the complete quadrihedral consisting of six lines on $\mathbb{P}^2$ in a special position. All ball quotient arrangements supported by A can be found in 4.11. On the one hand this example demonstrates the use of molecular graphs. On the other hand we know by some work of PICARD, TERADA, MOSTOW, DELIGNE that we get really ball quotient surfaces with monodromy lattices corresponding to certain classical systems of differential equations of PICARD-FUCHS-GAUSS-MANIN type. In the last section we interpret the Finiteness Theorem as existence criterion for (at most finitely many) ball metrics on a given open algebraic surface.

A further field of application is the theory of PICARD modular surfaces. Such surfaces are defined as (compactified) quotients $\mathbb{B}/\Gamma$ by certain arithmetic ball lattices Γ connected with imaginary quadratic number fields. They have rather complicated branch arrangements. The corresponding elliptic points and reflection discs together with stationary groups have been found by a hard number theoretic work of FEUSTEL during several years. The proofs would fill another monograph. So we present only the results and apply them to the rough KODAIRA classification of PICARD modular surfaces. Knowing the arrangements and heights we are able to calculate the CHERN numbers. The heights being essentially volumes of fundamental domains are calculated in the Appendix 5.A with number theoretic methods. It gives also an impression of the local height part of proofs in FEUSTEL'S work. We start in the first sections demonstratively with the fine classification of the PICARD modular surfaces of EISENSTEIN and GAUSS numbers. In section 5.4 we prove that almost all PICARD modular surfaces are of general type with the fractional invariant c_1^2/c_2 coming arbitrary near to the value 3, which is extremely possible in surface theory. Examples of rational, elliptic and $K3$ surfaces are presented in section 5.5 using suitable sublattice of the full PICARD modular groups.

The historical order is the following: We started with 5.2, 5.3. Then 4.1 - 4.9 have been developed in close connection with chapter 1 and 5.3 - 5.5. Then the idea was born to generalize the multiplicative property of volumes of fundamental domains of lattices with respect to sublattices to orbital surface invariants with respect to GALOIS coverings. It leads to the introduction of our heights, arrangements, orbital surfaces. This general surface language, independently of discs and balls, enabled us to attack the theorems of 4.9 - 4.12. The next step was to extend the functorial ideas of heights to a more general class of orbital surfaces (with rational coefficients at arrangenent components) with the first aim to present explicit HURWITZ formulas for CHERN invariants. This program is fullfilled in sections 6.1 - 6.9 for finite coverings of algebraic surfaces with (at most) cyclic singularities. In the last section 6.10 we present the following application with a glance to the well-known BOGOMOLOV-MIYAOKA-YAU Theorem: $c_1^2/c_2 \leq 3$ for smooth compact complex algebraic surfaces of general type. We prove that the sharper inequality $c_1^2/c_2 \leq 2$ holds for all surfaces allowing a finite projection onto a ruled surface or onto $\mathbb{P}^2$ with a nice branch locus consisting of smooth transversally intersecting curves.

Suggestions. The author believes that the functorial ideas of constructing orbital heights can be inductively extended also to higher dimensions. In [HOV], chapter 6 we established precise dimension formulas for spaces of cusp forms of ball lattices Γ with a proof in the spirit of chapters 4 and 5. These dimensions can be read off also in a universal manner for each weight from the molecular graph of the arrangement $A(\Gamma)$ knowing additionally the volumes of fundamental domains. There are good reasons to assume that a functorial transfer of the dimension formulas in connection with logarithmic pluricanonical genera of open surfaces in the style of the chapters 1, 2, 3 and 6 is possible. It would produce an infinite series of heights and, maybe, explicit HURWITZ formulas for (logarithmic) pluricanonical genera with respect to finite surfaces coverings. At the end we ask for possibilties to construct orbital surfaces and heights for surfaces over fields of finite characteristic p and for arithmetical surfaces.

1 Abelian Points

1.1 Cyclic Points

We consider complex representations of finite groups G

$$r : G \longrightarrow \mathbb{G}\mathrm{l}(V), \quad \dim_{\mathbb{C}}(V) = 2,$$

of rank 2.

The representation r is called *faithful*, if $\ker(r) = 1$. $\gamma \in G$ is called a *reflection* of r, if $\gamma \notin \ker(r)$ and there exists a line L (one-dimensional subspace) in V such that $r(\gamma)|_L = id_L$. The subgroup $S(r)$ of G generated by all reflections of r and by $\ker(r)$ is a normal subgroup. We will consider sometimes the vector space V as affine space isomorphism to $\mathbb{A}^2(\mathbb{C})$.

Then the quotient space $V/r(G)$ is well-defined. It is an affine algebraic variety with at most one singularity (the image of the point $O \in V$). Surface singularities which are isomorphic (in the analytic or algebraic sense) to a singularity of a suitable quotient space $V/r(G)$ are called *quotient singularities.*

Lemma 1.1.1 (CHEVALLEY [CHV]) .
$V/r(G)$ is smooth if and only if G is generated by r-reflections and by $\ker(r)$, *that means that $S(r) = G$. In this case there is an (algebraic) isomorphism $V \xrightarrow{\sim} V/r(G)$.*

Any of our representations r induces a faithful representation

$$r' : G' = G/\ker(r) \longrightarrow \mathbb{G}\mathrm{l}(V).$$

We call r a *small representation*, if $S(r) = 1$. Each small representation is faithful. To any of our representations r we correspond a small representation $\overline{r}$ in the following manner. We have an exact sequence

$$1 \longrightarrow S(r) \longrightarrow G \longrightarrow \overline{G} = G/S(r) \longrightarrow 1.$$

The quotient space $\overline{V} = V/S(r)$ is smooth by CHEVALLEY's lemma and can be endowed with the structure of a two-dimentional vector space such that $\overline{G}$ acts linearly on V. The corresponding representation is denoted by $\overline{r}$. It is called the *reduction* of r:

$$\overline{r} : \overline{G} \longrightarrow \mathbb{G}l(\overline{V}). \tag{1.1.1}$$

The representation r is called *cyclic (abelian)* or *central* if $r(G)$ is a cyclic (abelian) subgroup of $\mathbb{G}l_2(V)$ or $r(G)$ lies in the center $Zl(V)$ of $\mathbb{G}l(V)$, respectively:

$$Zl(V) \cong Zl_2(\mathbb{C}) = Zl(\mathbb{C}^2) = \left\{ \begin{pmatrix} \zeta & 0 \\ 0 & \zeta \end{pmatrix} \in \mathbb{G}l_2(\mathbb{C}); \zeta \in \mathbb{C}^* \right\}$$

Lemma 1.1.2 *Let G be a finite abelian group and $r : G \longrightarrow \mathbb{G}l(V)$ a small representation of rank two. Then r is a cyclic representation. Consequently, G is a cyclic group.*

Proof Since r is faithful we can w.l.o.g. assume that G is a subgroup of $\mathbb{G}l(V) \cong \mathbb{G}l_2(\mathbb{C})$. Any abelian representation is diagonalizable, that means that $V \cong V_1 \oplus V_2$ splits into two G-invariant lines V_1, V_2 and $r = r_1 + r_2$, $r_i : G \longrightarrow \mathbb{G}l(V_i) \cong \mathbb{G}l_1(\mathbb{C})$. The images $r_i(\gamma), \gamma \in G$, can be identified with unit roots. We compare the cyclic groups $r_1(G) = \langle\zeta_d\rangle$ and $r_2(G) = \langle\zeta_k\rangle$, where ζ_m denotes a primitive unit root of order m. Since r is faithful there is a group isomorphism

$$\begin{array}{ccc} G & \xrightarrow{\sim} & (\mathbb{Z}/d\mathbb{Z}) \times (\mathbb{Z}/k\mathbb{Z}) \\ g & \longmapsto & (m,n) \bmod (d,k) \\ & \searrow \quad (\zeta_d^m, \zeta_k^n) \quad \swarrow & \end{array}$$

Since r is small it cannot happen that $\zeta_d^m = 1$ and $\zeta_k^n \neq 1$. Consequently k divides d (and d divides k), hence

$$(\zeta_d, \zeta_k) = (\zeta_d, \zeta_d^e) = \left(\zeta_k^{e'}, \zeta_k\right) = \left((\zeta_d^e)^{e'}, \zeta_d^e\right) \tag{1.1.2}$$

and $r(G)$ is cyclic. □

Definition 1.1.3 Two representations $r : G \to \mathbb{G}l(V)$ and $s : H \to \mathbb{G}l(W)$ are said to be *bi-equivalent*, if there exist an isomorphism of groups $h \colon G \xrightarrow{\sim} H$ and a linear isomorphism $\lambda : V \xrightarrow{\sim} W$ such that all diagrams:

$$\begin{array}{ccc} V & \xrightarrow[\lambda]{\sim} & W \\ {\scriptstyle r(\gamma)}\downarrow & & \downarrow{\scriptstyle s(h(\gamma)),\ \gamma \in G} \\ V & \xrightarrow[\lambda]{\sim} & W \end{array} \tag{1.1.3}$$

are commutative. If $H = G$, $h = \mathrm{id}_G$ and the corresponding diagrams (1.1.3) are commutative, then r and s are said to be *equivalent.*

For $G = H$ and $V = W$ the representations r and s are equivalent if and only if $r(G)$ and $s(G)$ are conjugated in $\mathbb{G}\mathrm{l}(V)$. Let $\mathfrak{G}$ be a set of representatives of all isomorphy classes of finite groups. The (discrete) space of all (complex) faithful representations of rank 2 of elements of $\mathfrak{G}$ is denoted by $\widehat{\mathfrak{G}}$, and $\widehat{\mathfrak{G}}(V)$ is the subspace of representations in V. On $\widehat{\mathfrak{G}}(V)$ acts $\mathbb{G}\mathrm{l}(V)$ by conjugation. We have the following biunivoque correspondence:

$$\begin{aligned} \widehat{\mathfrak{G}}/\text{equ.} &\Longleftrightarrow \widehat{\mathfrak{G}}(V)/\text{equ.} \Longleftrightarrow \widehat{\mathfrak{G}}(V)/\mathbb{G}\mathrm{l}(V) \\ &\Longleftrightarrow \widehat{\mathfrak{G}}(\mathbb{C}^2)/\mathbb{G}\mathrm{l}_2(\mathbb{C}) \\ &\Longleftrightarrow \left\{ \begin{array}{c} \text{conjugacy classes of} \\ \text{finite subgroups of } \mathbb{G}\mathrm{l}_2(\mathbb{C}) \end{array} \right\} \end{aligned} \tag{1.1.4}$$

The correspondences restrict to finite abelian and finite cyclic groups $Z_d = \mathbb{Z}/d\mathbb{Z}, d \in \mathbb{N}$. The set of all cyclic groups Z_d is denoted by $\mathfrak{Z}$, and $\widehat{\mathfrak{Z}}$ denotes the corresponding subspace in $\widehat{\mathfrak{G}}$ of all faithful representations of rank 2 of the cyclic groups Z_d; $\widehat{\mathfrak{G}}^{\text{small}}$ and $\widehat{\mathfrak{Z}}^{\text{small}}$ are the subspaces of small representations. With (1.1.4) and the proof of Lemma 1.1.2 we obtain the following correspondences:

$$\begin{aligned} \widehat{\mathfrak{Z}}/\text{equ.} &\Longleftrightarrow \widehat{\mathfrak{Z}}(V)/\mathbb{G}\mathrm{l}(V) \Longleftrightarrow \widehat{\mathfrak{Z}}(\mathbb{C}^2)/\mathbb{G}\mathrm{l}_2(\mathbb{C}) \\ &\Longleftrightarrow \{\text{pairs } \{\zeta, \zeta^e\}; \zeta \text{ unit root}, 0 \le e < \operatorname{ord}\zeta\} \\ &\qquad \text{(pairs of eigenvalues of a fixed generator)} \\ \widehat{\mathfrak{Z}}^{\text{small}}/\text{equ.} &\Longleftrightarrow \widehat{\mathfrak{Z}}^{\text{small}}(\mathbb{C}^2)/\mathbb{G}\mathrm{l}_2(\mathbb{C}) \\ &\Longleftrightarrow \left\{ \begin{array}{c} \text{pairs } \{\zeta, \zeta^e\};\ \zeta \text{ unit root}, \\ 0 \le e < \operatorname{ord}\ \zeta, \gcd(e, \operatorname{ord}\ \zeta) = 1 \end{array} \right\} \\ \widehat{\mathfrak{Z}}/\text{bi-equ.} &\Longleftrightarrow \text{group autom.} \setminus \widehat{\mathfrak{Z}}(\mathbb{C}^2)/\mathbb{G}\mathrm{l}_2(\mathbb{C}) \\ &\Longleftrightarrow \{\text{pairs } (d, e); d, e \in \mathbb{N}, d \ge 1, \end{aligned}$$

$$
\begin{aligned}
& 0 \le e < d\} / e \cdot e' \qquad \equiv 1 \bmod d \\
\widehat{\mathfrak{Z}}^{\text{small}}/\text{bi-equ.} \Longleftrightarrow & \{(d,e); d,e \in \mathbb{N}, d \ge 1, 0 \le e < d, \\
& \gcd(d,e) = 1\} / e \cdot e' \equiv 1 \bmod d
\end{aligned} \tag{1.1.5}
$$

The equivalence $/e \cdot e' \equiv 1 \bmod d$ means that two pairs (d,e) and (d',e') have to be identified if and only if $d = d'$ and $e \cdot e' \equiv 1 \bmod d$ (cp. with 1.1.2). We want to remove this double equivalence by introduction of distinguished representations in the following sense:

Definition 1.1.4 Let A be a finite abelian group. An *arranged representation* of A (complex, rank 2) is a pair $\rho = (r, \mathbb{C}\mathfrak{a})$ consisting of a faithful representation $r : A \longrightarrow \mathbb{G}\mathrm{l}(V)$ of rank 2 and a line (eigenline) through an eigenvector $0 \neq \mathfrak{a} \in V$ of the representation.

Let $\mathcal{A} \subset \mathcal{G}$ be a complete representative set of representations of isomorphism classes of finite abelian groups and $\widehat{\mathcal{A}} \subset \widehat{\mathcal{G}}$ the space of faithful representations of rank 2 of (elements of) $\mathcal{A}$. The space of arranged representations of $\mathcal{A}$ is denoted by $2\widehat{\mathcal{A}}$ because we have in general two possibilities to arrange a representation $r \in \widehat{\mathcal{A}}$.

Definition 1.1.5 Two arranged representations $(r, \mathbb{C}\mathfrak{a})$ and $(s, \mathbb{C}\mathfrak{b})$ are said to be ***bi-equivalent*** (***equivalent***) if and only if r and s are bi-equivalent (equivalent) in the sense of Definition (1.1.3) and the linear isomorphism λ in the diagrams (1.1.3) sends $\mathbb{C}\mathfrak{a}$ to $\mathbb{C}\mathfrak{b}$.

Remark 1.1.6 Central representations have not been doubled. In this case each line of the representing space V is an eigenline and each linear automorphism of V is compatible with the representation. All other representations have been doubled.

The notations $2\widehat{\mathcal{A}}^{\text{ small}}, 2\widehat{\mathfrak{Z}}, 2\widehat{\mathfrak{Z}}^{\text{ small}}, 2\widehat{\mathcal{A}}(V), 2\widehat{\mathfrak{Z}}^{\text{ small}}(V), \ldots$ are used in obvious manner. The following biunivoque correspondences are immediate from (1.1.5):

$$
\begin{aligned}
2\widehat{\mathfrak{Z}}/\text{equ.} \Longleftrightarrow & \; 2\widehat{\mathfrak{Z}}(V)/\mathbb{G}\mathrm{l}(V) \Longleftrightarrow 2\widehat{\mathfrak{Z}}(\mathbb{C}^2)/\mathbb{G}\mathrm{l}_2(\mathbb{C}) \\
\Longleftrightarrow & \; \{(\zeta, \zeta^e);\ \zeta \text{ unit root, } 0 \le e < \operatorname{ord} \zeta\}
\end{aligned}
$$

$$
2\widehat{\mathfrak{Z}}^{\text{small}}/\text{equ.} \Longleftrightarrow 2\widehat{\mathfrak{Z}}(\mathbb{C}^2)/\mathbb{G}\mathrm{l}_2(\mathbb{C}) \Longleftrightarrow
$$

$$\Longleftrightarrow \left\{ \begin{array}{c} (\zeta, \zeta^e);\ \zeta \text{ unit root}, 0 \leq e < \operatorname{ord}\zeta, \\ \gcd(e, \operatorname{ord}\zeta) = 1 \end{array} \right\}$$

$$\begin{aligned} 2\widehat{\mathfrak{Z}}/\text{bi-equ.} &\Longleftrightarrow \text{group autom.} \setminus 2\widehat{\mathfrak{Z}}(\mathbb{C}^2)/\mathbb{G}\mathrm{l}_2(\mathbb{C}) \\ &\Longleftrightarrow \{(d, e); d \geq 1, 0 \leq e < d\} \end{aligned}$$

$$2\widehat{\mathfrak{Z}}^{\text{small}}/\text{bi-equ.} \Longleftrightarrow \{(d, e); d \geq 1, 0 \leq e < d, \gcd(d, e) = 1\} \tag{1.1.6}$$

The quotient interpretations on the right hand sides allow to consider the space of (bi-) equivalence classes on the left-hand sides as discrete analogon of quotient spaces. We will consider functions on $2\widehat{\mathfrak{Z}}$ with integral values which are invariant w.r.t. the group action (like automorphic functions), and we will be able to define derivatives of them.

Definition 1.1.7 The elements of the spaces $\widehat{\mathfrak{Z}}^{\text{small}}/\text{bi-equ.}$, $2\widehat{\mathfrak{Z}}^{\text{small}}/\text{bi-equiv.}$ are called *cyclic points* or *arranged cyclic points*, respectively. The elements of $\widehat{\mathfrak{Z}}\left(2\widehat{\mathfrak{Z}}, \widehat{\mathfrak{Z}}^{\text{small}}, 2\widehat{\mathfrak{Z}}^{\text{small}}\right)$ are called (*small, arranged*) *cyclic representations*.

Remark 1.1.8 It is useful to correspond $2\widehat{Z}^{\text{small}}/\text{bi-equ.}$ via (1.1.6) with the rational boundary points of the POINCARÉ upper half plane $\mathbb{H}$ modulo $\mathbb{Z}$:

$$2\widehat{Z}^{\text{small}}/\text{bi-equ.} \Longleftrightarrow \partial_{\mathbb{Q}}\mathbb{H}/\mathbb{Z}\left((d, e) \mapsto \frac{e}{d}\right).$$

There will be a cooperation with DEDEKIND's η-function (see sections 1.8 - 1.11).

1.2 Graphs of Abelian Points

Let $\rho = (r, L)$ be an arranged cyclic representation. By (1.1.6) we correspond to ρ, via bi-equivalence in a unique manner a pair (d, e) of natural numbers, $0 \leq e < d$. Preparing geometric interpretations we correspond to ρ the weighted arrow

$$\underset{\langle d,e\rangle}{\longrightarrow\!\circ} \quad \text{or} \quad \underset{\langle d,e\rangle}{\circ\!\longrightarrow\!\circ} \tag{1.2.1}$$

endowed with (a) small circle(s) at the end(s). This arrow is called the *graph* of ρ or of the corresponding arranged cyclic point. By (1.1.6) we have a biunivoque correspondence between arranged cyclic points and graphs (1.2.1).

We want to define now opposite arranged representations. For each representation $r : A \longrightarrow \mathrm{Gl}(V)$ of finite groups there exists a (positive definit) hermitian metric on V such that the representation is a unitary one. Let A be abelian, r as above and $\rho = (r, L)$ an arranged representation. We denote by $L^\perp$ the orthogonal complement of L in V, $\dim V = 2$. Then $\rho^\perp = (r, L^\perp)$ is called the *opposite arranged representation* of ρ. The definition is unique, if r is not central. It is unique up to equivalence in any case. If (1.2.1) is the graph of the arranged cyclic representation ρ, then the graph of the opposite arranged representation $\rho^\perp$ is:

$$\underset{\langle d, e' \rangle}{\circ\!\!\longleftarrow} , \quad e \cdot e' \equiv 1 \bmod d \tag{1.2.2}$$

which is called the *opposite graph* of (1.2.1).

Going down to bi-equivalence classes the map $\perp$ defines an involution on $2\hat{Z}^{\text{small}}$/bi-equ., $2\hat{Z}$/bi-equ. or, more generally, on the space $2\hat{A}$/bi-equ. of arranged abelian points:

$$\mathbf{P} \mapsto \mathbf{P}^\perp$$

We want to define now the graph of an arranged abelian representation $\rho = (r, L)$, $r : A \to \mathrm{Gl}(V)$. We remember to the exact sequence

$$1 \longrightarrow S(A) \longrightarrow A \longrightarrow \overline{A} \longrightarrow 1, \tag{1.2.3}$$

where $S(A)$ is the subgroup generated by the r-reflections of A (see (1.1.1)). The representation r induces a small representation $\overline{r} : \overline{A} \longrightarrow \mathrm{Gl}(\overline{V}), \overline{V} = V/r(S(A))$, which is cyclic by Lemma 1.1.2. The image $\overline{L} \subset \overline{V}$ of L arranges $\overline{r}$ to $\overline{\rho} = (\overline{r}, \overline{L})$. Let

$$\underset{\langle d, e \rangle}{\longrightarrow\!\!\circ}$$

be the graph of $\overline{\rho}$. Then we define the *graph of* ρ by

$$\rho: \ \underset{v \ \langle d, e \rangle \ w}{\circ\!\!\longrightarrow\!\!\circ}$$

We have to explain v and w. We denote by $S_L(A)$ the group of L-reflections:

$$S_L(A) = \{\gamma \in S(A); \gamma | L = id_L\}$$

Then v, w are the orders of $S_L(A)$ or of $S_{L^\perp}(A)$, respectively. Obviously, the graph of the opposite arranged abelian representation (point) is the following one:

$$\text{Graph of } \rho^{\perp}: \underset{v\ \langle d,e'\rangle\ w}{\circ\!\longleftarrow\!\!\!-\!\!-\!\!\circ}\ , \quad e\cdot e' \equiv 1 \bmod d.$$

Forgetting the arranging line we correspond to abelian point $\mathbf{R} \in \widehat{\mathcal{A}}/$ bi-equ. represented by r the

$$\text{Graph of } \mathbf{R} = \text{Graph of } r: \underset{v\ \langle d,e\rangle\ w}{\circ\!\!-\!\!-\!\!-\!\!\circ} = \underset{v\ \langle d,e'\rangle\ w}{\circ\!\!-\!\!-\!\!-\!\!\circ}$$

$$(e\cdot e' \equiv 1 \bmod d) \qquad \| \qquad\qquad \|$$

$$\underset{w\ \langle d,e\rangle\ v}{\circ\!\!-\!\!-\!\!-\!\!\circ} = \underset{w\ \langle d,e'\rangle\ v}{\circ\!\!-\!\!-\!\!-\!\!\circ}$$

All the four versions have to be identified.

Looking back to (arranged) cyclic representations we have the following list of identifications:

$$\underset{\langle d,e\rangle}{\circ\!\!-\!\!-\!\!\!\longrightarrow\!\circ} = \underset{v\ \langle \bar{d},\bar{e}\rangle}{\circ\!\!-\!\!-\!\!\!\longrightarrow\!\circ} = \underset{v\ \langle \bar{d},\bar{e}\rangle\ 1}{\circ\!\!-\!\!-\!\!\!\longrightarrow\!\circ} \quad \text{if } S_L(A) \cong Z_v, \ {\scriptstyle v=\text{g.c.d.}(d,e),\ d=v\bar{d},\ e=v\bar{e};}$$

$$\underset{\langle d,e\rangle}{\circ\!\!-\!\!-\!\!\!\longrightarrow\!\circ} = \underset{\langle \bar{d},\bar{e}\rangle\ v}{\circ\!\!-\!\!-\!\!\!\longrightarrow\!\circ} = \underset{1\ \langle \bar{d},\bar{e}\rangle\ v}{\circ\!\!-\!\!-\!\!\!\longrightarrow\!\circ} \quad \text{if } S_{L^{\perp}}(A) \cong Z_v$$

$$\underset{\langle d,e\rangle}{\circ\!\!-\!\!-\!\!-\!\!\circ} = \underset{v\ \langle \bar{d},\bar{e}\rangle}{\circ\!\!-\!\!-\!\!-\!\!\circ} = \underset{v\ \langle \bar{d},\bar{e}\rangle\ 1}{\circ\!\!-\!\!-\!\!-\!\!\circ} = \underset{1\ \langle \bar{d},\bar{e}\rangle\ v}{\circ\!\!-\!\!-\!\!-\!\!\circ} = \underset{\langle \bar{d},\bar{e}\rangle\ v}{\circ\!\!-\!\!-\!\!-\!\!\circ}$$

$$\underset{\langle v\bar{d},v\bar{e}'\rangle}{\circ\!\!-\!\!-\!\!-\!\!\circ} = \underset{v\ \langle \bar{d},\bar{e}'\rangle}{\circ\!\!-\!\!-\!\!-\!\!\circ} = \underset{v\ \langle \bar{d},\bar{e}'\rangle\ 1}{\circ\!\!-\!\!-\!\!-\!\!\circ} = \underset{1\ \langle \bar{d},\bar{e}'\rangle\ v}{\circ\!\!-\!\!-\!\!-\!\!\circ} = \underset{\langle \bar{d},\bar{e}'\rangle\ v}{\circ\!\!-\!\!-\!\!-\!\!\circ}$$

with $\bar{e}\cdot\bar{e}' \equiv 1 \bmod \bar{d}$.

For (arranged) cyclic points coming from small cyclic representations v is equal to 1, and it can be omitted in the graphs above.

1.3 Geometric Interpretation

We correspond to a cyclic point $\mathbf{R}$ represented by $r : Z_d \longrightarrow \mathrm{Gl}(V)$ a special quotient singularity called *cyclic singularity* or HIRZEBRUCH-JUNG *singularity* $Q(\mathbf{R}) = Q(r)$. By Definition 1.1.7 $r(Z_d)$ is a small cyclic subgroup of $\mathrm{Gl}(V)$. The singularity $Q(r)$ is defined to be the image point of $O \in V$ along the quotient map

$$\begin{array}{ccc} V & \longrightarrow & V/r\,(Z_d)\,, \\ \Psi & & \Psi \\ O & \longmapsto & Q(r) = Q(\mathbf{R}) \end{array}$$

If r' is another representative of $\mathbf{R}$, then $r(Z_d)$ and $r'(Z_d)$ are conjugated in $\mathbb{Gl}(V)$. Therefore, the analytic spaces (affine algebraic varieties) $V/r(Z_d)$ and $V/r'(Z_d)$ are isomorphic. We will identify (analytically) isomorphic singularities. This justifies the notation $Q(\mathbf{R})$.

Proposition 1.3.1 (see e.g. [BRI]) *Two (two-dimensional) cyclic singularities are (analytically or algebraically) isomorphic if and only if they come from conjugated small cyclic subgroups of* $\mathbb{Gl}(V) \cong \mathbb{Gl}_2(\mathbb{C})$.

□

The two subgroups $r(Z_d)$ and $r'(Z_d)$ of small cyclic representations of Z_d are conjugated in $\mathbb{Gl}(V)$ if and only if r and r' are bi-equivalent. So there is a biunivoque correspondence between the set of cyclic singularities and the space $\mathfrak{Z}^{\text{small}}/\text{bi-equ.}$ of cyclic points. Moreover, $(\partial_{\mathbb{Q}}\mathbb{H}/\mathbb{Z})/\bot$ appears as "moduli space" of cyclic singularities by Proposition 1.3.1 and Remark (1.1.8)

Now we can attach more generally to each faithful abelian representation $r \in \widehat{\mathcal{A}}(V)$ a cyclic singularity in the same manner as above:

$$V \longrightarrow V/r(A),$$
$$O \mapsto Q(r) = Q(\mathbf{R}), \mathbf{R} = r/\text{bi-equ.}, r : A \to \mathbb{Gl}(V).$$

The exact sequence (1.2.3) decomposes our quotient map into

$$V \longrightarrow \overline{V} = V/S(A) \longrightarrow \overline{V}/\overline{r}(\overline{A}) \tag{1.3.1}$$

with the notations used above. Since $\overline{r}$ is a small cyclic representation it follows that $Q(r)$ is really a cyclic singularity. We have the commutative reduction diagram:

$$\begin{array}{ccc} \widehat{\mathfrak{A}} & \xrightarrow{\text{reduction}} & \widehat{\mathfrak{Z}}^{\text{small}} \\ \downarrow & & \downarrow \\ \widehat{\mathfrak{A}}/\text{bi-equ.} & \longrightarrow & \widehat{\mathfrak{Z}}^{\text{small}}/\text{bi-equ.} \\ \downarrow & & \downarrow \\ \{\text{cyclic singularities}\} & \Longleftrightarrow & \{\text{cyclic points}\} \\ \Updownarrow & & \Updownarrow \\ \left\{\begin{array}{l}\text{graphs of cyclic} \\ \text{singularities}\end{array}\right\} & \underset{\text{Df}}{=} & \left\{\begin{array}{l}\text{graphs of cyclic} \\ \text{points}\end{array}\right\} \end{array} \tag{1.3.2}$$

A cyclic singularity is called *of type* $\langle d, e\rangle$, if $\overset{\circ\longrightarrow\circ}{1\langle d,e\rangle 1}$ is the graph of the corresponding cyclic point. Each cyclic singularity has (in general) two types, namely $\langle d, e\rangle$ and $\langle d, e'\rangle, e \cdot e' \equiv 1 \bmod d$. The cyclic singularity of type $\langle 1, 0\rangle$ is the regular point.

Now, we want to find a natural geometric interpretation of arranged representations. Let $\rho = (r, L) \in 2\widehat{A}, r : A \to \mathbb{G}l(V)$ be a faithful abelian representation. The eigenlines L and $L^\perp$ are understood as *ordered* coordinate axis $(L, L^\perp)$ in V. So they define a coordinate system uniquely up to scaling factors, if r is not central. If r is central then we can work with arbitrary coordinate axis and we will pick out one. The reduction $V \to \overline{V}$ (see (1.3.1)) corresponds to $(L, L^\perp)$ a coordinate system $(\overline{L}, \overline{L}^\perp)$ on $\overline{V}$. We will endow the lines $\overline{L}, \overline{L}^\perp$ with weights:

$$v = v(\overline{L}) = \text{ order of } S_L(A), w = v\left(\overline{L}^\perp\right) = \text{ order of } S_{L^\perp}(A)$$

The weighted coordinate system is denoted by $(v\overline{L}, w\overline{L}^\perp)$. The lines $\overline{L}$ and $\overline{L}^\perp$ project along the second quotient map of (1.3.1) onto curves C or $C^\perp$, respectively, intersecting each other in the cyclic singularity $Q(r)$. We call $(vC, wC^\perp)$ the *weighted singular coordinate system* of r at $Q(r)$. Let $U(r)$ be an open analytic neighbourhood of $Q(r)$. The quadruple $(U(r), Q(r); vC', wC'^\perp)$, $C' = C \cap U(r)$, $C'^\perp = C^\perp \cap U(r)$, is called *arranged abelian point attached to* ρ.

We will identify arranged abelian points attached to r, if we change the neighbourhood $U(r)$. Moreover, we call any quadruple $\mathbf{Q} = (U, Q; vD, wD')$, which is analytically and weight preserving isomorphic to an arranged abelian point attached to a suitable $\rho \in 2\widehat{A}$, an *arranged abelian point.* We identify also the analytic isomorphic objects with the same weights and call, more precisely, the corresponding equivalence classes *arranged abelian points.*

The following picture illustrates how to get an arranged abelian point from a faithful representation of rank 2 of a finite abelian group A (cp. with (1.3.1)):

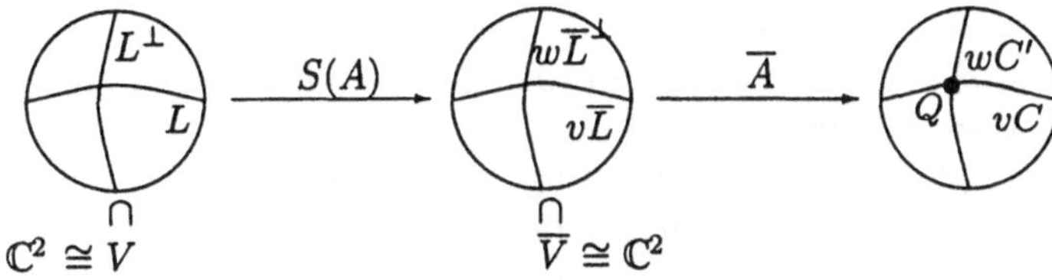

Figure 1.3.1

The following commutative diagram is closely connected with diagram 1.3.2:

$$\begin{array}{ccc}
2\widehat{\mathfrak{A}} & \xrightarrow{\text{forget arranging}} & \widehat{\mathfrak{A}} \\
\downarrow & & \downarrow \\
2\widehat{\mathfrak{A}}/\text{bi}-\text{equ.} & \longrightarrow & \widehat{\mathfrak{A}}/\text{bi}-\text{equ.} \\
\Updownarrow & & \Updownarrow \\
\left\{\begin{array}{l}\text{arranged}\\ \text{abelian points}\end{array}\right\} & \longrightarrow & \{\text{abelian points}\} \\
\Updownarrow & & \Updownarrow \\
\left\{\begin{array}{l}\text{graphs of arranged}\\ \text{abelian points}\end{array}\right\} & \xrightarrow[\text{(vertex of arrow)}]{\text{forget the order}} & \left\{\begin{array}{l}\text{graphs of}\\ \text{abelian points}\end{array}\right\} \\
\underset{v \quad \langle d,e\rangle \quad w}{\circ\!\!\longrightarrow\!\!\circ} & \longmapsto & \underset{v \quad \langle d,e\rangle \quad w}{\circ\!\!-\!\!\!-\!\!\!-\!\!\circ}
\end{array} \tag{1.3.3}$$

Remark 1.3.2 We come to the same end with finite abelian subgroups of the analytic automorphism groups of a two-dimensional complex analytic manifold fixing a point 0 because by a theorem of H. CARTAN [CAR] there exists an analytic coordinate system around 0 such that the group acts linearly with respect to these coordinates.

Remark 1.3.3 We dispose now on two interpretations of (arranged) abelian points: a geometric one as illustrated in Figure 1.3.1 and via representation $\left(2\widehat{\mathcal{A}} \text{ or } \widehat{\mathcal{A}}/\text{bi-equ.}\right)$. For both kinds of these points we will use the same name. From the text it will be clear what kind of point we work with. Both interpretations will be important and should be kept in mind by the reader.

1.4 Derived Representations

We have given a precise definition of bi-equivalence classes of representations. So we will identify frequently and tacitely bi-equivalent representations. Especially we will often restrict our considerations to finite abelian subgroups A of $\mathbb{G}l_2(\mathbb{C})$ and the representing space $\mathbb{C}^2$ with the natural coordinate axis $L_x(x-axis)$, $L_y(y-axis)$ for arranging. Then A appears as subgroup of the diagonal group $D_2(\mathbb{C}) = \left\{\begin{pmatrix} a & 0 \\ 0 & b \end{pmatrix}; a, b \in \mathbb{C}^*\right\}$. The reader should be familiar with the σ–process (see e.g. [SC2], IV, § 3).

Let $\rho = (r, L_x)$ be an arranged small cyclic representation, say of type $\langle d,e\rangle$. We want to define its derivations ρ_x and ρ_y. We identify r with the subgroup $Z_{d,e} = \left\langle \begin{pmatrix} \zeta_d & 0 \\ 0 & \zeta_d^e \end{pmatrix} \right\rangle$ of $D_2(\mathbb{C})$, ζ_d a primitive unit root of order d.

Step 1

We consider the groups $Z = \left\langle \begin{pmatrix} \zeta_d & 0 \\ 0 & \zeta_d \end{pmatrix} \right\rangle$ and $S = \left\langle \begin{pmatrix} 1 & 0 \\ 0 & \zeta_e \end{pmatrix} \right\rangle$ and define the abelian group $A = S \cdot Z$ generated by S and Z. The corresponding arranged representation on $\mathbb{C}^2$ is denoted by ρ_A.

Step 2

We blow up the point $0 \in \mathbb{C}^2$ to the exceptional line L:

$$\begin{array}{ccc} \tilde{\mathbb{C}}^2 & \longrightarrow & \mathbb{C}^2 \\ \cup & & \Cup \\ L & \longrightarrow & O \end{array}$$

The proper transforms of the axis are denoted by L_x or L_y, respectively, as before, and their intersection points with L by 0_x or 0_y, respectively.

Step 3

We extend the representation ρ_A to an action of A on $\tilde{\mathbb{C}}^2$. A acts on L, L_x and L_y. Looking at the tangent spaces in 0_x and 0_y the action produces two representations ρ_0 and ρ_1 around the points 0_x and 0_y arranged by L_x or L, respectively.

Lemma-Definition 1.4.1 *The reduction $\overline{\rho}_A$ of ρ_A is equal to the arranged representation ρ we started with. The reduction $\overline{\rho}_1$ of ρ_1 is of type $\langle e, -d\rangle$.*
It is called the derived representation of ρ and is denoted by ρ_x.

Proof Using coordinate systems $(x, \frac{y}{x})$ and $\left(\frac{x}{y}, y\right)$ in 0_x or 0_y, respectively, we find the following presentation of the indicated generators of S and Z:

$$\begin{pmatrix} 1 & 0 \\ 0 & \zeta_e \end{pmatrix}, \begin{pmatrix} \zeta_d & 0 \\ 0 & 1 \end{pmatrix} \text{ in } 0_x, \begin{pmatrix} \zeta_e^{-1} & 0 \\ 0 & \zeta_e \end{pmatrix}, \begin{pmatrix} 1 & 0 \\ 0 & \zeta_d \end{pmatrix} \text{ in } 0_y. \tag{1.4.1}$$

We denote the arranged cyclic (quotient) points by fat letters. With obvious notations we have the following commutative diagram:

$$
\begin{array}{ccccccc}
 & \mathbf{O}_y \in & \widetilde{\mathbb{C}}^2 & \longrightarrow & \mathbb{C}^2 & \ni & \mathbf{O} \\
 & & \cup\!\!\!\!\cup & & & & \\
 & & \mathbf{O}_x & & & & \\
\downarrow Z & \downarrow S & \downarrow S & \searrow & \downarrow & & \downarrow S \\
\overline{\mathbf{O}}{}'_y & \overline{\mathbf{O}}_y & \overline{\mathbf{O}}_x \in \widetilde{\mathbb{C}}^2/S & \xrightarrow{\sigma_e} & \mathbb{C}^2/S & \ni & \overline{\mathbf{O}} \\
\downarrow S & \downarrow Z & \downarrow Z & \swarrow & \downarrow & & \downarrow Z \\
 & & \mathbf{Q}_x & & & & \\
 & & \cap & & & & \\
\longrightarrow & \mathbf{Q}_y \in & \widetilde{\mathbb{C}}^2/A & \xrightarrow{\overline{\sigma}_e} & \mathbb{C}^2/A & \ni & \mathbf{P}
\end{array}
\qquad (1.4.2)
$$

The diagram of the cyclic points looks like

$$
\begin{array}{ccccc}
 & & \circ\!\!\longleftarrow\!\!\bullet\!\!\longrightarrow\!\!\circ & \xrightarrow{\sigma} & \longrightarrow\!\!\circ \\
 & & \langle 1,0\rangle \quad \langle 1,0\rangle & & \langle 1,0\rangle \\
 & \nearrow Z & \downarrow S & & \downarrow S \\
\circ\!\!\longleftarrow & & \circ\!\!\longleftarrow\!\!\bullet\!\!\longrightarrow\!\!\circ & \xrightarrow{\sigma_e} & \longrightarrow\!\!\circ \\
\langle 1,0\rangle & & \langle e,e-1\rangle \quad \langle 1,0\rangle & & \langle 1,0\rangle \\
 & \searrow S & \downarrow Z & & \downarrow Z \\
 & & \circ\!\!\longleftarrow\!\!\bullet\!\!\longrightarrow\!\!\circ & \xrightarrow{\overline{\sigma}_e} & \longrightarrow\!\!\circ \\
 & & \langle e,-d\rangle \quad \langle 1,0\rangle & & \langle d,e\rangle
\end{array}
\qquad (1.4.3)
$$

The filled circle represents L, their quotient $L_e = L/S$ or $\overline{L}_e = L/A$, respectively. An arrow is joined with $\circ$ if the corresponding point lies on one of these curves. We have to find the types of the quotient points indicated in diagram (1.4.3). All regular points represented by the graphs $\underset{\langle 1,0\rangle}{\longrightarrow\!\!\circ}$ come from quotients by reflection groups (CHEVALLEY's Lemma 1.1.1). They are immediately discovered by means of (1.4.1). In $\overline{\mathbf{O}}{}'_y$ and $\overline{\mathbf{O}}$ we have the coordinate systems $\left(\frac{x}{y}, y^d\right)$ or (x, y^e), respectively, represented by invariant functions. Now the types are immediately determined. □

We called in [HO1] the map σ_e a σ_e-process. It can be applied to any smooth surface point $0 \in X$. The point is substituted by a smooth projective line $L_e \simeq \mathbb{P}^1(\mathbb{C})$ and on this line there exists exactly one surface singularity Q_e (if $e > 1$), which is a cyclic singularity of type $\langle e, e-1\rangle$:

$$\begin{array}{ccc} X_e & \longrightarrow & X \\ \cup & & \Cup \\ Q_e \in L_e & \longrightarrow & O \end{array} \tag{1.4.4}$$

The σ_1-process is the σ-process.

The map $\overline{\sigma}_e$ will be used for resolving cyclic singularitues. The type $\langle e, -d\rangle$ of Q_y is simpler than the type $\langle d, e\rangle$ of Q because $e < d$, and we have to understand $-d$ as $-d \bmod e$.

Definition 1.4.2 Let ρ be a small arranged cyclic representation, ρ_x its derivative, $\mathbf{P}$, $\mathbf{P}_x$ the correspondingly arranged cyclic points and $\rho^\perp(\mathbf{P}^\perp)$ the opposite representation (point). The *derivatives* ρ_y, $\mathbf{P}_y$ are defined to be

$$\rho_y = \left((\rho^\perp)_x\right)^\perp, \ \mathbf{P}_y = \left((\mathbf{P}^\perp)_x\right)^\perp. \tag{1.4.5}$$

Now we can define also *higher derivatives* of ρ and $\mathbf{P}$ by induction:

$$\rho_{z_1 \cdots z_n z_{n+1}} = (\rho_{z_1\cdots z_n})_{z_{n+1}}, z_i = x \text{ or y} .$$

So we get a triangle (1.4.6)

$$\begin{array}{c} \rho_x \\ \rho_{xx}\ \rho_{xy}\ \rho_{yy} \\ \rho_{xxx}\ \rho_{xxy}\ \rho_{xyy}\ \rho_{yyy} \\ \text{-----------} \\ \rho_{xx\cdots xx} \quad \cdots \quad \rho_{x\cdots xy\cdots y} \quad \cdots \quad \rho_{y\cdots y} \end{array} \tag{1.4.6}$$

which is finite because the order of the corresponding cyclic groups become lower and lower by Lemma 1.4.1. It would be more precise to write the triangle with $\mathbf{P}_{x\cdots xy\cdots y}$ instead of $\rho_{x\cdots xy\cdots y}$. By our convention identifying bi-equivalent representations both triangles are understood to be equal. The triangle (1.4.6) has the following properties:

(i) All mixed derivatives $\rho_{z_1\cdots z_k}, z_i = x$ or y, occur in the triangle

(ii) There is a number l such that the l-th row consists only of the trivial representations $1 = Z_1 \longrightarrow \mathrm{Gl}(\mathbb{C}^2)$ corresponding to the smooth point, and the row before does not contain the trivial representation.

So, derivatives of ρ of higher degree $m > l$ are trivial. We omit them and say that the triangle (1.4.6) has the *height* l and ρ has the *lenght* l.

The first property follows from

Lemma 1.4.3 *The derivations*

$$\begin{array}{rcl}\delta_x,\ \delta_y:\ 2\widehat{Z}/\text{bi-equ.} & \longrightarrow & 2\widehat{Z}/\text{bi-equ.} = \{\text{arranged cyclic points}\} \\ \rho & \mapsto & \rho_x \text{ or } \rho_y, \text{respectively}\end{array}$$

commute with each other.

Proof We have to show that

$$\rho_{xy} = \rho_{yx}. \tag{1.4.7}$$

First we assume that ρ is not of type $\langle 1,0\rangle$ or $\langle d,1\rangle$. Identifying representations or, more precisely, their bi-equivalence classes with their types, we start with

$$\rho = \langle d,e\rangle \text{ and } \rho^\perp = \langle d,e'\rangle,\ e,e' > 1.$$

We calculate ρ_{yx} by means of Lemma 1.4.1

$$\begin{aligned}\rho_y &= \left(\left(\rho^\perp\right)_x\right)^\perp = \langle e',-d\rangle^\perp = \langle e',g\rangle \\ \rho_{yx} &= \langle g,-e'\rangle = \langle g,h\rangle, \\ g(-d) &\equiv 1 \bmod e',\ 0 \le g < e'; \\ h &\equiv -e' \bmod g,\ 0 \le h < g.\end{aligned} \tag{1.4.8}$$

The number g can taken away from the equation

$$ee' + g(-d) = 1,\ 0 < g < e,e' \tag{1.4.9}$$

which is uniquely solvable because $ee' \equiv 1 \bmod \mathrm{d}$. The estimations are automatically satisfied. Now we calculate ρ_{xy}.

$$\begin{aligned}\rho_x &= \langle e,-d\rangle = \langle e,f\rangle, \\ (\rho_x)^\perp &= \langle e,f'\rangle = \langle e,g\rangle, \\ f &\equiv -d \bmod e \\ f'f &\equiv f'\cdot(-d) \equiv 1 \bmod e\end{aligned} \tag{1.4.10}$$

$$\rho_{xy} = \left(\left((\rho_x)^\perp\right)_x\right)^\perp = (\langle e,g\rangle_x)^\perp = \langle g,-e\rangle^\perp = \langle g,k\rangle = \langle g,h\rangle$$
$$k\cdot(-e) \equiv 1 \mod g,\ 0 \leq k < g$$

The number k and h must be equal because they are smaller than g and both inverse to $-e$ mod g in $\mathbb{Z}/g\mathbb{Z}$, namely $h \equiv e'$ mod g , hence

$$h\cdot(-e) \equiv e\cdot e' \equiv 1 \equiv k\cdot(-e) \mod g.$$

Now (1.4.7) follows from (1.4.8) and (1.4.10).

If $\rho = \langle 1,0\rangle$, then $\rho, \rho^\perp$ and all derivatives correspond to the smooth (cyclic) point. For the non-trivial central representations $\rho = \langle d,1\rangle$, $d > 1$, the derivatives ρ_x and ρ_y and all higher derivatives are trivial. □

We introduce the notation $\rho_{x^iy^j} = \rho_{x\cdots xy\cdots y}$. The property (ii) follows from the following statement:

(ii') If $\rho_{x^my^n} = \langle 1,0\rangle$, then $\rho_{x^iy^j} = \langle 1,0\rangle$ for all i, j such that $i+j = l := m+n$.

Proof We can assume that $l \geq 1$. Then we look at the neighbours of $\rho_{x^my^n}$ in the (l-1)-th row of the triangle (1.4.6). It is easy to see that they have to be central, that means of types $\langle d,1\rangle$. The derivations of central representations are trivial. Consequently the neighbours of $\rho_{x^my^n}$ in the l-th row are of trivial type $\langle 1,0\rangle$. This proves the statements (ii') and (ii). □

1.5 The Differential Relation

We look for invariant functions $f(\rho)$ for $\rho \in 2\widehat{Z}^{\text{small}}$ with respect to the $\mathbb{Gl}_2$ -action and automorphisms of the represented cyclic groups Z_d (see 1.1.5). Like automorphic functions they are considered also as functions on $2\widehat{Z}^{\text{small}}/\text{bi-equ}$. The most simplest one is the order $d(\rho) = d$ of the represented cyclic group, $\rho\colon Z_d \hookrightarrow \mathbb{Gl}_2(\mathbb{C})$. The derived representations allow to define derivatives of such functions, especially of $d(\rho)$.

Definition 1.5.1

$$d_{z_1\cdots z_n}(\rho) = d(\rho_{z_1\cdots z_n}),\ z_i = x \text{ or } y.$$

Since $\rho_{z_1\cdots z_n} = \rho_{x\cdots xy\cdots y} = \rho_{x^m y^n}$ by (4.10.) we have also

$$d_{z_1\cdots z_n}(\rho) = d_{x\cdots xy\cdots y}(\rho) = d_{x^m y^n}(\rho) = (\delta_{x^m}\delta_{y^n}(d))(\rho)$$

on $2\widehat{Z}^{\text{small}}$/bi-equ., where we introduced the derivations δ_x and δ_y of our function $d(\rho)$. Obviously

1.5.2 *The derivations δ_x, δ_y commute with each other.*
If we fix ρ, then we write $d_{x^m x^n}$ instead of $d_{x^m x^n}(\rho)$. With this notation each $\rho \in 2\widehat{Z}$ defines a triangle:

$$\begin{array}{lc} 0-\text{th row:} & d \\ & d_x \; d_y \\ & d_{xx} \; d_{xy} \; d_{yy} \\ & d_{x^3} \; d_{x^2y} \; d_{xy^2} \; d_{y^3} \\ & \vdots \qquad\qquad \vdots \\ l\text{-th row:} & 1 = d_{x^l} \; \ldots \; 1 \; \ldots \; 1 \; \ldots d_{y^l} = 1 \end{array} \tag{1.5.1}$$
$$d_{x^iy^j} = 0 \text{ for } i+j > l,\; d_{x^iy^j} > 1 \text{ for } i+j < l,\; l = l(\rho)$$

derived from the triangle (1.4.6).

We want to express now the derivative $d_y(\rho)$ of $d(\rho)$ by means of higher x-derivatives of $d(\rho)$. For this purpose we work with the "logarithmic derivative" $\frac{d_y(\rho)}{d(\rho)}$.

Proposition 1.5.3 *It holds that*

$$\frac{d_y(\rho)}{d(\rho)} = \sum_{i=1}^{l(\rho)} \frac{1}{\delta_x^{i-1}d(\rho)\cdot\delta_x^i d(\rho)} \tag{1.5.2}$$

for any small arranged cyclic representation ρ.

Proof If $l = 0$, then ρ is the trivial representation of type $\langle 1,0\rangle$ and both sides of (1.5.2) are equal to 0. Now let ρ be non-trivial. Then $l = l(\rho) \geq 1$ and we go on by induction over l. By inductive hypothesis we start with the identity

$$\frac{(d_x)_y}{d_x} = \sum_{i=1}^{l-1} \frac{1}{d_{x x^{i-1}} d_{x x^i}}. \tag{1.5.3}$$

Now we consider for $\rho \neq \langle 1,0\rangle$ the matrix

$$M(\rho) = \begin{pmatrix} d_x & d_{xy} \\ d & d_y \end{pmatrix} \in \mathbb{S}\mathrm{l}_2(\mathbb{Z}) \tag{1.5.4}$$

This matrix belongs really yo $\mathbb{S}\mathrm{l}_2(\mathbb{Z})$. Namely, for $\rho = \langle d, e\rangle$, $\rho^\perp = \langle d, e'\rangle$ we have $\rho_y = \langle ee', g\rangle$, $\rho_{xy} = \langle g, h\rangle$ with the notations of the proof of Lemma 1.4.3. Therefore, we can write

$$M(\rho) = \begin{pmatrix} e & g \\ d & e' \end{pmatrix} = \begin{pmatrix} d_x & d_{xy} \\ d & d_y \end{pmatrix}. \tag{1.5.5}$$

The relation (1.4.9) shows that the determinant of $M(\rho)$ is equal to 1. This relation can now be written as

$$d_x d_y - d \cdot d_{xy} = 1 \text{ or } \frac{d_{xy}}{d_x} = \frac{d_y}{d} - \frac{1}{d \cdot d_x}.$$

The latter identity combined with 1.5.3 yields 1.5.2. The proposition is proved. □

Lemma 1.5.4 *For $\rho \in 2\widehat{Z}^{\mathrm{small}}$ and $i,\ j \in \mathbb{N}, i+j \le l = l(\rho)$, the derivative $\rho_{x^i y^j}$ is of type $\langle d_{x^i y^j}, d_{x^{i+1} y^j}\rangle$.*

Proof Let ρ be of type $\langle d, e\rangle$ and $i,\ j = 0$. Then

$$\rho = \langle d, e\rangle = \langle d, d_x\rangle \tag{1.5.6}$$

by (1.5.5). Now we assume that $i + j + 1 \ge l$ and that

$$\rho_{x^i y^j} = \langle d_{x^i y^j}, d_{x^{i+1} y^j}\rangle.$$

We set $\tau = \rho_{x^i y^j} = \langle f, h\rangle$. Then it holds that $h = f_x$ by (1.5.6) and $f = d_{x^i y^j}$ by Definition 1.5.1, hence

$$\begin{aligned} \rho_{x^{i+1} y^j} &= \left(\rho_{x^i y^j}\right)_x = \langle f_x, (f_x)_x\rangle \\ &= \langle d_{x^i y^j x}, d_{x^i y^j x x}\rangle = \langle d_{x^{i+1} y^j}, d_{x^{i+2} y^j}\rangle. \end{aligned} \tag{1.5.7}$$

For calculating $\rho_{x^i y^{j+1}}$ we look first at the matrix $M(\rho)$. From (1.5.5) it follows that

$$\rho_y = \langle e', g\rangle = \langle d_y, d_{xy}\rangle \tag{1.5.8}$$

We substitute ρ by τ and obtain

$$\rho_{x^iy^{j+1}} = \left(\rho_{x^iy^j}\right)_y = \tau_y = \langle (d_{x^iy^j})_y, (d_{x^iy^i})_{xy}\rangle$$
$$= \langle d_{x^iy^{j+1}}, d_{x^{i+1}y^{j+1}}\rangle.$$

So we proved the lemma by induction. □

A counter part to Lemma 1.5.4 is

Lemma 1.5.5 *With the notations of Lemma 1.5.4 and the usual identifications it holds that*

$$\left(\rho^{\perp}\right)_{x^iy^j} = \left(\rho_{x^jy^i}\right)^{\perp} = \langle d_{x^jy^i}, d'_{x^{j+1}y^i}\rangle,$$

where $d'_{x^{j+1}y^i}$ is defined by

$$d'_{x^{j+1}y^i}\cdot d_{x^{j+1}y^i} \equiv 1 \bmod d_{x^jy^i}, 0 \le d'_{x^{j+1}y^i} < d_{x^jy^i}$$

Proof For $i,\ j = 0$, ρ of type $\langle d, e\rangle$, $\rho^{\perp}$ of type $\langle d, e'\rangle$ we have $\left(\rho^{\perp}\right)_{x^iy^j} = \rho^{\perp} = \langle d, e'\rangle = \langle d, d'_x\rangle$ because $e\cdot e' \equiv 1 \bmod d$ and $e = d_x$.

The definition (1.4.5) of ρ_y can be written as $(\rho_y)^{\perp} = (\rho^{\perp})_x$. We substitute ρ by $\rho^{\perp}$ and obtain $(\rho^{\perp})_y = (\rho_x)^{\perp}$. Both identities show that the first identity of (1.5.9) holds for $i + j = 1$. They imply also the general case by induction:

$$\begin{aligned}(\rho^{\perp})_{x^{i+1}y^j} &= ((\rho^{\perp})_{x^iy^j})_x = (\rho_{x^jy^i})^{\perp})_x = ((\rho_{x^jy^i})_y)^{\perp} = \rho_{x^jy^{i+1}},\\ (\rho^{\perp})_{x^iy^{j+1}} &= ((\rho^{\perp})_{x^iy^j})_y = ((\rho_{x^jy^i})^{\perp})_y = ((\rho_{x^jy^i})_x)^{\perp} = (\rho_{x^{j+1}y^i})^{\perp}.\end{aligned} \tag{1.5.9}$$

The second identity follows from $\rho_{x^jy^i} = \langle d_{x^jy^i}, d_{x^{j+1}y^i}\rangle$ (Lemma 1.5.4) and (1.2.2). □

As a summary we illustrate the relations by means of the following type triangles (1.5.10), (1.5.11) of ρ or $\rho^{\perp}$, respectively.

$$\begin{array}{ccc} & \rho & \\ & \| & \\ & \langle d, d_x \rangle & \\ & \langle d_x, d_{xx} \rangle \ \langle d_x, d_{xx} \rangle & \\ & \cdots & \\ & \langle d_{x^{l-1}}, d_{x^l} \rangle \ \ldots \ \langle d_{y^{l-1}}, d_{xy^{l-1}} \rangle & \\ \langle d_{x^l}, 0 \rangle & \cdots & \langle d_{y^l}, 0 \rangle \\ \| & & \| \\ \langle 1, 0 \rangle & & \langle 1, 0 \rangle \end{array} \tag{1.5.10}$$

$$\begin{array}{ccc} & \rho^\perp & \\ & \| & \\ & \langle d, d'_x \rangle & \\ & \langle d_y, d'_{xy} \rangle \ \langle d_x, d'_{xx} \rangle & \\ & \cdots & \\ & \langle d_{y^{l-1}}, d'_{xy^{l-1}} \rangle \ \ldots \ \langle d_{x^{l-1}}, d'_{x^l} \rangle & \\ \langle d_{y^l}, 0 \rangle & \cdots & \langle d_{x^l}, 0 \rangle \\ \| & & \| \\ \langle 1, 0 \rangle & & \langle 1, 0 \rangle \end{array} \tag{1.5.11}$$

1.6 Stepwise Resolutions of Cyclic Points

We change over now to the geometric interpretation of arranged cyclic points as singularities with a singular coordinate system $(U, Q; C, C') = \mathbf{Q}$ (see Figure 1.3.1). We restrict ourselves to arranged small representations ρ of cyclic groups $A \cong Z_d$. Let ρ and $\mathbf{Q} = \mathbf{Q}(\rho)$ be of type $\langle d, e \rangle$. The σ_e-process induces a resolution step $\overline{\sigma}_e$ introduced in the proof of 1.4.1 and described in:

$$\begin{array}{ccc} L & \longrightarrow & O \\ \cap & & \Cap \\ W_e & \xrightarrow{\sigma_e} & W \\ \downarrow{\scriptstyle \langle d,e\rangle} & & \downarrow{\scriptstyle \langle d,e\rangle} \\ U_1 & \xrightarrow{\overline{\sigma}_e} & U \\ \cup & & \Cup \\ \overline{L} & \longrightarrow & Q \end{array} \tag{1.6.1}$$

where W is a suitable neighbourhood of O in $\mathbb{C}^2$ and $\langle d, e\rangle$ indicate quotient morphisms by the corresponding cyclic subgroups. The diagram (1.6.1) is illustrated in the following picture:

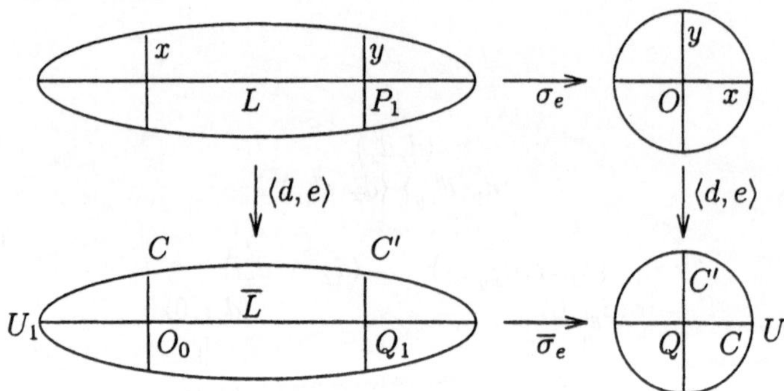

Figure 1.6.1

The only singular points are:

$$\begin{aligned} \mathbf{P}_1 : \ \langle e, e-1\rangle, \ \mathbf{Q}_1 = \rho_x : \ \langle e, -d\rangle &= \langle d_x, d_{xx}\rangle \\ \mathbf{Q} : \ \langle d, e\rangle &= \langle d, d_x\rangle 1.6.3 \end{aligned} \tag{1.6.2}$$

(see (1.4.2) and (1.4.3)). We adapt from (1.4.3) the diagram description:

$$\begin{array}{ccc} \overset{L_1 = L}{\circ\!\!\longleftarrow\!\!\bullet\!\!\longrightarrow\!\!\circ} & \xrightarrow{} & \longrightarrow\!\!\circ \\ \langle 1,0\rangle \ \ \langle e, e-1\rangle & \sigma_e & \langle 1,0\rangle \\ \downarrow & & \downarrow \\ \overset{\overline{L}_1 = \overline{L}}{\circ\!\!\longleftarrow\!\!\bullet\!\!\longrightarrow\!\!\circ} & \xrightarrow{} & \longrightarrow\!\!\circ \\ \langle 1,0\rangle \ \ \langle d_x, d_{xx}\rangle & \overline{\sigma}_e & \langle d, d_x\rangle \end{array} \tag{1.6.3}$$

The same procedure applies to $\mathbf{Q}_1$. The arranged cyclic point $\mathbf{Q}_1$ splits into two arranged cyclic points $\mathbf{Q}_1$ and $\mathbf{Q}_2$ joined by a projective line $\overline{L_2}$. The types of these points are $\langle 1,0\rangle$ or $\langle d_{xx}, d_{xxx}\rangle$, respectively. In our diagram language the lower row of (1.6.3) has been extended to:

$$\overset{C}{\circ}\!\longleftarrow\!\overset{\overline{L}_1}{\bullet}\!\longrightarrow\!\overset{\overline{L}_2}{\bullet}\!\longrightarrow\!\underset{\langle d_{x^2}, d_{x^3}\rangle}{\overset{C'}{\circ}} \;\Longrightarrow\; \circ\!\longrightarrow\!\overset{\overline{L}_1}{\bullet}\!\longrightarrow\!\underset{\langle d_x, d_{xx}\rangle}{\circ} \;\Longrightarrow\; \longrightarrow\!\underset{\langle d, d_x\rangle}{\circ} \qquad (1.6.4)$$

The diagram part $\overset{\overline{L}_1}{\bullet}\!\longrightarrow\!\overset{\overline{L}_2}{\bullet}$ means that the two curves $\overline{L_1}$, $\overline{L_2}$ intersect each other transversally in a regular point. The symbol $\langle 1,0\rangle$ for a regular point has been omitted. After $l = l(\rho)$ steps of the same kind we obtain a linear graph (1.6.5). We write L_i instead of $\overline{L_i}$.

$$\overset{C}{\circ}\!\longrightarrow\!\overset{L_1}{\bullet}\!\longrightarrow\!\overset{L_2}{\bullet}\cdots\longrightarrow\!\overset{L_{l-1}}{\bullet}\!\longrightarrow\!\overset{L_l}{\bullet}\!\longrightarrow\!\overset{C'}{\circ} \qquad (1.6.5)$$

because $\langle d_{x^l}, d_{x^{l+1}}\rangle = \langle 1,0\rangle$ (regular point). This graph corresponds to the configuration Figure 1.6.2 of curves

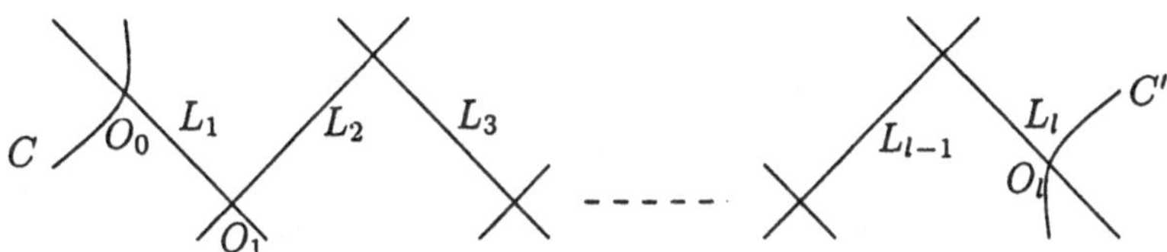

Figure 1.6.2

consisting of l projective lines together with the proper transforms of the singular coordinate axis C, C' interesecting successively and transversally in regular points O_j. They lie on an open analytic surface U_l. There exists an analytic morphism $U_l \longrightarrow U$ contracting $L_1 + \cdots + L_l$ to the starting point Q. Altogether we found a resolution of the singularity $\mathbf{Q}$ by a stepwise procedure which is called the ***stepwise resolution of*** $\mathbf{Q}$.

1.7 Continued Fractions and Selfintersection Numbers

We introduce further bi-equivalent invariants on $\widehat{23}^{small}$. For $r \in \mathbb{R}$ we denote by $\lceil r \rceil$ the smallest integer which is not smaller than r:

$$\lceil r \rceil \in \mathbb{Z}, r \leq \lceil r \rceil < r+1.$$

Let ρ be an arranged small cyclic representation and $d_{x^iy^j} = d_{x^iy^j}(\rho)$. We set

$$\begin{aligned} b_1 = b_1(\rho) = \left\lceil \frac{d}{d_x} \right\rceil = \frac{d}{d_x} + \frac{d_{xx}}{d_x} \quad (\rho \neq \langle 1,0 \rangle) \\ b_i = b_i\ (\rho) = b_1(\rho_{x^{i-1}}) \geq 2, 1 \leq i \leq l(\rho) \end{aligned} \tag{1.7.1}$$

From $\rho_x = \langle d_x, d_{xx} \rangle = \langle d_x, -d \rangle$ we know that $d_{xx} \equiv -d \mod d_x$. Therefore, $\frac{d}{d_x} + \frac{d_{xx}}{d_x}$ is a natural number. Since $0 \leq d_{xx} < d_x$ it is the smallest natural number not smaller than d/d_x, and it is not smaller than 2. The identity (1.7.1) can be written as

$$\frac{d}{d_x} = b_1 - \frac{1}{d_x/d_{xx}} = b_1 - \frac{1}{d(\rho_x)/d_x(\rho_x)} \tag{1.7.2}$$

substituting ρ by ρ_x we get

$$d(\rho_x)/d_x(\rho_x) = b_2 - \frac{1}{d(\rho_{xx})/d_x(\rho_{xx})} \tag{1.7.3}$$

Extending this procedure we can write the inverse logarithmic derivative d/d_x as continued fraction

$$\frac{d}{d_x} = b_1 - \frac{1}{b_2} - \frac{1}{b_3} - \ldots - \frac{1}{b_{l-1}} - \frac{1}{b_l} \tag{1.7.4}$$

Proposition 1.7.1 (HIRZEBRUCH,[HI1]) *The numbers $-b_1, ..., -b_l$ are the selfintersection numbers of $L_1, ..., L_l$ in the minimal resolution Figure 1.6.2 of the arranged cyclic singularity $\mathbf{Q} = \mathbf{Q}(\rho)$.*

Proof First we consider the case $l = 1$, that means that ρ is of type $\langle d, e \rangle = \langle d, d_x \rangle = \langle d, 1 \rangle$. The σ_e-process is nothing else than the σ-process. We have a commutative diagram:

$$\begin{array}{ccccc} L \subset & \widetilde{\mathbb{C}}^2 & \xrightarrow{\sigma} & \mathbb{C}^2 & \ni O \\ & \downarrow Z & & \downarrow Z & \downarrow \\ L_1 = \overline{L} \subset & \widetilde{\mathbb{C}}^2/Z & \xrightarrow{\overline{\sigma}} & \mathbb{C}^2/Z & \ni Q \end{array} \tag{1.7.5}$$

$$Z = \left\langle \begin{pmatrix} \zeta_d & 0 \\ 0 & \zeta_d \end{pmatrix} \right\rangle \subset Zl_2(\mathbb{C})$$

The smooth surface $\tilde{\mathbb{C}}^2/Z$ is the minimal singularity resolution of $\mathbb{C}^2/Z$; and $\tilde{p} : \tilde{\mathbb{C}}^2 \longrightarrow \mathbb{C}^2/Z$ is a purely branched finite covering, branched along L with ramification index d. The exceptional line L has selfintersection number $(L^2) = -1$. Now we apply the degree formula (see e.g. [HAR], App.A)

$$\left(\overline{L}^2\right) = (\deg \tilde{p}) \cdot (L^2)$$

and find that $(L_1^2) = \left(\overline{L}^2\right) = -d = -\frac{d}{1} = -b_1(\rho)$.

In the next step we prove that the result of the stepwise resolution described in Figure 1.6.2 is the minimal resolution of $\mathbf{Q}$. We have to show that all intersection numbers (L_i^2) are smaller than -1. This is true for all ρ of length 1. Assume that $l \geq 2$, and the statement holds for all small cyclic representations of length $l-1$. Then for ρ of length l the derived representation ρ_x has length $l-1$.

By the first resolution step the singularity $\mathbf{Q}(\rho_x) = \mathbf{Q}_1$ sits on L_1 (see Figure 1.6.1). Resolving $\mathbf{Q}_1$ we know from our induction hypothesis that $(L_2^2), ..., (L_l^2) \leq -2$. So we have only to prove that $(L_1^2) < -1$ in Figure 1.6.2. If $(L_1^2) = -1$, then we could blow down L_1 and successively $L_2, L_3, L_4, ..., L_k$ if, additionally, $L_2^2 = ... = L_k^2 = -2, L_{k+1} < -2$. The remaining configuration is described in Figure 1.7.1:

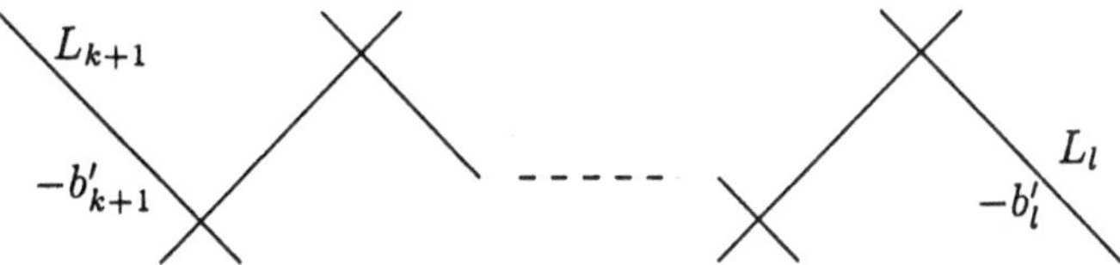

Figure 1.7.1

On the one hand this is the minimal resolution of $\mathbf{Q}$ by construction. On the other hand it is the minimal singularity resolution of the arranged cyclic point $\mathbf{S}$ of type $\langle t, u\rangle$, where t, u are taken from the continued fraction

$$\frac{t}{u} = b'_{k+1} - \frac{1}{b'_{k+2}} - \dots - \frac{1}{b'_l}$$

For this conclusion we used inductively the statement of our proposition. Cyclic points and their minimal resolutions are in bijective correspondence (see e.g. [BRI]). So $\mathbf{Q}$ has to be equal to $\mathbf{S}$. But $\mathbf{S}$ has length $l-(k+1) < 1$ by induction hypothesis. Comparing the lengths of $\mathbf{Q}$ and $\mathbf{S}$ we get a contradiction.

In our last step we assume again that $l \geq 2$ and the proposition is true for all our representations of length $l-1$. Looking at $\mathbf{Q} = \langle d_x, d_{xx} \rangle = \rho_x$ and at (1.7.3) we know the last $l-1$ selfintersection numbers in the resolution of $\mathbf{Q}$, see Figure 1.7.2:

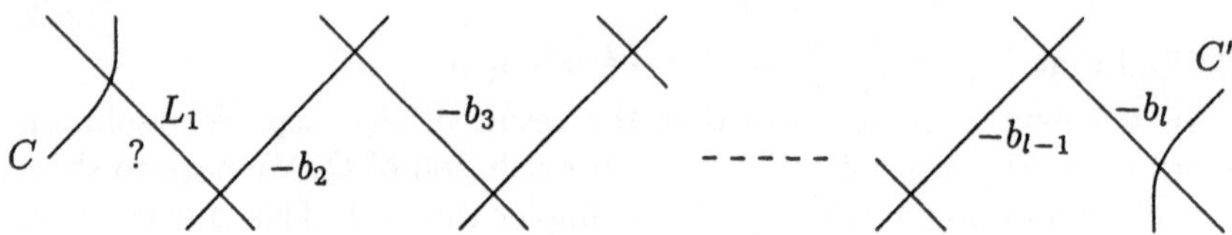

Figure 1.7.2

The cyclic point corresponding to the opposite representation $\rho^\perp$ is the same as that for ρ because we interchanged only the roles of C and C'. Hence $\mathbf{Q}$ and $\mathbf{Q}^\perp$ have the same minimal singularity resolution. So (L_1^2) can be read of from the resolution of $\mathbf{Q}^\perp$ at the last resolving line L', see Figure 1.7.3:

Figure 1.7.3

Thus, the proposition is proved if we have checked the following

Lemma 1.7.2 *With the notations of (1.7.4) for the continued fraction of ρ the continued fraction of $\rho^\perp$ is*

$$\frac{d(\rho^\perp)}{d_x(\rho^\perp)} = b_l - \frac{1}{b_{l-1}} - \ldots - \frac{1}{b_2} - \frac{1}{b_1} \tag{1.7.6}$$

Proof The relation (1.7.2) extends to the matrix decomposition

$$M'(\rho) = \begin{pmatrix} d, & -d_y \\ d_x, & -d_{xy} \end{pmatrix} = \begin{pmatrix} b_1, & -1 \\ 1, & 0 \end{pmatrix} \begin{pmatrix} d_x, & -d_{xy} \\ d_{xx}, & * \end{pmatrix}$$
$$= \begin{pmatrix} b_1, & -1 \\ 1, & 0 \end{pmatrix} \begin{pmatrix} d_x, & -d_{xy} \\ d_{xx}, & -d_{xxy} \end{pmatrix}$$

We used that all occurring matrices lie in $\mathbb{Sl}_2(\mathbb{Z})$ (see (1.5.4)). We apply the decomposition to the last matrix $M'(\rho_x)$ and so on. Finally we get the decomposition

$$M'(\rho) = \begin{pmatrix} d, & -d_y \\ d_x, & -d_{xy} \end{pmatrix} = \begin{pmatrix} b_1, & -1 \\ 1, & 0 \end{pmatrix} \begin{pmatrix} b_2, & -1 \\ 1, & 0 \end{pmatrix} \circ \ldots \circ \begin{pmatrix} b_l, & -1 \\ 1, & 0 \end{pmatrix}$$

Transposition and conjugation with $\begin{pmatrix} -1 & 0 \\ 0 & 1 \end{pmatrix}$ yield

$$M'(\rho^{\perp}) = \begin{pmatrix} d, & -d_x \\ d_y, & -d_{xy} \end{pmatrix} = \begin{pmatrix} b_l & 1 \\ 1 & 0 \end{pmatrix} \cdot \ldots \cdot \begin{pmatrix} b_1 & -1 \\ 1 & 0 \end{pmatrix}$$

The stepwise re-interpretation of this decomposition yields the continued fraction (1.7.6).

Definition 1.7.3 With the notation of Proposition 1.7.1 we call

$$\underset{-b_1}{\bullet} \longrightarrow \underset{-b_2}{\bullet} \cdots \longrightarrow \underset{-b_{l-1}}{\bullet} \longrightarrow \underset{-b_l}{\bullet}$$

the *resolution graph* of the cyclic point $\mathbf{Q} = \mathbf{Q}(\rho)$.

From the graph one obtains back the type $\langle d, d_x \rangle$ by means of the continued fraction (1.7.4). The type triangles (1.5.10), (1.5.11) are translated in this way into the graph triangles (1.7.7), (1.7.8).

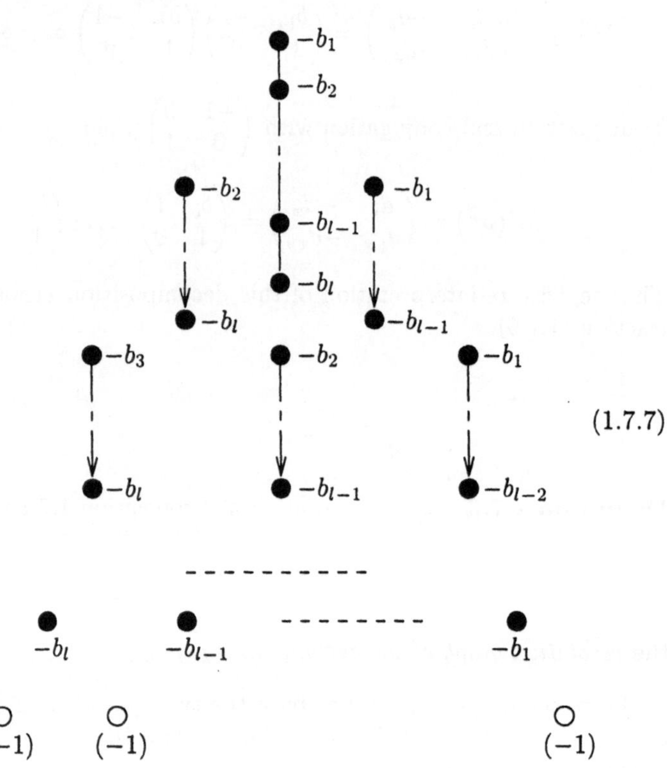

(1.7.7)

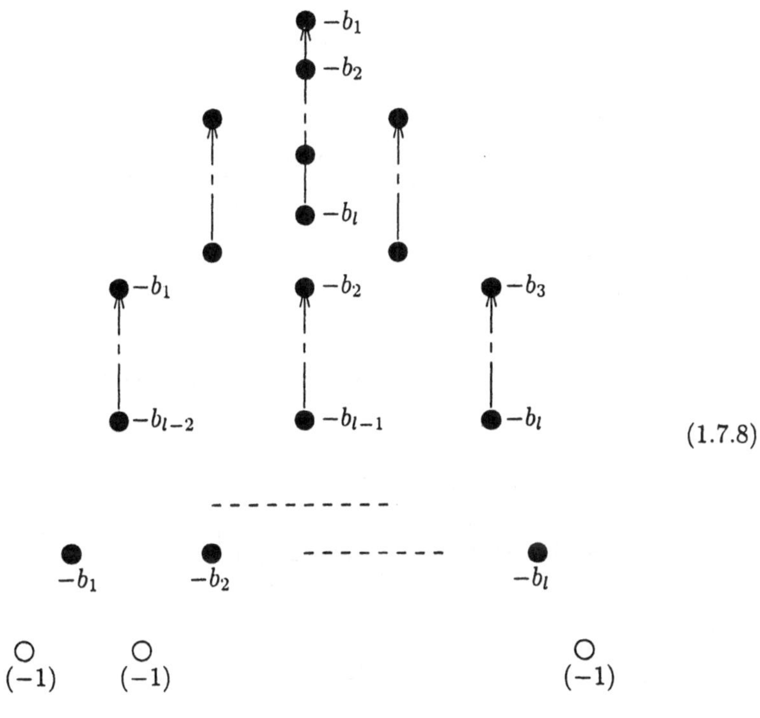

(1.7.8)

○ (−1) corresponds to the regular point.

1.8 Reciprocity Law for Geometric Sums

Let ρ be again a small (arranged) cyclic representation with corresponding cyclic point $\mathbf{Q} = \mathbf{Q}(\rho) = (U, Q; C, C')$, $\varphi : \tilde{U} \longrightarrow U$ the minimal resolution of the singularity $Q \in U$ and $\varphi^{-1}(Q) = L_1 \cup \cdots \cup L_l, l = l(\rho) \geq 1$. By Proposition 1.7.1 and Figure 1.6.2 the intersection matrix of the curve $\varphi^{-1}(Q)$ on $\widetilde{U}$ is

$$I(\rho) = I(\mathbf{Q}) = \begin{pmatrix} -b_1 & 1 & & & \\ 1 & -b_2 & 1 & & \bigcirc \\ & 1 & \ddots & \ddots & \\ \bigcirc & & \ddots & -b_{l-1} & 1 \\ & & & 1 & -b_l \end{pmatrix}$$

We define the *trace* of ρ (of $\mathbf{Q}$) by

$$Tr(\rho) = Tr(\mathbf{Q}) = Tr(I(\rho)) = -b_1 - b_2 - \ldots - b_l. \tag{1.8.1}$$

Definition 1.8.1 The number

$$\mathsf{S}(\rho) = \mathsf{S}(\mathbf{Q}) = \frac{1}{12}\left(-3l - Tr(\rho) + \frac{d_x}{d} + \frac{d_y}{d}\right)$$

is called the *geometric sum* of ρ (of $\mathbf{Q}$) .

Notice that $Tr(\rho) = Tr(\rho^\perp)$, hence $\mathsf{S}(\rho) = \mathsf{S}(\rho^\perp)$, that means $\mathsf{S}(\mathbf{Q})$ depends only on the singularity $Q \in U$. Let ρ be of type $\langle d, e \rangle = \langle d, d_x \rangle$, $\rho^\perp$ of type $\langle d, e' \rangle = \langle d, d_y \rangle$. Knowing the type we can calculate $\mathsf{S}(\rho)$ immediately by means of the continued fraction (1.7.4):

$$\mathsf{S}(\rho) = \mathsf{S}(\mathbf{Q}) = \frac{1}{12}\left(-3l + b_1 + \ldots + b_l + \frac{e}{d} + \frac{e'}{d}\right), e \cdot e' \equiv 1 \bmod d \tag{1.8.2}$$

In order to discover a reciprocity law we set

$$\mathsf{S}(e, d) = \mathsf{S}(\langle d, e \rangle) = \mathsf{S}(\rho).$$

We extend the definition to

$$\mathbf{S} : (\mathbb{Z} \times \mathbb{N})_1 \longrightarrow \mathbb{Q},$$

where $(\mathbb{Z} \times \mathbb{N})_1$ denotes the set of all pairs $(g, n) \in \mathbb{Z} \times \mathbb{N}$ with g.c.d.$(g, n) = 1$ setting

$$\begin{aligned} \mathbf{S}(e + kd, d) &= \mathbf{S}(e, d), k \in \mathbb{N}, 0 \leq e < d \\ \mathbf{S}(e - kd, d) &= -\mathbf{S}(kd - e, d) \end{aligned} \tag{1.8.3}$$

Proposition 1.8.2 *(Reciprocity law for geometric sums).*
For $(e, d) \in (\mathbb{N} \times \mathbb{N})_1, e \neq 0$, *it holds that*

$$\mathbf{S}(e, d) + \mathbf{S}(d, e) = \frac{1}{12}\left(-3 + \frac{d}{e} + \frac{e}{d} + \frac{1}{ed}\right). \tag{1.8.4}$$

Proof Let $(e, d) \in (\mathbb{N} \times \mathbb{N})_1, 0 < e < d$ and ρ of type $\langle d, e\rangle$. Consider the derived representation ρ_x. We want to relate $\mathbf{S}(\rho_x)$ with $\mathbf{S}(\rho)$. For this purpose we use the differential relation (1.5.2). Then we can write

$$\mathbf{S}(\rho) = \frac{1}{12}\left(-3l + b_1 + \ldots + b_l + \frac{d_x}{d} + \sum_{i=1}^{l} (d_{x^{i-1}} \cdot d_{x^i})^{-1}\right),$$

$$\mathbf{S}(\rho_x) = \frac{1}{12}\left(-3(l-1) + b_2 + \ldots + b_l + \frac{d_{xx}}{d_x} + \sum_{i=1}^{l-1} (d_{x^i} \cdot d_{x^{i+1}})^{-1}\right).$$

The difference is

$$\mathbf{S}(\rho) - \mathbf{S}(\rho_x) = \left(-3 + b_1 + \frac{e}{d} + \frac{1}{de} - \frac{d_{xx}}{d_x}\right)/12.$$

Together with (1.8.3), 1.4.1 and (1.7.2) we see that

$$\begin{aligned} \mathbf{S}(e, d) + \mathbf{S}(d, e) &= \mathbf{S}(e, d) - \mathbf{S}(-d, e) = \mathbf{S}(\rho) - \mathbf{S}(\rho_x) \\ &= \left(-3 + \frac{e}{d} + \frac{d}{e} + \frac{1}{de}\right)/12. \end{aligned}$$

The reciprocity law is proved for $0 < e < d$, but also for $0 < d < e$ by symmetry. Additionally one finds that

$$\mathbf{S}(1, 1) = \mathbf{S}(0, 1) = 0. \tag{1.8.5}$$

So (1.8.4) is also satisfied for $0 < e = d (= 1)$. □

Geometric sums can now be calculated in the most effective manner by the reciprocity law. One has only to connect it with the Euclidean algorithm.

The reciprocity law was found here by a kind of discrete function theory imitating automorphic function theory. Natural and rational numbers appeared as values of such discrete functions, discrete derivations of them or of their combinations. It seems not to be a piece of bad luck, that the reciprocity law (1.8.4) is already known more than 100 years ago in the theory of modular functions. It will be important to connect our discrete function theory with analytic function theory on this place.

DEDEKIND'S *η-function* is defined by

$$\eta(\tau) = q^{1/24} \prod_{\nu=1}^{\infty} (1 - q^{\nu}), q = e^{2\pi i \tau}, \tau \in \mathbb{H} = \{z \in \mathbb{C}, \operatorname{Im} z > 0\}.$$

We change over to the logarithmic η-function $\log \eta(\tau)$, where one takes the branch with real value $\log \eta(i), i = \sqrt{-1}$. The group $\mathbb{G}\mathrm{l}_2^+(\mathbb{R}) = \{\gamma \in \mathbb{G}\mathrm{l}_2(\mathbb{R}); \det \gamma > 0\}$ acts on $\mathbb{H}$ by linear fractions.

Proposition 1.8.3 *For* $\gamma = \begin{pmatrix} e' & f \\ d & e \end{pmatrix} \in \mathbb{S}\mathrm{l}_2(\mathbb{Z}), d \neq 0,$ *it holds that*

$$\log \eta(\gamma\tau) = \log \eta(\tau) + \left(\frac{1}{2} \log \frac{d\tau + e}{di} + \frac{1}{4} \log d^2\right)$$
$$+ \frac{2\pi i}{24}\left(\left(\frac{e}{d} + \frac{e'}{d}\right) - 12 S_{Ded}(e, d)\right)$$

with a function $S_{Ded} : (\mathbb{Z} \times \mathbb{Z})_1 \longrightarrow \mathbb{Q}$.

Definition 1.8.4 The value $S_{Ded}(e, d)$ is called the DEDEKIND sum of (e, d).

Proposition 1.8.5 *The* DEDEKIND *sum has the following properties:*

(i)	$S_{Ded}(-e, -d)$	$= S_{Ded}(e, d)$
(ii)	$S_{Ded}(-e, d)$	$= -S_{Ded}(e, d)$
(iii)	$S_{Ded}(e, -d)$	$= -S_{Ded}(e, d)$
(iv)	$S_{Ded}(e, d)$	$= S_{Ded}(\overline{e}, d)$ *for* $\overline{e} \equiv e \bmod d$
(v)	$S_{Ded}(0, 1)$	$= S_{Ded}(e, 1) = 0$
(vi)	$S_{Ded}(e, d)$	$= S_{Ded}(e', d)$ *if* $e \cdot e' \equiv 1 \bmod d$
(vii)	$12 S_{Ded}(e, d) + 12 S_{Ded}(d, e)$	$= -3 + \frac{e}{d} + \frac{d}{e} + \frac{1}{ed}$ *for* $e, d > 0$.

The last property is called the Reciprocity Law for DEDEKIND *sums.*

For the proofs of the statements we refer the reader to Chapter IX of [LA2]. An original reference is DEDEKIND'S article [DED]. Comparing the reciprocity laws (vii), (1.8.4) and the properties (ii), (iv) with (1.8.3) we see via Euclidean algorithm that the geometric sum and the DEDEKIND sum coincide on $(\mathbb{Z} \times \mathbb{N})_1$. We used also the normalizing conditions (1.8.5) and (v), which are consequences of the reciprocity laws and the congruence property. So the DEDEKIND sum extends the definition of the geometric sum to $\mathbf{S} : (\mathbb{Z} \times \mathbb{Z})_1 \longrightarrow \mathbb{Q}$.

Corollary 1.8.6 *For all* $(e, d) \in (\mathbb{Z} \times \mathbb{Z})_1$ *it holds that*

$$\mathbf{S}(e, d) = S_{Ded}(e, d).$$

1.9 Explicit Dedekind Sums

Now we present DEDEKIND'S explicit definition of S_{Ded}. The odd periodic (piecewise linear) function

$$((x)) : \mathbb{R} \longrightarrow (-1/2, 1/2) \subset \mathbb{R}$$

is defined by

$$((x)) = \begin{cases} \text{residue of } x - \frac{1}{2} \bmod \mathbb{Z} \text{ in } (-1/2, 1/2) & \text{if } x \notin \mathbb{Z} \\ 0 & \text{if } x \in \mathbb{Z} \end{cases}$$

The graph of the function $((x))$ is described in picture Figure 1.9.1.

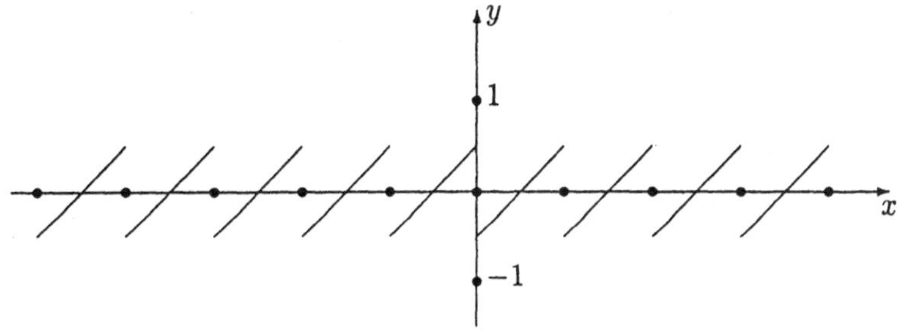

Definition 1.9.1 For $(e,d) \in (\mathbb{Z} \times \mathbb{N})_1, d \neq 0$, we define the *explicit* DEDEKIND *sum* by

$$s(e,d) = \sum_{k=1}^{d} \left(\left(\frac{k}{d}\right)\right) \cdot \left(\left(\frac{ek}{d}\right)\right) \tag{1.9.1}$$

Following DEDEKIND'S idea we want to prove that both the implicit DEDEKIND sum $S_{Ded}(e,d)$ and the explicit DEDEKIND sum $s(e,d)$ coincide. Obviously we have an analogon of 1.8.5 (iv):

$$s(e,d) = s(\overline{e},d) \text{ for } \overline{e} \equiv e \bmod d$$

It suffices to check the coincidence for $(e,d) \in (\mathbb{N} \times \mathbb{N})_1$. This follows from DEDEKIND'S reciprocity law 1.8.5 (vii) for $S_{Ded}(e,d)$ and from

Lemma 1.9.2 *(*DEDEKIND'S *Reciprocity Law for $s(e,d)$).*

$$s(e,d) + s(d,e) = \frac{1 + d^2 + e^2 - 3de}{12de}, \quad (e,d > 0). \tag{1.9.2}$$

There is a quite elementary proof of (1.9.2). We present the main steps of this proof. The details are left as elementary exercise to the reader (see [H-Z1])

$$\sum_{k=1}^{d-1} \left(\left(\frac{ke}{d}\right)\right)^r = \sum_{k=1}^{d-1} \left(\left(\frac{k}{d}\right)\right)^r \text{ for } r \in \mathbb{N} \tag{1.9.3}$$

$$\sum{}' \left(\left(\frac{ke}{d}\right)\right)^r = \begin{cases} 0 & \text{if } r = 1 \\ (d^2 - 3d + 2)/12d^2 & \text{if } r = 2 \end{cases} \tag{1.9.4}$$

where $\sum'$ means $\sum\limits_{k=1}^{d-1}$.

Hint: Substitute $((\frac{k}{d}))$ by $\frac{k}{d} - 1/2$.

$$\sum{}' \left[\frac{ke}{d}\right] = (d-1)(e-1)/2 \tag{1.9.5}$$

Hint: Use

$$\left[\frac{ke}{d}\right] = \frac{ke}{d} - \left(\left(\frac{ke}{d}\right)\right) - \frac{1}{2} \tag{1.9.6}$$

and (1.9.4).

$$\sum{}'\left[\frac{ke}{d}\right]^2 - 2\frac{e}{d}\sum{}' k\left[\frac{ke}{d}\right] = (2d-1)(d-1)\left(1-e^2\right)/6d \tag{1.9.7}$$

Hint: Consider $\sum'\left(\frac{ke}{d} - \left[\frac{ke}{d}\right] - \frac{1}{2}\right)^2 = \sum'\left(\left(\frac{ke}{d}\right)\right)^2$. We define now the sum of natural numbers

$$s_d(e) = \sum{}' k\left[\frac{ke}{d}\right]. \tag{1.9.8}$$

Then

$$6ds(e,d) = (d-1)\left(2de - e - \frac{3}{2}d\right) - 6s_d(e) \tag{1.9.9}$$

Hint: Express in the defining sum (1.9.1) for $s(e,d)$ the summands $((k/d))$ by their correct values $\frac{k}{d} - 1/2$ and apply (1.9.3), (1.9.4), (1.9.6) and the definition (1.9.8) of $s_d(e)$.
Now we consider $s_e(d)$.

$$s_e(d) = \frac{1}{4}(e-1)(d-1)(2e-1) - \frac{1}{2}\sum_{k=1}^{l-1}\left[\frac{ke}{d}\right]^2 \tag{1.9.10}$$

Proof

$$\begin{aligned} s_e(d) &= \sum_{l=1}^{e-1} l\left[\frac{ld}{e}\right] = \sum_{l=1}^{e-1} l\cdot\left(\sum_{0<k<ld/e} 1\right) = \sum_{0<\frac{k}{d}<\frac{l}{e}<1} l \\ &= \sum_{k=1}^{d-1}\sum_{l=[ke/d]+1}^{e-1} l = \sum{}'\left(\frac{e^2-e}{2} - \frac{[ke/d]^2 + [ke/d]}{2}\right) \end{aligned}$$

Now apply (1.9.5).

The relation (1.9.7) connects also $s_d(e)$ with $\sum'\left[\frac{ke}{d}\right]^2$. A comparison with (1.9.10) yields

$$ds_e(d) + es_d(e) = (d-1)(e-1)(8de - d - e - 1)/12.$$

The reciprocity law for $s(e,d)$ follows now after substituting $s_d(e)$ and $s_e(d)$ by means of (1.9.9).

□

1.10 Eisenstein Sums

We started with representations of finite abelian groups in order to define the geometric sum. The eigenvalues of the represented groups are unit roots. So it is quite natural to ask for a cyclotomic expression of $S(e,d)$. We know that $S(e,d)$ is the DEDEKIND sum and look at DEDEKIND'S explicit definition 1.9.1 It was EISENSTEIN [EIS] who found a cyclotomic expression for $((\frac{k}{d})), ((\frac{ke}{d}))$ in the proofs. We follow the treatment of HIRZEBRUCH, ZAGIER [H-Z1]

Definition 1.10.1 The EISENSTEIN *sum* is defined by

$$S_{Eis}(e,d) = -\frac{1}{4d}\sum_{\zeta^d=1}{}' \left(\frac{\zeta+1}{\zeta-1}\cdot\frac{\zeta^e+1}{\zeta^e-1}\right), \ (e,d) \in (\mathbb{Z}\times\mathbb{N})_1 \tag{1.10.1}$$

where the sum runs over all non-trivial d-th unit roots.

Proposition 1.10.2 *It holds that*

$$S(e,d) = S_{Ded}(e,d) = s(e,d) = S_{Eis}(e,d), \ (e,d) \in (\mathbb{Z}\times\mathbb{N})_1.$$

The first two identities have already been proven by reciprocity. As mentioned above we express $s(e,d)$ by means of unit roots. We need the FOURIER expansion of the function $((x))$ due to EISENSTEIN.

Lemma 1.10.3 *The odd periodic function* $((x))$ *has the* FOURIER *expansion*

$$((x)) = -\sum_{n=1}^{\infty}\frac{1}{\pi n}\sin(2\pi n x)$$

Proof We have to determine the FOURIER coefficients a_n in

$$((x)) = \sum_{n=1}^{\infty} a_n \sin(2\pi n x).$$

A straightforward calculation yields

$$a_n = 2\int_0^1 ((x))\sin(2\pi n)dx = \int_0^1 (2x-1)\sin(2\pi n x)dx$$

$$= \int_0^1 2x \sin(2\pi n x)dx - \int_0^1 \sin(2\pi n x)dx$$

$$= \int_0^1 x \sin(2\pi n x)dx + \left[\frac{1}{2\pi n}\cos(2\pi n x)\right]_0^1$$

$$= 2\left[\frac{\sin(2\pi n x)}{(2\pi n)^2} - \frac{x}{2\pi n}\cos(2\pi n x)\right]_0^1 + 0$$

$$= -\frac{1}{\pi n}[x\cos(2\pi n x)]_0^1 = -\frac{1}{\pi n}.$$

□

Corollary 1.10.4 *For* $(e,d) \in (\mathbb{Z} \times \mathbb{N})_1, d \neq 0$, *it holds that*

$$\left(\left(\frac{e}{d}\right)\right) = -\frac{1}{2d}\sum_{\zeta^d=1}{}' \zeta^e \cdot \frac{\zeta+1}{\zeta-1}$$

where the sum is taken over all non-trivial d-th unit roots.

Proof We need again elementary function theory. The cotangent function has simple poles in $\mathbb{Z}\pi$. It can be expressed in the following manner (see e.g. [PRI], IV, § 2(30)):

$$\cot(z) = \frac{1}{2} + \sum_{k=-\infty}^{\infty}{}' \left(\frac{1}{2-k\pi} + \frac{1}{k\pi}\right).$$

Therefore, we have

$$\pi \cot(\pi x) = \lim_{M\to\infty} \sum_{m=-M}^{M} \frac{1}{x+m}$$

This identity allows to calculate

$$\lim_{N\to\infty}\left(\sum_{\substack{0<n<N \\ n\equiv k \bmod d}} \frac{1}{n} - \sum_{\substack{0<n<N \\ n\equiv -k \bmod d}} \frac{1}{n}\right)$$

$$= \lim_{M\to\infty} \sum_{-M<m<M} \frac{1}{k+dm} = \frac{1}{d}\lim_{M\to\infty}\sum_m \frac{1}{(k/d)+m}$$

$$= \frac{\pi}{d} \cdot \cot\left(\pi \frac{k}{d}\right) = \frac{\pi}{d} \cdot \frac{\cos(\pi k/d)}{\sin(\pi k/d)} \tag{1.10.2}$$

$$= \frac{\pi}{d} \cdot \frac{Re\left(e^{\pi i k/d}\right)}{\mathrm{Im}\left(e^{\pi i k/d}\right)} = i\frac{\pi}{d} \cdot \frac{\zeta_{2d}^k + \zeta_{2d}^{-k}}{\zeta_{2d} - \zeta_{2d}^{-k}}$$

$$= i\frac{\pi}{d}\frac{\zeta^k + 1}{\zeta^k - 1}, \zeta = \zeta_d = e^{2\pi i/d}$$

By substitution we obtain

$$\left(\left(\frac{e}{d}\right)\right) = -\frac{1}{\pi}\sum_{n=1}^{\infty}\frac{1}{n}\sin(2\pi n e/d)$$

$$= -\frac{1}{\pi}\sum_{k=1}^{d-1}\lim_{N\to\infty}\left(\sum_{\substack{0<n<N\\ n\equiv k \bmod d}}\frac{1}{n}\right)\sin(2\pi k e/d)$$

$$= -\frac{1}{2\pi}\sum_{k=1}^{d-1}\lim_{N\to\infty}\left(\sum_{\substack{0<n<N\\ n\equiv k \bmod d}}\frac{1}{n} - \sum_{\substack{0<n<N\\ n\equiv -k \bmod d}}\frac{1}{n}\right)\sin(2\pi k e/d)$$

$$= -\frac{i}{2d} = -\frac{1}{2d}\sum_{\zeta^d=1}{}'\frac{\zeta+1}{\zeta-1}\cdot\left(\frac{\zeta^e}{2} - \frac{\overline{\zeta}^e}{2}\right) \tag{1.10.3}$$

$$= -\frac{1}{2d}\left(\sum_{\zeta}{}'\frac{\zeta^e}{2}\cdot\frac{\zeta+1}{\zeta-1} - \sum_{\zeta}{}'\frac{\overline{\zeta}^e}{2}\cdot\frac{\zeta+1}{\zeta-1}\right)$$

$$= -\frac{1}{2d}\left(\sum_{\zeta}{}'\frac{\zeta^e}{2}\cdot\frac{\zeta+1}{\zeta-1} + \sum_{\zeta}{}'\frac{\overline{\zeta}^e}{2}\cdot\frac{\overline{\zeta}+1}{\overline{\zeta}-1}\right) = -\frac{1}{2d}\sum_{\zeta}{}'\zeta^e\frac{\zeta+1}{\zeta-1}$$

□

The nextTh relation we use is well-known (see e.g. [BOS])

$$\sum_{k=1}^{d}\zeta^k = \begin{cases} 0 & \zeta \text{ a non trivial } d\text{ - th unit root} \\ d & \zeta = 1 \end{cases} \tag{1.10.4}$$

Proof of Proposition 1.10.2 Using Corollary 1.10.4 and (1.10.4) one receives

$$s(e,d) = \sum_{k=1}^{d} \left(\left(\frac{k}{d}\right)\right) \cdot \left(\left(\frac{ke}{d}\right)\right)$$
$$= \frac{1}{4d^2} \sum_{k=1}^{d} \sum_{\mu^d=1}{}' \sum_{\zeta^d=1}{}' \frac{\mu+1}{\mu-1} \cdot \frac{\zeta+1}{\zeta-1}$$
$$= \frac{1}{4d^2} \sum_{\mu}{}' \sum_{\zeta}{}' \left(\sum_{k=1}^{d} \mu^k \zeta^{ek} \right) \frac{\mu+1}{\mu-1} \cdot \frac{\zeta+1}{\zeta-1}$$

$$= \frac{1}{4d} \sum_{\zeta}{}' \frac{\zeta^{-e}+1}{\zeta^{-e}-1} \cdot \frac{\zeta+1}{\zeta-1} = S_{Eis}(e,d)$$

□

1.11 Hirzebruch's Sum

We remember that $S(e,d)$ has been originally defined as geometric sum $S(\rho)$ of an arranged small representation of type $\langle d, e \rangle$ (see Definition 1.8.1). We know that

$$S(\rho) = S_{Eis}(\rho) = -\frac{1}{4d} \sum_{\zeta^d=1}{}' \frac{\zeta+1}{\zeta-1} \cdot \frac{\zeta^e+1}{\zeta^e-1} \tag{1.11.1}$$

Now we change over from cyclic representations to faithful (arranged) abelian representations $\rho \in \widehat{\mathcal{A}}\left(\text{or } 2\widehat{\mathcal{A}}\right), \rho : A \longrightarrow \mathbb{G}\mathrm{l}(V), \dim V = 2, A$ a finite abelian group. Each $\rho(g)$ has a unique pair of eigenvalues ζ_g, μ_g. The following definition generalizes EISENSTEIN sums.

Definition 1.11.1 Let ρ be an arranged abelian representation and identify A with its image in $\mathbb{G}\mathrm{l}(V)$. The number

$$\mathrm{S}(\rho) = S_{Hir}(\rho) = S_{Hir}(A) = -\frac{1}{4d} \sum_{g \in A}{}^{*} \frac{\zeta_g+1}{\zeta_g-1} \cdot \frac{\mu_g+1}{\mu_g-1}$$

is called the HIRZEBRUCH sum of ρ.

The sum is taken over all non-reflections of A such that all summands are defined. It is defined to be 0 if the sum is void.

HIRZEBRUCH called $-4d\mathrm{S}(\rho)$ the *signature defect* of A, if $A \subset \mathrm{Gl}_2(\mathbb{C})$ because it appeared in the signature theorem of equivariant K-theory (see [H-Z1]). It is important to connect the sum $\mathrm{S}(\rho)$ with the sum $\mathrm{S}(\overline{\rho})$ of the reduction $\overline{\rho} \in 2\widehat{Z}^{\mathrm{small}}$ of ρ, defined in (1.1.1). We follow again HHIRZEBRUCH'S and ZAGIER'S line of reasoning in [H-Z1].

Proposition 1.11.2 *Let $\Sigma(\rho)$ be the subgroup of A generated by all reflections with respect to the faithful representation*

$$\rho : A \longrightarrow \mathrm{Gl}_2(\mathbb{C}),\ \overline{A} = A/\Sigma,\ \Sigma = \Sigma(\rho),$$

and $\overline{\rho} : \overline{A} \longrightarrow \mathrm{Gl}_2(\mathbb{C})$ the reduction of ρ, say of type $\langle d, e\rangle$. Then it holds that

$$\mathrm{S}(\rho) = |\Sigma| \cdot \mathrm{S}(\overline{\rho}),\ (|\Sigma| = \mathit{card}\ \Sigma). \tag{1.11.2}$$

For the proof we need some elementary function theory again.

Lemma 1.11.3 *For $b \in \mathbb{C}^*, d \in \mathbb{N}_+$ it holds that*

$$\frac{1}{d}\sum_{a^d=b} \frac{az+1}{az-1} = \frac{bz^d+1}{bz^d-1} \tag{1.11.3}$$

Proof Both rational functions in (1.11.3) have the same value at $z = 0$ and they have all their (simple) poles in the d-th roots of $1/b$. It suffices to prove that the residue coincide there. We calculate the residue in $z_o = \frac{1}{a}, a^d = b$:

$$\mathrm{Res}_{z_o} f(z) = \lim_{z\to 1/a}\left(z - \frac{1}{a}\right) f(z)$$

Obviously the residue at z_o of the function on the left-hand side of (1.11.3) is equal to $2/ad$. For $f(z) = (bz^d+1)/(bz^d-1)$ we find by means of de l'HOSPITAL'S rule that

$$\begin{aligned}\mathrm{Res}_{z_o} f(z) &= \lim_{z\to\frac{1}{a}}\left(z-\frac{1}{a}\right)\frac{z^d+\frac{1}{b}}{z^d-\frac{1}{b}} = \lim_{z\to\frac{1}{a}} \frac{d(z-\frac{1}{a})+z^d+\frac{1}{b}}{dz^{d-1}} \\ &= \lim_{z\to\frac{1}{a}} \frac{(d+1)z^d - \frac{1}{a}dz^{d-1}+\frac{1}{b}}{dz^{d-1}} = \lim_{z\to\frac{1}{a}}\left(\frac{d+1}{d}z - \frac{1}{a} + \frac{z/b}{dz^d}\right) \\ &= \frac{d+1}{ad} - \frac{1}{a} + \frac{1}{ad} = \frac{2}{ad}.\end{aligned}$$

□

Corollary 1.11.4

$$\sum_{a^d=b,(a\neq 1)} \frac{a+1}{a-1} = \begin{cases} d\frac{b+1}{b-1} & \text{if } b\neq 1 \\ 0 & \text{if } b=1 \end{cases} \tag{1.11.4}$$

Proof We set $z=1$ in (1.11.3). For $b\neq 1$ the result is immediate. For $b=1$ the sum is equal to

$$\begin{aligned}\lim_{z\to 1}\left(d\frac{z^d+1}{z^d-1}-\frac{z+1}{z-1}\right) &= \lim_{x=z-1\to 0}\left(d\frac{(1+x)^d+1}{(1+x)^d-1}-\frac{2+x}{x}\right)\\ &= \lim_{x\to 0}\left(\frac{2+dx+x^2p(x)}{x+\frac{1}{2}(d-1)x^2+x^3q(x)}-\frac{2+x}{x}\right).\\ &= \lim_{x\to 0}\frac{x^3(p(x)+\ldots)}{x^2+\frac{1}{2}(d-1)x^3+x^4q(x)}=0.\end{aligned}$$

□

Proof of Proposition 1.11.2. Without loss of generality we can assume that $A\subset D_2(\mathbb{C})\subset \mathbb{Gl}_2(\mathbb{C})$. We consider the exact sequence

$$\begin{array}{ccccccccc} 1 & \longrightarrow & \Sigma & \longrightarrow & A & \longrightarrow & \overline{A} & \longrightarrow & 1. \\ & & & & g & \mapsto & \overline{g} & & \end{array} \tag{1.11.5}$$

Σ is the product of two cyclic groups ${}_v\Sigma$ and Σ_w of orders v or w consisting of the elements of the form $\begin{pmatrix} * & 0\\ 0 & 1\end{pmatrix}$ or $\begin{pmatrix} 1 & 0\\ 0 & *\end{pmatrix}$, respectively. With obvious notations, ζ a primitive d-th unit root, we get

$$\begin{aligned}-4d\mathbf{S}(\rho) &= {\sum_{g\in A}}^{*}\frac{\zeta_g+1}{\zeta_g-1}\cdot\frac{\mu_g+1}{\mu_g-1} = \sum_{\overline{g}\in\overline{A}}{\sum_{h\in\overline{g}}}^{*}\frac{\zeta_h+1}{\zeta_h-1}\frac{\mu_h+1}{\mu_h-1}\\ &= \sum_{k=1}^{d}\left({\sum_{\lambda^v=\zeta^k}}'{\sum_{\nu^w=\zeta^{ek}}}'\frac{\lambda+1}{\lambda-1}\cdot\frac{\nu+1}{\nu-1}\right)\\ &= \sum_{k=1}^{d-1} vw\frac{\zeta^k+1}{\zeta^k-1}\cdot\frac{\zeta^{ek}+1}{\zeta^{ek}-1} \\ &= -4dvw\mathbf{S}(\overline{\rho}) = -4d|\Sigma|\mathbf{S}(\overline{\rho})\end{aligned} \tag{1.11.6}$$

□

1.12 Geometric Interpretation

We introduce the category of (arranged) abelian points. The objects are the (arranged) abelian points $\mathbf{P} = (P, U; uC, u'C')$ introduced in section 3. A *quasi-finite morphism* $\mathbf{f} : \mathbf{P} \longrightarrow \mathbf{Q}, \mathbf{Q} = (Q, V; vD, v'D')$ is a finite analytic morphism (germ) $U \longrightarrow V$ sending P to Q and (C, C') to (D, D') for suitably chosen open analytic neighbourhoods U, V of P or Q, respectively. A *morphism* $\mathbf{f} : \mathbf{P} \to \mathbf{Q}$ is represented by an analytic morphism $U \longrightarrow V$ with finite restriction $U \setminus C' \longrightarrow V \setminus D'$ sending C onto D and P to Q (C' may be contracted). The quotient maps

$$O \underset{A}{\longrightarrow} \mathbf{Q} = (Q,\ \mathbb{C}^2/A;\ vL_x,\ v'L_y)\ :\ \underset{v\quad \langle d,e\rangle \quad v'}{\circ\!\!\longrightarrow\!\!\circ} \tag{1.12.1}$$

L_x, L_y the images of the coordinate axis diagonalizing the finite abelian group $A \subset \mathbb{G}l_2(\mathbb{C})$, are called ***uniformizations.*** We will sometimes use the notation $\mathbf{Q} = O/A$, where O denotes the reduced regular point $O \in \mathbb{C}^2$. More generally, $\mathbf{Q}$ with weights $v = v' = 1$ is called a ***reduced point*** . Reduced points will be frequently denoted by thin letters. So we write simply Q instead of $\mathbf{Q}$ if $\mathbf{Q}$ is reduced. The exact sequence (1.11.5) yields a decomposition (1.12.2) of our uniformization (1.12.1):

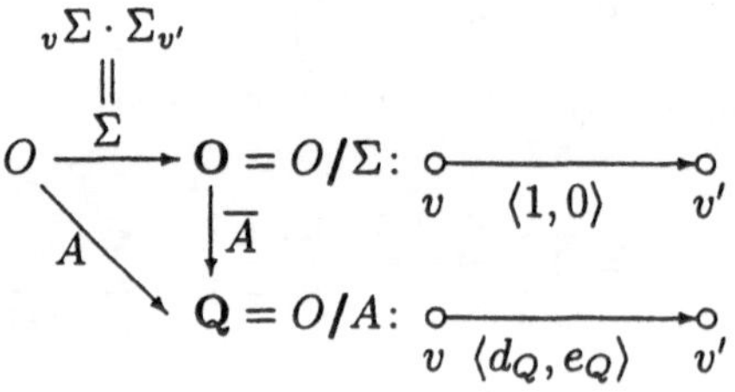

$$\begin{array}{ccll} & {}_v\Sigma \cdot \Sigma_{v'} & & \\ & \| & & \\ O & \xrightarrow{\ \Sigma\ } & \mathbf{O} = O/\Sigma : & \underset{v\quad \langle 1,0\rangle \quad v'}{\circ\!\!\longrightarrow\!\!\circ} \\ & {}_{A}\searrow & \downarrow \bar{A} & \\ & & \mathbf{Q} = O/A : & \underset{v\quad \langle d_Q,e_Q\rangle \quad v'}{\circ\!\!\longrightarrow\!\!\circ} \end{array} \tag{1.12.2}$$

A morphism $\mathbf{f} : \mathbf{P} \to \mathbf{Q}$ is said to be *finite* , if there is a commutative diagram:

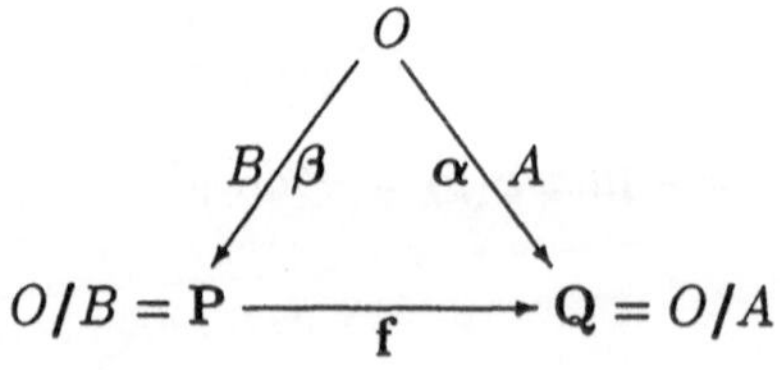

$$\begin{array}{ccccc} & & O & & \\ & {}^{B/\beta}\swarrow & & \searrow^{\alpha\backslash A} & \\ O/B = \mathbf{P} & & \xrightarrow{\quad \mathbf{f}\quad} & & \mathbf{Q} = O/A \end{array} \tag{1.12.3}$$

with uniformizations α, β.

Lemma 1.12.1 *Let*

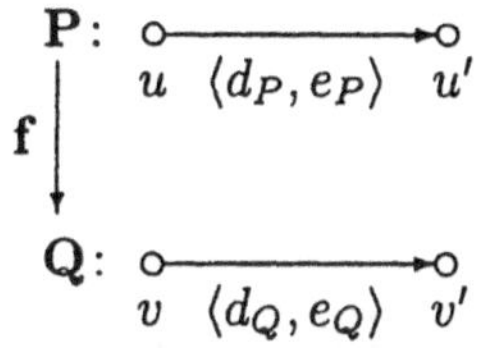

$$\begin{array}{ll} \mathbf{P}: & \circ\!\longrightarrow\!\circ \\ & u \;\; \langle d_P, e_P \rangle \;\; u' \\ \mathbf{f}\big\downarrow & \\ \mathbf{Q}: & \circ\!\longrightarrow\!\circ \\ & v \;\; \langle d_Q, e_Q \rangle \;\; v' \end{array}$$

be a quasifinite morphism. The following two conditions are equivalent:

(i) $\mathbf{f}$ *is finite.*

(ii) $u|v$ *and* $u'|v'$*, and the quotients* v/u*,* $v/u, v'/u' \in \mathbb{N}$ *are the ramification indices of the underlying quotient map* $f : U \to V$ *along* C *or* C'*, respectively.*

Proof We first construct a "co-uniformization" $Q \longrightarrow \mathbf{O}$ of the reduced point Q via diagram:

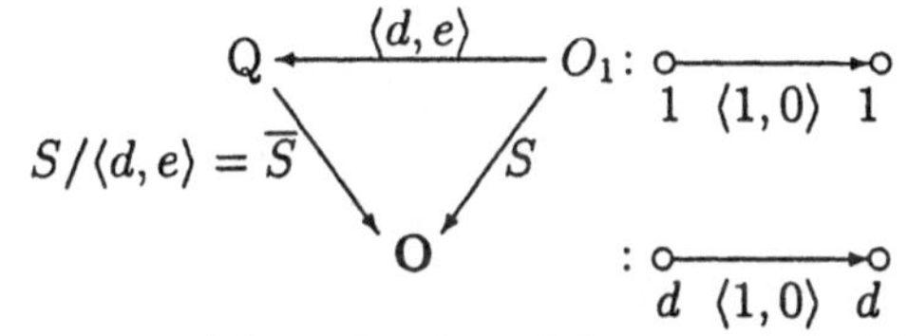

$$\begin{array}{ccccl} Q & \xleftarrow{\langle d,e \rangle} & & O_1: & \circ\!\longrightarrow\!\circ \quad 1 \;\langle 1,0\rangle\; 1 \\ S/\langle d,e\rangle = \overline{S} \searrow & & \swarrow S & & \\ & \mathbf{O} & & : & \circ\!\longrightarrow\!\circ \quad d \;\langle 1,0\rangle\; d \end{array} \tag{1.12.4}$$

with subgroups $S = \left\langle \begin{pmatrix} \zeta & 0 \\ 0 & 1 \end{pmatrix}, \begin{pmatrix} 1 & 0 \\ 0 & \zeta \end{pmatrix} \right\rangle, \langle d, e \rangle = \left\langle \begin{pmatrix} \zeta & 0 \\ 0 & \zeta^e \end{pmatrix} \right\rangle \subset S \subset$ $\mathrm{Gl}_2(\mathbb{C})$, ζ a primitive d-th unit root. The morphism $O_1 \to Q$ is the uniformization of Q.

Let $U_O = D^2 = D \times D$ be a small open bidisc around O and $U_O^{**} = D^* \times D^*, D^* = D \setminus \{O\}$. We denote the preimages of U_O^{**} along $P \to Q \to O$ by U_P^{**}, U_Q^{**}. Then $U_P^{**} \to U_Q^{**} \to U_O^{**}$ is a tower of unramified finite coverings because we have thrown out the branch and ramification loci. The universal covering of $D^* \times D^*$ is

$$\begin{array}{ccc} \mathbb{H}^2 & \xrightarrow{\mathbb{Z} \times \mathbb{Z}} & D^{*2} \\ (z_1, z_2) & \longmapsto & (x_1, x_2) = \left(e^{2\pi i z_1}, e^{2\pi i z_2}\right), \end{array}$$

$\mathbb{H}$ the POINCARÉ upper half plane. Our unramified finite coverings yield subgroups $\Gamma_O \subset \Gamma_Q \subset \Gamma_P$ of finite index in $\mathbb{Z} \times \mathbb{Z}$ and a commutative diagram of quotient maps:

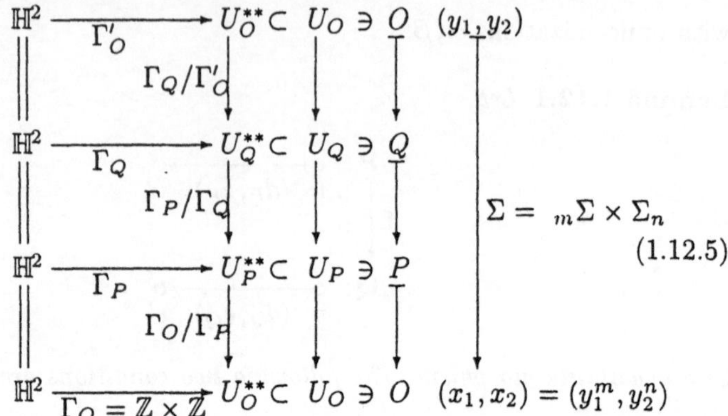

(1.12.5)

The tower on the right hand side consists of subquotient maps of $U_O \xrightarrow{\Sigma} U_O$. We have some freedom to choose $\Gamma_0' = m\mathbb{Z} \times n\mathbb{Z} \subset \Gamma_Q$. The diagram suggests to take the biggest splitting subgroup $\Gamma_0'' = m'\mathbb{Z} \times n'\mathbb{Z}$ of Γ_Q. This should be done, but then one has to change over to $\Gamma_0' = (um'\mathbb{Z}) \times (u'n'\mathbb{Z})$ in order to get the correct weights at Q and P we started with. It remains to check that the underlying map $f : U \longrightarrow V$ of $\mathbf{f}$ is analytically equivalent to $U_Q \longrightarrow U_P$ in diagram (1.12.5). This is true for the restrictions $U^{**} \longrightarrow V^{**}, U_Q^{**} \longrightarrow U_P^{**}$ by construction. Now one has only to apply the following

Extension Theorem 1.12.2 (Stein, [G-R]) *Let X be a complex manifold, $Z \subset X$ an analytic submanifold, $X^o = X \setminus Z$, $\pi^o : Y^o \longrightarrow X^o$ an unramified finite covering. Then there exists a unique finite branched covering $\pi : Y \longrightarrow X$ extending π^o, see:*

$$\begin{array}{ccc} Y & \overset{\pi}{\dashrightarrow} & X \\ \uparrow & & \uparrow \\ Y^0 & \xrightarrow{\pi^0} & X^0 \end{array} \tag{1.12.6}$$

□

Corollary 1.12.3 *(of 1.12.1). Each finite morphism $\mathbf{f} : \mathbf{P} \longrightarrow \mathbf{Q}$ is a quotient map by a finite abelian group C, notation: $\mathbf{Q} = \mathbf{P}/C$.*

This means that up to analytic equivalence the underlying map

$$f : U_P \longrightarrow V_Q$$

is a finite abelian quotient map for suitable analytic neighbourhoods of P or Q, respectively, branched along the singular coordinate systems, and the ramification indices are the weight quotients v/u or v'/u', respectively, with the notations of Lemma 1.12.1.

We introduce the operation of *weight shifting* for abelian points

$\mathbf{Q} : \underset{v \;\; \langle d,e \rangle \;\; v'}{\circ\!\!\longrightarrow\!\!\circ}$. If $w|v$ and $w'|v'$, say $v = dw, v' = d'w'$, then we denote by $\mathbf{Q}(d,d')$ the abelian point with graph

$$\mathbf{Q}(d,d'): \quad \underset{w = \frac{v}{d} \;\; \langle d,e \rangle \;\; \frac{v'}{d'} = w'}{\circ\!\!\longrightarrow\!\!\circ}$$

The weight shifting extends in obvious manner to morphisms

$$\mathbf{f} : \mathbf{P} \longrightarrow \mathbf{Q},$$

whenever $\mathbf{P}(d,d')$ and $\mathbf{Q}(d,d')$ are defined. The shifted morphism is denoted by $\mathbf{f}(d,d')$.

Definitions 1.12.4 Let $\mathbf{f} : \mathbf{P} \longrightarrow \mathbf{Q}$ be a finite morphism. With the notations of Lemma 1.12.1 we call

$$\mathbf{f}(u,u'): P = \mathbf{P}(u,u') \longrightarrow \mathbf{Q}(u,u')$$
$$\underset{1 \;\; \langle d_P, e_P \rangle \;\; 1}{\circ\!\!\longrightarrow\!\!\circ} \qquad \underset{\frac{v}{u} \;\; \langle d_Q, e_Q \rangle \;\; \frac{v'}{u'}}{\circ\!\!\longrightarrow\!\!\circ},$$

the *reduction* of $\mathbf{f}$. It is denoted by $red(\mathbf{f})$. $\mathbf{f}$ is called *reduced*, if $red(\mathbf{f}) = \mathbf{f}$. $\mathbf{f}$ is called *completely reduced*, if it is reduced and $\mathbf{Q}$ is also reduced.

From the criterion (ii) of Lemma 1.12.1 it follows that any quasifinite morphism $P \longrightarrow \mathbf{Q}$ with reduced starting point P is finite, hence reduced.

Let $\mathbf{f} : \mathbf{P} \longrightarrow \mathbf{Q} = \mathbf{P}/C$ be a finite morphism, $f : U_P \longrightarrow U_Q$ the underlying finite map. The *degree* of $\mathbf{f}$ is defined by

$$\deg(\mathbf{f}) = \deg(f). \tag{1.12.7}$$

The *degree of* $\mathbf{Q} = O/A$ is the degree of its uniformization:

$$\deg(\mathbf{Q}) = \deg(O \underset{A}{\longrightarrow} \mathbf{Q}) = v_Q v'_Q d_Q = |A|, \;\; \mathbf{Q}: \underset{v_Q \;\; \langle d_Q, e_Q \rangle \;\; v'_Q}{\circ\!\!\longrightarrow\!\!\circ} \tag{1.12.8}$$

With obvious notations and those of diagram (1.12.3) we have altogether

$$
\begin{aligned}
[\mathbf{P}:\mathbf{Q}] = \deg(\mathbf{P}/\mathbf{Q}) = \deg(\mathbf{f}) &= \frac{\deg(\mathbf{Q})}{\deg(\mathbf{P})} \\
&= \frac{v_Q \cdot v'_Q \cdot d_Q}{v_P \cdot v'_P \cdot d_P} = \frac{|A|}{|B|} \qquad (1.12.9)\\
&= |C| = \deg(\mathrm{red}\,(\mathbf{f})).
\end{aligned}
$$

Lemma 1.12.5 *For two reduced abelian points P, Q the following conditions are equivalent:*

(i) There exists a (completely reduced) finite morphism $f : P \longrightarrow Q$.

(ii) The subgroup $\langle d_Q, e_Q\rangle = \left\langle \begin{pmatrix} \zeta_d & 0 \\ 0 & \zeta_d^e \end{pmatrix} \right\rangle$ of $\mathbb{Gl}_2(\mathbb{C})$, ζ_d a primitive d-th unit root, contains the group $\langle d_P, e_P\rangle$ belonging to P.

If both conditions are satisfied, then d_P divides d_Q and $\deg(f) = d_Q/d_P$.

Proof Since $P \longrightarrow Q$ is finite there is a uniformizing composition $O \longrightarrow P \longrightarrow Q$ of completely reduced morphisms (see Diagram (1.12.3)), $Q = O/A$, $P = O/B$, $A = \langle d_Q, e_Q\rangle \subseteq B = \langle d_P, e_P\rangle$. The opposite implication and the last statement are trivial. □

Definition 1.12.6 Let $\mathbf{Q} = O/A$ be an abelian point. The number

$$\mathbf{S}(\mathbf{Q}) = S_{Hir}(A) \in \mathbb{Q}$$

is called the (HIRZEBRUCH) sum of $\mathbf{Q}$ (cp. with Definition 1.11.1). The following proposition means that the reduction preserves quotients of HIRZEBRUCH sums of abelian points along finite morphisms.

Proposition 1.12.7 *For any finite morphism $\mathbf{P} \longrightarrow \mathbf{Q}$ of abelian points with reduction $P \longrightarrow \overline{\mathbf{Q}}$ it holds that*

$$
\begin{aligned}
&(i)\quad \mathbf{S}(\mathbf{Q}) : \mathbf{S}(\mathbf{P}) = \mathbf{S}(\overline{\mathbf{Q}}) : \mathbf{S}(P) = \deg(\overline{\mathbf{Q}}) \cdot (\mathbf{S}(Q) : \mathbf{S}(P))\\
&(ii)\quad \mathbf{S}(\mathbf{Q}) = v_Q \cdot v'_Q \mathbf{S}(Q) = v_Q \cdot v'_Q \cdot \mathbf{S}(e_Q, d_Q)
\end{aligned}
$$

where $\mathbf{S}(Q)$ *is the geometric sum of the cyclic point Q which is equal to the* HIRZEBRUCH *sum of Q.*

Proof We first prove property (ii). The reflection subgroup Σ of $A, \mathbf{Q} = O/A$, yields a decomposition $O \longrightarrow \mathbf{O} \xrightarrow{\langle d_Q, e_Q \rangle} \mathbf{Q}$ of the uniformization of $\mathbf{Q}$ with $\mathbf{O} : \underset{v_Q}{\circ} \xrightarrow{\langle 1,0 \rangle} \underset{v'_Q}{\circ}$ (see diagram (1.12.2)). Now we refer to the Propositions 1.11.2, 1.10.2 and to (1.11.1).

Applying (ii) to $\mathbf{P}, \mathbf{Q}, \overline{\mathbf{Q}}$ we get

$$\mathbf{S}(\mathbf{P}) = v_P \cdot v'_P \mathbf{S}(P), \ \mathbf{S}(\overline{\mathbf{Q}}) = v_{\overline{Q}} v'_{\overline{Q}} \mathbf{S}(\overline{Q})$$
$$\mathbf{S}(\mathbf{Q}) = v_Q \cdot v'_Q \mathbf{S}(Q), \ \mathbf{S}(\overline{Q}) = \mathbf{S}(Q).$$

We have $v_{\overline{Q}} = v_Q / v_P, v'_{\overline{Q}} = v'_Q / v'_P$ by the definition of reductions 1.12.4 Now (i) follows immediately. □

Remark 1.12.8 *If* $\mathbf{S}(\mathbf{P}) = S(P) = 0$, *then the quotients in (i) are interpreted as* ∞ *or* $0:0$.

We remember that $-4d \cdot \mathbf{S}(A), d$ the order of the reduction $\overline{A}$ of A, is the signature defect of A. We define for $\mathbf{Q} = O/A$ the *defect* of $\mathbf{Q}$ by

$$\operatorname{def}(\mathbf{Q}) = \operatorname{def}(A) = -4d_Q \cdot \mathbf{S}(\mathbf{Q}) \tag{1.12.10}$$

Corollary 1.12.9 *With the notations of the proposition it holds that*

$$\operatorname{def}(\mathbf{Q}) : \operatorname{def}(\mathbf{P}) = \operatorname{def}(\overline{\mathbf{Q}}) : \operatorname{def}(P) = \mathit{deg}\,(\overline{\mathbf{Q}}) \cdot (\operatorname{def}(Q) : \operatorname{def}(P)).$$

□

The property (ii) of Proposition 1.12.7 and its analogon for the defect can also be understood as multiplicative properties. The reduction index of a finite morphism $\mathbf{f} : \mathbf{P} \longrightarrow \mathbf{Q}$ or of $\mathbf{P}$ is the number

$$\operatorname{ind}_{\mathbf{red}}(\mathbf{f}) = \operatorname{ind}_{\mathbf{red}}(\mathbf{P}) = v_P \cdot v'_P.$$

Corollary 1.12.10

$$\mathbf{S}(\mathbf{P}) = \mathit{ind}_{red}(\mathbf{P}) \cdot \mathbf{S}(P)$$
$$\operatorname{def}(\mathbf{P}) = \mathit{ind}_{red}(\mathbf{P}) \cdot \operatorname{def}(P)$$

1.13 Quotients and Coverings of Modifications

Let $\mathbf{P} = \left(U_P, P; v_P C_P, v'_P C'_P\right), \mathbf{Q} = \left(V_Q, Q; v_Q C_Q, v'_Q C'_Q\right)$ be abelian points. A *modification* $\varphi : \mathbf{P} \longrightarrow \mathbf{Q}$ of abelian points is supported by a morphism $\varphi : U_P \longrightarrow U_Q$ such that $U_P^* = U_P \setminus C'_P \longrightarrow U_P^\bullet = U_Q \setminus \{Q\}$ is bijective. We call also $\mathbf{P}$ a *modification* of $\mathbf{Q}$. We work with algebraic morphisms. Then φ is a birational morphism contracting C'_P.

There is a simple method to produce modifications. Let $Q \in U$ be an arranged cyclic point and $\widetilde{U}_Q \longrightarrow U_Q$ the minimal resolution of Q. The preimage of Q is a linear tree of projective lines $E_Q = L_1 + \ldots + L_l$ as described in Figure 1.6.2. Now we blow down $L_1, \ldots, L_{k-1}$ to a cyclic point P and $L_{k+1}, \ldots, L_l$ to a cyclic point P'. The resulting surface is denoted by U_P. We have a commutative diagram:

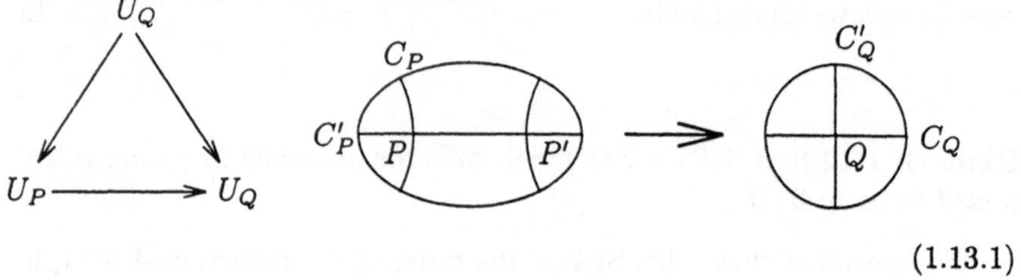

(1.13.1)

with natural identifications $C_P = C_Q$, $C'_P = L_k$. The local restriction $\mathbf{P} \longrightarrow \mathbf{Q}$ is a modification. Instead of the minimal resolution of the cyclic singularity one can use more generally arbitrary resolutions with a linear configuration of the resolving curve as described in Figure 1.6.2 by blowing up intersection points of the curves in the minimal resolution and so on. For example, we can start with the regular point Q. We blow up Q, then the intersection point of the exceptional line with C'_Q, then again the intersection point of C'_Q with the resolution curve and so on, say e times. By abuse of language we say that the point Q has been *blown up e times at* C'_Q. Now, we contract simultaneously all components of the resolution curves not intersecting C'_Q, that means all −2–lines. We get a commutative diagram pictured in Figure 1.13.1: Adding arbitrary weights we obtain a morphism $\mathbf{P} \longrightarrow \mathbf{Q}$ of abelian points.

The morphism $V \longrightarrow U$ is nothing else than the σ_e*–process* (at C'_Q) described in diagram (1.4.4). We have only to check that the cyclic singularity P in Figure 1.13.1 is of type $\langle e, e-1\rangle$. The graphs of the resolution curves of P and of the cyclic singularity Q_e of type $< e, e-1 >$ coincide because of the continued fraction

$$\frac{e-1}{e} = 2 - \frac{1|}{|2} - \ldots - \frac{1|}{|2} \tag{1.13.2}$$

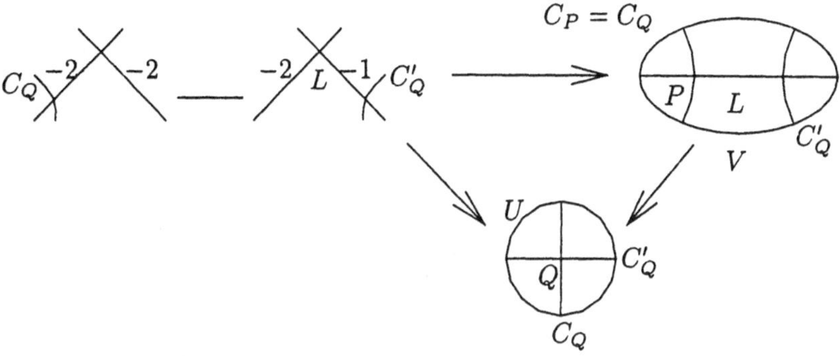

Figure 1.13.1

of length $e-1$ and HIRZEBRUCH'S Proposition 1.7.1

We want to study quotients of modifications. A quotient of the σ_e-process has been already introduced in Figure 1.6.1. Restricting this diagram to neighbourhoods of P_1 and Q_1 we obtain a commutative diagram

$$\begin{array}{ccc} P_1 & \xrightarrow{\sigma_e} & O \\ {\scriptstyle \langle d,e\rangle}\downarrow & & \downarrow{\scriptstyle \langle d,e\rangle} \\ \mathbf{Q}_1 & \longrightarrow & Q \end{array}$$

of abelian points induced from σ_e by quotients. The three points P_1, O, Q are reduced. The quotient point $\mathbf{Q}_1$ is endowed with a non-trivial branch weight along $\overline{L}$.

Now consider more generally the following situation. Let $P \longrightarrow \mathbf{Q}$ be a reduced quotient morphism, $\mathbf{Q} = P/A, A$ a finite abelian group, and $R \longrightarrow P$ a modification such that the group action can be extended to U_R. The quotient point R/A is denoted by $\mathbf{S}$. We obtain commutative diagrams:

$$\begin{array}{ccc} R & \xrightarrow{\psi} & P \\ A\downarrow\mathbf{g} & & A\downarrow\mathbf{f} \\ \mathbf{S} & \xrightarrow{\varphi} & \mathbf{Q}, \end{array} \qquad \begin{array}{ccc} U_R & \xrightarrow{\psi} & U_P \\ A\downarrow & & \downarrow A \\ U_S & \xrightarrow{\varphi} & U_Q \end{array} \tag{1.13.3}$$

Definition 1.13.1 In the situation of (1.13.3) we call φ a *quotient* of ψ, ψ a *finite covering* (*quasi-uniformization*) of φ, and $\mathbf{g}$ a *modification* of $\mathbf{f}$.

Definition 1.13.2 A *local linear arrangement* is a quadrupel $\mathbf{E} = (U, E; \mathbf{C}, \mathbf{C}')$, where $U = U_E$ is a surface (germ along E), E is a linear tree of weighted projective lines on U; $\mathbf{C}, \mathbf{C}'$ are weighted curve germs on U intersecting E in the first or last component, respectively, and only there such that all intersection points of C, C' and components of E are arranged cyclic points arranged by the intersecting curves, see Figure 1.13.2

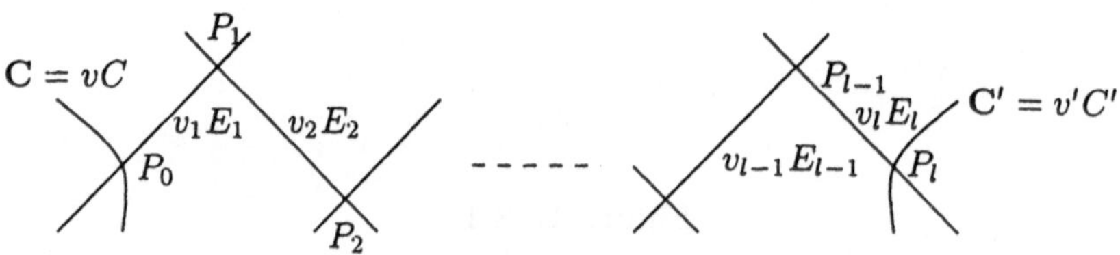

$E = E_1 + \ldots + E_l$, v, v', $v_i \in \mathbb{N}_+$

Figure 1.13.2

The only admitted surface singularities on U_E are the points $P_o, \ldots, P_l$. The graph of E is defined to be

$$\overset{}{\underset{v_0 = v}{\circ}} \xrightarrow{\langle d_0, e_0\rangle} \overset{s_1}{\underset{v_1}{\bullet}} \xrightarrow{\langle d_1, e_1\rangle} \overset{s_2}{\underset{v_2}{\bullet}} \cdots \longrightarrow \overset{s_{l-1}}{\underset{v_{l-1}}{\bullet}} \xrightarrow{\langle d_{l-1}, e_{l-1}\rangle} \overset{s_l}{\underset{v_l}{\bullet}} \xrightarrow{\langle d_l, e_l\rangle} \underset{v' = v_{l+1}}{\circ} \tag{1.13.4}$$

where s_i is the selfintersection number of E_i on U and $\langle d_i, e_i\rangle$ is the type of the arranged reduction P_i of the abelian point $\mathbf{P}_i = (U_{P_i}, P_i; v_i E_i, v_{i+1} E_{i+1})$ lying on $\mathbf{E}$ (E_0, E_{l+1} means C or C', respectively). $\mathbf{E}$ is called a *resolution* of $\mathbf{P} = (U_P, P; vC_P, v'C'_P)$, if E can be contracted to P sending C, C' to C_P or C'_P, respectively. This resolution is denoted by $\mathbf{E} \longrightarrow \mathbf{P}$. Together with the realizing morphism $U \longrightarrow U_P$ it is also called a *contraction* with contraction point $\mathbf{P}$. $\mathbf{E}$ is called a *singularity resolution* (of $\mathbf{P}$), if U_E does not support any surface singularity, that means that the points P_i are regular.

Let P be a reduced (arranged) abelian point and $E \longrightarrow P$ be a reduced resolution of P, that means that E is a P–resolving linear arrangement $\mathbf{E}$ with trivial weights $v_i = 1$, $i = 0, \ldots, l+1$.

A finite group action on P, more precisely on U_P, C, C', can be sometimes extended to an action on U_E, for example, if E is the minimal resolution of P, $U_E \setminus E \simeq U_P \setminus \{P\}$. Then the corresponding finite group A acts around each component of E. With obvious notations we get a commutative diagram

$$\begin{array}{ccc} E = E_P & \longrightarrow & P \\ \downarrow A & & \downarrow A \\ E_P/A = \mathbf{E_Q} & \longrightarrow & \mathbf{Q} = P/A \end{array} \tag{1.13.5}$$

The weights of the components of $\mathbf{E}_Q$ are uniquely defined as ramification indices of the finite covering $U_E \longrightarrow U_E/A$ along the components of E. The resolution $\mathbf{E_Q} \longrightarrow \mathbf{Q}$ is called a *quotient* of $E_P \longrightarrow P$, and $E_P \longrightarrow P$ is called a *finite (reduced, abelian) cover* of $\mathbf{E_Q} \longrightarrow \mathbf{Q}$. We wish to calculate the data of $\mathbf{E_Q}$ by means of those of E_P in the special case when E_P is a singularity resolution of P. In the following graph diagram (1.13.6) corresponding to (1.13.5) we omit as usual trivial weights and types:

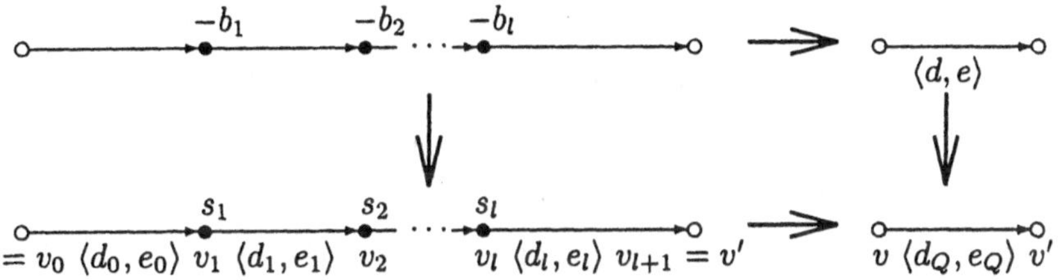

$$\begin{array}{ccc} \circ\!\!\longrightarrow\!\!\overset{-b_1}{\bullet}\!\!\longrightarrow\!\!\overset{-b_2}{\bullet}\cdots\!\!\longrightarrow\!\!\overset{-b_l}{\bullet}\!\!\longrightarrow\!\!\circ & \Longrightarrow & \circ\!\!\xrightarrow[\langle d,e\rangle]{}\!\!\circ \\ \Big\downarrow & & \Big\downarrow \\ \underset{v=v_0}{\circ}\xrightarrow[\langle d_0,e_0\rangle]{}\underset{v_1}{\overset{s_1}{\bullet}}\xrightarrow[\langle d_1,e_1\rangle]{}\underset{v_2}{\overset{s_2}{\bullet}}\cdots\longrightarrow\underset{v_l}{\overset{s_l}{\bullet}}\xrightarrow[\langle d_l,e_l\rangle]{}\underset{v_{l+1}=v'}{\circ} & \Longrightarrow & \underset{v}{\circ}\xrightarrow[\langle d_Q,e_Q\rangle]{}\underset{v'}{\circ} \end{array} \tag{1.13.6}$$

In order to dispose on short notations for the occurring curves and points we write the local linear arrangements as sum of curves and points:

$$\begin{aligned} E_P &= L_1 + L_2 + \ldots + L_1 + P_0 + P_1 + +P_l \\ \mathbf{E_Q} &= v_1 E_1 + v_2 E_2 + \ldots + v_l E_l + \mathbf{Q}_0 + \mathbf{Q}_1 + \cdots + \mathbf{Q}_l \\ P_i &= L_{i-1} \cap L_i \ (L_0 = C, L_{l+1} = C'), \\ \mathbf{Q}_i &= P_i/A,\ E_i = L_i/A, L_i \simeq \mathbb{P}^1. \end{aligned} \tag{1.13.7}$$

C. We consider the quotient surface (germ) $(X/A, C/A)$ along C/A. The quotient curve C/A is smooth but it supports finitely many cyclic surface singularities of X/A. The action of A on X is assumed to be effective, that means that $A \subseteq \operatorname{Aut} X$, the automorphism group of X. In general A does not act effectively on C. The ineffective kernel is denoted by Z. There is an exact sequence

$$1 \longrightarrow Z \longrightarrow A \longrightarrow \overline{A} \longrightarrow 1$$

of finite abelian groups. The group $\overline{A}$ acts effectively on C and $C/A = C/\overline{A} = \overline{C}$. The subgroup Z of A is cyclic because it is a reflection subgroup of A w.r.t. one axis of a coordinate system on the tangent space $T_P(X) \simeq \mathbb{C}^2, P \in C$, diagonalizing A. With a similar argument we see that also the stationary subgroups $\overline{A}_P$ of $\overline{A}$ are cyclic groups. The orders of Z and $\overline{A}_P$ are denoted by v or d_P, respectively. The graphs $\underset{v}{\circ}\xrightarrow[\langle \overline{d}_P, \overline{e}_P\rangle]{}\underset{v'_P}{\circ}$ of the abelian points $\overline{\mathbf{P}} = P/A_P = \left(X/A, \overline{P}; v\overline{C}, v'_P\overline{C}'\right), A_P$ the stationary group of A at $P \in C$, fix further notations. It holds that

$$|A_P| = |Z| \cdot |\overline{A}_P| = v \cdot d_P = vv'_P\overline{d}_P, e_P = v'_P\overline{e}_P \tag{1.14.1}$$

because of the cancellation rule $O/\langle d_P, e_P\rangle \cong O/\left\langle \frac{d_P}{v'_P}, \frac{e_P}{v'_P}\right\rangle$.

The aim of this section is to relate the selfintersection numbers $(C^2) = (C^2)_X$ and $\left(\overline{C}^2\right) = \left(\overline{C}^2\right)_{\widetilde{X/A}}$ taken on the minimal singularity resolution $\widetilde{X/A}$ of X/A. In two special cases (a), (b), below, the relation we look for can be directly derived from the degree formula

$$(p^*(D))^2 = \deg(p) \cdot (D^2) \tag{1.14.2}$$

for finite coverings $p : X \longrightarrow Y$, of smooth surfaces, D a smooth compact curve on Y (see [KUR])

(a) $A = Z$

(b) $Z = 1$ and all stationary groups $A_P, P \in C$, are generated by "vertical" reflections (w.r.t. C)

Lemma 1.14.1 *In the two cases above it holds that*

$$\left(\overline{C}^2\right) = |Z| \cdot (C^2), \tag{1.14.3.a}$$

$$(C^2) = |A| \cdot \left(\overline{C}^2\right), \tag{1.14.3.b}$$

respectively.

Proof In both cases the quotient points $\overline{\mathbf{P}} = P/A_P$ on $\overline{C} \subset X/A$ are regular points. The projection $p : X \longrightarrow X/A$ is finite. So we can apply the degree formula. The curve $p^*(\overline{C})$ has only one component because A acts on C, that means that A does not move C to another curve. In case (a) one checks easily that $p^*(\overline{C}) = |Z| \cdot C$. Now (1.14.3.a) follows from the degree formula (1.14.2). For the second case we have only to check that $p^*(\overline{C}) = C$. Looking at local parameters at a general point P of C and its image point $\overline{P}$ it is easy to see that $\overline{C}$ has the same local equation $f = 0, f \in \mathbb{C}(X/A) \subset \mathbb{C}(X)$, at $\overline{P}$ as C has at P. This proves our lemma. □

In the general situation we push off from $\overline{C}$ the cyclic singularities $\overline{P}$ in order to obtain a smooth surface germ along $\overline{C}$ which allows to apply the degree formula for the calculation of selfinteresection numbers. The pushing procedure is prepared on X by means of σ_{e_P}–processes at all points $P \in C$ with honestly singular quotient points $\overline{P} = P/A_P$.

The action of A extends to an action on the arising surface X_n by the following reasoning. It suffices to assume that there is only one surface singularity $\overline{P}$ on $\overline{C} \subset \overline{X}$, say of type $\langle d, e\rangle$. The σ_e–process at a preimage P of $\overline{P}$ is received by blowing up e times the point P at C and contracting the –2–lines to the point P_e (see Figure 1.13.1). The action of A_P around P can be extended step by step to the e times blown up surface X_P because of a universal property of the σ–process. The group A_P acts also on the –2–curves, hence on the complement of these curves in X_P. Therefore, it acts on the complement of P_e in the contracted surface X_e. By HARTOG'S theorem (see e.g. [HER]) the action can be extended to X_e. We can do the same simultaneously at any preimage of $\overline{P}$. For equivariant reasons we obtain an action of A on X_n and a commutative diagram:

$$\begin{array}{ccc} X_n & \longrightarrow & X \\ \downarrow & & \downarrow \\ X_n/A & \longrightarrow & X/A \end{array} \tag{1.14.4}$$

We assume now that $A = \overline{A}$, $Z = 1$. Locally the quotient diagram is described in Figure 1.6.1. The cyclic singularities have been pushed off from C/A and the new singularities are the derived singularities of the original ones (see (1.6.3)). The morphism $X_n/A \longrightarrow X/A$ is the first step of the stepwise resolution of X/A (see (1.6.4), (1.6.5)). Therefore, we have a decomposition of the minimal singularity resolution: $\widetilde{X/A} \longrightarrow X_n/A \longrightarrow$

X/A. The next steps of the resolution do not change anything along the smooth curve $C \subset X_n/A$.

The quotient morphism $X_n \longrightarrow X_n/A$ is of type (b) described before in Lemma 1.14.1. Namely, X_n/A is smooth along C/A. Therefore, the stationary groups $A_P, P \in C$, are generated by reflections. This follows from CHEVALLEY's result 1.1.1 Under our assumption $Z = 1$ there are no "horizontal" reflections in A_P. Consequently A_P is generated by "vertical" reflections. Now we can apply the formula (1.14.3.b). It follows that

$$(C^2)_{X_n} = |\overline{A}| \cdot \left(\overline{C}^2\right)_{X_n/A} = |\overline{A}| \cdot \left(\overline{C}^2\right)_{\widetilde{X/A}} = |\overline{A}| \cdot \left(\overline{C}^2\right). \qquad (1.14.5)$$

We have to compare $(C^2)_{X_n}$ and $(C^2)_X$. Looking at C'_Q in diagram Fig. 1.13.1 we learn that each σ_e–process decreases the selfintersection number of C by e. It follows that

$$(C^2)_{X_n} = (C^2) - \sum_{\mathbf{P}} \sum_{P \in C, P \to \mathbf{P}} e_P = (C^2) - \sum_{\mathbf{P}} \left[\overline{A} : \overline{A}_P\right] \cdot e_P \qquad (1.14.6)$$

Comparing (1.14.5) and (1.14.6) we get together with (1.14.1) the relation

$$\left(\overline{C}^2\right) = \frac{(C^2)}{|\overline{A}|} - \sum_{\overline{P}} \frac{\overline{e}_P}{\overline{d}_P}. \qquad (1.14.7)$$

Now it is easy to derive the formula for the non-reduced case $A \neq \overline{A}$, $Z \neq 1$. We divide first X by Z and consider the decomposition
$X \longrightarrow X/Z = X' \longrightarrow X/A = X'/\overline{A}$.

From (1.14.3.a) it follows that

$$|Z| \cdot (C^2)_X = (C^2)_{X'}$$

Together with (1.14.7) we proved the following

Proposition 1.14.2 *Let (X, C) be a smooth surface germ along a smooth compact curve C and A a finite abelian group acting effectively on X, and on C with ineffective kernel Z. Then the selfintersection number (C^2) of $\overline{C} = C/A$ on the minimal resolution $\widetilde{X/A}$ of singularities of X/A can be calculated by the formula*

$$\left(\overline{C}^2\right) = \left(\overline{C}^2\right)_{\widetilde{X/A}} = \frac{|Z|^2 \cdot \left(C^2\right)_X}{|A|} - \sum_{\overline{P}} \frac{\overline{e}_P}{\overline{d}_P},$$

$$\overline{\mathbf{P}}: \underset{v\ \langle \overline{d}_P, \overline{e}_P\rangle\ v_P}{\circ\!\!-\!\!\!-\!\!\!-\!\!\!-\!\!\circ}, \quad v = |Z| \tag{1.14.8}$$

where the sum is taken over all abelian points $\overline{\mathbf{P}} = P/A$.

1.15 The Bridge Algorithm

We continue our investigations of Section 13. Especially we want to determine the type and weights of the singularity $\mathbf{Q} = P/A$ if the type of P is known. It will be sufficient to know additionally the ramification index of $P \longrightarrow \mathbf{Q}$ along one of the singular coordinate axis at P, say $v_0 = v_Q$. We consider an arbitrary equivariant singularity resolution $E_P \longrightarrow P$, for example the minimal one. The factorization by A yields a commutative diagram (1.13.5). The corresponding graph diagram (1.13.6) and (1.13.7) fix further notations. The bridge algorithm is a stepwise transport of data from one point $\mathbf{P}_{i-1}$ of $\mathbf{E}_Q$ to the next one $\mathbf{P}_i$. More precisely, the algorithm allows to calculate effectively the data v_i, d_i, e_i, s_i knowing $v_{i-1}, d_{i-1}, e_{i-1}, b_i$.

1.15.1 Steps of the Bridge Algorithm:

$$v_{i-1}, d_{i-1}, e_{i-1}, b_i \Longrightarrow v_i, d_i, e_i, s_i, i = 1, \ldots, l$$

In terms of the graph diagram (1.13.6) the whole algorithm is described in the following scheme:

1.15.2 Bridge Algorithm:

$$\boxed{\underset{v_0\ \langle d_0, e_0\rangle}{\circ\!\!\longrightarrow}} \Longrightarrow \boxed{\overset{s_1}{\underset{v_1\ \langle d_1, e_1\rangle}{\bullet\!\!\longrightarrow}}} \Longrightarrow \cdots \Longrightarrow \boxed{\overset{s_l}{\underset{v_l\ \langle d_l, e_l\rangle\ v_{l+1}}{\bullet\!\!\longrightarrow\!\!\circ}}}$$

For the algorithm we need the following three identities:

$$e_{i-1} \cdot e_{i-1}' \equiv 1 \bmod d_{i-1}, i = 1, \ldots, l+1 \tag{1.15.1}$$

$$|A| = |A_{P_i}| = v_{i-1}v_i d_{i-1}, E_P \ni P_i \longrightarrow P_i/A = \mathbf{Q}_i \tag{1.15.2}$$

$$s_i = \frac{v_i^2}{|A|} \cdot (-b_i) + \frac{e'_{i-1}}{d_{i-1}} + \frac{e_i}{d_i} \tag{1.15.3}$$

The numbers e'_i come from the opposite points $\mathbf{Q}_i^{\perp}$: $\underset{v_{i-1}}{\circ}\!\xleftarrow[\langle d_i, e'_i\rangle]{}\!\!\underset{v_i}{\circ}$ of the abelian points $\mathbf{Q}_i$.

The relation (1.15.2) can be adapted from (1.12.8). The identity (1.15.3) follows from Proposition 1.14.2 applied to the covering $(U_{E_P}, L_i) \underset{A}{\longrightarrow} (U_{E_Q}, E_i)$. Namely, Q_{i-1} and Q_i are the only singular points of U_{E_Q} on E_i. The corresponding arranged points $P_{i-1}/A, P_i/A$ are $\mathbf{Q}_{i-1}^{\perp}$ and $\mathbf{Q}_i$ because they have to be arranged by E_i in order to apply (1.14.8) for $C = L_i$, $\overline{C} = C/A = E_i$.

Now we are able to operate the algorithm step 1.15.1 With e_{i-1}, d_{i-1} we know also e'_{i-1} by (1.15.1) and the condition $0 \leq e'_{i-1}/d_{i-1} < 1$. The number v_i can be taken from (1.15.2). Now, in (1.15.3) only the numbers $\frac{e_i}{d_i}$ and s_i are unknown. We know that s_i is an integer because it is the self-intersection number of E_i. Since $0 \leq \frac{e_i}{d_i} < 1$ the fraction e_i/d_i is uniquely determined only by this property. With e_i/d_i we know also s_i. The natural numbers e_i, d_i are relative prime. Therefore they are both uniquely determined by e_i/d_i. The bridge algorithm has been established.

Especially we proved the following

Lemma 1.15.3 *Let $E_P \longrightarrow P$ be the minimal singularity resolution of the arranged reduced cyclic point P and A a finite abelian group acting around P. Then the linear resolution $\mathbf{E_Q} = E_P/A$ of $\mathbf{Q} = P/A$ and the abelian point $\mathbf{Q}$ itself are uniquely determined by the type $\langle d_P, e_P\rangle$ of P and the representation type ${}_{v_o}\langle d_o, e_o\rangle$ of A at Q_o.*

Corollary 1.15.4 *If A is a cyclic group of order n and $v = v_o = n$ (or $v' = v_{l+1} = n$), then $\mathbf{E_Q} = E_P/A$, hence $\mathbf{Q}$, is uniquely determined by the type $\langle d_P, e_P\rangle$ of P.*

The reader is invited to check the following

Example 1.15.5 Let P be of type $\langle 6, 5\rangle$, A cyclic of order 6 and $v = v_o = 6$. Using continued fractions (see (1.13.2)) one finds for the minimal singularity resolution of P the graph

$$E_P : \circ\!\!-\!\!\!\!\underset{}{\overset{-2}{\bullet}}\!\!-\!\!\overset{-2}{\bullet}\!\!-\!\!\overset{-2}{\bullet}\!\!-\!\!\overset{-2}{\bullet}\!\!-\!\!\overset{-2}{\bullet}\!\!-\!\!\circ$$

The bridge algorithm yields

$$\underset{6\ \langle 1,0\rangle}{\circ\!\!\longrightarrow} \Longrightarrow \overset{-1}{\underset{1\ \langle 3,2\rangle}{\bullet\!\!\longrightarrow}} \Longrightarrow \overset{-2}{\underset{2\ \langle 1,0\rangle}{\bullet\!\!\longrightarrow}} \Longrightarrow \overset{-3}{\underset{3\ \langle 1,0\rangle}{\bullet\!\!\longrightarrow}}$$

$$\Longrightarrow \overset{-2}{\underset{2\ \langle 3,2\rangle}{\bullet\!\!\longrightarrow}} \Longrightarrow \overset{-1}{\underset{1\ \langle 1,0\rangle\ 6}{\bullet\!\!\longrightarrow\!\!\circ}}, \tag{1.15.4}$$

$$\mathbf{E_Q}: \underset{6\ \langle 1,0\rangle}{\circ\!\!\longrightarrow} \overset{-1}{\underset{1\ \langle 3,2\rangle}{\bullet\!\!\longrightarrow}} \overset{-2}{\underset{2\ \langle 1,0\rangle}{\bullet\!\!\longrightarrow}} \overset{-3}{\underset{3\ \langle 1,0\rangle}{\bullet\!\!\longrightarrow}} \overset{-2}{\underset{2\ \langle 3,2\rangle}{\bullet\!\!\longrightarrow}} \overset{-1}{\underset{1\ \langle 1,0\rangle\ 6}{\bullet\!\!\longrightarrow\!\!\circ}}$$

If we resolve the singular points $Q_1, Q_4 \in E_Q$ of type $\langle 3,2\rangle$ and forget the weights we get the minimal resolution $\tilde{E}_Q$ of the reduced linear arrangement E_Q with graph

$$-1 \quad -2 \quad -2 \quad -2 \quad -3 \quad -2 \quad -2 \quad -2 \quad -1 \tag{1.15.5}$$

Consequently $\tilde{E}_Q$, hence E_Q, is contractible to the regular point,

$$\mathbf{Q}: \underset{6\ \langle 1,0\rangle\ 6}{\circ\!\!\longrightarrow\!\!\circ}$$

This is not surprising because of the following

Remark 1.15.6 If A is cyclic of order n, P of type $\langle n,e\rangle$ and $v_P = n$ (or $v'_P = n$), then

$$\mathbf{Q} = P/A: \underset{n\ \langle 1,0\rangle\ n}{\circ\!\!\longrightarrow\!\!\circ}$$

Proof By Corollary 1.15.4 it suffices to check one special case, where the hypothesis of 1.15.6 is satisfied. Let Σ be the abelian reflection subgroup of $\mathbb{Gl}_2(\mathbb{C})$ generated by two reflections of order n. It contains the subgroup $\langle n,e\rangle$, *g.c.d.*$(n,e) = 1$. Then we have a decomposition

$$O \xrightarrow{\langle n,e\rangle} P = O/\langle n,e\rangle \xrightarrow{A} \mathbf{Q} = P/A, \ A = \Sigma/\langle n,e\rangle$$
$$= O/\Sigma$$

$$\underset{1\ \langle n,e\rangle\ 1}{\circ\!\!\longrightarrow\!\!\circ} \Longrightarrow \underset{n\ \langle 1,0\rangle\ n}{\circ\!\!\longrightarrow\!\!\circ}$$

□

Also the symmetry in $\mathbf{E_Q}$ and $\tilde{E}_Q$ (see (1.15.4) and (1.15.5)) is not surprising:

Remark 1.15.7 If (the graph of) P is symmetric and $v_P = v'_P = n = |A|$, then also $\mathbf{Q} = P/A$ is symmetric.

Proof We have to show that $\mathbf{Q}^\perp = \mathbf{Q}$ or

$$\underset{v'}{\circ}\xrightarrow{\langle d_Q, e'_Q\rangle}\underset{v}{\circ} \quad = \quad \underset{v}{\circ}\xrightarrow{\langle d_Q, e_Q\rangle}\underset{v'}{\circ}$$

For a resolution $\mathbf{E_Q} \longrightarrow \mathbf{Q}$ we introduce the *opposite resolution* as $\mathbf{E}_\mathbf{Q}^\perp \longrightarrow \mathbf{Q}^\perp$ of $\mathbf{Q}^\perp$. We inverse all arrows in the graph of $\mathbf{E_Q}$ see (1.13.6) and substitute $\langle d_i, e_i\rangle = \langle d_{P_i}, e_{P_i}\rangle$ by $\langle d_{P_i}, e'_{P_i}\rangle = \langle d_i, e'_i\rangle$ Now let $E_P \longrightarrow P$ be the minimal resolution of P. The symmetry of P means that $e_P = e'_P$. Consequently d_P/e_P and d_P/e'_P have the same continued fraction, hence the minimal resolutions of P and $P^\perp$ coincide by HIRZEBRUCH'S Proposition 1.7.1, $E_{P^\perp} = E_P = E_P^\perp$. The bridge algorithm for $\mathbf{E_Q}$ starts at $\mathbf{Q}_o = P_o/A$. The graph of $\mathbf{Q}_o$ is $\underset{n}{\circ}\xrightarrow{\langle 1,0\rangle}\underset{1}{\circ}$ because $v_o = v_P = |A|$, hence A is a cyclic reflection group of order n at P_o. By assumption we have the same initial data at $Q_l^\perp$, the starting point of the bridge algorithm for $\mathbf{E}_{\mathbf{Q}^\perp}$. So both bridge algorithms 1.15.2 produce step by step the same data. Forgetting branch weights we see that the minimal resolutions $\tilde{E}_Q$ and $\tilde{E}_{Q^\perp}$ of E_Q or $E_{Q^\perp}$, respectively, coincide. Now blow down step by step -1–curves as long as possible. We obtain the minimal resolutions of Q and $Q^\perp$. Both are the same, hence $Q = Q^\perp$, $e_Q = e'_Q$. Since also $v = n = v' = v_o = v_{l+1}$ it follows that $\mathbf{Q} = \mathbf{Q}^\perp$. □

1.16 First Orbital Properties

The *relative category* of the *category of abelian points* is defined as follows. Consider morphisms $\varphi : \mathbf{R} \longrightarrow \mathbf{P}$ of the category of abelian points. For abbreviation we denote such an object by $\mathbf{R/P}$ and call it sometimes a *relative object*. Morphisms $\mathbf{F} : \mathbf{R/P} \longrightarrow \mathbf{S/Q}$ are commutative diagrams

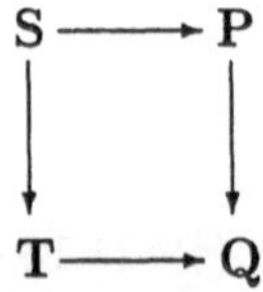

in the category of abelian points.

The *catgegory of modifications of abelian points* is the complete subcategory with modifications of abelian points as objects. A modification $\mathbf{S}/\mathbf{P}$ is called *reduced* if the weights of $\mathbf{S}$ and $\mathbf{P}$ are trivial. For reduced objects we write S/P. A *finite reduced covering* of $\mathbf{S}/\mathbf{Q}$ is a commutative quotient diagram:

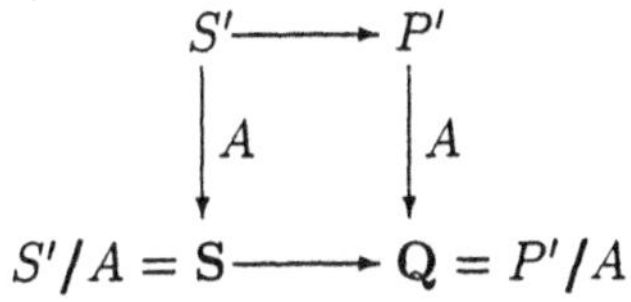

$$\begin{array}{ccc} S' & \longrightarrow & P' \\ \downarrow{\scriptstyle A} & & \downarrow{\scriptstyle A} \\ S'/A = \mathbf{S} & \longrightarrow & \mathbf{Q} = P'/A \end{array} \tag{1.16.1}$$

with reduced modification S'/P' and finite abelian group A. A morphism $\mathbf{F} : \mathbf{S}/\mathbf{P} \longrightarrow \mathbf{T}/\mathbf{Q}$ is called *finite* , if there exists a commutative diagram:

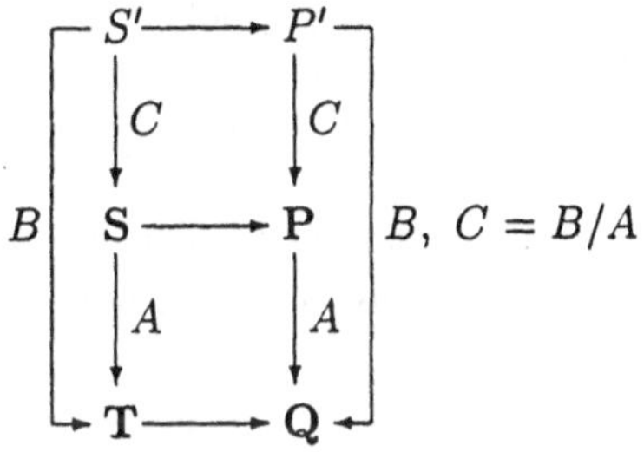

$$\begin{array}{ccc} S' & \longrightarrow & P' \\ \downarrow{\scriptstyle C} & & \downarrow{\scriptstyle C} \\ \mathbf{S} & \longrightarrow & \mathbf{P} \\ \downarrow{\scriptstyle A} & & \downarrow{\scriptstyle A} \\ \mathbf{T} & \longrightarrow & \mathbf{Q} \end{array} \qquad B:\ S' \to \mathbf{T},\quad B:\ P' \to \mathbf{Q},\quad C = B/A \tag{1.16.2}$$

with finite reduced cover S'/P' of $\mathbf{S}/\mathbf{P}$ and of $\mathbf{T}/\mathbf{Q}$. Notice that

$$v_S = v_P, v_T = v_Q \tag{1.16.3}$$

holds in this situation because the first coordinate curves are not blown down along the modifications. Hence the action of the groups on this curve is the same in S' as well as in P'.

Let $\mathfrak{C}$ be a category containing a special class of morphisms called *finite morphisms*. A *degree function* corresponds to any morphism f a number $d(f)$ with the property

$$d(g \cdot f) = d(g) \cdot d(f). \tag{1.16.4}$$

Let γ correspond to each object C of $\mathfrak{C}$ a rational number $\gamma(C)$ such that $\gamma(C) = \gamma(C')$ for isomorphic objects C, C'. We say that γ is *orbital with respect to the degree function* d, if for any finite morphism $f : B \longrightarrow C$ the "degree formula"

$$\gamma(B) = d(f) \cdot \gamma(C) \tag{1.16.5}$$

is satisfied. In this case we call the numbers $\gamma(C)$, $C \in Ob(\mathfrak{C})$, *orbital invariants.* We correspond also to a morphism $f : B \longrightarrow C$ a rational number $\gamma(f)$ by the formula

$$\gamma(f) = \gamma(B) - d(f) \cdot \gamma(C) \tag{1.16.6}$$

Immediately, we obtain the additive relation

$$\gamma(g \cdot f) = \gamma(f) + d(f) \cdot \gamma(g) \tag{1.16.7}$$

For the proof let $g : C \longrightarrow D$ be a morphism. Applying (1.16.6) we get

$$\gamma(g) = \gamma(C) - d(g) \cdot \gamma(D).$$

The multiplication with $d(f)$ yields

$$d(f) \cdot \gamma(g) = d(f) \cdot \gamma(C) - d(g \cdot f) \cdot \gamma(D) \tag{1.16.8}$$

according to (1.16.4). Now we have only to add (1.16.6) and (1.16.8) and to look at the definition of $\gamma(g \cdot f)$.

The "degree formula" (1.16.5) allows to define derived orbital invariants as orbital invariants on the relative category $\mathfrak{C}'$ with "degree formula". The objects of $\mathfrak{C}'$ are the morphism of $\mathfrak{C}$. The morphisms $F = (f_1, f_2) : g' = S/X \longrightarrow g = T/Y$ are commutative diagrams:

$$\begin{array}{ccc} S & \xrightarrow{g'} & X \\ \downarrow f_1 & & \downarrow f_2 \\ T & \xrightarrow{g} & Y \end{array} \tag{1.16.9}$$

The morphism F is called *finite*, if f_1 and f_2 are finite of the same degree $d(F) = d(f_1) = d(f_2)$. The multiplicative property (1.16.4) is obviously satisfied for finite morphisms of $\mathfrak{C}'$. Let $\gamma : Ob(\mathfrak{C}) \longrightarrow \mathbb{Q}$ be orbital with respect to d : Mor $(\mathfrak{C}) \longrightarrow \mathbb{Q}$. We already defined $\gamma(T/Y)$ for relative objects $g : T \longrightarrow Y$. We have to prove the analogon of (1.16.5):

$$\gamma(S/X) = d(F) \cdot \gamma(T/Y) \tag{1.16.10}$$

for finite relative morphisms $F = (f_1, f_2)$. Using the notations of diagram (1.16.9) we find that

$$\begin{aligned} \gamma(S/X) &= \gamma(g') = \gamma(g') + 0 = \gamma(g') + d(f_1) \cdot 0 \\ &= \gamma(g') + d(f_1) \cdot \gamma(f_1) = \gamma(f_1 \cdot g') = \gamma(g \cdot f_2) \\ &= \gamma(f_2) + d(f_2) \cdot \gamma(g) = 0 + d(F) \cdot \gamma(g) \\ &= d(F) \cdot \gamma(g) = d(F) \cdot \gamma(T/Y). \end{aligned}$$

We used that γ is trivial on the subclass of finite morphisms of $\mathfrak{C}$. This follows from the definition (1.16.6) and the "degree formula" (1.16.5). We applied also the relation (1.16.7).

Let $\mathfrak{C}$ be a category with degree function d. A *modification* is a morphism of degree 1. A morphism $G = (g, g')$ in $\mathfrak{C}'$ is a *relative modification*, if there is a commutative diagram (1.16.9) and g, g' are modifications in $\mathfrak{C}$. In this case we call also S/T a *modification* of X/Y and, sometimes, X/Y a *contraction* of S/T. The relative objects X/Y and X'/Y' are called *equivalent with respect to modifications*, if there exists a series of relative objects

$$X/Y = X_0/Y_0,\ X_1/Y_1,\ \dots\ , X_n/Y_n = X'/Y'$$

such that X_i/Y_i is a modification or a contraction of X_{i-1}/Y_{i-1} for $i = 1, ..., n$. An important role in all chapters of this monograph plays the following simple lemma. It gives a method for discovering orbital invariants on a category $\mathfrak{C}$ by means of orbital invariants on the category $\mathfrak{C}_1'$ whose objects are the morphisms of $\mathfrak{C}$ of degree 1.

Shifting Lemma 1.16.1 *Let $\mathfrak{C}$ be a category with degree function d, finite morphisms and a function $\gamma : Ob(\mathfrak{C}) \longrightarrow \mathbb{Q}$. The function γ is extended to morphisms, especially to $\mathfrak{C}_1'$ by means of (1.16.6). Assume that*

(i) γ is orbital on $\mathfrak{C}_1'$.

(ii) For any finite morphism f in $\mathfrak{C}'$ there exists a finite morphism f' equivalent to f with respect to modifications such that $\gamma(f') = 0$.

Then γ is orbital w.r.t. the degree function d.

Proof We have to show that γ is trivial on the class of finite morphisms, that means that $\gamma(f) = 0$ for any finite morphisms f. By assumption (ii) it suffices to prove that this triviality is preserved under modifications. With the notations of diagram (1.16.9) we assume that g, g' are modifications and f_1, f_2 are finite morphisms. It suffices to show that

$$\gamma(f_1) = 0 \text{ if and only if } \gamma(f_2) = 0. \tag{1.16.11}$$

We first remark that f_1 and f_2 have the same degree. This follows from (1.16.4) and our assumptions:

$$\begin{aligned} d(f_1) &= 1 \cdot d(f_1) = d(g) \cdot d(f_1) = d(g \circ f_1) \\ &= d(f_2 \circ g') = d(f_2) \cdot d(g') = d(f_2). \end{aligned}$$

So $F = (f_1, f_2)$ is a finite morphism in $\mathfrak{C}_1'$ of degree $d(F) = d(f_1) = d(f_2)$. By assumption (i) we have

$$\gamma(g') = \gamma(S/X) = d(F) \cdot \gamma(T/Y) = d(f_1)\gamma(g). \tag{1.16.12}$$

From $f_2 \circ g' = g \circ f_1$ and (1.16.7) applied to these products it follows that

$$\gamma(g') + \gamma(f_2) = \gamma(f_1) + d(f_1) \cdot \gamma(g).$$

Together with (1.16.12) we see that

$$\gamma(f_1) = \gamma(f_2). \tag{1.16.13}$$

This proves (1.16.11) and our Shifting Lemma. □

Remark 1.16.2 . For the proof of the lemma it is not necessary that the degree function is defined everywhere. It suffices to work with a degree function defined on the class of finite morphisms and on a second class of morphisms called modifications. The latter morphisms must have degree 1. The axiom (1.16.4) has to be satisfied on the class of all compositions of modifications and finite morphisms. So the shifting lemma can be applied to the category of abelian points.

The name of the lemma becomes clear from formula (1.16.13). It says that the value $\gamma(f)$ of finite morphisms f is preserved under modifications. More specially, from the statement (1.16.11) it follows that the degree formula (1.16.5) can be shifted from one finite morphism to another along modifications.

Corollary 1.16.3 *Let $\mathfrak{C}$ be a category with finite morphisms, modifications a degree function for finite morphisms and $\gamma : Ob(\mathfrak{C}) \longrightarrow \mathbb{Q}$ a function on $\mathfrak{C}$. If the derived function γ on the relative category $\mathfrak{C}_1'$ is orbital and $f_1 = S/T \longrightarrow X/Y = f_2$ is a modification of finite relative objects f_1, f_2, then*

$$\gamma(f_1) = \gamma(f_2).$$

□

Now we come to the first example.

Let $\mathbf{P} : \underset{v_P}{\circ}\!\xrightarrow[\langle d_P, e_P\rangle]{}\!\underset{v'_P}{\circ}$ be an abelian point. We call

$$e^1(\mathbf{P}) = 1 - \frac{1}{v'_P d_P} \tag{1.16.14}$$

the *first* EULER *contribution* of $\mathbf{P}$. The degree formula $e^1(\mathbf{P}) = d(f) \cdot e^1(\mathbf{Q})$ for finite morphisms $\mathbf{f} : \mathbf{P} \longrightarrow \mathbf{Q}$ is in general not satisfied. For example, if $\mathbf{f}$ is a uniformization, $\mathbf{P} = O\colon \underset{1\ \langle 1,0\rangle\ 1}{\circ\!\!\longrightarrow\!\!\circ}$ then $e^1(\mathbf{P}) = 0$. But $\mathbf{Q}$ can be arbitrary and $e^1(\mathbf{Q})$ does not vanish in general. In spite of that we will see that the derived function is orbital on the relative category of abelian points. We introduce the following degree function in the category $\mathfrak{C}$ of abelian points. Let $\mathbf{f}$ in diagram:

$$\begin{array}{ll} \mathbf{P}: & \underset{v_P\quad \langle d_P, e_P\rangle \quad v_{P'}}{\circ\!\!\longrightarrow\!\!\circ} \\ \mathbf{f}\big\downarrow A & \\ \mathbf{Q}: & \underset{v_Q\quad \langle d_Q, e_Q\rangle \quad v'_Q}{\circ\!\!\longrightarrow\!\!\circ} \end{array} \tag{1.16.15}$$

be a finite morphism, A the realizing finite abelian group (see (1.12.3)). We set

$$d(\mathbf{f}) = \deg_{\mathrm{red}}(\mathbf{f}) = |A| \cdot v_P/v_Q = \deg(\mathbf{f}) \cdot v_P/v_Q \tag{1.16.16}$$

and call it the *reduced degree* of $\mathbf{f}$. The function d is really a degree function for finite morphisms. For the proof we consider a second finite morphism $\mathbf{g} : \mathbf{Q} \underset{B}{\rightarrow} \mathbf{R}$ induced by a quotient morphism with finite abelian group B. We uniformize $\mathbf{P}$ and get a tower corresponding to finite abelian subgroups $D \subseteq E \subseteq F$ of $\mathbb{Gl}_2(\mathbb{C})$. The abelian tower can be constructed by uniformization as in the proof of Lemma 1.12.1 (see diagram (1.12.5) and Corollary 1.12.3). Now we can calculate

$$\begin{aligned} d(\mathbf{g} \circ \mathbf{f}) &= \deg(g \circ f) \cdot v_P/v_R = |C| \cdot v_P/v_R \\ &= |A| \cdot |B| \cdot v_P/v_R = \deg(\mathbf{f}) \cdot \deg(\mathbf{g}) \cdot v_P/v_R \\ &= (\deg(\mathbf{f}) \cdot v_P/v_Q) \cdot (\deg(\mathbf{g}) \cdot v_Q/v_R) = d(\mathbf{g}) \cdot d(\mathbf{f}). \end{aligned}$$

If $\mathbf{G} : \mathbf{S}/\mathbf{T} \longrightarrow \mathbf{P}/\mathbf{Q}$ is a modification of $\mathbf{f} = \mathbf{f}_2$, then we have, by definition, a commutative diagram

$$\begin{array}{ccc} \mathbf{S} & \xrightarrow{\mathbf{g}'} & \mathbf{P} \\ \mathbf{f}_1\big\downarrow A & & A\big\downarrow \mathbf{f}_2 \\ \mathbf{T} & \xrightarrow{\mathbf{g}} & \mathbf{Q} \end{array} \tag{1.16.18}$$

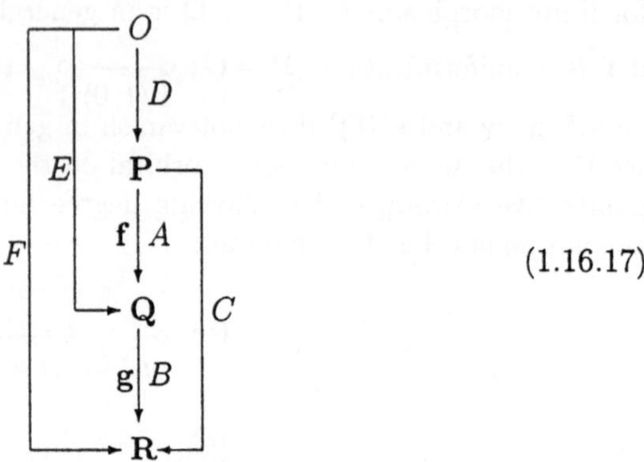

(1.16.17)

We assume that $\mathbf{F} = (\mathbf{f}_1, \mathbf{f}_2) : \mathbf{S}/\mathbf{P} \longrightarrow \mathbf{T}/\mathbf{Q}$ is finite (see Diagram (1.16.2). We remember that $v = v_P$, $v_T = v_Q$ (see (1.16.3)). It follows that

$$d(\mathbf{f}_1) = \deg_{\mathrm{red}}\ (\mathbf{f}_1) = \deg_{\mathrm{red}}\ (\mathbf{f}_2) = d(\mathbf{f}_1). \tag{1.16.19}$$

Namely, $d(\mathbf{f}_1) = |A| \cdot v_S/v_T = |A| \cdot v_P/v_Q = d(\mathbf{f}_2)$. So the *reduced degree* $d(\mathbf{F}) = \deg_{\mathrm{red}}\ (\mathbf{F})$ is correctly defined as $d(\mathbf{f}_1) = d(\mathbf{f}_2)$. Now we can prove that the derived function e^1 is orbital:

Lemma 1.16.4 *Let* $\mathbf{F} : \mathbf{S}/\mathbf{P} \longrightarrow \mathbf{T}/\mathbf{Q}$ *be a finite morphism in the category of modifications of abelian points as described in diagram (1.16.18). Then it holds that*

$$e^1(\mathbf{S}/\mathbf{P}) = \deg_{red}\ (\mathbf{F}) \cdot e^1(\mathbf{T}/\mathbf{Q}) \tag{1.16.20}$$

Proof By the definitions (1.16.6), (1.16.14) and $d(g) = d(g') = 1$ we have

$$\begin{aligned} e^1(\mathbf{T}/\mathbf{Q}) &= e^1(\mathbf{T}) - e^1(\mathbf{Q}) = \frac{1}{v'_Q d_Q} - \frac{1}{v'_T d_T} \\ e^1(\mathbf{S}/\mathbf{P}) &= e^1(\mathbf{S}) - e^1(\mathbf{P}) = \frac{1}{v'_P d_P} - \frac{1}{v'_S d_S} \end{aligned} \tag{1.16.21}$$

We multiply the first identity with

$$d(\mathbf{F}) = d(\mathbf{f}_2) = |A| \cdot v_P/v_Q = d(\mathbf{f}_1) = |A| \cdot v_S/v_T$$

and obtain

$$d(\mathbf{F}) \cdot e^1(\mathbf{T}/\mathbf{Q}) = |A| \cdot \left(\frac{v_P}{v_Q v_Q' d_Q} - \frac{v_S}{v_T v_T' d_T} \right) \tag{1.16.22}$$

Looking at the uniformizing diagram (1.16.17) we remember that

$$|A| = |E|/|D| = \frac{v_Q v_Q' d_Q}{v_P v_P' d_P}$$

and, analogously,

$$|A| = \frac{v_T v_T' d_T}{v_S v_S' d_S}$$

By substitution into (1.16.22) we receive with (1.16.21)

$$d(\mathbf{F}) \cdot e^1(\mathbf{T}/\mathbf{Q}) = \frac{1}{v_P' d_P} - \frac{1}{v_S' d_S} = e^1(\mathbf{T}/\mathbf{Q}).$$

The lemma is proved. □

As we have already remarked the orbital property (1.16.20) does not discover the orbital property of the numbers $e^1(\mathbf{P})$. The reason is that the property (ii) of the Shifting Lemma 1.16.1 is not satisfied. Later we will see that the degree formula (1.16.20) discovers orbital invariants in a more global category, the category of orbital curves. These numbers are connected with our local Euler numbers by means of a geometric Local-Global-Principle (see section 2.3 below).

1.17 Local Orbital Euler Numbers

In order to find further orbital invariants we consider the following geometric situation. Let Y be a compact complex algebraic surface with exactly one singularity Q. This point is assumed to be a cyclic singularity of type $\langle d_Q, e_Q \rangle$. Furthermore, we assume that Q is the only intersection point of two smooth curves D, D' forming a singular coordinate system at Q. We endow D, D' with weights v or v', respectively, such that $\mathbf{Q} = (X, Q; vD, v'D')$ is an abelian point. The formal sum

$$\mathcal{A} = \mathbf{D} + \mathbf{D}' + \mathbf{Q},\ \mathbf{D} = (vD, \mathbf{Q}),\ \mathbf{D}' = (v'D', \mathbf{Q})$$

is the first simple example of an *arrangement on* Y, and $\mathbf{Y} = (Y,\ \mathbf{D}+\mathbf{D}'+\mathbf{Q})$ is the first simple example of an *orbital surface*. For the general definitions we refer to sections 3.1, 6.2. We introduce the *local* EULER *contribution* of $\mathbf{Q} : \underset{v}{\circ}\xrightarrow[\langle d_Q, e_Q\rangle]{}\underset{v'}{\circ}$ by

$$e(\mathbf{Q}) = -1 + \frac{1}{v} + \frac{1}{v'} - \frac{1}{vv'd_Q} = -1 + \frac{1}{v} + \frac{1}{v'} - \frac{1}{\deg(\mathbf{Q})} \tag{1.17.1}$$

The *orbital* EULER *height* of the orbital surface $\mathbf{Y}$ is defined as

$$\mathbf{e}(\mathbf{Y}) = e(Y) - \left(1 - \frac{1}{v}\right)e(D) - \left(1 - \frac{1}{v'}\right)e(D') - e(\mathbf{Q}), \tag{1.17.2}$$

where $e(Y), e(D), e(D')$ are the EULER numbers of Y, D or D', respectively.

Now we consider a linear resolution of the point Q. Globally it is a compact complex surface Y' together with a birational morphism

$$\varphi : Y' \longrightarrow Y$$

such that φ induces an isomorphism $Y' \setminus \varphi^{-1}(Q) \xrightarrow{\sim} Y \setminus \{Q\}$ and the preimage $\varphi^{-1}(Q)$ is a linear tree of projective lines E_i whose intersection points Q_i are cyclic points arranged by the intersecting lines, see Figure 1.17.1:

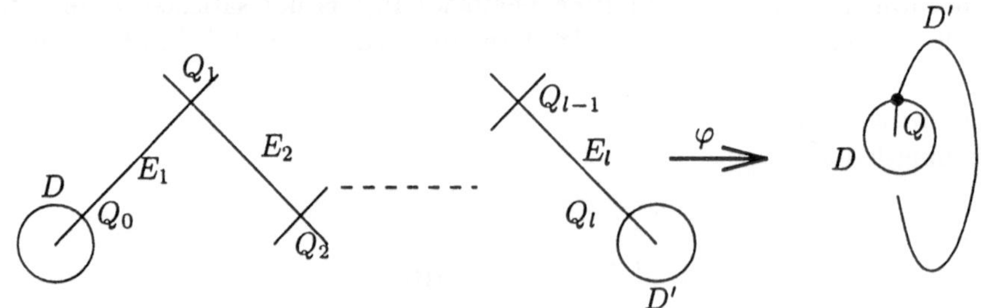

Figure 1.17.1

The points Q_0, Q_l are also assumed to be cyclic and arranged by $(D,\ E_1)$ or $(E_l,\ D')$, respectively. We endow the components E_i of $\varphi^{-1}(Q)$ with weights v_i. Then $\mathbf{Q}_i = (Y',\ Q_i;\ v_iE_i,\ v_{i+1}E_{i+1})$ are abelian points, $(E_0 = D,\ E_{l+1} = D',\ v_0 = v,\ v_{l+1} = v')$, and $\mathbf{E}_i = (v_iE_i,\ \mathbf{Q}_{i-1},\ \mathbf{Q}_i)$, $i = 1, \dots, l$, are *orbital lines*.

The formal sum

$$\mathbf{A}' = \mathbf{D} + \mathbf{E}_1 + \ldots + \mathbf{E}_l + \mathbf{D}' + \mathbf{Q}_0 + \ldots + \mathbf{Q}_l$$

is a further simple example of an arrangement. Let us call it an *extended linear arrangement.* The pair $\mathbf{Y}' = (Y, \mathbf{A}')$ is a further simple example of an orbital surface. The graph of the arrangement $\mathbf{A}'$ is defined as

$$\begin{array}{ccccccccccc} s_0 & & s_1 & & s_2 & & & s_l & & s_{l+1} & \\ \bullet & \longrightarrow & \bullet & \longrightarrow & \bullet & \cdots & \bullet \longrightarrow & \bullet & \longrightarrow & \bullet & \\ v = v_0 & \langle d_0, e_0\rangle & v_1 & \langle d_1, e_1\rangle & v_2 & & \langle d_{l-1}, e_{l-1}\rangle & v_l & \langle d_l, e_l\rangle & v_{l+1} & = v' \end{array} \tag{1.17.3}$$

where $s_i = (E_i^2)$ is the selfintersection number of E_i on $Y, i = 0, ..., l+1$, and $\mathbf{Q}_i : \underset{v_i\ \langle d_i, e_i\rangle v_{i+1}}{\circ\!\longrightarrow\!\circ}$, $i = 0, \ldots, l$. With a glance at (1.17.2) it is consequent to define the *orbital* EULER *height* of $\mathbf{Y}'$ by

$$\mathbf{e}(\mathbf{Y}') = e(Y') - \sum_{i=0}^{l+1}\left(1 - \frac{1}{v_i}\right)e(E_i) - \sum_{i=0}^{l} e(\mathbf{Q}_i). \tag{1.17.4}$$

The morphism $\boldsymbol{\varphi} : \boldsymbol{Y}' \longrightarrow \boldsymbol{Y}$ supported by $\varphi : Y' \longrightarrow Y$ is the first example of an orbital morphism. The only distinction between $\boldsymbol{\varphi}$ and φ comes from the difference of the objects joint by these morphisms. Now we are able to correspond to $\boldsymbol{\varphi}$ in a natural manner the derived *orbital* EULER *invariant* $\mathbf{e}(\boldsymbol{\varphi})$. Obviously $\boldsymbol{\varphi}$ has degree 1 because φ has. So the definition (1.16.6) suggests to define

$$\mathbf{e}(\boldsymbol{\varphi}) = \mathbf{e}(\mathbf{Y}') - \mathbf{e}(\mathbf{Y}). \tag{1.17.5}$$

We want to express $\mathbf{e}(\boldsymbol{\varphi})$ explicitly in terms of the data of the graph (1.17.3) of $\mathbf{A}'$ and of the graph of $\mathbf{A} : \underset{v\ \langle d, e\rangle\ v'}{\bullet\!\longrightarrow\!\bullet}$, $d = d_Q$, $e = e_Q$. We have

$$e(E_i) = e(\mathbb{P}^1) = 2,\ i = 1, \ldots, l, \tag{1.17.6}$$

$$e(\mathbf{Q}_i) = 1 - \frac{1}{v_i} - \frac{1}{v_i'} + \frac{1}{v_i v_i' d_i} = 1 - \frac{1}{v_i} - \frac{1}{v_{i+1}} + \frac{1}{\deg(\mathbf{Q}_i)}. \tag{1.17.7}$$

One knows that the EULER number ascends by the number of projective lines if we change over from a surface Y to a resolution surface Y' by blowing up a cyclic point. Therefore,

$$e(Y') - e(Y) = l \tag{1.17.8}$$

The relations (1.17.1), (1.17.6), (1.17.7) and (1.17.8) allow to express $\mathbf{e}(\varphi)$ subtracting (1.17.2) from (1.17.4):

$$\mathbf{e}(\varphi) = l - 2\sum_{i=1}^{l}\left(1 - \frac{1}{v_i}\right) - \sum_{i=0}^{l}\left(-1 + \frac{1}{v_i} + \frac{1}{v_{i+1}} - \frac{1}{\deg(\mathbf{Q}_i)}\right)$$
$$+ \left(-1 + \frac{1}{v_0} + \frac{1}{v_{l+1}} - \frac{1}{\deg(\mathbf{Q})}\right),$$

$$\mathbf{e}(\varphi) = \sum_{i=0}^{l}\frac{1}{\deg(\mathbf{Q}_i)} - \frac{1}{\deg(\mathbf{Q})}, \tag{1.17.9}$$

$$\deg(\mathbf{Q}_i) = v_i v_{i+1} d_i,\ \deg(\mathbf{Q}) = v_Q v_Q' d_Q.$$

This is a local invariant. The global EULER numbers $e(Y)$, $e(Y')$, $e(D)$, $e(D')$ disappeared. It allows to define $\mathbf{e}(\varphi)$ for local morphisms φ. Instead of $\mathbf{Y}'$ we can take a linear arrangement $\mathbf{E} = (U, E; \mathbf{D}, \mathbf{D}')$ defined in 1.13.2 Instead of the global morphism φ considered above we take a contraction ("resolution" of $\mathbf{Q}$)

$$\varphi : \mathbf{E} \longrightarrow \mathbf{Q}$$

defined in section 13. The *relative orbital* EULER *height* $\mathbf{e}(\varphi)$ is defined by (1.17.9). This number can be read off from the graph presentation described see in (1.17.10):

$$v = v_0\ \langle d_0, e_0\rangle\ v_1\ \langle d_1, e_1\rangle\ v_2\ \cdots\ v_l\ \langle d_l, e_l\rangle\ v_{l+1} = v' \Longrightarrow v\ \langle d_Q, e_Q\rangle\ v' \tag{1.17.10}$$

The resolutions of abelian points form a category. The morphisms are commutative diagrams (1.17.11):

$$\begin{array}{ccc} \mathbf{L} & \xrightarrow{\psi} & \mathbf{P} \\ \downarrow{\scriptstyle \mathbf{g}} & & \downarrow{\scriptstyle \mathbf{f}} \\ \mathbf{E} & \xrightarrow{\varphi} & \mathbf{Q} \end{array} \tag{1.17.11}$$

$$\mathbf{L} = (U_L, \mathbf{L}_1 + \ldots + \mathbf{L}_r + \mathbf{P}_o + \ldots + \mathbf{P}_r), \mathbf{P} = \left(U_P, P; v_P C, v_P' C'\right),$$
$$\mathbf{E} = (U_E, \mathbf{E}_1 + \ldots + \mathbf{E}_l + \mathbf{Q}_o + \ldots + \mathbf{Q}_l), \mathbf{Q} = \left(U_Q, Q; v_Q D, v_Q' D'\right),$$

$\mathbf{f}$ a morphism of abelian points and $\mathbf{g}$ a morphism supported by a morphism $g : U_L \longrightarrow U_E$ sending $L = \cup L_i$ to $E = \cup E_i$. We extend the definition of finite morphisms given in section 13. The morphism $\mathbf{F} = (\mathbf{g}, \mathbf{f})$ is called a *quasi-uniformization,* if $\mathbf{L}$ and $\mathbf{P}$ are reduced, that means that all weights are trivial, and $\mathbf{g}, \mathbf{f}$ are quotient morphisms by the same finite abelian group A, see (1.17.12):

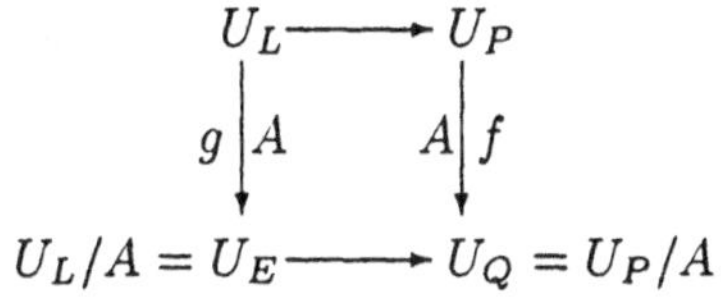

$$\begin{array}{ccc} U_L & \longrightarrow & U_P \\ g \downarrow A & & A \downarrow f \\ U_L/A = U_E & \longrightarrow & U_Q = U_P/A \end{array} \tag{1.17.12}$$

This condition includes that $r = l$, $L = E/A$ is the quotient curve, $\mathbf{E}_i = L_i/A$, $\mathbf{Q}_i = P_i/A$ endowed with the quotient weights defined by the corresponding ramification indices.

Definition 1.17.1 A morphism $\mathbf{F} = (\mathbf{g}, \mathbf{f})$ in the category of resolutions of abelian points is called *finite*, if there is a commutative diagram:

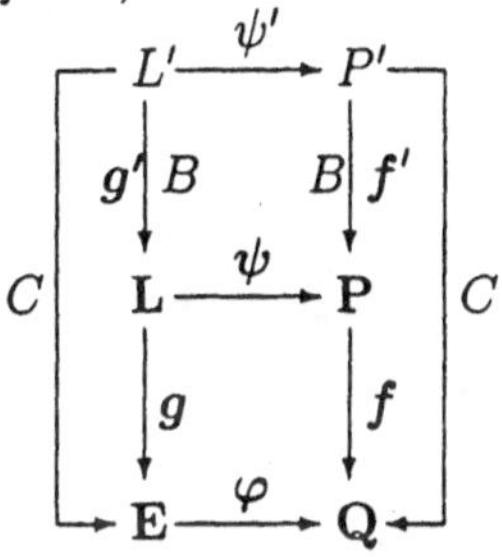

$$\begin{array}{ccc} L' & \xrightarrow{\psi'} & P' \\ g' \downarrow B & & B \downarrow f' \\ \mathbf{L} & \xrightarrow{\psi} & \mathbf{P} \\ \downarrow g & & \downarrow f \\ \mathbf{E} & \xrightarrow{\varphi} & \mathbf{Q} \end{array} \quad (C \text{ on both outer arrows } L' \to \mathbf{E},\ P' \to \mathbf{Q}) \tag{1.17.13}$$

with quasi-uniformization $L'/P' \longrightarrow \mathbf{E}/\mathbf{Q}$ and $L'/P' \longrightarrow \mathbf{L}/\mathbf{P}$. The (analytic) *degree* of $\mathbf{F}$ is the quotient of group orders $|C|/|B|$.

Lemma 1.17.2 *The relative* EULER *numbers* $\mathbf{e}(\mathbf{E}/\mathbf{Q}) = \mathbf{e}(\varphi)$ *are orbital invariants in the category of resolutions of abelian points.*

Proof We have to prove the degree formula

$$\mathbf{e}(\psi) = \deg(\mathbf{F}) \cdot \mathbf{e}(\varphi) \tag{1.17.14}$$

for finite morphisms $\mathbf{F} : \mathbf{L}/\mathbf{P} \longrightarrow \mathbf{E}/\mathbf{Q}$. We first assume that $\mathbf{F}$ is a quasi-uniformization. Uniformizing P we have a composition of quotient maps

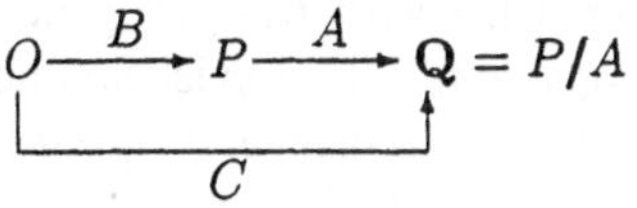

$$O \xrightarrow{B} P \xrightarrow{A} \mathbf{Q} = P/A, \qquad O \xrightarrow{C} \mathbf{Q}$$

with finite abelian groups A, B, C, where $B = \langle d_P, e_P \rangle$. According to diagram (1.17.13) we find that

$$|A| = \deg(f) = |C|/|B| = v_Q v_Q' d_Q/d_P = \deg(\mathbf{Q})/\deg(P)$$

Since also for $\mathbf{Q}_i = P_i/A$ it holds that

$$|A| = \deg(f) = \deg(\mathbf{Q}_i)/\deg(P_i)$$

we can calculate with definition (1.17.9)

$$\begin{aligned} \mathrm{e}(\psi) &= \sum_{i=0}^{l} \frac{1}{\deg(P_i)} - \frac{1}{\deg(P)} = \sum_{i=0}^{l} \frac{|A|}{\deg(\mathbf{Q}_i)} - \frac{|A|}{\deg(\mathbf{Q})} \\ &= |A| \cdot \mathrm{e}(\varphi) = \deg(\mathbf{F}) \cdot \mathrm{e}(\varphi). \end{aligned}$$

If $\mathbf{F}$ is finite with quasi-uniformizations described in diagram (1.17.13) then

$$\mathrm{e}(\psi) = \frac{1}{|B|}\mathrm{e}(\psi') = \frac{|C|}{|B|}\mathrm{e}(\varphi) = \deg(\mathbf{F}) \cdot \mathrm{e}(\varphi)$$

□

A *modification* $\varphi : \mathbf{E}' \longrightarrow \mathbf{E}$ *of linear arrangements* is a birational morphism $\varphi : U_E' \longrightarrow U_E$ with $\varphi^{-1}(E) = E'$, φ isomorphic outside of E', E. Morphisms $\mathbf{L}'/\mathbf{L} \longrightarrow \mathbf{E}'/\mathbf{E}$ in the category of these modifications are commutative diagrams

$$\begin{array}{ccc} \mathbf{L}' & \xrightarrow{\psi} & \mathbf{L} \\ \downarrow g' & & \downarrow g \\ \mathbf{E}' & \xrightarrow{\varphi} & \mathbf{E} \end{array} \qquad \begin{array}{ccc} U_{L'} & \xrightarrow{\psi} & U_L \\ \downarrow g' & & \downarrow g \\ U_{E'} & \xrightarrow{\varphi} & U_E \end{array}$$

The linear arrangement $\mathbf{L}$ is *reduced*, if all weights v_i of the components L_i of $\mathbf{L}$ are trivial ($v_i = 1$). The *modification* $\mathbf{L}'/\mathbf{L}$ is *reduced*, if both $\mathbf{L}'$ and $\mathbf{L}$ are reduced. If $\mathbf{L}$ or $\mathbf{L}'/\mathbf{L}$ is reduced then we write often L or L'/L instead of $\mathbf{L}$ or $\mathbf{L}'/\mathbf{L}$, respectively. Let $\mathbf{L}'/\mathbf{L} = L'/L$ be reduced. Then $\mathbf{G} = (\mathbf{g}', \mathbf{g}) : L'/L \longrightarrow \mathbf{E}'/\mathbf{E}$ is called a *quasi-uniformization*, if there is a finite abelian group A acting on $\mathbf{L}'$ and $\mathbf{L}$, more precisely on U_L', U_L, such that A fixes each intersection point of the components L_i' or L_j of $\mathbf{L}'$ or $\mathbf{L}$, respectively, φ is compatible with the group action, $\mathbf{E}' = L'/A, \mathbf{E} = L/A$ are the quotient arrangements and φ, φ is the quotient morphism of ψ or ψ, respectively. The relative morphism $\mathbf{F} = (\mathbf{g}', \mathbf{g}) : \mathbf{L}'/\mathbf{L} \longrightarrow \mathbf{E}'/\mathbf{E}$ is called *finite* , if there exists a commutative diagram:

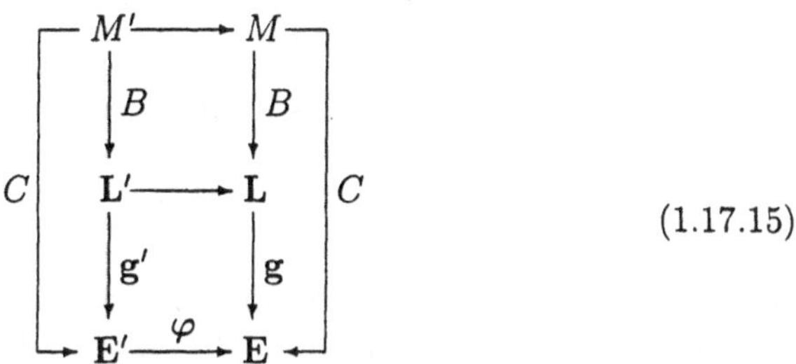

(1.17.15)

with finite abelian groups B, C belonging to quasi-uniformizations $M'/M \to \mathbf{L}'/\mathbf{L}$ or $M'/M \longrightarrow \mathbf{E}'/\mathbf{E}$, respectively. The morphism $\mathbf{g} : \mathbf{L} \longrightarrow \mathbf{E}$ is *finite* if there exists a commutative diagram (1.17.15). Let $\mathbf{F}$ and $\mathbf{g}$ be finite with diagram (1.17.15). Then we call the quotient $|C|/|B|$ the (analytic) *degree* of $\mathbf{F}$ or $\mathbf{g}$, respectively, and denote them by $\deg(\mathbf{F}) = \deg(\mathbf{g}) = \deg(\mathbf{g}')$. The degree is intrinsically determined. Namely, the quotient group $A = C/B$ acts around each intersection point P_i of two neighbour components of $L, \mathbf{Q}_i = \mathbf{P}_i/A$ is an abelian point on $\mathbf{E}$ and $\deg(\mathbf{F}) = \deg(\mathbf{g}) = |A| = \deg(\mathbf{P}_i \longrightarrow \mathbf{Q}_i)$.

Definition 1.17.3 The *orbital* EULER *heights* of $\mathbf{E}$ and $\mathbf{E}'/\mathbf{E}$ are

$$\mathbf{e}(\mathbf{E}) = \sum_{i=0}^{l} \frac{1}{\deg(\mathbf{Q}_i)}, \quad \mathbf{E} = \mathbf{E}_1 + \ldots + \mathbf{E}_l + \mathbf{Q}_o + \ldots + \mathbf{Q}_l$$

$$\mathbf{e}(\varphi) = \mathbf{e}(\mathbf{E}'/\mathbf{E}) = \mathbf{e}(\mathbf{E}') - \mathbf{e}(\mathbf{E}).$$

Lemma 1.17.4 *The relative orbital* EULER *height* $\mathbf{e}(\mathbf{E}'/\mathbf{E})$ *is an orbital invariant in the category of modifications of local linear arrangements.*

Proof The proof is essentially the same as that of Lemma 1.17.2. One has to check the degree formula

$$\mathbf{e}(\mathbf{L}'/\mathbf{L}) = \deg(\mathbf{F}) \cdot \mathbf{e}(\mathbf{E}'/\mathbf{E}) \qquad (1.17.16)$$

for any finite relative morphism $\mathbf{F} : \mathbf{L}'/\mathbf{L} \longrightarrow \mathbf{E}'/\mathbf{E}$. The repetition of the arguments used in the proof of Lemma 1.17.2 is left to the reader. □

At the end of this section we introduce for reduced finite morphisms $\mathbf{f} : P \longrightarrow \mathbf{Q} = P/A$ of abelian points the local orbital EULER height $\mathbf{e}(\mathbf{f})$. A *(linear) resolution* of $\mathbf{f}$ is a commutative diagram

$$\begin{array}{ccc} L & \xrightarrow{\psi} & P \\ \Big\downarrow{\scriptstyle \mathbf{g}} & & \Big\downarrow{\scriptstyle \mathbf{f}} \\ \mathbf{E} & \xrightarrow{\varphi} & \mathbf{Q} \end{array} \tag{1.17.17}$$

with (linear) resolutions $\mathbf{E} \longrightarrow \mathbf{Q}$ and $L \longrightarrow P$, L reduced. The resolution is called *finite* if $\mathbf{g}$ is finite, that means a quasi-uniformization of $\mathbf{E}$, with the same group A as above acting around $L, \mathbf{E} = L/A, \psi$ is compatible with A and φ is the quotient morphism. For any reduced finite morphism $\mathbf{f}$ there exist finite resolutions. For example, one can take the minimal singularity resolution $L = E_P$ of P or the σ-process $L \longrightarrow P$, if $P = O$ is the regular point. Then the group action around P extends to a group action around L (on U_L) and we can establish a quotient diagram (1.17.17).

Definition 1.17.5 We set

$$\mathbf{e}(\mathbf{f}) = \mathbf{e}(L/\mathbf{E})$$

$L/\mathbf{E}$ a finite resolution of $\mathbf{f} : P \longrightarrow \mathbf{Q} = P/A$.

The definition is correct, that means that it does not depend on the choice of the finite resolution of $\mathbf{f}$. It suffices to prove that $\mathbf{e}(L/\mathbf{E}) = \mathbf{e}(E_P/\mathbf{E}_Q)$ where E_P is the minimal singularity resolution of P and $\mathbf{E}_Q = E_P/A$. (If P is regular, then we take the σ-process $E_P \longrightarrow P$). Let $L' \longrightarrow L$ be the minimal resolution of singularities on L and $\mathbf{E}' = L'/A$. Then we have a commutative diagram

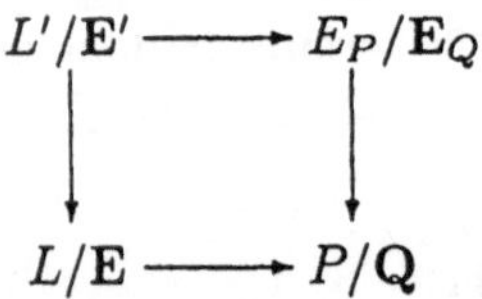

$$\begin{array}{ccc} L'/\mathbf{E}' & \longrightarrow & E_P/\mathbf{E}_Q \\ \Big\downarrow & & \Big\downarrow \\ L/\mathbf{E} & \longrightarrow & P/\mathbf{Q} \end{array}$$

of relative modifications and contractions. The Corollary 1.16.3 of the Shifting Lemma 1.16.1 yields

$$\mathbf{e}(L/\mathbf{E}) = \mathbf{e}(L'/\mathbf{E}') = \mathbf{e}(E_P/\mathbf{E}_P).$$

So $\mathbf{e}(\mathbf{f})$ is correctly defined.

1.18 Absorptive Numbers

The *(modified) selfintersection number* of a smooth curve C on a normal compact complex algebraic surface Y is defined by

$$\left(C^2\right) = \left(C^2\right)_Y =_{Df} \left(C^2\right)_{\tilde{Y}} \tag{1.18.1}$$

where $\tilde{Y} \longrightarrow Y$ is the minimal resolution of the singularities of Y. Assume that Y has only one singularity Q, Q is cyclic and C, C' are two smooth curves on Y intersecting each other in Q and only there, such that $\mathbf{Q} = (Y, Q; D, D')$ is a (reduced) abelian point denoted by Q again. Let $\varphi : Y' \longrightarrow Y$ be a linear resolution of the singularity P. The situation has been already described in Figure 1.17.1.

The proper transforms of C and C' on Y' are denoted by the same symbols. The minimal singularity resolution of Y' is in general not the same as that of Y. Therefore, the selfintersection numbers $(C^2)_{Y'}$ and $(C^2)_Y$ are not equal.

Definition 1.18.1 The difference

$$a^-(\varphi) = \left(C^2\right)_Y - \left(C^2\right)_{Y'} \tag{1.18.2}$$

is called the *left absorptive number* of φ. The *right absorptive number* of φ is defined by

$$a^+(\varphi) = \left(C'^2\right)_Y - \left(C'^2\right)_{Y'}, \tag{1.18.3}$$

and

$$a(\varphi) = a^+(\varphi) + a^-(\varphi) \tag{1.18.4}$$

is called the *absorptive number* of φ.

The left-right order of C, C' comes from the order of C, C' in the (arranged) abelian point Q. It is more precise to write $a^+_Q(\varphi), a^-_Q(\varphi)$ instead of $a^+(\varphi)$ or $a^-(\varphi)$, respectively.

1.18.2 *The absorptive numbers depend only on the reduced linear arrangement E resolving the abelian point Q.*

Proof In order to prove this we consider the graph of the extended (reduced) linear arrangement $C + E + C'$

$$\overset{s_0}{\bullet} \longrightarrow \overset{s_1}{\bullet} \longrightarrow \bullet \cdots \longrightarrow \overset{s_l}{\bullet} \longrightarrow \overset{s_{l+1}}{\bullet}$$
$$\langle d_0, e_0\rangle \quad \langle d_1, e_1\rangle \qquad \langle d_l, e_l\rangle$$

$$s_0 = \left(C^2\right)_{Y'}, \; s_{l+1} = \left(C'^2\right)_{Y'}, \; s_i = \left(E_i^2\right)_{Y'}, \; i = 1, \ldots, l.$$

The trivial weights (equal to 1) have been omitted as usual. Setting $s = (C^2)_Y, s' = ({C'}^2)_Y$ we have to compare s_0 with s and s_{l+1} with s'. Let $\tilde{E} \longrightarrow E$ and $E_Q \longrightarrow Q$ be the minimal singularity resolutions of E or Q, respectively. There is a commutative diagram:

$$\begin{array}{ccc} \tilde{E} & \xrightarrow{\tilde{\varphi}} & E_Q \\ \downarrow & & \downarrow \\ E & \xrightarrow{\varphi} & Q \end{array} \qquad (1.18.5)$$

The modification $\tilde{E} \longrightarrow E$ resolves in the minimal manner the singularities $Q_i = E_i \cap E_{i+1}$ of type $\langle d_i, e_i \rangle$, $i = 0, ..., l$, $E_0 = C, E_{l+1} = C'$. The graph of the extended arrangement $C + \tilde{E} + C'$ looks like

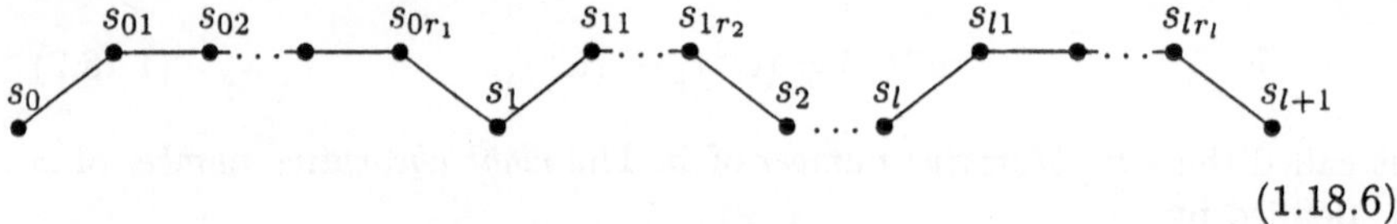

(1.18.6)

The trivial types $\langle 1, 0 \rangle$ corresponding to the smooth intersection points of the components have been omitted. Successively we blow down (-1)-curves of $\tilde{E}$. On this way we come down to E_Q. The composition of the corresponding σ–processes is $\tilde{\varphi}$ in diagram (1.18.5).

We distinguish three types of these σ–processes described in the following diagrams:

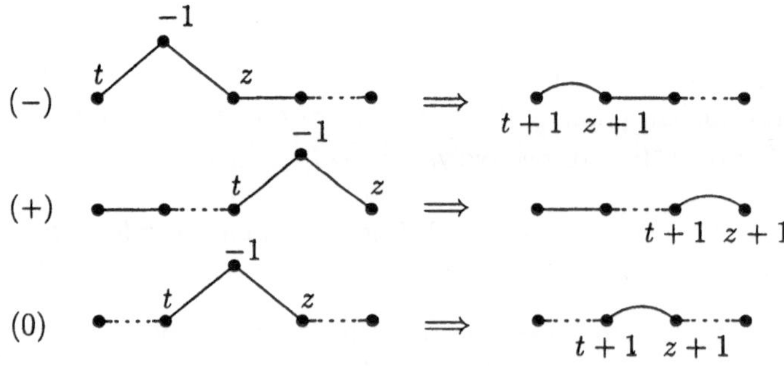

The first kind of σ–processes descreases (C^2) by 1, the second kind descreases $({C'}^2)$ by 1, and the type (0) does not alter (C^2) and $({C'}^2)$. So we get

$$\begin{aligned} a^-(\varphi) &= s - s_0 = \left(C^2\right)_{Y} - \left(C^2\right)_{Y'} \\ &= |\{\text{steps of type } (-)\}| \end{aligned} \qquad (1.18.7)$$

$$a^+(\varphi) = s' - s_{l+1} = \left(C'^2\right)_Y - \left(C'^2\right)_{Y'}$$
$$= |\{\text{steps of type } (+)\}| \quad (1.18.8)$$

$$a(\varphi) = a^-(\varphi) + a^+(\varphi)$$
$$= |\{\text{steps of type } (-) \text{ or } (+)\}| \quad (1.18.9)$$

Therefore $a^-(\varphi), a^+(\varphi), a(\varphi)$ depend only on $\tilde{E}$, hence on E. Moreover, knowing E, these numbers can be calculated explicitly because the minimal resolutions of $Q_i \in E$ can be calculated explicitly by means of continued fractions (HIRZEBRUCH'S Proposition 1.7.1). □

Now we can forget the extended linearly arranged global surfaces Y' and Y. For any linear arrangement $\mathbf{E}$ with reduction E and minimal singularity resolution $\tilde{E} \longrightarrow E$ we can define $a^-(\mathbf{E}), a^+(\mathbf{E}), a(\mathbf{E})$ by the right sides of (1.18.7), (1.18.8) or (1.18.9), respectively, where the "steps" are σ–processes decomposing $\tilde{E} \longrightarrow E_Q$, E_Q the minimal resolution of Q, Q the contraction point of $\tilde{E}$ or E, which has to be an abelian point.

Definition 1.18.3 The number $(a^-(\mathbf{E}) = a^-(E), a^+(\mathbf{E}) = a^+(E))$ $a(\mathbf{E}) = a(E)$ is called the *(left, right) absorptive number* of $\mathbf{E}$ or E, respectively.

By the way we introduce a kind of trace of an abelian point $\mathbf{P}$ which does not depend on the resolution of $\mathbf{P}$. Let $\mathbf{E} \longrightarrow \mathbf{P}$ be a linearly arranged resolution of $\mathbf{P}, l(\mathbf{E})$ the number of components of $\mathbf{E}$ and $\tilde{E} \longrightarrow E$ the minimal resolution of E.

Definition 1.18.4 The number

$$Th(\mathbf{P}) = Th(P) = Th(\mathbf{E}) = Th(\tilde{E}) = Th(E)$$
$$= 3l(\tilde{E}) + Tr(\tilde{E}) + a(\mathbf{E})$$

is called the *homogeneous trace* of $\mathbf{P}, P, \mathbf{E}, \tilde{E}$ or E, respectively, where $Tr(\tilde{E})$ is the sum of selfintersection numbers of the components of $\tilde{E}$.

In order to prove that the definition is correct it suffices to check that $Th(\tilde{E}) = Th(E_P)$, $E_P \longrightarrow P$ the minimal singularity resolution of P. We decompose $\tilde{E} \longrightarrow E_P$ into σ–processes. The first of the σ–processes yields a decomposition $\tilde{E} \rightarrow_\sigma E' \longrightarrow E_P$. We consider the case when σ is of type

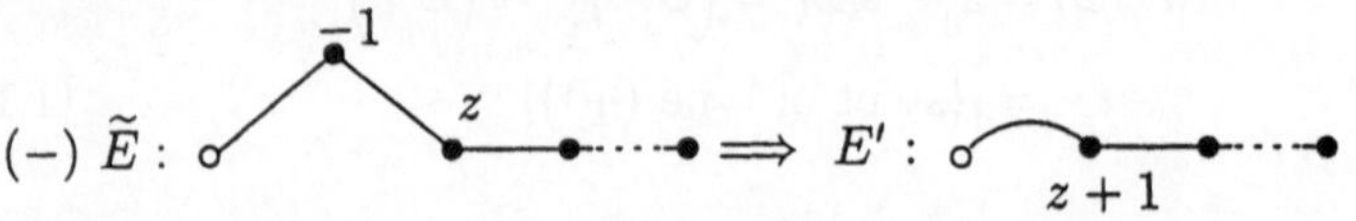

Then

$$a^-\left(\tilde{E}\right) = a^-(E') - 1, Tr\left(\tilde{E}\right) = Tr(E') - 2, 3l\left(\tilde{E}\right) = 3l(E') + 3.$$

Adding together we obtain $Th(\tilde{E}) = Th(E')$. The same identity holds, if σ is of type (+) or of type (0). After $l(\tilde{E}) - l(E_P)$ σ-processes we come to E_P and to the identity

$$Th(\mathbf{E}) = Th\left(\tilde{E}\right) = Th(E_P) = 3l_p + Tr(P) = 3l(E_P) + Tr(E_P). \tag{1.18.10}$$

Let $\mathbf{f} : P \underset{A}{\rightarrow} \mathbf{Q} = P/A$ be a reduced quotient morphism of abelian points. We set $\mathbf{E} = E_P/A$, E_P the minimal resolution of P again. We have a commutative diagram:

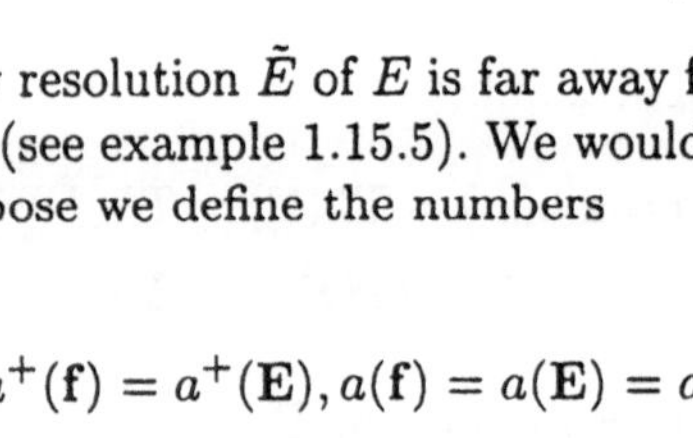

(1.18.11)

The minimal singularity resolution $\tilde{E}$ of E is far away from being the minimal resolution E_Q of Q (see example 1.15.5). We would like to measure the difference. For this purpose we define the numbers

$$a^-(\mathbf{f}) = a^-(\mathbf{E}), a^+(\mathbf{f}) = a^+(\mathbf{E}), a(\mathbf{f}) = a(\mathbf{E}) = a^-(\mathbf{f}) + a^+(\mathbf{f}) \tag{1.18.12}$$

and call them the *(left, right) absorptive numbers* of $\mathbf{f}$. These absorptive numbers can be expressed explicitly by means of the graph data of P and $\mathbf{Q}$.

Proposition 1.18.5 *In the situation of diagram (1.18.11) it holds that*

$$a^-(\mathbf{f}) = \frac{e_P v_Q^2}{d_P \cdot |A|} + \frac{e_o}{d_o} - \frac{e_Q}{d_Q}, a^+(\mathbf{f}) = \frac{e'_P v'^2_Q}{d_P \cdot |A|} + \frac{e'_l}{d_l} - \frac{e'_Q}{d_Q}$$

$$a(\mathbf{f}) = \left(e_P v_Q^2 + e'_P v'^2_Q\right)/d_P|A| + \left(e_o/d_o + e'_l/d_l'\right) + (e_Q + e_Q')/d_Q,$$

where d_o, e_o *are taken away from the graph of*

$$\mathbf{E}: \; v_Q \; \langle d_0, e_0 \rangle \; v_1 \; \langle d_1, e_1 \rangle \; v_2 \; \cdots \; v_l \; \langle d_l, e_l \rangle \; v'_Q$$

and $\langle d_l, e'_l \rangle, \langle d_Q, e'_Q \rangle$ *are the opposite types of* $\langle d_l, e_l \rangle$ *or* $\langle d_Q, e_Q \rangle$*, respectively.*

The proof is delegated to the next chapter (see 2.4.9).

2 Orbital Curves

2.1 Point Arrangements on Curves

Let (U_C, C) be an analytic surface germ along a smooth compact complex curve C. Working with a surface representant $U = U_C$ we assume that U is a normal algebraic surface and has only cyclic singularities on C. Furthermore, we can assume that for any cyclic singularity $P \in C$ there exists a smooth curve germ C'_P on U_C through P such that $(U_C, P; C, C'_P)$ is a reduced abelian point. A *weighted analytic surface germ* along C is a pair $(U, vC), v \neq 0$ a natural number. We say that the abelian point $\mathbf{P} = (U_C, P; wC, v'C'_P)$ *lies on* (U, vC), if $w = v$, with C'_P, U_C as above.

Definition 2.1.1 The formal sum of abelian points $\mathbf{P}_1 + ... + \mathbf{P}_k$ is called a *(point) arrangement* on (U, vC), if

(i) all $\mathbf{P}_i$ lie on $(U,\ vC)$, $i = 1, \ldots k$;

(ii) $P_i \neq P_j$ if $i \neq j$ (as points of the surface U);

(iii) If P is a non-trivial cyclic singularity on C, then there exists a number i, $1 \leq i \leq k$, such that $P = P_i$.

The tripel $\mathbf{C} = (U, vC; \mathbf{P}_1 + ... + \mathbf{P}_k)$ is called an *orbital curve* .

Remark 2.1.2 Later, in the more general context of Chapter VI, we will admit rational coefficients at the points $\mathbf{P}_i$ (orbital curves with rational weights).

Supplement 2.1.3 The phrases "$\mathbf{P}$ lies on (U, vC)" and "$\mathbf{P}$ lies on $\mathbf{C}$" will be used synonymously. In this case we write sometimes $\mathbf{P} \in \mathbf{C}$ or $\mathbf{P} \in (U, vC)$. The sum $\mathbf{P}_1 + ... + \mathbf{P}_k$ in the definition is also called an *arrangement of abelian points* (on (U, vC)). We call (U, vC) *reduced* , if $v = 1$. The point arrangement $\mathbf{P}_1 + ... + \mathbf{P}_k$ is reduced if $k \geq 1$ and all occurring abelian points $\mathbf{P}_i$ are reduced. The orbital curve $\mathbf{C}$ is called reduced if (U, vC) and $\mathbf{P}_1 + ... + \mathbf{P}_k$ defining $\mathbf{C}$ are reduced. We admit the case $k = 0$. In this case the arrangement is called void and is denoted by $\emptyset$.

An orbital curve can be illustrated by the following figure:

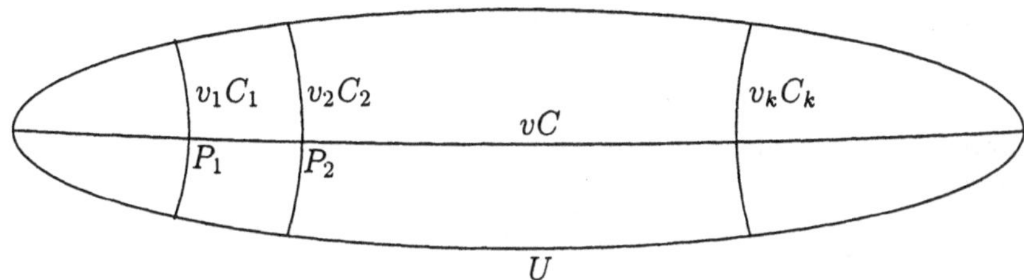

Figure 2.1.1

with curve germs $C_i = C'_{P_i}$. The *graph of the orbital curve* **C** is defined by the following figure:

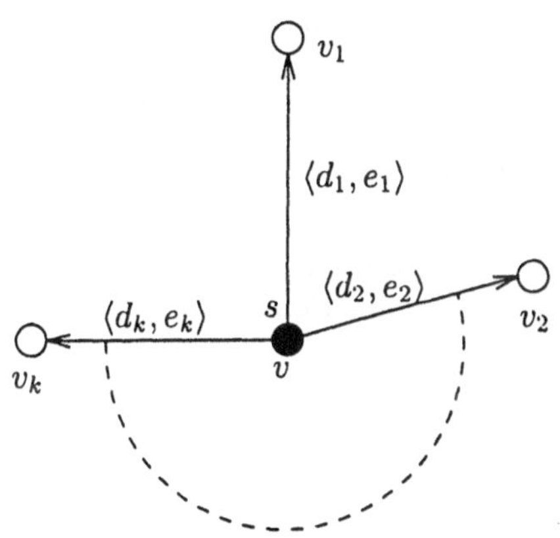

Figure 2.1.2

where

$$s = (C^2) = (C^2)_{U_C} = (C^2)_{\tilde{U}_C} \tag{2.1.1}$$

is the selfintersection number of C on the minimal resolution $\tilde{U}_C$ of singularities of U_C and $\underset{v\ \langle d_i, e_i\rangle\ v_i}{\circ\!\!\longrightarrow\!\!\circ}$ is the graph of the abelian points $\mathbf{P}_i$ of the arrangement.

In [HO6] we called (2.1.2) the *atomic graph* of the orbital curve $\mathbf{C}$ and the graphs of $\mathbf{P}_i$ the *electronic graphs* of these abelian points because the graph (2.1.2) look like a model of an atom kernel with electrons around it. Two atomic graphs are called *equivalent* , if they coincide up to some "trivial" electronic parts $\underset{v}{\circ}\xrightarrow[\langle 1,0\rangle]{}\underset{1}{\circ}$ corresponding to regular points of U_C on C.

Two arrangements on (U, vC) are *equivalent* if the corresponding orbital curves have equivalent graphs, and the supports of the non-trivial points coincide.

A *morphism* $\mathbf{f} : \mathbf{C} \longrightarrow \mathbf{D}$ of orbital curves with point arrangements $\mathbf{Q}_1 + \ldots + \mathbf{Q}_k$, or $\mathbf{P}_1 + ... + \mathbf{P}_m$ on (U_D, wD), (U_C, vC) defining $\mathbf{D}$ or $\mathbf{C}$, respectively, is a generically finite map $f : U_C \longrightarrow U_D$ of normal algebraic surfaces with finite restriction

$$f_C : C \longrightarrow D \tag{2.1.2}$$

sending each P_i to one of the points $Q_{j(i)}$ such that the *localizations*

$$\mathbf{f}_{P_i} : \mathbf{P}_i \longrightarrow \mathbf{Q}_{j(i)}$$

of $\mathbf{f}$ at P_i are morphisms of abelian points.

2.2 Euler Heights of Orbital Curves

We look for orbital invariants in the category of orbital curves. For this purpose we need a suitable degree function.

Definition 2.2.1 Let $\mathbf{f} : \mathbf{C} \longrightarrow \mathbf{D}$ be a morphism of orbital curves. The *reduced degree* of $\mathbf{f}$ is the degree

$$\deg_{\mathbf{red}}(\mathbf{f}) = \deg(f_C)$$

of the generically finite restriction $f_C : C \longrightarrow D$ of f to C.

Let A be a finite group acting on U_C and on $C \subset U_C$. We say that this action is *locally abelian* on (U_C, C) or on the corresponding reduced orbital curve denoted by C again, if all stationary groups $A_P, P \in C$, are abelian. The group A is not assumed to be abelian. Let v be the ramification index of the finite GALOIS covering $U_C \longrightarrow U_C/A$ at C and $D = C/A$, $U_C/A = U_D$ the quotients. For any point $P \in C$ we consider the quotient point P/A_P lying on (U_D, vD). There are only finitely many such quotient points which are not of trivial type $\underset{v}{\circ}\xrightarrow[\langle 1,0\rangle]{}\underset{1}{\circ}$ because $U_D = U_C/A$ is normal, supporting only finitely many singularities, and there exist only finitely many curve germs on U_C with non-trivial ramification index. Up to equivalence we define the quotient arrangement on (U_D, D) of $U_C \xrightarrow{A} U_D$ as the sum $\sum P/A_P$ of all non-trivial quotient points $P/A_P, P \in C$.

Definitions 2.2.2 $\mathbf{D} = C/A = (U_D, vD; \sum P/A_P)$ is called the *orbital quotient curve* of the GALOIS covering $f : U_C \longrightarrow U_C/A = U_D$. The corresponding morphism $\mathbf{f} : C \longrightarrow \mathbf{D} = C/A$ is called the *quotient morphism* belonging to f. For a given arranged curve $\mathbf{D}$ we call

$$\mathbf{f} : \mathbf{C} \longrightarrow \mathbf{D}$$

a *quasi-uniformization* if $\mathbf{C}$ is reduced, $\mathbf{D} = C/A$ and $\mathbf{f}$ is the corresponding quotient morphism. A quasi-uniformization $\mathbf{f} : C \longrightarrow \mathbf{D}$ is a *uniformization* if U_C is smooth along C. A morphism $\mathbf{f} : \mathbf{C} \longrightarrow \mathbf{D}$ is called GALOIS-*finite* if there is a commutative diagram:

$$\begin{array}{ccc} & E & \\ {}^{\mathbf{h}}\swarrow & & \searrow^{\mathbf{g}} \\ \mathbf{C} & \xrightarrow[\mathbf{f}]{} & \mathbf{D} \end{array} \qquad (2.2.1)$$

with quasi-uniformizations $\mathbf{h}$ and $\mathbf{g}$.

In this chapter the *class of finite morphisms* will consist of the GALOIS-finite morphisms. A morphism $\mathbf{f} : \mathbf{C} \longrightarrow \mathbf{D}$ is a *modification* if $f : U_C \longrightarrow U_D$ is birational and the weights v_D, v_C of D and C in $\mathbf{D}$ or $\mathbf{C}$, respectively, coincide. In the category of relative objects $\mathbf{C}/\mathbf{D}$ the GALOIS-*finite morphisms* $\mathbf{C}'/\mathbf{C} \longrightarrow \mathbf{D}'/\mathbf{D}$ are defined by commutative diagrams:

$$\begin{array}{ccc} \mathbf{E}' & \xrightarrow[\chi]{} & \mathbf{E} \\ {\scriptstyle B}\big\downarrow{\scriptstyle \mathbf{h}'} & & {\scriptstyle B}\big\downarrow{\scriptstyle \mathbf{h}} \\ A\ \mathbf{C}' & \xrightarrow[\psi]{} & \mathbf{C}\ A \\ \big\downarrow{\scriptstyle \mathbf{f}'} & & \big\downarrow{\scriptstyle \mathbf{f}} \\ \mathbf{D}' & \xrightarrow[\varphi]{} & \mathbf{D} \end{array} \qquad (2.2.2)$$

(with outer arrows $\mathbf{g}' : \mathbf{E}' \to \mathbf{D}'$ and $\mathbf{g} : \mathbf{E} \to \mathbf{D}$)

with modifications χ, ψ, φ and quasiuniformizations $\mathbf{g}, \mathbf{h}, \mathbf{g}', \mathbf{h}'$ such that the corresponding group actions of A and B on U'_E are extensions of the actions of the same groups on U_E.

We correspond to each orbital curve a rational number in the following manner.

Definition 2.2.3 Let $\mathbf{D}$ be an orbital curve with arrangement $\mathbf{Q}_1 + \ldots + \mathbf{Q}_m$, $e(D)$ the EULER number of D and $e^1(\mathbf{Q}_i) = 1 - \frac{1}{v_i d_i}$ the first EULER contribution of $\mathbf{Q}_i$: $\underset{v}{\circ} \xrightarrow{\langle d_i, e_i \rangle} \underset{v_i}{\circ}$ The rational number

$$\mathbf{e}(\mathbf{D}) = e(D) - \sum_{i=1}^{m} e^1(\mathbf{Q}_i) = e(D) - \sum_i \left(1 - \frac{1}{v_i d_i}\right) \tag{2.2.3}$$

is called the *orbital* EULER *height* of $\mathbf{D}$.

Theorem 2.2.4 *The* EULER *heights of orbital curves are orbital invariants with respect to the degree function* $\deg_{\text{red}}$ *and the class of* GALOIS*-finite morphisms in the category of orbital curves.*

Proof The Shifting Lemma 1.16.1 gives us the criterion for the orbital property of $\mathbf{e}$. First we look at criterion (ii). We have to find "sufficiently many" GALOIS-finite morphisms $\mathbf{f} : \mathbf{C} \longrightarrow \mathbf{D}$ satisfying the degree formula

$$\mathbf{e}(\mathbf{C}) = \deg_{\mathbf{red}}\ (\mathbf{f}) \cdot \mathbf{e}(\mathbf{D}). \tag{2.2.4}$$

Our first candidates are uniformizations. Let $\mathbf{f} : \mathbf{C} \underset{\mathbf{A}}{\rightarrow} \mathbf{D} = \mathbf{C}/A$ be a uniformization. The exact sequence

$$1 \longrightarrow Z_A(C) \longrightarrow A \longrightarrow A_C \longrightarrow 1$$

defines the cyclic *inertia group* of order v

$$Z = Z_A(C) = \{a \in A; a|_C = id_C\}$$

and $\overline{A} = A_C = A/Z$ acting effectively on C. The quotient curve C/A_C is normal, hence smooth. Therefore the curves D and C/A_C are equal. Now we have only to apply the

HURWITZ' Genus Formula 2.2.5 (see e.g. [LA1])

$$e(C) = |\overline{A}| \cdot \left(e(D) - \sum_{P \in C} \left(1 - \frac{1}{|\overline{A}_P|}\right)\right),$$

where $\overline{A}_P$ is the isotropy group of $\overline{A}$ at a preimage point $P \in D$ of Q along the quotient map $C \longrightarrow D = C/\overline{A}$.

All non-trivial summands in the sum of 2.2.5 come from the quotient arrangement $\sum_{i=1}^{m} \mathbf{Q}_i$, $\mathbf{Q}_i = P_i/A$: $\underset{v}{\circ}\xrightarrow[\langle d_i, e_i\rangle]{}\underset{v_i}{\circ}$ of $\mathbf{D} = C/A$. We have

$$|\overline{A}| = |A|/|Z| = |A|/v = \frac{\deg(f)}{v} = \deg_{\mathrm{red}}(\mathbf{f}), \tag{2.2.5}$$

$$|A_{P_i}| = v \cdot v_i \cdot d_i, |\overline{A}_{P_i}| = |A_{P_i}|/v = v_i \cdot d_i \tag{2.2.6}$$

Therefore, the degree formula (2.2.4) for uniformizations follows directly from HURWITZ' genus formula 2.2.5.

Now we change over to special GALOIS-finite coverings.

Definition 2.2.6 A GALOIS-finite covering $\mathbf{f} : \mathbf{C} \longrightarrow \mathbf{D}$ is called *uniformizable*, if the commutative diagram (2.2.1) can be chosen in such a manner that $\mathbf{g}$ and $\mathbf{h}$ are uniformizations, that means that U_E is a smooth surface (germ) along the curve E.

2.2.7 *For uniformizable morphisms* $\mathbf{f} : \mathbf{C} \longrightarrow \mathbf{D}$ *the degree formula* (2.2.4) *holds.*

Proof Since $\mathbf{g}$ and $\mathbf{h}$ are uniformizations we know that

$$\mathbf{e}(\mathbf{E}) = \deg_{\mathrm{red}}(\mathbf{h}) \cdot \mathbf{e}(\mathbf{C})$$
$$\mathbf{e}(\mathbf{E}) = \deg_{\mathrm{red}}(\mathbf{g}) \cdot \mathbf{e}(\mathbf{D})$$

Now the degree formula (2.2.4) follows from the multiplicative property of reduced degrees

$$\deg_{\mathrm{red}}(\mathbf{g}) = \deg_{\mathrm{red}}(\mathbf{f} \cdot \mathbf{h}) = \deg_{\mathrm{red}}(\mathbf{f}) \cdot \deg_{\mathrm{red}}(\mathbf{h}). \tag{2.2.7}$$

□

2.2.8 *The condition (ii) of the Shifting-Lemma 1.16.1 is satisfied.*

Proof Indeed, if $\mathbf{f} : \mathbf{C} \longrightarrow \mathbf{D}$ is a GALOIS-finite morphism of orbital curves we can construct a commutative diagram (2.2.2) such that $\mathbf{g}'$ and $\mathbf{h}'$ are uniformizations. We have only to resolve the cyclic surface singularities of U_E in a minimal manner and to divide the arising surface $U'_E = U_{E'}$ by the groups A and B. □

For the proof of Theorem 2.2.4 it remains to check the property (i) of the Shifting-Lemma 1.16.1. For this purpose we need a Local-Global Principle. This principle will be used not only for EULER heights but also for other invariants. We will describe it in the following section.

2.3 The Geometric Local-Global Principle

Let $\mathcal{C}, \mathcal{P}$ be the categories of orbital curves or abelian points, respectively. A *punctured orbital curve* is a pair $(\mathbf{C}, \mathbf{P})$, $\mathbf{C}$ an orbital curve and $\mathbf{P}$ an abelian point lying on $\mathbf{C}$. A *morphism* $\mathbf{f} : (\mathbf{C}, \mathbf{P}) \longrightarrow (\mathbf{D}, \mathbf{Q})$ is a morphism $\mathbf{f} : \mathbf{C} \longrightarrow \mathbf{D}$ of orbital curves with $Q = f(P)$ such that $\mathbf{f_P} : \mathbf{P} \longrightarrow \mathbf{Q}$ is a morphism in the category of abelian points induced by f, and $\mathbf{f_P}$ is called the *localization* of $\mathbf{f}$ at $\mathbf{P}$. For quotient morphisms (quasi-uniformizations) $\mathbf{C} \underset{A}{\to} \mathbf{D} = \mathbf{C}/A$ the "image point" $\mathbf{Q}$ is uniquely determined by $\mathbf{P}$ and $\mathbf{f_P} : \mathbf{P} \longrightarrow \mathbf{Q} = \mathbf{P}/A_P$ is the corresponding quotient morphism of abelian points. The same is true for GALOIS-finite morphisms (see diagram (2.2.1)).

The category of punctured orbital curves is denoted by $\mathcal{CP}$. We dispose on localization functors

$$\mathcal{CP} \longrightarrow \mathcal{P} : (\mathbf{C}, \mathbf{P}) \longrightarrow \mathbf{P}, \mathbf{f} \mapsto \mathbf{f_P};\ (\mathbf{CP})' \longrightarrow \mathbf{P}' : (\mathbf{C}, \mathbf{P})/(\mathbf{D}, \mathbf{Q}) \mapsto \mathbf{P}/\mathbf{Q} :$$

$$\begin{array}{ccccccc} (\mathbf{C}', \mathbf{P}') & \longrightarrow & (\mathbf{C}, \mathbf{P}) & & \mathbf{P}' & \longrightarrow & \mathbf{P} \\ \downarrow \mathbf{f}' & & \downarrow \mathbf{f} & \longmapsto & \downarrow \mathbf{f}'_{\mathbf{P}'} & & \downarrow \mathbf{f_P} \\ (\mathbf{D}', \mathbf{Q}') & \longrightarrow & (\mathbf{D}, \mathbf{Q}) & & \mathbf{Q}' & \longrightarrow & \mathbf{Q} \end{array} \tag{2.3.1}$$

on the categories of relative objects/morphisms, and $(\mathcal{CP})'_1 \longrightarrow \mathcal{P}'_1$ as above, restricted to modifications. Obviously these functors transform GALOIS-finite morphisms and global modifications to local ones and they preserve the reduced degree.

We consider numerical functions $\mathbf{c} : \mathrm{Ob}\ (\mathcal{C}) \longrightarrow \mathbb{Q}$ with the canonical extension to the category $\mathcal{C}'$ of relative objects and the restriction to $\mathcal{C}'_1$ (see section 1.16). We look for a numerical function γ on $\mathcal{P}, \mathcal{P}'$ or $\mathcal{P}'_1$ such that we have for any modification $\varphi = \mathbf{C}'/\mathbf{C}$ a local decomposition

$$\mathbf{c}(\varphi) = \sum_{\mathbf{P}'} \gamma(\varphi_{\mathbf{P}'}), \tag{2.3.2}$$

where the sum runs over all points $\mathbf{P}' \in \mathbf{C}'$ which belong to the point arrangement of $\mathbf{C}'$. We assume that the sum does not change if we change over to an equivalent arrangement.

Definition 2.3.1 Under the above conditions the numerical function γ is called a *localization* of $\mathbf{c}$ (on $\mathcal{C}_1'$).

For a localization of $\mathbf{c}$ we will frequently use the same symbol $\mathbf{c}$ as for the global numerical function or a similar one (e.g. $\mathbf{e}^1$). Then (2.3.2) can be written as

$$\mathbf{c}(\varphi) = \sum_{\mathbf{P}'} \mathbf{c}(\varphi_{\mathbf{P}'}). \tag{2.3.3}$$

The crucial point is the following:

Geometric Local-Global Principle 2.3.2 *If* $\mathbf{c}$ *is a numerical function on* $\mathcal{C}$ *with localization and this localization is (relative) orbital on* $\mathcal{P}'_1$ *then* $\mathbf{c}$ *is orbital on the relative category* $\mathcal{C}_1'$.

Proof For GALOIS-finite coverings $\mathbf{F} : \mathbf{C}'/\mathbf{C} \longrightarrow \mathbf{D}'/\mathbf{D}$ of modifications we have to prove that

$$\mathbf{c}(\mathbf{C}'/\mathbf{C}) = \deg_{\text{red}}(\mathbf{F}) \cdot \mathbf{c}(\mathbf{D}'/\mathbf{D}). \tag{2.3.4}$$

$\mathbf{F} = (\mathbf{f}', \mathbf{f})$ is a commutative diagram:

$$\begin{array}{ccc} \mathbf{C}' & \xrightarrow{\psi} & \mathbf{C} \\ \downarrow{\mathbf{f}'} & & \downarrow{\mathbf{f}} \\ \mathbf{D}' & \xrightarrow{\varphi} & \mathbf{D} \end{array} \tag{2.3.5}$$

with modifications ψ, φ. We first assume that $\mathbf{F}$ is a quasiuniformization with GALOIS group A. With corresponding notations we find

$$\begin{aligned} \mathbf{c}(\psi) &= \sum_{P'} \mathbf{c}(\psi_{P'}) = \sum_{Q'} \sum_{P' \to Q'} \mathbf{c}(\psi_{P'}) = \sum_{Q'} |f'^{-1}(Q')| \mathbf{c}(\psi_{P'}) \\ &= \sum_{Q'} [A : A_{P'}] \cdot \deg_{\text{red}}(\mathbf{f}) \mathbf{c}(\varphi_{Q'}) \\ &= \sum_{Q'} [A : A_{P'}] \frac{|A_{P'}|}{v} \mathbf{c}(\varphi_{Q'}) \\ &= \sum_{Q} \frac{|A|}{v} \mathbf{c}(\varphi_{Q'}) = \deg_{\text{red}}(\mathbf{F}) \mathbf{c}(\varphi). \end{aligned}$$

We used the identity

$$\mathbf{c}(\psi_{P'}) = \deg_{\text{red}}(\mathbf{f}) \cdot \mathbf{c}(\varphi_{Q'}), \tag{2.3.6}$$

which comes from the local relative orbital assumption about **c**. If $\mathbf{C}'/\mathbf{C} \longrightarrow \mathbf{D}'/\mathbf{D}$ is GALOIS-finite then we go through diagram (2.2.2) with quasi-uniformizations $\mathbf{G} = (\mathbf{g}', \mathbf{g})$ and $\mathbf{H} = (\mathbf{h}', \mathbf{h})$:

$$\mathbf{c}(\psi) = \frac{1}{\deg_{\mathrm{red}}(\mathbf{H})}\mathbf{c}(\chi) = \frac{\deg_{\mathrm{red}}(\mathbf{G})}{\deg_{\mathrm{red}}(\mathbf{H})}\mathbf{c}(\varphi) = \deg_{\mathrm{red}}(\mathbf{F})\mathbf{c}(\varphi).$$

Thus the geometrical Local-Global Principle 2.3.2 has been established. □

Proof of Property 1.16.1(i) for e. We have to look for local EULER contributions satisfying (2.3.3) (local decomposition) and (2.3.6) (local orbital). These numbers have already been introduced in (1.16.14):

$$e^1(\mathbf{P}) = 1 - \frac{1}{v'_P d_P}, \quad \mathbf{P} : \underset{v \ \langle d_P, e_P \rangle \ v'}{\circ\!\!\longrightarrow\!\!\circ}$$

The local degree formula (2.3.6) has been proved in chapter 1, Lemma 1.16.4.
If $\varphi_{\mathbf{Q}} : \mathbf{Q} \longrightarrow \mathbf{P}$ is a morphism of abelian points, then we set

$$\mathbf{e}^1(\varphi_{\mathbf{Q}}) = -e^1(\varphi_{\mathbf{Q}}) = \frac{1}{v'_Q d_Q} - \frac{1}{v'_P d_P} \tag{2.3.7}$$

Since we only changed the sign, the degree formula (2.3.6) for $\mathbf{e}^1$, applied to GALOIS-finite covering of local modifications, is valid. Furthermore we have the decomposition

$$\mathbf{e}(\varphi) = \sum_{\mathbf{P}'} \mathbf{e}^1(\varphi_{\mathbf{P}'}), \ \varphi : \mathbf{C}' \longrightarrow \mathbf{C}$$

because of $\mathbf{e}(\varphi) = \mathbf{e}(\mathbf{C}') - \mathbf{e}(\mathbf{C})$, the definition 2.2.3 and $e(C') = e(C)$ for modifications φ.

□

End of the Proof of Theorem 2.2.4: With the geometric Local-Global Principle we checked condition (i) of the Shifting-Lemma 1.16.1. Also the condition (ii) is satisfied by 2.2.8. Now Theorem 2.2.4 follows from the conclusion of the Shifting-Lemma.

□

2.4 Signature Heights of Orbital Curves

Definition 2.4.1 Let **C** be an orbital curve with point arrangement $\mathbf{P}_1 + ... + \mathbf{P}_m$ and atomic graph (Fig. 2.1.2). We set

$$\tau^1(\mathbf{P}_i) = -\frac{e_i}{3vd_i}, \quad \mathbf{P}_i : \underset{v}{\circ} \xrightarrow{\langle d_i, e_i \rangle} \underset{v_i}{\circ} \tag{2.4.1}$$

$$\tau(\mathbf{C}) = s/3v = (C^2)/3v, \tag{2.4.2}$$

$$\tau(\mathbf{C}) = \tau(\mathbf{C}) - \sum_i \tau^1(\mathbf{P}_i) = \left(s + \sum_i e_i/d_i\right)/3v, \tag{2.4.3}$$

and call $\tau(\mathbf{C})$ the *signature height* of **C**.

As usual one has to take the selfintersection number $s = (C^2)$ on the minimal resolution $\tilde{U}_C$ of surface singularities of U_C. In analogy with Theorem 2.2.4 we prove the following

Theorem 2.4.2 *The signature heights of orbital curves are orbital invariants with respect to the degree function* $\deg_{\text{red}}$ *and the class of* GALOIS*-finite morphisms in the category of orbital curves.*

The Shifting-Lemma (1.16.1) outlines again the proof. We look first at uniformizations and uniformizable GALOIS-finite morphisms of orbital curves.

2.4.3 *The degree formula*

$$\tau(\mathbf{C}) = \deg_{red}(\mathbf{f}) \cdot \tau(\mathbf{D}) \tag{2.4.4}$$

is valid for uniformizable morphisms $\mathbf{f} : \mathbf{C} \longrightarrow \mathbf{D}$.

Proof We start with a uniformization **f** and $\mathbf{D} = \mathbf{C}/A$. Since U_C is smooth along C and the v_C is equal to 1 we have

$$\tau(\mathbf{C}) = \frac{1}{3}(C^2) = \frac{1}{3}(C^2)_{U_C}$$

by definition (2.4.3). Let $v = v_D$ be the weight of **D**. Proposition 1.14.2 yields with the above notations

$$v \cdot \tau(\mathbf{D}) = \frac{v^2 \cdot (C^2)}{3|A|},$$

hence

$$\tau(\mathbf{D}) = \frac{1}{3}(C^2)/\deg_{\mathrm{red}}(\mathbf{f}) = \tau(\mathbf{C})/\deg_{\mathrm{red}}(\mathbf{f}), \tag{2.4.5}$$

which is the degree formula (2.4.4). The formula can be extended in obvious manner to uniformizable morphism $\mathbf{C} \longrightarrow \mathbf{D}$ via diagram (2.2.1) with uniformizations $\mathbf{h}$ and $\mathbf{g}$. □

2.4.4 *In each class of* GALOIS*-finite morphisms of modifications there exists at least one* GALOIS*-finite covering satisfying the degree formula* (2.4.4).

Proof This is an immediate consequence of 2.4.3 because each GALOIS-finite morphism can be modified to a uniformizable one by resolving the singularities of U_E in (2.2.1). □

So we know that

2.4.5 *The condition (ii) of the Shifting-Lemma 1.16.1 is satisfied for the signature heights of orbital curves and the class of* GALOIS*-finite morphisms.*

As for EULER heights we want to check condition (i) of the Shifting-Lemma via the geometric Local-Global-Principle 2.3.2. We need a localization of τ on the relative category $\mathcal{P}_1'$ of abelian points. It is clear that for τ the differences of selfintersection numbers play a role. So it is quite natural to work with absorptive numbers.

Let $\varphi : \mathbf{P} \longrightarrow \mathbf{Q}$ be a modification of abelian points. Forgetting the weights we can draw a commutative diagram:

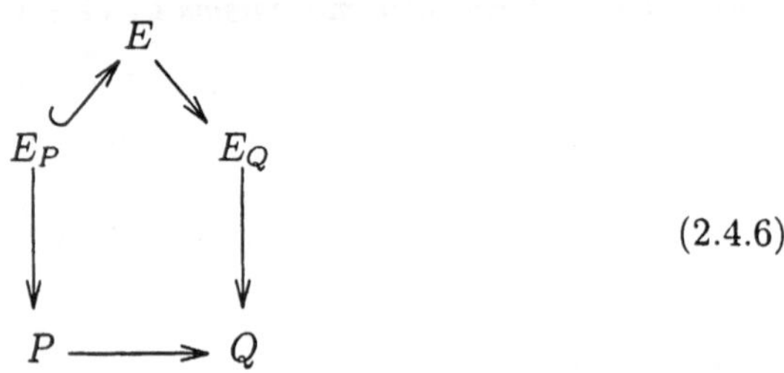

(2.4.6)

with minimal singularity resolutions E_P and E_Q of P or Q, respectively. $E_P \subset E$ means that E_P is the first part of a greater linear arrangement E, which is not minimal in general. $E \longrightarrow E_Q$ is received by contracting step by step (-1)-curves. We choose E uniquely in a minimal manner contracting, if necessary, step by step all (-1)-curves such that the first part E_P is preserved. Now we define

$$s(\varphi_{\mathbf{P}}) = -a^-(E), \varphi_{\mathbf{P}} : \mathbf{P} \longrightarrow \mathbf{Q} \tag{2.4.7}$$

where $a^-(E)$ is the left absorptive number of E (see 1.18.3). By the definition of $a^-(E)$ it is clear that for a modification $\varphi : \mathbf{C} \longrightarrow \mathbf{D}$ of orbital curves degenerating only at $\mathbf{P} \in \mathbf{C}$ it holds that

$$(C^2) = (D^2) + s(\varphi_{\mathbf{P}})$$

For arbitrary modifications $\varphi : \mathbf{C} \longrightarrow \mathbf{D}$ it follows that

$$(C^2) = (D^2) + \sum_P s(\varphi_P) \tag{2.4.8}$$

For modification $\varphi_{\mathbf{P}} : \mathbf{P} \longrightarrow \mathbf{Q}$ of abelian point we set

$$\begin{aligned} \tau(\varphi_{\mathbf{P}}) &= \frac{1}{3v} s(\varphi_{\mathbf{P}}) - \tau^1(\mathbf{P}) + \tau^1(\mathbf{Q}) \\ &= \left(s(\varphi) + \frac{e_P}{d_P} - \frac{e_Q}{d_Q} \right) /3v, \ v = v_P = v_Q, \end{aligned} \tag{2.4.9}$$

and call it the *(relative) signature height of* $\varphi_{\mathbf{P}}$. Now let $\varphi : \mathbf{C} \longrightarrow \mathbf{D}$ be a modification of orbital curves. We remember that $\tau(\varphi)$ is defined as difference $\tau(\mathbf{C}) - \tau(\mathbf{D})$. With definition (2.4.3), (2.4.9) and (2.4.8) we get the

Local Decomposition 2.4.6

$$\tau(\varphi) = \sum_{\mathbf{P} \in C} \tau\left(\varphi_{\mathbf{P}}\right), \ \varphi : \mathbf{C} \longrightarrow \mathbf{D} \text{ in } \mathcal{C}_1'$$

In order to apply the Local-Global Principle we have to show that

$$\tau\left(\psi_{\mathbf{P}'}\right) = \deg_{\text{red}}(\mathbf{F}) \cdot \tau\left(\varphi_{\mathbf{Q}'}\right) \tag{2.4.10}$$

for any GALOIS-finite covering $\mathbf{F} = (\mathbf{f}', \mathbf{f})$, see:

$$\begin{array}{ccc} \mathbf{C}' & \xrightarrow{\psi} & \mathbf{C} \\ \downarrow{\scriptstyle \mathbf{f}'} & & \downarrow{\scriptstyle \mathbf{f}} \\ \mathbf{D}' & \xrightarrow{\varphi} & \mathbf{D} \end{array} \qquad \begin{array}{ccc} \mathbf{P}' & \longrightarrow & \mathbf{P} \\ \downarrow & & \downarrow \\ \mathbf{Q}' & \longrightarrow & \mathbf{Q} \end{array} \tag{2.4.11}$$

We assume first that $\mathbf{F}$ is a quasiuniformization with Galois group A. Furthermore we assume that each of $\mathbf{C}'$ and $\mathbf{C}$ support at most $k \geq 1$ singularities and proceed by induction over k.

Inductive Hypothesis 2.4.7 Let $\beta : \mathbf{B} \longrightarrow \mathbf{E}$ be a quasi-uniformization of the orbital curve $\mathbf{E}$ such that $\mathbf{B}$ supports at most $k-1 \geq 0$ (cyclic surface) singularities. Then the degree formula

$$\tau(\mathbf{B}) = \deg_{\text{red}}(\beta) \cdot \tau(\mathbf{E})$$

holds.

Now we resolve in equivariant manner the singularities P' and P on $\mathbf{C}'$ or $\mathbf{C}$, respectively, and their A-orbits. We divide the arising (trivially) arranged curves $\tilde{\mathbf{C}}'$and $\tilde{\mathbf{C}}$ by A and get the following commutative quotient diagrams:

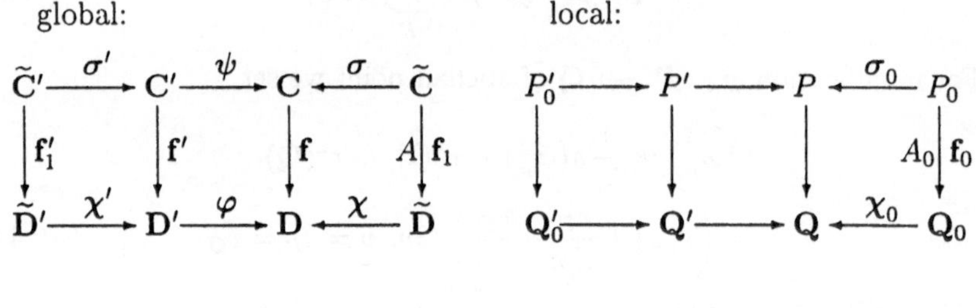

(2.4.12)

For the special quasi-uniformization $\tilde{\mathbf{C}}/\mathbf{C} \underset{A}{\rightarrow} \tilde{\mathbf{D}}/\mathbf{D}$ we are able to prove the local degree formula

$$\tau(\sigma_0) = \frac{|A_0|}{v} \cdot \tau(\chi_0), \qquad (2.4.13)$$
$$v = v_C = v_Q = v_{Q_0}, A_0 = A_{P_0}.$$

It is easy to see that $-s(\chi_0)$ is nothing else than the left absorptive number $a^-(\mathbf{f}_0)$. We refer the reader to the definitions (1.18.12) and (2.4.7). With definition (2.4.9) and Proposition 1.18.5 we get

$$\tau(\chi_0) = \left(-a^-(\mathbf{f}_0) + \frac{e_0}{d_0} - \frac{e_Q}{d_Q}\right)/3v = -\frac{e_P v^2}{d_p\,|A_0|}/3v = -\frac{e_P v}{d_p\,|A_0|} \qquad (2.4.14)$$

$\mathbf{Q}_0 :$ $v \; \langle d_0, e_0 \rangle \; w_0$

Since $\sigma_0 : P_0 \longrightarrow P$ is a minimal singularity resolution we find that

$$\tau(\sigma_0) = \left(s(\sigma_0) + \frac{0}{1} - \frac{e_P}{d_P} \right) /3 = -\frac{e_P}{3d_P}.$$

Comparing with (2.4.14) we see that (2.4.13) is satisfied. We conclude that

$$\tau(\sigma) = \deg_{\text{red}}(\mathbf{f}) \cdot \tau(\chi) \qquad (2.4.15)$$

One has only to take into account the $|A|/|A_0|$ preimages of Q or Q_0 on $\mathbf{C}$ or $\tilde{\mathbf{C}}$, respectively. All produce the same local number $\tau(\sigma_0)$, hence

$$\tau(\sigma) = \sum_{P \to Q} \tau(\sigma_0) = \frac{|A|}{|A_0|} \cdot \tau(\sigma_0) = \frac{|A|}{|A_0|} \cdot \frac{|A_0|}{v} \tau(\chi_0) = \frac{|A|}{v} \tau(\chi_0).$$

From the shifting principle (see Corollary 1.16.3) and the degree formula (2.4.13) it follows that $\tau(\mathbf{f}) = \tau(\mathbf{f}_0)$. The inductive hypothesis 2.4.7 and the assumption that $\mathbf{C}$ supports at most k singularities imply that for $\mathbf{f}_0$ the degree formula holds because we removed at least one singularity of $\mathbf{C}$. This means that $\tau(\mathbf{f}_0) = 0$, hence $\tau(\mathbf{f}) = 0$.

Remark 2.4.8 The proof of Theorem 2.4.2 can be already finished on this place by induction over k. Namely the case $k = 0$ has been already checked in the proof of (2.4.3). So the degree formula is valid for all quasi-uniformizations $\mathbf{C} \longrightarrow \mathbf{D}$. It is easily extended to GALOIS-finite coverings via diagram (2.2.1) in the usual manner.

For the sake of completeness we bring the proof of the local degree formula (2.4.10) to the end. We know that $\tau(\mathbf{f}) = \tau(\mathbf{f}') = 0$. As in the proof of (1.16.10) it follows that

$$\tau(\psi) = \deg(\mathbf{F}) \cdot \tau(\varphi) \qquad (2.4.16)$$

with the notations of diagram (2.4.12). (The orbital property of τ on C implies the same property on C_1').

If φ would be isomorphic outside of $\mathbf{Q}' \longrightarrow \mathbf{Q}$, then the right side of (2.4.16) would be $\deg(\mathbf{F}) \cdot \tau\left(\varphi_{\mathbf{Q}'}\right)$ by the formula 2.4.6 of local decomposition. The left side would be

$$\tau(\psi) = \sum_{P' \to Q'} \tau\left(\psi_{\mathbf{P}'}\right) = [A : A_{P'}]\tau\left(\psi_{\mathbf{P}'}\right) = \frac{|A| \cdot v}{v \cdot |A_{P'}|} \tau\left(\psi_{\mathbf{P}'}\right)$$

The comparison of both expressions yields the local degree formula (2.4.10). So we need only to realize ψ'_P and $\varphi_{Q'} = \psi_{P'}/A$ as localizations of a uniformization $\mathbf{C}'/\mathbf{C} \longrightarrow \mathbf{D}'/\mathbf{D}$ such that these are the only non-trivial localizations of $\mathbf{C}'/\mathbf{C}$ or $\mathbf{D}'/\mathbf{D}$, respectively. It is easy to see that $\varphi : \mathbf{D}' \longrightarrow \mathbf{D}$ can be decomposed into simple modifications degenerating at only one point. Especially we get a decomposition $\mathbf{D}' \longrightarrow \mathbf{D_Q} \longrightarrow \mathbf{D}$ where the second modification degenerates only over $\mathbf{Q}$ such that the localization there is $\varphi_{Q'}$. We can do the same at all preimages of $\mathbf{Q}$ on $\mathbf{C}$ in an A-equivariant manner. With obvious notations we obtain a commutative diagram (2.4.17) with localizations we looked for.

$$\begin{array}{ccccccc} \mathbf{C}_Q & \longrightarrow & \mathbf{C} & \qquad & \mathbf{P}' & \xrightarrow{\psi_{P'}} & \mathbf{P} \\ \downarrow A & & \downarrow A & & \downarrow & A_{P'} = A_P & \downarrow \\ \mathbf{D}_Q & \longrightarrow & \mathbf{D} & & \mathbf{Q}' & \xrightarrow{\varphi_{Q'}} & \mathbf{Q} \end{array} \tag{2.4.17}$$

This proves the degree formula (2.4.10) for quasiuniformizations of local modifications. It can be easily extended to uniformizable local modifications via diagram (2.2.2).

Forgetting the Remark 2.4.8 we proved the Theorem 2.4.2 by means of the Local-Global-Principle 2.3.2. It works with the geometric local decomposition 2.4.6 and the local degree formula (2.4.10) and implies condition (i) of the Shifting-Lemma 1.16.1. The condition (ii) has already been checked by means of uniformizable (GALOIS-finite) coverings (see 2.4.5 and 2.4.4). The Shifting-Lemma implies that the signature heights $\tau(\mathbf{C})$ of orbital curves are orbital invariants.

We have to fill one gap, namely the

2.4.9 Proof of Proposition 1.18.5. We start with a reduced finite local quotient morphism $\mathbf{f}_P : P \longrightarrow \mathbf{Q} = P/A$, A a finite abelian group. We look for a reduced orbital curve $\mathbf{C}$ such that

$$(\mathbf{C}, P) : \quad \underset{1}{\overset{s}{\circ}} \xrightarrow[\langle d_P, e_P \rangle]{} \underset{1}{\circ}, \quad s = (C^2), \tag{2.4.18}$$

is a reduced punctured arranged curve and A extends to an action on $\mathbf{C}$, that means that up to analytic equivalence the localization of $\mathbf{f} : \mathbf{C} \longrightarrow \mathbf{D} = \mathbf{C}/A$ at $\mathbf{P}$ is the given point morphism $\mathbf{f}_P : P \longrightarrow \mathbf{Q}$. Additionally the following conditions should be satisfied:

(a) P is the only surface singularity on $\mathbf{C}$

(b) $\mathbf{f} : \mathbf{C} \longrightarrow \mathbf{D}$ is uniformizable.

Now we resolve the singularity P in the minimal manner. By factorization we obtain a commutative diagram:

$$\begin{array}{ccc} (\mathbf{C}, P_0) & \longrightarrow & (\mathbf{C}, P) \\ \mathbf{f}' \downarrow A & & A \downarrow \mathbf{f} \\ (\mathbf{D}', \mathbf{Q}_0) & \longrightarrow & (\mathbf{D}, \mathbf{Q}) \end{array} \qquad \begin{array}{ccc} \overset{s}{\bullet}\!\!\longrightarrow\!\!\circ & \Longrightarrow & \overset{s}{\bullet}\!\!\longrightarrow\!\!\circ \\ \langle 1,0\rangle \ 1 & & \langle d_P, e_P\rangle \ 1 \\ \Downarrow & & \Downarrow \\ \overset{t'}{\bullet}\!\!\longrightarrow\!\!\circ & \Longrightarrow & \bullet\!\!\longrightarrow\!\!\circ \\ v \ \langle d_0, e_0\rangle \ w_0 & & v \ \langle d_Q, e_Q\rangle \ w \end{array}$$

(2.4.19)

Taking into account the minimal resolution $E_P \longrightarrow P$, we have also a commutative diagram (1.18.11) of resolutions. The left absorptive number $a^-(\mathbf{f}_Q)$ measures the difference of $(D'^2) = t'$ and $(D^2) = t$. So we have to prove that

$$t - t' = (D^2) - \left(D'^2\right) = \frac{e_P v^2}{|A| \cdot d_P} + \frac{e_o}{d_o} - \frac{e_Q}{d_Q} \tag{2.4.20}$$

The formula for $a^+(\mathbf{f}_P)$ in Proposition 1.18.5 can be proved analogeously. One has only to change the orientation of the arranged abelian point P. More directly it follows from (2.4.20) and $a^+(\mathbf{f}_P) = a^-(\mathbf{f}_{P^\perp})$, $P^\perp$ the opposite point of P. The morphism $\mathbf{f}$ is uniformizable by (b), and $\mathbf{f}'$ is a uniformization by assumption (a). From Proposition 1.14.2 we know that for $\mathbf{f}$ and $\mathbf{f}'$ the degree formula holds. The corresponding translation of (1.14.8) into the degree formula was given in (2.4.5). Taking into account Definition (2.4.3) we have

$$\begin{aligned} \frac{|A|}{v}\tau(\mathbf{D}') = \tau(\mathbf{C}') = \frac{1}{3}\left(C'^2\right) = \frac{1}{3}(C^2) = \tau(\mathbf{C}) - \frac{e_P}{3 \cdot d_P} \\ = \frac{|A|}{v}\tau(\mathbf{D}) - \frac{e_P}{3 \cdot d_P}, \end{aligned}$$

hence

$$\tau(\mathbf{D}') - \tau(\mathbf{D}) = -\frac{e_P v}{3 d_P |A|} \tag{2.4.21}$$

The orbital curves $\mathbf{D}'$ and $\mathbf{D}$ are isomorphic outside of $\mathbf{Q}_0 \longrightarrow \mathbf{Q}$. Therefore the difference (2.4.21) is equal to

$$\left[\left(D'^2\right) - (D^2) + \frac{e_o}{d_o} - \frac{e_Q}{d_Q}\right] / 3v = -\frac{e_P v}{3 d_P |A|},$$

hence

$$-a^-(\mathbf{f}_Q) = \left(D'^2\right) - \left(D^2\right) = -\frac{e_P v^2}{d_P|A|} - \frac{e_o}{d_o} + \frac{e_Q}{d_Q}.$$

This is the formula of Proposition 1.18.5 which we wanted to prove. It remains to find the situation of diagram (2.4.19) with conditions (a) and (b). For the corresponding construction we first uniformize the given point morphism

$$\begin{array}{c} \overset{B}{\overbrace{\qquad\qquad\qquad\qquad}} \\ O \xrightarrow{\langle d_P, e_P\rangle} P \xrightarrow{A} \mathbf{Q} \end{array}$$

For the existence of B we use Corollary 1.12.3. We consider B and $\langle d_P, e_P\rangle$ as subgroups of $\mathbb{Gl}_2(\mathbb{C})$. So we dispose on an affine globalization

$$\mathbb{C}^2 \longrightarrow \mathbb{C}^2/\langle d_P, e_P\rangle \xrightarrow{A} \mathbb{C}^2/B$$

The action of B on $\mathbb{C}^2$ extends to $\mathbb{P}^2 = \mathbb{P}^2(\mathbb{C}) \supset \mathbb{C}^2$ in a natural manner. Let L be the closure of the x-axis of $\mathbb{C}^2$ in $\mathbb{P}^2$. The reduced orbital curve $(\mathbb{P}^2, L)$ is denoted by $\mathbf{L}$. A small finite abelian subgroup diagonalized by the coordinate axis has only two fixed points on L, namely 0 and ∞, $\{\infty\} = L \setminus \mathbb{C}^2$. Now we consider the tower

$$\mathbf{L} \longrightarrow \mathbf{L}/\langle d, e\rangle = \mathbf{C}_1 \xrightarrow{A} \mathbf{D}_1 = \mathbf{L}/B.$$

The localization at 0 is $O \longrightarrow P \underset{\mathbf{f}_P}{\rightarrow} \mathbf{Q}$ as we wanted to construct. But $\mathbf{C}_1$ may support two singularities. In order to remove the singularity $\infty/\langle d, e\rangle$ we apply the σ_{-e}-process at $\infty \in \mathbf{L}$. Observe that the action of $\langle d, e\rangle$ at $\infty \in \mathbf{L}$ is of type $\langle d, -e\rangle = \langle d, a\rangle$, $a \equiv -e \mod d$, $0 \le a < d$. So σ_{-e} means σ_a. Then we have a commutative diagram:

$$\begin{array}{ccccccc} & & \mathbf{L}' & \xrightarrow{\sigma_a} & \mathbf{L} & & \\ & & \downarrow{\scriptstyle\langle d,e\rangle} & & \downarrow{\scriptstyle\langle d,e\rangle} & & \\ B & & \mathbf{C} & \longrightarrow & \mathbf{C}_1 & & B \\ & & \mathbf{f}\downarrow{\scriptstyle A} & & \downarrow{\scriptstyle A} & & \\ & & \mathbf{D} & \longrightarrow & \mathbf{D} & & \end{array} \qquad (2.4.22)$$

with localization $\infty' \longrightarrow \infty$ of σ_a at ∞. The quotient point $\infty'/\langle d, e\rangle$ is smooth. To see this, the reader is referred back to the diagrams (1.6.1) and figure 1.6.1. Now it is easy to see that $\mathbf{f} : \mathbf{C} \longrightarrow \mathbf{D}$ defined in diagram (2.4.22) satisfies the conditions (a) and (b). the proof of Proposition 1.18.5 is finished.

□

3 Orbital Surfaces

3.1 Regular Arrangements on Surfaces

We work in the category of all compact complex normal algebraic surfaces with (at most) singularities of HIRZEBRUCH-JUNG type. A *regular arrangement* on such a surface X is a formal sum

$$\mathbf{A} = \mathbf{C}_1 + \ldots + \mathbf{C}_k + \mathbf{P}_1 + \ldots + \mathbf{P}_m,$$

where $\mathbf{C}_i = (X, v_i C_i; ...)$ is a (smooth) orbital curve on X and $\mathbf{P}_j$ is an (arranged) abelian point on X, $i = 1, ..., k$, $j = 1, ..., m$. The following axioms have to be satisfied:

(i) $C_i \neq C_j$ and $P_i \neq P_j$ for $i \neq j$

(ii) If P is an intersection point of C_i and C_j, then P can be arranged by C_i and C_j, $\mathbf{P} = (X, P; v_i C_i, v_j C_j) \in \mathbf{C}_i$, $\mathbf{P}^\perp \in \mathbf{C}_j$ and the corresponding abelian point belongs to the *point arrangement* $\mathbf{P}_1 + ... + \mathbf{P}_m$ on X.

(iii) If P_i lies on C_j but not on any other curve C_k, then $\mathbf{P}_i$ lies on $\mathbf{C}_j$ and has the graph $\underset{v_j \, \langle d_{P_i}, e_{P_i} \rangle 1}{\circ\!\!-\!\!-\!\!-\!\!\circ}$.

(iv) Each $\mathbf{P}_i$ not lying on the 1–*support* $\mathrm{supp}^1(A) = C_1 \cup \ldots \cup C_k$ of $\mathbf{A}$ is a reduced abelian point, $P_i : \underset{1 \, \langle d_{P_i}, e_{P_i} \rangle 1}{\circ\!\!-\!\!-\!\!-\!\!\circ}$.

(v) The set $Sg(X)$ of all singularities of X belongs to the 0–*support* $\mathrm{supp}^0(\mathbf{A}) = \{P_1, \ldots, P_m\}$ of $\mathbf{A}$.

(vi) If $\mathbf{P}$ is part of the point arrangement of the orbital curve $\mathbf{C}_i$, then $\mathbf{P}$ belongs to $\{\mathbf{P}_1, ... \mathbf{P}_m\}$. .

The *support* of $\mathbf{A}$ is the set

$$\mathrm{supp}(\mathbf{A}) = \mathrm{supp}^1(\mathbf{A}) \cup \mathrm{supp}^0(\mathbf{A}) = C_1 \cup \ldots \cup C_k \cup \{P_1, \ldots, P_m\}.$$

We allow that $k = m = 0$, that means that $\mathbf{A}$ is the zero-arrangement $\emptyset$. also $k = 0, m \neq 0$ and $k \neq 0, m = 0$.

Definition 3.1.1 A *regularly arranged surface* $\mathbf{X}$ is a pair $\mathbf{X} = (X, \mathbf{A})$, X a surface as above, and $\mathbf{A}$ a (locally abelian) arrangement on X as described above. It is also called an *orbital surface* (with regular arrangement).

Convention 3.1.2 Until we define new kinds of arrangements we will omit the attribute "regular". An arrangement $\mathbf{A}$ is called *connected* , if supp($\mathbf{A}$) is connected. It is called *reduced* , if all 1–*components* $\mathbf{C}_i$ of $\mathbf{A}$, hence also all 0–*components* $\mathbf{P}_j$, are reduced.

We can correspond to each arrangement $\mathbf{A}$ a *molecular graph* joining the atomic graph's of the 1–components at common points and adding the graphs of the isolated 0–components consisting of the points P_i not lying on $\mathrm{supp}^1(\mathbf{A})$. We give an example, which clarifies the general concept.

$$\mathbf{A} = \mathbf{C}_1 + \mathbf{C}_2 + \mathbf{C}_3 + \mathbf{P}_1 + \mathbf{P}_2 + \mathbf{P}_3 + \mathbf{P}_4 + \mathbf{P}_5 + \mathbf{P}_6$$

We omitted as usually the trivial weights 1 on the electronic parts. A

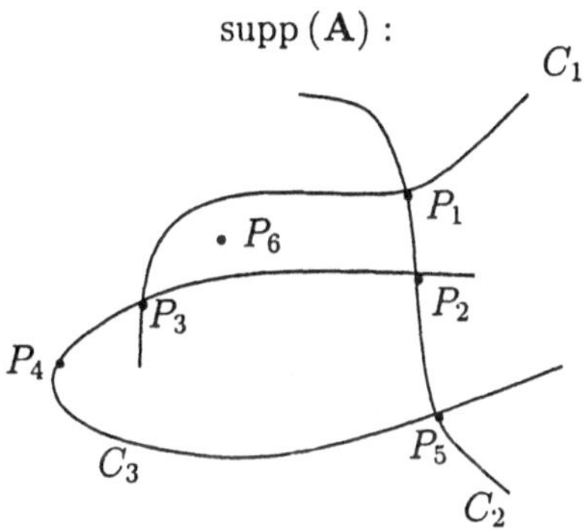

Figure 3.1.1

shorter form of graphs connecting directly the curve components $\overset{v_i}{\underset{s_i}{\bullet}}$ at intersection points is more useful: Clearly, the graph Figure 3.1.2 can be reconstructed from the latter one. The arrangement $\mathbf{A}$ is connected if and only if its molecular graph is connected.

We come to equivalent graphs if we add trivial "arrows" $\xrightarrow[\langle 1,0\rangle]{}\circ$ at curve components $\overset{s_i}{\underset{v_i}{\bullet}}$, trivial isolated "arrows" $\circ\xrightarrow[\langle 1,0\rangle]{}\circ$ or trivially

Molecular graph of **A** :

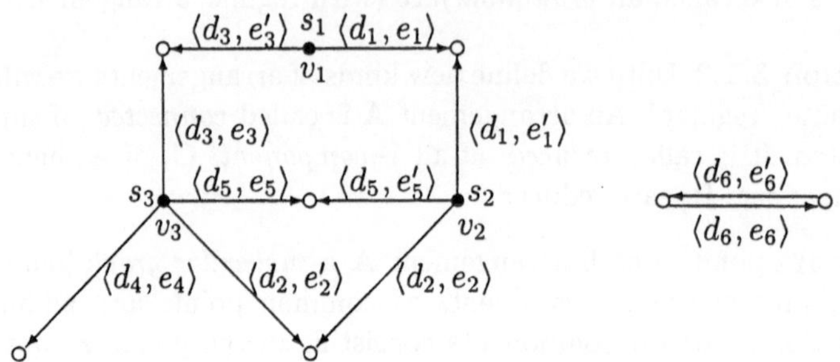

Figure 3.1.2

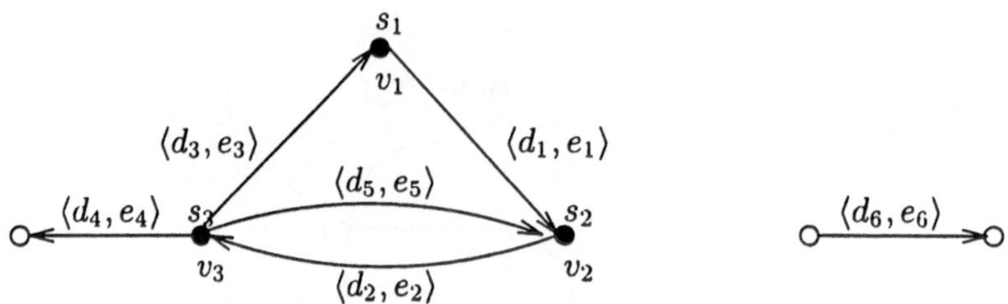

Figure 3.1.3

weighted components $\overset{s}{\underset{1}{\bullet}}$ These manipulations define an equivalence relation of molecular graphs. Two arrangements $\mathbf{A}, \mathbf{A}'$ are called *equivalent* , if their graphs are equivalent.

Two orbital surfaces $\mathbf{X}, \mathbf{X}'$ are called *equivalent* if they are supported by the same surface X and the corresponding arrangements are equivalent. Later we will see that equivalence implies numerical equivalence. Sometimes we change over tacitly from a given arrangement to a suitable equivalent one adding trivially weighted curves and smooth points, for example in the next

Definition 3.1.3 A *morphism* $\mathbf{f}: \ \mathbf{X} = (X, \mathbf{A}) \longrightarrow \mathbf{Y} = (Y, \mathbf{B})$ is a generically finite morphism of algebraic surfaces $f: X \longrightarrow Y$ such that

$f(\mathrm{supp}(\mathbf{A})) \subseteq (\mathrm{supp}\,\mathbf{B})$ and f induces morphisms $\mathbf{f_C} : \mathbf{C} \longrightarrow \mathbf{D}$, $\mathbf{f_P} : \mathbf{P} \longrightarrow \mathbf{Q}$ on the components (orbital curves, abelian points) of $\mathbf{A}$ or $\mathbf{B}$, respectively.

This definition is generalized by changing over to an equivalent bigger arrangement $\mathbf{B}'$ of $\mathbf{Y}$ such that $f(\mathrm{supp}(\mathbf{A})) \subseteq \mathrm{supp}(\mathbf{B}')$, if $f(\mathrm{supp}(\mathbf{A})) \not\subseteq \mathrm{supp}(\mathbf{B})$ or to a smaller arrangement $\mathbf{A}'$ equivalent to $\mathbf{A}$.

The category of orbital surfaces is denoted by $\mathfrak{S}$. We look for orbital invariants on $\mathfrak{S}$. For this purpose we need a class $\mathfrak{F}$ of morphisms playing the role of "finite morphisms". Let X be a surface and G a finite group acting on X. We say that G acts *locally abelian* at $P \in X$, if the isotropy group G_P is abelian. The quotient map $X \longrightarrow X/G$ induces a quotient morphism $P \longrightarrow P/G_P$ of abelian points. The action of G is called *locally abelian,* if G acts locally abelian at any point of X. Let $\mathbf{X} = (X, Sg(X))$ be the *trivially arranged surface* X, where $Sg(X)$ denotes the sum of all reduced singular points on X. Assume that G acts locally abelian on X and the components of the branch curve of the quotient morphism $f : X \longrightarrow X/G$ are smooth. Then we say that G acts *regularly* on X and we define the quotient arrangement $\mathbf{B}$ on $Y = X/G$ in the following manner:

$$\mathrm{supp}(\mathbf{B}) = (\text{branch locus of f}) \cup SgY;$$
$$\mathbf{B} = \mathbf{D}_1 + \ldots + \mathbf{D}_k + \mathbf{Q}_1 + \ldots + \mathbf{Q}_l,$$

where the D_i's are the 1–components of the branch locus and $\mathbf{D}_i = C_i/N_G(C_i)$, C_i an irreducible component of $f^{-1}(D_i)$,

$$N_G(C) = \{g \in G; g(C) = C\} \text{ (decomposition group)}$$
$$\mathbf{Q}_j = P_j/G_{P_j}, P_j \in f^{-1}(Q_j). \tag{3.1.1}$$

The orbital surface $\mathbf{Y} = (X/G; \mathbf{B})$ is called the *quotient* of X by G and $\mathbf{B}$ the *quotient arrangement* . We will write frequently $\mathbf{Y} = X/G$ with a fat / in order to indicate an orbital object. The morphism $\mathbf{f} : X \longrightarrow X/G$ defined by f is called a *quotient morphism* . Quotient morphisms are also called *quasi-uniformizations* . A *uniformization* is a quotient morphism $X \longrightarrow X/G$ with smooth covering surface X. The morphism $\mathbf{f} : \mathbf{X} \longrightarrow \mathbf{Y}$ is called *quasi-uniformizable,* if there exists a commutative diagram:

$$\begin{array}{ccc} & W & \\ {}^{\mathbf{h}}\swarrow_{H} & & {}_{G}\searrow^{\mathbf{g}} \\ \mathbf{X} & \xrightarrow{\mathbf{f}} & \mathbf{Y} \end{array} \quad , H \subseteq G \subseteq \mathrm{Aut}(W) \tag{3.1.2}$$

with quasi-uniformizations $\mathbf{h}$ and $\mathbf{g}$. If we can find such uniformizations $\mathbf{h}, \mathbf{g}$, then $\mathbf{f}$ is called *uniformizable.*

The class $\mathfrak{F}$ of morphism of arranged surfaces for which we want to prove degree formulas for certain numerical functions $Ob(\mathfrak{S}) \longrightarrow \mathbb{Q}$ consists of all quasi-uniformizable morphisms. For this reason we call them GALOIS-*finite coverings* or GALOIS-*finite morphisms* because GALOIS groups have been used to define them.

We have two kinds of localizations: the localizations along curves and the point localizations. They are defined in an obvious manner. Let $\mathbf{f} : \mathbf{X} \longrightarrow \mathbf{Y}$ be a morphism, $\mathbf{C} \subset \mathbf{X}$ an orbital curve on $\mathbf{X}$ with support C not contracted by f and $\mathbf{P} \in \mathbf{X}$ an abelian point on $\mathbf{X}$. The images $D = f(C)$, $Q = f(P)$ determine the orbital curve $\mathbf{D} \subset \mathbf{Y}$ or the abelian point $\mathbf{Q} \in \mathbf{Y}$, respectively. The obvious restrictions of $\mathbf{f}$ are denoted by

$$\mathbf{f}_C = \mathbf{f}_{\mathbf{C}} : \mathbf{C} \longrightarrow \mathbf{D} \text{ or } \mathbf{f}_P = \mathbf{f}_{\mathbf{P}} : \mathbf{P} \longrightarrow \mathbf{Q},$$

respectively, and are called the *localizations* or *restrictions* of $\mathbf{f}$ at $\mathbf{C}$ or $\mathbf{P}$, respectively. It is clear that uniformizations, quasi-uniformizations, GALOIS-finite morphisms and modifications restrict to the same classes of morphisms in lower dimension. Furthermore, any commutative diagram in $\mathfrak{S}$ restricts to local commutative diagrams:

$$\begin{array}{ccccccccc} \mathbf{X}' & \xrightarrow{\psi} & \mathbf{X} & \mathbf{C}' & \xrightarrow{\psi_{C'}} & \mathbf{C} & \mathbf{P}' & \xrightarrow{\psi_{P'}} & \mathbf{P} \\ \downarrow{\scriptstyle \mathbf{f}'} & & \downarrow{\scriptstyle \mathbf{f}} & \downarrow{\scriptstyle \mathbf{f}'_{\mathbf{C}'}} & & \downarrow{\scriptstyle \mathbf{f}_{\mathbf{C}}} & \downarrow{\scriptstyle \mathbf{f}'_{\mathbf{P}'}} & & \downarrow{\scriptstyle \mathbf{f}_{\mathbf{P}}} \\ \mathbf{Y}' & \xrightarrow{\varphi} & \mathbf{Y} & \mathbf{D}' & \xrightarrow{\varphi_{D'}} & \mathbf{D} & \mathbf{Q}' & \xrightarrow{\varphi_{Q'}} & \mathbf{Q} \end{array} \tag{3.1.3}$$

for $\mathbf{P} \in \mathbf{C} \subset \mathbf{X}$ under the assumption that $D = (\varphi \circ f')(C)$ is not a point.

3.2 Basic Invariants and Fixed Point Theorem

Two basic invariants of a normal compact complex algebraic surface X are the *arithmetic genus*

$$\chi(X) = \chi(X, \mathcal{O}_X) = \sum_{i=0}^{2} (-1)^i \dim H^i(X, \mathcal{O}_X)$$

and the EULER *number*

$$e(X) = c_2(X) = \sum_i (-1)^i \dim H^i(X, \mathbb{C}_X),$$

where $\mathcal{O}_X$ denotes the structure sheaf of X and $\mathbb{C}_X = X \times \mathbb{C}$ is the trivial vector bundle of rank 1 on X. We restrict our attention to *quasi-smooth surfaces.* By definition, these surfaces have only quotient singularities. In the class of quasismooth surfaces it holds that

3.2.1 $\chi(X)$ is a *birational invariant* (FREITAG, [FRE]).

The reason is that quotient singularities are rational (see BRIESKORN, [BRI]).

The EULER number is not birationally invariant. For smooth surfaces the following invariants are also well known:

$$c_1^2(X) = (K_X^2), K_X \text{a canonical divisor on X,} \tag{3.2.1}$$

$$\tau(X) = \tau^+(H_2(X,\mathbb{R})) - \tau^-(H_2(X,\mathbb{R})) \tag{3.2.2}$$

The first one is the *selfintersection number of the canonical class.* The second one is the *signature* of the surface X. The homology group $H_2(X,\mathbb{R})$ has a natural non-degenerate intersection product. If V is a real finite-dimensional vector space with a symmetric bilinear form Q, then we can find a basis on V, which diagonalizes Q. Then $\tau^+(V)$ denotes the number of positive entries in the corresponding diagonal matrix and $\tau^-(V)$ the number of negative entries. The *signature* (τ^+,τ^-) *of* (V,Q) is uniquely determined by Q (SYLVESTER). From RIEMANN-ROCH theory the following relations are well-known:

$$\chi(X) = \left(e(X) + c_1^2(X)\right)/12 \quad \text{(NOETHER's formula)} \tag{3.2.3}$$

$$\tau(X) = \left(c_1^2(X) - 2e(X)\right)/3 \quad \text{(Index Theorem)} \tag{3.2.4}$$

(see [HI3] IV, § 20, Introduction (9); II, Thm. 8.2.2; I, Thm. 4.6.1). The invariants $c_1^2(X), c_2(X)$ and some linear combinations are also known as CHERN *numbers* of X. They can be expressed as volumes of X :

$$c_1^2(X) = \int_X c_1^2 = \int_X \gamma_1 \wedge \gamma_1, c_2(X) = \int_X c_2 = \int_X \gamma_2, \tag{3.2.5}$$

where γ_1 or γ_2 is a real closed 2–form or 4–form on X representing the CHERN class c_1 or c_2 of X, respectively.

We extend the definition of τ to quasismooth surfaces X. Let

$$\varphi : \tilde{X} \longrightarrow X$$

be a singularity resolution of X and E the corresponding *exceptional divisor* on $\tilde{X}$ whose components are those irreducible curves which are contracted to a point. The subspace of $H_2\left(\tilde{X}, \mathbb{R}\right)$ generated by the components of E is denoted by $\langle E \rangle$. The restriction of the intersection form Q on $H_2\left(\tilde{X}, \mathbb{R}\right)$ to $\langle E \rangle$ is negative definit by a well-known theorem, see for instance MUMFORD [MU1]. Therefore it holds that

$$-\tau(\langle E \rangle) = |\{\text{components of E }\}|$$

In order to define $\tau(X)$ we look at $H_2\left(\tilde{X}, \mathbb{R}\right)$ and omit the part coming from $\langle E \rangle$. More precisely we define

$$\tau(X) = \tau(\langle E \rangle^{\perp}) = \tau\left(\tilde{X}\right) + |\{\text{components of E }\}| \tag{3.2.6}$$

where $\langle E \rangle^{\perp}$ is the orthogonal complement of $\langle E \rangle$ in $H_2\left(\tilde{X}, \mathbb{R}\right)$. We have to check that the definition is correct, that means that it does not depend on the choice of the resolution. It suffices to prove that

$$\tau(X) = 4\chi(X) - e(X) \tag{3.2.7}$$

because the right side is well-defined.

The EULER number has a nice additive property. In our special case $\tilde{X} \longrightarrow X$ with rational components of the exceptional divisor E (resolving quotient singularities) it appears as

$$e\left(\tilde{X}\right) = e(X) + |\{\text{components of E }\}| \tag{3.2.8}$$

(see [B-K]). The comparison with (3.2.6) yields

$$e(X) + \tau(X) = e\left(\tilde{X}\right) + \tau\left(\tilde{X}\right) = 4\chi\left(\tilde{X}\right) = 4\chi(X).$$

We used (3.2.3), (3.2.4) for $\tilde{X}$ and 3.2.1.

Now we let X be a smooth surface again and G a finite group acting effectively on X. Then $Y = X/G$ is a quasismooth surface. We look for relations between the invariants of X and Y. They have been found with the methods of equivariant K-theory, which can be understood as equivariant RIEMANN-ROCH theory. This theory emits three fixed point theorems, which we will present without proof. The reader is referred to the nice lecture note of SHANAHAN [SHA] and/or to [H-Z1].

Lefschetz Fixed Point Formula 3.2.2 ([H-Z1] p. 53, (37); [SHA], p. 127)

$$|G|\, e\,(X/G) = e(X) + \sum_C |Z_G(C)|\, e_G(C) + \sum_x |G_x|\, e_G(x)$$

where the sums run over all irreducible curves C on X or over all points $x \in X$, respectively, and

$$Z_G(C) = \{g \in G; g|_C = id_C\}.$$

$$e_G(C) = \frac{1}{|Z_G(C)|} \sum_{g \in Z_G(C)^*} e_g(C),\ e_g(C) = e(C), \tag{3.2.9}$$

$$e_G(x) = \frac{1}{|G_x|} \sum_{g \in G_x^{**}} e_g(x),\ e_g(x) = 1, \tag{3.2.10}$$

where

$$\begin{aligned} &G_x^{**} = \{g \in G;\ x \text{ is isolated fixed point of } g\}, \\ &H^* = H \setminus \{1\} \text{for any group } H \end{aligned} \tag{3.2.11}$$

So we have

$$e_G(C) = \left(1 - \frac{1}{|Z_G(C)|}\right) e(C),\ e_G(x) = \frac{|G_x^{**}|}{|G_x|}, \tag{3.2.12}$$

and the LEFSCHETZ fixed point formula can be written as

$$|G| \cdot e\,(X/G) = e(X) + \sum_C (|Z_G(C)| - 1) e(C) + \sum_x |G_x^{**}| \tag{3.2.13}$$

Holomorphic Lefschetz Theorem 3.2.3 ([SHA], p.567; [AS2], p.138)

$$|G| \cdot \chi\,(X/G) = \chi(X) + \sum_C |Z_G(C)| \cdot \chi_G(C) + \sum_x |G_x|\, \chi_G(x),$$

where

$$\chi_G(C) = \frac{1}{|Z_G(C)|} \sum_{g \in Z_G(C)^*} \chi_g(C), \tag{3.2.14}$$

$$\chi_g(C) = \left(\frac{1}{2}e(C) - \frac{\zeta_g}{1-\zeta_g}(C^2)\right)\frac{1}{1-\zeta_g} \tag{3.2.15}$$

$$\chi_G(x) = \frac{1}{|G_x|} \sum_{g \in G_x^{**}} \chi_g(x), \; \chi_g(x) = \frac{1}{(1-\mu_g)(1-\zeta_g)}. \tag{3.2.16}$$

We have to explain μ_g and ζ_g. Let P be a point on X. It can be arranged by means of a coordinate system through P diagonalizing the finite abelian group $\langle g \rangle$ generated by $g \in G_x^*$. With respect to these coordinates g has a unique diagonal form $\left(\begin{smallmatrix}\mu_g & 0\\ 0 & \zeta_g\end{smallmatrix}\right)$. If C is a curve left pointwise fixed by an element $g \neq 1$, then we let P be a general point and arrange it by $C : \mathbf{P} = (X, P; C, C'_P)$. The curve germ C'_P can be chosen such that $g \in Z_G(C)$ has diagonal form $\left(\begin{smallmatrix}1 & 0\\ 0 & \zeta_g\end{smallmatrix}\right)$ with respect to C, C'_P.

We would like to simplify the sum in (3.2.14) removing the unit roots. We will use the following

Lemma 3.2.4 *Let ζ be a primitive n-th unit root. Then it holds that*

(a) $\displaystyle\sum_{i=1}^{n-1} \frac{1}{1-\zeta^i} = \frac{n-1}{2},$

(b) $\displaystyle\sum_{i=1}^{n-1} \frac{1}{|1-\zeta^i|^2} = \frac{1}{12}(n^2-1)$

Proof We apply the logarithmic derivation:

$$\sum_{i=1}^{n-1} \frac{1}{1-\zeta^i} = \left[\sum_{i=1}^{n-1} \frac{1}{z-\zeta^i}\right]_{z=1} = \left[\frac{d}{dz}\sum_i ln(z-\zeta^i)\right]_{z=1}$$

$$= \left[\frac{d}{dz} ln \prod_{i=1}^{n-1}(z-\zeta^i)\right]_{z=1}$$

$$= \left[\frac{d}{dz} ln(z^{n-1} + z^{n-2} + \ldots + 1)\right]_{z=1}$$

$$= \left[\frac{(n-1)z^{n-2} + (n-2)z^{n-3} + \ldots + 1}{z^{n-1} + z^{n-2} + \ldots + z + 1}\right]_{z=1}$$

$$= \frac{n(n-1)/2}{n} = \frac{n-1}{2}.$$

For the proof of (b) we use (a) and

(c) $\sum_{i=1}^{n-1} \frac{1}{(1-\zeta^i)^2} = -\frac{1}{12}(n-1)(n-5)$:

$$\begin{aligned}\sum_{i=1}^{n-1} \frac{1}{|1-\zeta^i|^2} &= \sum_{i=1}^{n-1} \frac{1}{1-\zeta^i} - \sum_{i=1}^{n-1} \frac{1}{(1-\zeta^i)^2} \\ &= \frac{n-1}{2} + \frac{(n-1)(n-5)}{12} \\ &= \frac{(n-1)(n+1)}{12} = \frac{1}{12}(n^2-1).\end{aligned}$$

It remains to prove (c). We let

$$s_{n-3}(X_1,\ldots,X_{n-1}) = \sum_{i,j=1;i<j}^{n-1} X_1 \cdot \ldots \cdot \widehat{X}_i \cdot \ldots \cdot \widehat{X}_j \cdot \ldots \cdot X_{n-1}$$

be the $(n-3)$-th elementary function of $X_1, \ldots, X_{n-1}$ and set $\xi_i = 1-\zeta^i$, $i = 1, \ldots, n-1$, and $s_{n-3} = s_{n-3}(\xi_1, \ldots, \xi_{n-1})$. This number appears up to sign as coefficient of z^2 in the polynomial $\prod_{i=1}^{n-1}(z-\xi_i)$. With $y = 1-z$ we obtain

$$\begin{aligned}\sum_{i=1}^{n-1}(z-\xi_i) &= \prod_i(-y+\zeta^i) \\ &= (-1)^{n-1}(y^{n-1}+y^{n-2}+\ldots+y+1) \\ &= (-1)^{n-1}((1-z)^{n-1}+(1-z)^{n-2}+\ldots+(1-z)+1) \\ &= (-1)^{n-1}\left(\ldots+z^2\sum_{\nu=0}^{n-1}\binom{\nu}{2}+\ldots\right).\end{aligned}$$

It follows that

$$s_{n-3} = \sum_{\nu=0}^{n-1}\binom{\nu}{2} = \binom{n}{3}.$$

Using also (a) we get

$$\begin{aligned}\sum_{i=1}^{n-1} \frac{1}{(1-\zeta^i)^2} &= \left(\sum_{i=1}^{n-1} \frac{1}{1-\zeta^i}\right)^2 - 2\sum_{i<j} \frac{1}{(1-\zeta^i)(1-\zeta^j)} \\ &= \left(\frac{n-1}{2}\right)^2 - 2\frac{s_{n-3}}{\prod_{i=1}^{n-1}(1-\zeta^i)}\end{aligned}$$

$$= \frac{(n-1)^2}{4} - \frac{2s_{n-3}}{n}$$
$$= \frac{(n-1)^2}{4} - \frac{(n-1)(n-2)}{3}$$
$$= -\frac{1}{12}(n-1)(n-5).$$

□

Corollary 3.2.5

$$\chi_G(C) = \left(1 - \frac{1}{|Z_G(C)|}\right)\left(\frac{1}{4}e(C) + \frac{|Z_G(C)|+1}{12}(C^2)\right)$$

Proof One has only to substitute (3.2.15) in (3.2.14) and to calculate the coefficients of $e(C)$ and (C^2) by means of Lemma 3.2.4. □

Now we come to the third fixed point theorem. We consider the action of G on X as a "representation" $\rho : G \longrightarrow Aut(X)$. Via duality the intersection form on $H_2(X, \mathbb{R})$ yields a quadratic form on the cohomology group $H^2(X, \mathbb{R})$. The group G acts on $H^2(X, \mathbb{R})$. The subspace of G-invariant elements is denoted by $H^2(X, \mathbb{R})^G$.

Definition 3.2.6 (HIRZEBRUCH, see [H-Z1]) For $Y = X/G$ set

$$\tau_\rho(Y) = \tau_\rho(X/G) = \tau\left(H^2(X,\mathbb{R})^G\right)$$
$$= \tau^+\left(H^2(X,\mathbb{R})^G\right) - \tau^-\left(H^2(X,\mathbb{R})^G\right).$$

***G*-Signature Theorem of ATIYAH-SINGER 3.2.7** ([AS1], [H-Z1], p.31; [SHA], p.154)

$$|G|\tau_\rho(X/G) = \tau(X) + \sum_C |Z_G(C)| \cdot \tau_G(C) + \sum_x |G_x|\tau_G(x), \tag{3.2.17}$$

where

$$\tau_G(C) = \frac{1}{|Z_G(C)|} \sum_{g \in Z_G(C)^*} \tau_g(C), \tag{3.2.18}$$

$$\tau_g(C) = (C^2)/\sin^2\left(\frac{\arg \zeta_g}{2}\right) \tag{3.2.19}$$

$$\tau_G(x) = \frac{1}{|G_x|} \sum_{g \in G_x^{**}} \tau_g(x), \tag{3.2.20}$$

$$\begin{aligned} \tau_g(x) &= \frac{(\mu_g + 1) \cdot (\zeta_g + 1)}{(\mu_g - 1) \cdot (\zeta_g - 1)} \\ &= -\cot\left(\frac{\arg \mu_g}{2}\right) \cot\left(\frac{\arg \zeta_g}{2}\right). \end{aligned} \tag{3.2.21}$$

As we have done for $\chi_G(C)$ we can remove the unit root expressions in (3.2.18) by means of the elementary formula

$$\sum_{i=1}^{n-1} \frac{1}{\sin^2\left(\arg \frac{\zeta^i}{2}\right)} = \frac{n^2 - 1}{3},$$

where ζ is a primitive n-th unit root (see [H-Z1], p. 178, (15)). Together with (3.2.19) we obtain

$$\tau_G(C) = \frac{|Z_G(C)|^2 - 1}{3|Z_G(C)|}(C^2). \tag{3.2.22}$$

Lemma 3.2.8 (a) $\chi_G(C) = \frac{1}{4}(e_G(C) + \tau_G(C))$

(b) $\chi_G(x) = \frac{1}{4}(e_G(x) + \tau_G(x))$

Proof For (a) one has only to add (3.2.12) and (3.2.22). For the proof of (b) we use (3.2.10), (3.2.21) and (3.2.16). Assume that $g \in G_x^{**}$, $\mu = \mu_g$, $\zeta = \zeta_g$. Then

$$\begin{aligned} e_g(x) + \tau_g(x) &= 1 + \frac{(\mu + 1)(\zeta + 1)}{(\mu - 1)(\zeta - 1)} \\ &= \frac{2\mu\zeta + 2}{(\mu - 1)(\zeta - 1)} \\ &= \frac{2}{(1 - \mu)(1 - \zeta)} + \frac{2}{(1 - \overline{\mu})(1 - \overline{\zeta})} \\ &= 2\chi_g(x) + 2\chi_{g^{-1}}(x). \end{aligned}$$

Therefore,

$$|G_x| \cdot (e_G(x) + \tau_G(x)) = \sum_{g \in G_x^{**}} (e_g(x) + \tau_g(x))$$

$$= 2\sum_{G_x^{**}} (\chi_g(x) + \chi_{g^{-1}}(x))$$
$$= 4\sum_{G_x^{**}} \chi_g(x) = 4|G_x|\chi_G(x)$$

□

Corollary 3.2.9 *It holds that*

$$\tau_\rho(X/G) = 4\chi(X/G) - e(X/G). \tag{3.2.23}$$

So the signature $\tau_\rho(X/G)$ does only depend on $Y = X/G$ and is equal to $\tau(X/G)$ defined in (3.2.6).

Proof By the G-signature theorem and the formulas 3.2.8(a), 3.2.8(b) and (3.2.7) we find that

$$\begin{aligned}|G| \cdot \tau_\rho(X/G) =& \tau(X) + \sum_C |Z_G(C)|\tau_G(C) + \sum_x |G_x|\tau_G(x)\\ =& 4\chi(X) - e(X) + \sum_C |Z_G(C)|(4\chi_G(C) - e_G(C))\\ &+ \sum_x |G_x|(4\chi_G(x) - e_G(x)).\end{aligned}$$

Now we apply the LEFSCHETZ fixed point formula 3.2.2, the holomorphic LEFSCHETZ theorem 3.2.3 and the formula (3.2.7) for X/G. We receive

$$|G| \cdot \tau_\rho(X/G) = |G|(4\chi(X/G) - e(X/G)) = |G| \cdot \tau(X/G).$$

□

So the G-signature formula can be written as

$$|G| \cdot \tau(X/G) = \tau(X) + \sum_C |Z_G(C)|\tau_G(C) + \sum_x |G_x|\tau_G(x). \tag{3.2.24}$$

At the end of this section we assume that G acts regularly on the smooth surface X. The arranged abelian quotient point of $P \in X$ is denoted by $\mathbf{Q} = P/G_P$. With definitions 1.12.6, 1.11.1, (1.12.10) and (3.2.20) we have

$$|G_P| \cdot \tau_G(P) = -4d_Q \cdot \mathbf{S}(\mathbf{Q}) = \operatorname{def}(\mathbf{Q}) \tag{3.2.25}$$

Together with (3.2.22) we write (3.2.25) in the most manageable form

$$|G|\tau(X/G) = \tau(X) + \sum_C \frac{1}{3}\left(|Z_G(C)|^2 - 1\right)(C^2) + \sum_P \operatorname{def}(P/G_P). \tag{3.2.26}$$

3.3 EULER Heights

We work with compact complex normal algebraic surfaces X with (at most) HIRZEBRUCH-JUNG singularities and regular arrangements $\mathbf{B}$ on such surfaces. Any such arrangement has a decomposition

$$\mathbf{B} = \mathbf{B}^1 + \mathbf{B}^0 = (\mathbf{D}_1 + \ldots + \mathbf{D}_k) + (\mathbf{Q}_1 + \ldots + \mathbf{Q}_m)$$

in the 1-dimensional and the 0-dimensional part. We set

$$\lambda_e(\mathbf{D}) = \lambda_e(v) = \left(1 - \frac{1}{v}\right), \mathbf{D} \text{ an orbital curve of weight } v; \tag{3.3.1}$$

$$e(\mathbf{Q}) = -1 + \frac{1}{v} + \frac{1}{v'} - \frac{1}{vv'd}, \quad \mathbf{Q} : \underset{v\ \langle d,e\rangle\ v'}{\circ\!\!\longrightarrow\!\!\circ}; \tag{3.3.2}$$

$$e(\mathbf{D}) = e(D) = 2 - 2g(D) \text{ (EULER number)} \tag{3.3.3}$$

Definition 3.3.1 The EULER *height* of an orbital surface $\mathbf{Y} = (Y, \mathbf{B})$ with regular arrangement is defined by the

Geometric Decomposition 3.3.2

$$\mathbf{e}(\mathbf{Y}) = e(Y) - e(\mathbf{B}^1) - e(\mathbf{B}^0),$$

where

$$\begin{aligned} e(\mathbf{B}^1) &= \sum_{\mathbf{D}\subset\mathbf{Y}} \lambda_e(\mathbf{D})e(\mathbf{D}) = \sum_{i=1}^{k} \lambda_e(\mathbf{D}_i)e(\mathbf{D}_i), \\ e(\mathbf{B}^0) &= \sum_{\mathbf{Q}\in\mathbf{Y}} e(\mathbf{Q}) = \sum_{j=1}^{m} e(\mathbf{Q}_j) \end{aligned} \tag{3.3.4}$$

with the notations introduced above.

Explicitly we have

$$\begin{aligned} \mathbf{e}(\mathbf{Y}) = & e(Y) - \sum_{\mathbf{D}} \left(1 - \frac{1}{v_{\mathbf{D}}}\right) e(D) \\ & + \sum_{\mathbf{Q}} \left(1 - \frac{1}{v_Q} - \frac{1}{v'_Q} + \frac{1}{v_Q v'_Q d_Q}\right) \end{aligned} \tag{3.3.5}$$

We define the EULER *height* of an abelian point to be

$$\mathbf{e}(\mathbf{Q}) = \frac{1}{d}\left(1-\frac{1}{v}\right)\left(1-\frac{1}{v'}\right), \quad \mathbf{Q} : \underset{v\ \langle d,e\rangle\ v'}{\circ\!\longrightarrow\!\circ}; \tag{3.3.6}$$

We call $Sg(Y)$ the *trivial arrangement* of Y. If there is no danger of misunderstanding, then we will denote the trivially arranged orbital surface $(Y, Sg(X))$ by Y again. Such orbital surfaces and equivalent ones are called *reduced* . The EULER height of this reduced orbital surface is

$$\mathbf{e}(Y) = e(Y) - \sum_{Q\in Sg(Y)} \left(1 - \frac{1}{d_Q}\right). \tag{3.3.7}$$

The next simple result is very useful for proofs.

Lemma 3.3.3 (Orbital decomposition of EULER heights) *With the above notations it holds that*

$$\mathbf{e}(\mathbf{Y}) = \mathbf{e}(Y) - \mathbf{e}\left(\mathbf{B}^1\right) - \mathbf{e}\left(\mathbf{B}^0\right), \tag{3.3.8}$$

where

$$\mathbf{e}(\mathbf{B}^1) = \sum_{\mathbf{D}\subset\mathbf{Y}} \lambda_e(\mathbf{D})\mathbf{e}(\mathbf{D}) = \sum_{i=1}^{k} \lambda_e(\mathbf{D}_i)\mathbf{e}(\mathbf{D}_i)$$

$$\mathbf{e}(\mathbf{B}^0) = \sum_{\mathbf{Q}\in\mathbf{Y}} \mathbf{e}(\mathbf{Q}) = \sum_{j=1}^{m} \mathbf{e}(\mathbf{Q}_j)$$

We delegate the proof to the next section (proof of 3.4.2, below).

Proposition 3.3.4 *Let* $\mathbf{f} : X \underset{G}{\to} \mathbf{Y} = X/G$ *be a uniformization of an orbital surface with regular arrangement.*

Then the degree formula

$$\mathbf{e}(X) = \deg(\mathbf{f})\mathbf{e}(\mathbf{Y}) = |G|\mathbf{e}(\mathbf{Y}) \tag{3.3.9}$$

holds.

Corollary 3.3.5 *The degree formula holds for uniformizable morphisms.*

Proof We translate the LEFSCHETZ fixed point formula (3.2.13) into the langugage of orbital surfaces:

$$e(X) = |G| e(Y) - \sum_{C \subset X} \lambda_e(v_C) v_C e(C) - \sum_{P \in X} |G_P| e^{(0)}(P/G_P) \qquad (3.3.10)$$

where v_C is the ramification index of C and

$$e^{(0)}(\mathbf{Q}) = 1 - \frac{1}{vd} - \frac{1}{v'd} + \frac{1}{vv'd}, \quad \mathbf{Q} : \underset{v \; \langle d,e \rangle \; v'}{\circ \longrightarrow \circ}; \qquad (3.3.11)$$

Indeed, we have for abelian G_P and $\mathbf{Q} = P/G_P$ the relation

$$|G_P^{**}| = |G_P \setminus \{\text{reflections}\}| = v_Q v'_Q d_Q - v_Q - v'_Q + 1$$
$$= |G_P| \left(1 - \frac{1}{vd} - \frac{1}{v'd} + \frac{1}{vv'd}\right) = |G_P| e^{(0)}(\mathbf{Q}).$$

With (3.3.2) and (3.3.6) we get the relation

$$e^{(0)}(\mathbf{Q}) = e(Q) + \mathbf{e}(\mathbf{Q}). \qquad (3.3.12)$$

The restrictions

$$\mathbf{f}_C : C \longrightarrow \mathbf{D} = C/N_G(C)$$

are uniformizations of orbital curves. So we can apply the degree formula (see (2.2.4))

$$e(C) = \deg_{\text{red}}(\mathbf{f}_C) \cdot \mathbf{e}(\mathbf{D}) = \deg(\mathbf{f}_C) \cdot \mathbf{e}(\mathbf{D}) = \frac{|N_G(C)|}{v_D} \cdot \mathbf{e}(\mathbf{D}), \; v_D = v_C. \qquad (3.3.13)$$

Now (3.3.10) can be transformed into

$$\begin{aligned} e(X) = & |G| \cdot e(Y) - \sum_{\mathbf{D}} \sum_{C \to \mathbf{D}} \lambda_e(v_D) \cdot \deg_{\text{red}}(\mathbf{f}_C) \cdot v_D \cdot \mathbf{e}(\mathbf{D}) \\ & - \sum_{\mathbf{Q}} \sum_{P \to \mathbf{Q}} |G_P| (\mathbf{e}(\mathbf{Q}) + e(Q)) \\ = & |G| \cdot e(Y) - \sum_{\mathbf{D}} \lambda_e(\mathbf{D}) [G : N_G(C)] \cdot |N_G(C)| \cdot \mathbf{e}(\mathbf{D}) \\ & - \sum_{\mathbf{Q}} [G : G_P] \cdot |G_P| \cdot (\mathbf{e}(\mathbf{Q}) + e(Q)) \\ = & |G| \left(\mathbf{e}(Y) - \mathbf{e}(\mathbf{B}^1) - \mathbf{e}(\mathbf{B}^0)\right) = |G| \cdot \mathbf{e}(\mathbf{Y}). \end{aligned}$$

We used the orbital decomposition (3.3.8) and denoted the quotient arrangement of $\mathbf{Y} = X/G$ by $\mathbf{B} = \mathbf{B}^1 + \mathbf{B}^0$. The degree formula (3.3.9) for uniformizations is proved. □

Now we change over to GALOIS-finite morphisms $\mathbf{f} : \mathbf{X} \longrightarrow \mathbf{Y}$ (see diagram (3.1.2)). In order to prove that $\mathbf{e}$ is trivial on this morphism class, we will apply again the shifting principle (see Lemma 1.16.1) and a geometric Local-Global Principle for the category $\mathfrak{S}$ of orbital surfaces. We start with global orbital decompositions

$$\mathbf{c}(\varphi) = \mathbf{c}(\varphi) - \mathbf{c}(\varphi)^1 - \mathbf{c}(\varphi)^0 \tag{3.3.14}$$

for $\mathbf{c} = \mathbf{e}$ and modifications $\varphi : \mathbf{X}' \longrightarrow \mathbf{X}$, define *local orbital decompositions*

$$\mathbf{c}(\varphi_P) = \mathbf{c}(\varphi_P) - \mathbf{c}(\varphi_P)^1 - \mathbf{c}(\varphi)^0, \ P \in X, \tag{3.3.15}$$

and prove the formula of the

Local Orbital Decomposition for Modifications 3.3.6

$$\mathbf{c}(\varphi) = \sum_{\mathbf{P} \in \mathbf{X}} \mathbf{c}(\varphi_P)$$

We set

$$\begin{aligned} \mathbf{c}(\varphi) &= \mathbf{c}(X'/X) \\ &= c(X'/X) - \left(\sum_{P' \in SgX'} c(P') - \sum_{P \in SgX} c(P) \right); \end{aligned} \tag{3.3.16}$$

$$\mathbf{c}(\varphi)^1 = \sum_{\mathbf{C} \subset \mathbf{X}} \lambda_e(\mathbf{C})\mathbf{c}(\mathbf{C}'/\mathbf{C}) + \sum_{\mathbf{L} \subset \mathbf{E}} \lambda_e(\mathbf{L})\mathbf{c}(\mathbf{L}), \tag{3.3.17}$$

where C' denotes the proper image of C in X', $\mathbf{C}' \subset \mathbf{X}'$ the corresponding orbital curve, E is the excpetional locus of $\varphi : X' \longrightarrow X$, $\mathbf{E} \subset \mathbf{X}'$ the corresponding linear arrangement, and the sum runs over all components $\mathbf{L}$ of $\mathbf{E}$;

$$\mathbf{c}(\varphi)^0 = \sum_{\mathbf{P}' \in \mathbf{X}'} \mathbf{c}(\mathbf{P}') - \sum_{\mathbf{P} \in \mathbf{X}} \mathbf{c}(\mathbf{P}). \tag{3.3.18}$$

Together with Lemma 3.3.3 the relation (3.3.14) follows immediately. Explicitly we have

$$\begin{aligned}\mathbf{e}(\varphi) &= \sum_{P\in X} l_P + \sum_{P\in X}\left(e(P) - \sum_{\mathbf{P}'\in\mathbf{E}(\mathbf{P})} e(P')\right)\\ &= \sum_{P} l_P + \sum_{P\in X}\left(\left(1-\frac{1}{d_P}\right) - \sum_{\mathbf{P}'\in\mathbf{E}(\mathbf{P})}\left(1-\frac{1}{d_{P'}}\right)\right),\end{aligned} \tag{3.3.19}$$

where

$$\varphi_{\mathbf{P}} : \mathbf{E}(\mathbf{P}) \longrightarrow \mathbf{P}$$

is the arranged linear resolution of $\mathbf{P}$ along φ and $l_P = l(\mathbf{E}(\mathbf{P})) = l(E(P))$ denotes the length (number of components) of $\mathbf{E}(\mathbf{P}) \subset \mathbf{X}'$. With definition (2.2.3) we find that

$$\begin{aligned}\mathbf{e}(\varphi)^1 =& \sum_{\mathbf{C}}\sum_{\mathbf{P}\in\mathbf{C}}\left(1-\frac{1}{v_P}\right)\cdot\left(e^1(\mathbf{P}) - e^1(\mathbf{P}')\right)\\ &+ 2\sum_P l_P - 2\sum_P\sum_{\mathbf{L}\subset\mathbf{E}(\mathbf{P})}\frac{1}{v_{\mathbf{L}}}\\ &- \sum_P\sum_{\mathbf{L}\subset\mathbf{E}(\mathbf{P})}\left(1-\frac{1}{v_{\mathbf{L}}}\right)\left(e^1(\mathbf{P}_1) + e^1(\mathbf{P}_2)\right)\end{aligned} \tag{3.3.20}$$

because $e(\mathbf{L}) = e(\mathbb{P}^1) = 2$; $\mathbf{P}_1, \mathbf{P}_2$ denote the two intersection points of $\mathbf{L}$ with two components of $\mathbf{E}(\mathbf{P})$, or better of the arrangement $\mathbf{A}'$ of $\mathbf{X}'$, and P' is the preimage point of P on C' in the first sum, $\mathbf{P}' \in \mathbf{C}'$.

We decompose

$$\mathbf{e}(\varphi) = \sum_P \mathbf{e}(\varphi)_P,\ \mathbf{e}(\varphi)^1 = \sum_P \mathbf{e}(\varphi_P)^1,\ \mathbf{e}(\varphi)^0 = \sum_P \mathbf{e}(\varphi_P)^0 \tag{3.3.21}$$

with

$$\begin{aligned}\mathbf{e}(\varphi)_P =& l_P + 1 - \frac{1}{d_P} - \sum_{P'\in E(P')}\left(1-\frac{1}{d_{P'}}\right)\\ =& -\frac{1}{d_P} + \sum_{P'\in E(P')}\frac{1}{d_{P'}} = -\frac{1}{d} + \sum_{i=0}^{l}\frac{1}{d_i},\\ \mathbf{e}(\varphi_P)^1 =& -\left(1-\frac{1}{v}\right)e^1(\mathbf{P}_0/\mathbf{P}) - \left(1-\frac{1}{v'}\right)\left(e^1(\mathbf{P}_l^{\perp}/\mathbf{P}^{\perp})\right) + 2l\\ &- 2\sum_{i=1}^{l}\frac{1}{v_i} - \sum_{i=1}^{l}\left(1-\frac{1}{v_i}\right)\left(e^1(\mathbf{P}_i) + e^1(\mathbf{P}_{i-1}^{\perp})\right)\end{aligned}$$

$$= -\left(1-\frac{1}{v_o}\right)\left[\left(1-\frac{1}{v_1 d_0}\right)-\left(1-\frac{1}{v'd}\right)\right] \tag{3.3.22}$$
$$-\left(1-\frac{1}{v_{l+1}}\right)\left[\left(1-\frac{1}{v_l d_l}\right)-\left(1-\frac{1}{vd}\right)\right]$$
$$+2l-2\sum_{i=1}^{l}\frac{1}{v_i}-\sum_{i=1}^{l}\left(1-\frac{1}{v_i}\right)\left[\left(1-\frac{1}{v_{i+1}d_i}\right)-\left(1-\frac{1}{v_{i-1}d_{i-1}}\right)\right]$$

We used the notations of the graph (3.3.24) of

$$\varphi_P : \mathbf{E}=\mathbf{E}(\mathbf{P})=\sum_{i=1}^{l}\mathbf{E}_i \longrightarrow \mathbf{P}:$$

$$\underset{v=v_0\ \langle d_0,e_0\rangle\ v_1}{\overset{\mathbf{P}_0}{\circ\!\!\longrightarrow\!\!\bullet}}\cdots\underset{v_i\ \langle d_i,e_i\rangle\ v_{i+1}}{\overset{\mathbf{P}_i}{\bullet\!\!\longrightarrow\!\!\bullet}}\cdots\underset{v_l\ \langle d_l,e_l\rangle\ v_{l+1}=v'}{\overset{\mathbf{P}_l}{\bullet\!\!\longrightarrow\!\!\circ}}\Longrightarrow\underset{v\ \langle d,e\rangle\ v'}{\overset{\mathbf{P}}{\circ\!\!\longrightarrow\!\!\circ}} \tag{3.3.23}$$

$$\mathbf{e}(\varphi_P)^1 = 2l-2\sum_{i=1}^{l}\frac{1}{v_i}-\sum_{i=0}^{l}\left(1-\frac{1}{v_i}\right)\left(1-\frac{1}{v_{i+1}d_i}\right)$$
$$-\sum_{i=0}^{l}\left(1-\frac{1}{v_{i-1}d_i}\right)$$
$$+\left(1-\frac{1}{v}\right)\left(1-\frac{1}{v'd}\right)+\left(1-\frac{1}{v'}\right)\left(1-\frac{1}{vd}\right)$$

$$\mathbf{e}(\varphi_P)^1 = \sum_{i=0}^{l}\frac{1}{v_{i+1}d_i}+\sum_{i=0}^{l}\frac{1}{v_i d_i}-2\sum_{i=0}^{l}\frac{1}{v_i v_{i+1}d_i}$$
$$-\frac{1}{v'd}+\frac{1}{vv'd}-\frac{1}{vd}+\frac{1}{vv'd}, \tag{3.3.24}$$

$$\mathbf{e}(\varphi_P)^0 = \sum_{i=0}^{l}\mathbf{e}(\mathbf{P}_i)-\mathbf{e}(\mathbf{P}),$$
$$\mathbf{e}(\varphi_P)^0 = \sum_{i=0}^{l}\frac{1}{d_i}\left(1-\frac{1}{v_i}\right)\left(1-\frac{1}{v_{i+1}}\right)$$
$$-\frac{1}{d}\left(1-\frac{1}{v}\right)\left(1-\frac{1}{v'}\right) \tag{3.3.25}$$

With definition (3.3.15) of $\mathbf{e}(\varphi_P)$ the local decomposition (3.3.6) follows from (3.3.22). Furthermore, we can calculate $\mathbf{e}(\varphi_P)$ by means of (3.3.15), (3.3.22), (3.3.24) and (3.3.25). We receive

$$\mathrm{e}(\varphi_P) = \sum_{i=0}^{l} \frac{1}{v_i v_{i+1} d_i} - \frac{1}{vv'd} = -\frac{1}{\deg(\mathbf{P})} + \sum_{i=0}^{l} \frac{1}{\deg(\mathbf{P}_i)} \qquad (3.3.26)$$

This local number has already been defined in (1.17.9) in the category $\mathcal{P}_1'$ of modification of abelian points. The orbital property of these numbers in $\mathcal{P}_1'$ has already been proved with Lemma 1.17.2 and definition (1.17.9). In our context we need the following formulation:

Lemma 3.3.7 *Let* $\mathbf{F} : \mathbf{X}'/\mathbf{X} \longrightarrow \mathbf{Y}'/\mathbf{Y}$ *be a* GALOIS*-finite morphism with global and local base change diagrams described in (3.1.2). Then it holds for all* $P \in X$ *that*

$$\mathbf{e}_P(\mathbf{X}'/\mathbf{X}) = \deg_{\mathbf{P}}(\mathbf{F}) \cdot \mathbf{e}_Q(\mathbf{Y}'/\mathbf{Y}), \qquad (3.3.27)$$

where

$$\mathbf{e}_P(\mathbf{X}'/\mathbf{X}) = \mathbf{e}(\psi_P), \mathbf{e}_Q(\mathbf{Y}'/\mathbf{Y}) = \mathbf{e}(\varphi_Q), Q = f(P)$$

and

$$\deg_{\mathbf{P}}(\mathbf{F}) = \deg(\mathbf{f_P}).$$

The Global-Local Principle can now be realized. For later use we write $\mathbf{c}$ instead of $\mathbf{e}$.

Local-Global Principle 3.3.8 *If* $\mathbf{c} : \mathfrak{S} \longrightarrow \mathbb{Q}$ *is a function on the category of orbital surfaces, then we call a function* $\gamma : \mathcal{P}_1' \longrightarrow \mathbb{Q}$ *a localization of* $\mathbf{c}$ *if we have for any relative object* $\varphi = \mathbf{X}'/\mathbf{X}$ *the local decomposition*

$$\mathbf{c}(\mathbf{X}'/\mathbf{X}) = \sum_{P \in X} \mathbf{c}_P(\mathbf{X}'/\mathbf{X}),\ \mathbf{c}_P(\mathbf{X}'/\mathbf{X}) = \mathbf{c}_P(\varphi) \underset{\mathrm{Df}}{=} \gamma(\varphi_P). \qquad (3.3.28)$$

If the degree formula for uniformizations is true and the localization is orbital with respect to the class of GALOIS*-finite morphisms, that means that for any* GALOIS*-finite morphism* $\mathbf{F} = (\mathbf{f}', \mathbf{f}) : \mathbf{X}'/\mathbf{X} \longrightarrow \mathbf{Y}'/\mathbf{Y}$ *it holds that*

$$\mathbf{c}_P(\mathbf{X}'/\mathbf{X}) = \deg_P(\mathbf{F}) \cdot \mathbf{c}_{f(P)}(\mathbf{Y}'/\mathbf{Y}), \qquad (3.3.29)$$

then $\mathbf{c}$ *is orbital on* $\mathfrak{S}$.

Proof We have to show that $\mathbf{c}(\mathbf{X}/\mathbf{Y}) = 0$ for any GALOIS-finite covering $\mathbf{f} : \mathbf{X} \longrightarrow \mathbf{Y}$. There exists a uniformizable modification $\mathbf{f}' : \mathbf{X}' \longrightarrow \mathbf{Y}$. Namely, $\mathbf{f}$ is quasi-uniformizable by definition. Therefore, there exists a commutative diagram (3.1.2) which can be extended to a commutative diagram:

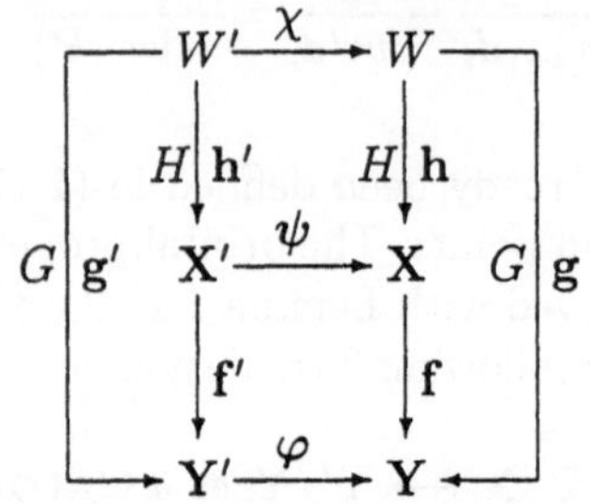

$$\begin{array}{ccc} W' & \xrightarrow{\chi} & W \\ H \downarrow \mathbf{h}' & & H \downarrow \mathbf{h} \\ \mathbf{X}' & \xrightarrow{\psi} & \mathbf{X} \\ \downarrow \mathbf{f}' & & \downarrow \mathbf{f} \\ \mathbf{Y}' & \xrightarrow{\varphi} & \mathbf{Y} \end{array} \qquad G\,|\,\mathbf{g}' : W' \to \mathbf{Y}', \quad G\,|\,\mathbf{g} : W \to \mathbf{Y} \tag{3.3.30}$$

by removing the singularities of W in an equivariant manner and taking quotients. Then $\mathbf{f}'$ is uniformizable, hence The Shifting Principle (see 1.16.1 or 1.16.2) allows to conclude that $\mathbf{c}(\mathbf{f}) = \mathbf{c}(\mathbf{f}') = 0$, if we know that

$$\mathbf{c}(\mathbf{X}'/\mathbf{X}) = \deg(\mathbf{F}) \cdot \mathbf{c}(\mathbf{Y}'/\mathbf{Y}). \tag{3.3.31}$$

According to diagram (3.3.30) it suffices to assume that $\mathbf{F}$ is a quasi-uniformization, say with GALOIS group G, because the degree formulas for χ/φ and χ/ψ imply (3.3.31). With $\deg_P(\mathbf{F}) = |G_P|$ and $f(P) = Q$ we find

$$\begin{aligned} |G| \cdot \mathbf{c}(\mathbf{Y}'/\mathbf{Y}) &= |G| \sum_{Q \in Y} \mathbf{c}_Q(\mathbf{Y}'/\mathbf{Y}) = \sum_{Q \in Y} |G| \cdot \mathbf{c}_Q(\mathbf{Y}'/\mathbf{Y}) \\ &= \sum_Q [G : G_P] |G_P| \mathbf{c}_Q(\mathbf{Y}'/\mathbf{Y}) \\ &= \sum_Q [G : G_P] \deg_P(\mathbf{F}) \cdot \mathbf{c}_Q(\mathbf{Y}'/\mathbf{Y}) \\ &= \sum_Q [G : G_P] \mathbf{c}_P(\mathbf{X}'/\mathbf{X}) = \sum_Q \sum_{P \to Q} \mathbf{c}_P(\mathbf{X}'/\mathbf{X}) \\ &= \sum_{P \in X} \mathbf{c}_P(\mathbf{X}'/\mathbf{X}) = \mathbf{c}(\mathbf{X}'/\mathbf{X}) \end{aligned}$$

using (3.3.28) and (3.3.29). The Local-Global Principle has been established. □

Proposition 3.3.9 *The* EULER *heights* $\mathbf{e}(\mathbf{X})$ *are orbital invariants on the category of regularly arranged orbital surfaces with respect to* GALOIS*-finite morphisms.*

Proof The assumptions (3.3.28) and (3.3.29) are satisfied, see (3.3.27) and (3.3.18). □

3.4 Signature Heights

We follow the line of the last section and define

$$\lambda_\tau(\mathbf{D}) = \lambda_\tau(v_\mathbf{D}) = \left(v_\mathbf{D} - \frac{1}{v_\mathbf{D}}\right), \quad \mathbf{D} \text{ an orbital curve;} \tag{3.4.1}$$

$$\tau(\mathbf{Q}) = -4\mathrm{S}(Q) + \frac{e}{3d}\left(1 - \frac{1}{v^2}\right) + \frac{e'}{3d}\left(1 - \frac{1}{v'^2}\right) \tag{3.4.2}$$

$$= \frac{1}{3}\left(3l_Q + Tr(Q) - \frac{e}{v'^2 d} - \frac{e'}{v'^2 d}\right), \mathbf{Q} : \underset{v\ \langle d,e\rangle\ v'}{\circ\!\!\longrightarrow\!\!\circ}$$

$$\tau(\mathbf{D}) = \frac{s}{3v} = (D^2)/3v, \ \mathbf{D}: \tag{3.4.3}$$

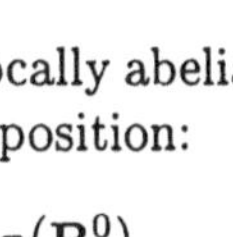

For the identity in (3.4.2) we refer to 1.8.1 and (1.8.2). The self-intersection (D^2) is taken as usually on the minimal singularity resolution of U_D.

Definition 3.4.1 The *signature height* of a (locally abelian) orbital surface $\mathbf{Y} = (Y, \mathbf{B})$ is defined by the geometric decomposition:

$$\tau(\mathbf{Y}) = \tau(Y) - \tau(\mathbf{B}^1) - \tau(\mathbf{B}^0), \tag{3.4.4}$$

where

$$\tau(\mathbf{B}^1) = \sum_{\mathbf{D}\subset\mathbf{Y}} \lambda_\tau(\mathbf{D})\tau(\mathbf{D}) = \sum_{i=1}^{k} \lambda_\tau(\mathbf{D}_i)\tau(\mathbf{D}_i), \tag{3.4.5}$$

$$\tau(\mathbf{B}^0) = \sum_{\mathbf{Q}\in\mathbf{Y}} \tau(\mathbf{Q}) = \sum_{j=1}^{m} \tau(\mathbf{Q}_j), \tag{3.4.6}$$

with

$$\mathbf{B} = \mathbf{B}^1 + \mathbf{B}^0 = \sum_{i=1}^{k} \mathbf{D}_i + \sum_{j=1}^{m} \mathbf{Q}_j.$$

Explicitly we can write

$$\begin{aligned} \tau(\mathbf{Y}) = & \tau(Y) - \frac{1}{3}\sum_{\mathbf{D}} \left(1 - \frac{1}{v_{\mathbf{D}}^2}\right)(D^2) \\ & - \frac{1}{3}\sum_{\mathbf{Q}} \left(3l_Q + Tr(Q) - \frac{e_Q}{v_Q^2 d_Q} - \frac{e'_Q}{v_Q'^2 d_Q}\right). \end{aligned} \tag{3.4.7}$$

Following the analogy with the EULER heights we define the *signature height of an abelian point* $\mathbf{Q}$ to be

$$\tau(\mathbf{Q}) = 0. \tag{3.4.8}$$

The signature height of the reduced orbital surface $(Y, Sg(Y))$ is

$$\begin{aligned} \tau(Y) &= \tau(Y, Sg(Y)) \\ &= \tau(Y) - \frac{1}{3} \sum_{Q\in Sg(Y)} \left(3l_Q + Tr(Q) - \frac{e_Q}{d_Q} - \frac{e'_Q}{d_Q}\right) \\ &= \tau(Y) + 4\mathbf{S}(Y) =_{\mathrm{Def}} \tau(Y) + 4\sum_{Q\in Y} \mathbf{S}(Q) \end{aligned} \tag{3.4.9}$$

Lemma 3.4.2 *(Orbital decomposition of signature heights)* .
With the above notations it holds that

$$\tau(\mathbf{Y}) = \tau(Y) - \tau(\mathbf{B}^1) - \tau(\mathbf{B}^0) = \tau(Y) - \tau(\mathbf{B}^1)$$

where

$$\begin{aligned} \tau(\mathbf{B}^1) &= \sum_{\mathbf{D}\subset\mathbf{Y}} \lambda_\tau(\mathbf{D})\tau(\mathbf{D}) = \sum_{i=1}^{k} \lambda_\tau(\mathbf{D}_i)\tau(\mathbf{D}_i), \\ \tau(\mathbf{B}^0) &= \sum_{\mathbf{Q}} \tau(\mathbf{Q}) = 0. \end{aligned}$$

Proof (of 3.4.2 and 3.3.3). Let c be one of the symbols e, τ. We use the following decomposition:

$$c(\mathbf{Q}) = \mathbf{c}(\mathbf{Q}) - \lambda_c(v)c^1(\mathbf{Q}) - \lambda_c(v')c^1(\mathbf{Q}^{\perp}) + c(Q),$$

$$\mathbf{Q} : \underset{v\ \langle d,e\rangle\ v'}{\circ\!\!\longrightarrow\!\!\circ}$$

The relation follows immediately from the definitions (see Table 3.A, B at the end of the chapter). Now we transform

$$\begin{aligned}
\mathbf{c}(\mathbf{Y}) &= c(Y) - c(\mathbf{B}^1) - c(\mathbf{B}^0) \\
&= c(Y) - \sum_D \lambda_c(\mathbf{D}) \cdot c(\mathbf{D}) - \sum_{\mathbf{Q}} c(\mathbf{Q}) \\
&= \left[c(Y) - \sum_Q c(Q)\right] - \sum_{\mathbf{D}} \lambda_c(\mathbf{D}) \left[c(\mathbf{D}) - \sum_{\mathbf{Q}\in\mathbf{D}} c^1(\mathbf{Q})\right] - \sum_{\mathbf{Q}} \mathbf{c}(\mathbf{Q}) \\
&= \mathbf{c}(Y) - \sum_{\mathbf{D}} \lambda_c(\mathbf{D})\mathbf{c}(\mathbf{D}) - \sum_{\mathbf{Q}} \mathbf{c}(\mathbf{Q}) = \mathbf{c}(Y) - \mathbf{c}(\mathbf{B}^1) - \mathbf{c}(\mathbf{B}^0)
\end{aligned}$$

□

Proposition 3.4.3 *Let* $\mathbf{f} : X \xrightarrow{G} \mathbf{Y} = X/G$ *be a uniformization of locally abelian orbital surfaces. Then the degree formula*

$$\tau(X) = \deg(\mathbf{f}) \cdot \tau(\mathbf{Y}) \tag{3.4.10}$$

holds.

Corollary 3.4.4 *The τ-degree formula holds for uniformizable morphisms.*

Proof We translate the G-Signature Theorem 3.2.7 into our language. The first step has been done with (3.2.26). We read this relation together with (3.2.25) as

$$\tau(X) = |G| \cdot \tau(Y) - \sum_C v_C \lambda_\tau(v)\tau(C) - \sum_P |G_P|\tau_G(P)$$

Using local orbital properties we find that

$$\begin{aligned}\tau(X) =&|G|\tau(Y) - \sum_{\mathbf{D}}\sum_{C\to\mathbf{D}} v_{\mathbf{D}}\cdot\lambda_\tau(\mathbf{D})\cdot \deg_{\mathrm{red}}(\mathbf{f_D})\cdot\boldsymbol{\tau}(\mathbf{D})\\ &-\sum_{\mathbf{Q}}\sum_{P\to\mathbf{Q}} \operatorname{def}(\mathbf{Q})\\ =&|G|\tau(Y) - \sum_{\mathbf{D}}[G:N_G(C)]\cdot v_{\mathbf{D}}\cdot\lambda_\tau(\mathbf{D})\frac{|N_G(C)|}{v_{\mathbf{D}}}\boldsymbol{\tau}(\mathbf{D})\\ &+4\sum_{\mathbf{Q}}[G:G_P]\cdot|G_P|\cdot\frac{1}{|G_P|}d_Q\mathbf{S}(\mathbf{Q})\\ =& |G|\tau(Y)-|G|\sum_{\mathbf{D}}\lambda_\tau(\mathbf{D})\boldsymbol{\tau}(\mathbf{D})+4|G|\cdot\sum_{\mathbf{Q}}\frac{1}{v_Qv_Q'}\mathbf{S}(\mathbf{Q})\\ =&|G|\left[\tau(Y)-\boldsymbol{\tau}(\mathbf{B}^1)+\sum_{Q}4\mathbf{S}(Q)\right] = |G|\boldsymbol{\tau}(\mathbf{Y}).\end{aligned}$$

In the last steps we used the relations of 1.12.10, (3.4.9) and (3.4.8). □

For $c = \tau$ we want to check now the decomposition formulas (3.3.14), (3.3.15), (3.3.6) for modifications $\varphi : \mathbf{X}' \to \mathbf{X}$ of orbital surfaces, $\boldsymbol{\tau}(\varphi), \boldsymbol{\tau}(\varphi)^1$ and $\boldsymbol{\tau}(\varphi)^0$ are defined by (3.3.16), (3.3.17) or (3.3.18), respectively. With regard to the orbital decompositions of $\boldsymbol{\tau}(\mathbf{X}')$ and $\boldsymbol{\tau}(\mathbf{X})$ (Lemma 3.4.2) it is clear that $\boldsymbol{\tau}(\varphi)$ has an orbital decomposition (3.3.14). We calculate the components explicitly and use notations of the previous section at the corresponding places.

$$\begin{aligned}\boldsymbol{\tau}(\varphi) =& -\sum_{P\in X} l_P + \sum_{P\in X}\left(\tau(P) - \sum_{\mathbf{P}'\in\mathbf{E}(\mathbf{P})}\tau(P')\right)\\ =& -\sum_{P\in X} l_P + \frac{1}{3}\sum_{P\in X}\left[\left(3l_P + Tr(P) - \frac{e_P}{d_P} - \frac{e_P'}{d_P}\right)\right.\\ &\left.-\sum_{\mathbf{P}'\in\mathbf{E}(P)}\left(3l_{P'} + Tr(P') - \frac{e_{P'}}{d_{P'}} - \frac{e_{P'}'}{d_{P'}'}\right)\right]\\ \boldsymbol{\tau}(\varphi)^1 =& \sum_{\mathbf{C}}\left(v_{\mathbf{C}} - \frac{1}{v_{\mathbf{C}}}\right)(\boldsymbol{\tau}(\mathbf{C}') - \boldsymbol{\tau}(\mathbf{C}))\\ &+\sum_{P}\sum_{\mathbf{L}\subset\mathbf{E}(\mathbf{P})}\left(v_{\mathbf{L}} - \frac{1}{v_{\mathbf{L}}}\right)\cdot\boldsymbol{\tau}(\mathbf{L})\end{aligned}$$

$$=\frac{1}{3}\sum_{\mathbf{C}}\left(1-\frac{1}{v_{\mathbf{C}}^2}\right)\cdot\left[\left(C'^2\right)-(C^2)-\sum_{\mathbf{P}\in\mathbf{C}}3\left(\tau^1(\mathbf{P}_0)-\tau^1(\mathbf{P})\right)\right]$$
$$+\frac{1}{3}\sum_{P}\sum_{\mathbf{L}\subset\mathbf{E}(\mathbf{P})}\left(1-\frac{1}{v_{\mathbf{L}}^2}\right)\left[(L^2)-3\left(\tau^1(\mathbf{P}_1)-\tau^1(\mathbf{P}_2)\right)\right].$$

The absorptive numbers allow to express

$$\left(C'^2\right)-(C^2)=-\sum_{\mathbf{P}\in\mathbf{C}}a^-(\mathbf{E}(\mathbf{P}))$$

So the sum over all $\mathbf{C}$ changes over into a sum over all $\mathbf{P},\mathbf{P}^\perp$. Since $a^-(\mathbf{E}(\mathbf{P}))=a^+(\mathbf{E}(\mathbf{P}^\perp))$ we can take the sum over all abelian points $\mathbf{P}$ forgetting their arranging. We receive

$$\tau(\varphi)^1=\frac{1}{3}\sum_{\mathbf{P}}\left[\left(1-\frac{1}{v_{\mathbf{P}}^2}\right)\left(-a^-(\mathbf{E}(\mathbf{P}))+\frac{e_{P_0}}{d_{P_0}}-\frac{e_P}{d_P}\right)\right]$$
$$+\left(1-\frac{1}{v_{\mathbf{P}}'^2}\right)\cdot\left(-a^+(\mathbf{E}(\mathbf{P}))-\frac{e_P'}{d_P}\right)$$
$$+\frac{1}{3}\sum_{P}\sum_{\mathbf{L}\subset\mathbf{E}(\mathbf{P})}\left(1-\frac{1}{v_{\mathbf{L}}^2}\right)\left[(L^2)-3\left(\tau^1(\mathbf{P}_1)-\tau^1(\mathbf{P}_2)\right)\right]$$

So we get the local decompositions

$$\tau(\varphi)=\sum_{P\in X}\tau(\varphi)_P$$
$$\tau(\varphi)^1=\sum_{P\in X}\tau(\varphi_P)^1 \tag{3.4.11}$$

where with the above notations

$$\tau(\varphi)_P=-l_P+\frac{1}{3}\left[3l_P+Tr(P)-\frac{e}{d}-\frac{e'}{d}-\sum_{i=0}^{l}\left(3l_i+Tr(P_i)-\frac{e_i}{d_i}-\frac{e_i'}{d_i}\right)\right] \tag{3.4.12}$$

$$\tau(\varphi_P)^1=\frac{1}{3}\left[\left(1-\frac{1}{v^2}\right)\cdot\left(a_P^-+\frac{e_o}{d_o}-\frac{e}{d}\right)+\left(1-\frac{1}{v_2'}\right)\cdot\left(a_P^++\frac{e_l'}{d_l}-\frac{e'}{d}\right)\right]$$
$$+\frac{1}{3}\sum_{i=1}^{l}\left(1-\frac{1}{v_i^2}\right)\left[(E_i^2)+\frac{e_i}{d_i}+\frac{e_{i-1}'}{d_{i-1}}\right] \tag{3.4.13}$$

Since τ is trivial for abelian points by (3.4.8) we set $\tau(\varphi)^0 = \tau(\varphi_P)^0 = 0$ and get the decompositions (3.3.15) and (3.3.6). The difference of (3.4.12) and (3.4.13) yields

$$\begin{aligned} 3\tau(\varphi_P) =& 3\tau(\varphi)_P - 3\tau(\varphi_P)^1 \\ =& [3l + Tr(P)] - \left[3l - a^- - a^+ + \sum_{i=1}^{l}(E_i^2) + \sum_{i=0}^{l}(3l_i + Tr(P_i))\right] \\ &+ \frac{1}{v^2}\left(-a^- + \frac{e_o}{d_o} - \frac{e}{d}\right) + \frac{1}{v'^2}\left(-a_P^+ + \frac{e_l'}{d_l'} - \frac{e'}{d'}\right) \\ &+ 3\sum_{i=1}^{l}\frac{1}{v_i}\tau(\mathbf{E}_i). \end{aligned} \tag{3.4.14}$$

The sums in the brackets are the homogeneous traces $Th(E_P)$ or $Th(\mathbf{E}(\mathbf{P}))$, respectively, where E_P is the minimal resolution of the singularity (see 1.18.4 and (1.18.10)). Both sums are equal to $Th(\mathbf{P})$ by definition 1.18.4. So we get finally

$$\begin{aligned} \tau(\varphi_P) =& \sum_{i=1}^{l}\frac{1}{v_i}\tau(\mathbf{E}_i) + \frac{1}{3v_P^2}\left(-a^-(\mathbf{E}(\mathbf{P})) + \frac{e_o}{d_o} - \frac{e_P}{d_P}\right) \\ &+ \frac{1}{3v_P'^2}\left(-a^+(\mathbf{E}(\mathbf{P})) + \frac{e'}{d_e'} - \frac{e'}{d_P'}\right) \\ \mathbf{E}(\varphi_P) =& \mathbf{E}(\mathbf{P}) = \sum_{i=1}^{l}\mathbf{E}_i + \sum_{i=0}^{l}\mathbf{P}_i. \end{aligned} \tag{3.4.15}$$

Proposition 3.4.5 *The function τ on the category of (locally abelian) orbital surfaces is orbital with respect to the class of* GALOIS*-finite morphisms.*

Proof With $\tau(\varphi_P)^0 = 0$,

$$\tau(\varphi_P) \underset{\mathrm{Df}}{=} \tau(\varphi)_P - \tau(\varphi_P)^1 - \tau(\varphi_P)^0 \tag{3.4.16}$$

we get the orbital decomposition

$$\tau(\varphi) = \tau(\varphi) - \tau(\varphi)^1 - \tau(\varphi)^0 \tag{3.4.17}$$

with $\tau(\varphi)^0 = 0$ via (3.4.11) and the local decomposition

$$\tau(\varphi) = \sum_{\mathbf{P}}\tau(\varphi_P) \tag{3.4.18}$$

for modifications $\varphi : \mathbf{X}' \longrightarrow \mathbf{X}$. We know also that the degree formula for uniformizations is true. Therefore, the Local-Global Principle reduces the proof to the local degree formula

$$\tau_P(\mathbf{X}'/\mathbf{X}) = \deg_P(\mathbf{F}) \cdot \tau_Q(\mathbf{Y}'/\mathbf{Y}) \tag{3.4.19}$$

for GALOIS-finite coverings

$$\mathbf{F} = (\mathbf{f}', \mathbf{f}) : \mathbf{X}'/\mathbf{X} \to \mathbf{Y}'/\mathbf{Y}$$

of relative objects of degree 1 (modifications). It suffices to prove (3.4.19) for quasi-uniformizations. We do it first for a special one.

Lemma 3.4.6 *If in the local base change diagram:*

$$\begin{array}{ccc} P_0 & \xrightarrow{\psi} & P \\ \downarrow A & & \downarrow \mathbf{f}_P \\ \mathbf{Q}_0 & \xrightarrow{\varphi} & \mathbf{Q} \end{array} \tag{3.4.20}$$

ψ is a minimal resolution of the reduced abelian singular point P, then (3.4.19) holds for the related base change diagram (3.4.21) of resolutions

$$\begin{array}{ccc} L & \xrightarrow{\psi_P} & P \\ \mathbf{f}_L \downarrow A & & A \downarrow \mathbf{f}_P \\ \mathbf{E} & \xrightarrow{\varphi_Q} & \mathbf{Q} \end{array} \qquad \begin{array}{l} L = E_P = \sum_{i=1}^{l} L_i + \sum_{j=0}^{l} P_j \\ \\ \mathbf{E} = \mathbf{E}(\varphi_Q) = \sum_{i=1}^{l} \mathbf{E}_i + \sum_{j=0}^{l} \mathbf{Q}_j \end{array} \tag{3.4.21}$$

$\mathbf{F} = (\mathbf{f}_L, \mathbf{f}_P), \deg_P(\mathbf{F}) = |A|$. *More precisely one has*

$$\tau(\psi_P) = |A| \cdot \tau(\varphi_Q). \tag{3.4.22}$$

Proof With obvious notations we deduce from (3.4.15)

$$\begin{aligned} |A|\,\tau(\varphi_Q) &= \sum_{i=1}^{l} \frac{|A|}{v_i} \tau(\mathbf{E}_i) + \frac{|A|}{3v_Q^2} \left(-a^-(\mathbf{E}) + \frac{e_0}{d_0} - \frac{e_Q}{d_Q} \right) \\ &+ \frac{|A|}{{v'_Q}^2} \left(-a^+(\mathbf{E}) + \frac{e'_l}{d'_l} + \frac{e'_Q}{d'_Q} \right) \end{aligned}$$

We remember that $\frac{|A|}{v_i} = \deg_{\text{red}}(L_i \to \mathbf{E}_i)$ and $a^\pm(\mathbf{E}) = a^\pm(\mathbf{f}_P)$ because $L = E_P$ is the minimal singularity resolution of P. For this notation we refer to section 1.18. With Proposition 1.18.5 we can calculate the absorptive numbers $a^-(\mathbf{f}_P)$ and $a^+(\mathbf{f}_P)$. The degree formula for the signature heights of orbital curves changes $\frac{|A|}{v_i}\tau(\mathbf{E}_i)$ to $\tau(L_i)$. So we obtain

$$|A|\tau(\varphi_Q) = \sum_{i=1}^{l} \tau(L_i) - \frac{|A|}{3v_Q^2} \cdot \frac{e_P v_Q^2}{d_P|A|} - \frac{|A|}{3v_Q'^2} \cdot \frac{e'_P v_Q'^2}{d_P|A|}$$

$$= \sum_{i=1}^{l} \tau(L_i) + \frac{1}{3}\left(0 - \frac{e_P}{d_P}\right) + \frac{1}{3}\left(0 - \frac{e'_P}{d_P}\right).$$

But this is exactly $\tau(\psi_P)$ by (3.4.15). The Lemma is proved. □

As in the proof of the orbital property of τ on the category of orbital curves we make the following

Inductive Hypothesis 3.4.7 Inductive Hypothesis The signature height is orbital on the category of (locally abelian) orbital surfaces $\mathfrak{S}$ with respect to the subclass $\mathfrak{F}_k$ of quasi-uniformizations $\mathbf{f} : X \overset{G}{\to} \mathbf{Y}$ whose starting objects X support at most $k \geq 0$ singularities.

Let $\mathbf{f} : X \overset{G}{\to} \mathbf{Y} = X/G$ be a quasi-uniformization such that X has $k+1$ singularities. We remove one of them, say P, in a minimal manner and do the same for all other points of the orbit GP. So we obtain commutative diagrams:

$$\begin{array}{ccc} X' & \xrightarrow{\psi} & X \\ \mathbf{f}' \downarrow G & & G \downarrow \mathbf{f} \\ X'/G = \mathbf{Y}' & \xrightarrow{\varphi} & \mathbf{Y} = X/G \end{array} \qquad \begin{array}{ccc} E_P & \xrightarrow{\psi_P} & P \\ G_P \downarrow & & G_P \downarrow \mathbf{f}_P \\ \mathbf{E} = E_P/A_P & \xrightarrow{\varphi_Q} & \mathbf{Q} = P/A_P \end{array} \qquad (3.4.23)$$

The surface X' has at most k singularities. The local decomposition (3.4.18) for φ and ψ and the local degree formula (3.4.22) yield

$$\tau(\psi) = \sum_{G_P} \tau(\psi_P) = [G : G_P]\tau(\psi_P) = [G : G_P] \cdot |G_P| \cdot \tau(\varphi_Q)$$

$$= |G| \cdot \tau(\varphi) \qquad (3.4.24)$$

By the inductive hypothesis we know that $\tau(\mathbf{f}') = 0$. The relation (3.4.24) allows to shift $\tau(\mathbf{f}')$ to the right side of the first diagram of (3.4.23), that means that $\tau(\mathbf{f}') = \tau(\mathbf{f}) = 0$ (see Corollary 1.16.3). The degree formula is proved for the quasi-uniformization $\mathbf{f}$. Since we know that the degree formula (3.4.10) is true for all uniformizations $(k = 0)$, it holds for all quasi-uniformizations by induction. The extension to all GALOIS-finite morphisms is evident. Thus τ is orbital on $(\mathfrak{S}, \mathfrak{F})$, $\mathfrak{F}$ the class of GALOIS-finite morphisms in $\mathfrak{S}$. The proof of Proposition 3.4.5 is already finished on this place. □

For the sake of completeness we continue to prove the local degree formula (3.4.19). First we can prove the global version:

3.4.8 *The function* τ *is orbital on the category of modifications of orbital surfaces with respect to* GALOIS*-finite coverings* $\mathbf{F} : \mathbf{X}'/\mathbf{X} \to \mathbf{Y}'/\mathbf{Y}$:

$$\tau(\mathbf{X}'/\mathbf{X}) = \deg(\mathbf{F}) \cdot \tau(\mathbf{Y}'/\mathbf{Y}) \tag{3.4.25}$$

This is formula (1.16.10) applied to our situation. It has been proved in section 1.16. Now we decompose the base change diagram of $\mathbf{F}$ in the style of diagram:

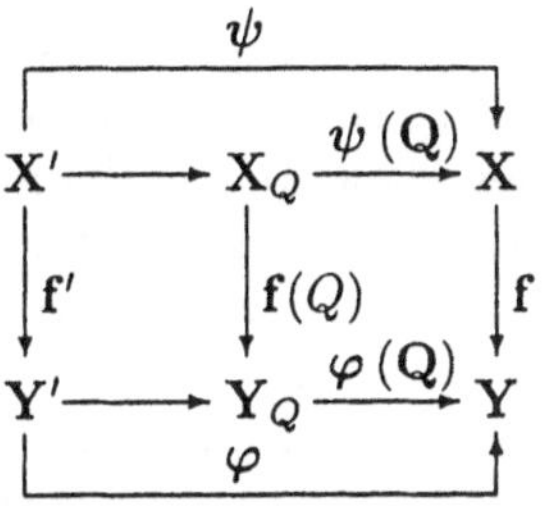

$$\begin{array}{ccccc} & & \psi & & \\ \mathbf{X}' & \longrightarrow & \mathbf{X}_Q & \xrightarrow{\psi(\mathbf{Q})} & \mathbf{X} \\ \downarrow \mathbf{f}' & & \downarrow \mathbf{f}(Q) & & \downarrow \mathbf{f} \\ \mathbf{Y}' & \longrightarrow & \mathbf{Y}_Q & \xrightarrow{\varphi(\mathbf{Q})} & \mathbf{Y} \\ & & \varphi & & \end{array} \tag{3.4.26}$$

where $\varphi(\mathbf{Q})$ resolves only $\mathbf{Q}$ and $\psi(\mathbf{Q})$ only the preimages $\mathbf{P} \in \mathbf{X}$ of $\mathbf{Q}$ such that we have the same resolutions $\psi(\mathbf{Q})_P = \psi_P$, $\varphi(\mathbf{Q})_Q = \varphi_Q$ as described in diagram (3.4.21). It suffices for the proof of (3.4.19) to assume that $\mathbf{F}$ is a quasi-uniformization, say with GALOIS group G. From the global degree formula (3.4.25) we get

$$\deg(\mathbf{f}) \cdot \tau(\varphi(\mathbf{Q})) = \tau(\psi(\mathbf{Q})),$$

hence via local decomposition (3.4.18)

$$|G| \cdot \tau(\varphi_Q) = \sum_{G_P} \tau(\psi_P) = [G : G_P] \cdot \tau(\psi_P),$$
$$\tau(\psi_P) = |G_P| \cdot \tau(\varphi_Q) = \deg_P(\mathbf{F}) \cdot \tau(\varphi_Q).$$

The local degree formula (3.4.19) and Proposition 3.4.5 are proved.

Altogether we found with $\mathbf{e}$ and τ two different orbital invariants on $\mathfrak{S}$. Each rational linear combination of two orbital invariants is obviously orbital. The formal $\mathbb{Q}$ -vector space $\mathbb{Q}e \oplus \mathbb{Q}\tau$ of dimension two is called the space of *rational* CHERN *symbols.* With regard to (3.2.3) and (3.2.4) we set

$$\chi = \frac{1}{4}(\tau + e), \; c_1^2 = 3\tau + 2e. \tag{3.4.27}$$

The first is called the *arithmetic genus symbol* , the second the *self-intersection symbol of canonical classes.* The name CHERN symbol comes from c_1^2 and $e = c_2$ generating $\mathbb{Q}e + \mathbb{Q}\tau$. On this way we dispose also on orbital invariants

$$\chi = \frac{1}{4}(\tau + \mathbf{e}), \; \mathbf{c}_1^2 = 3\tau + 2\mathbf{e} \tag{3.4.28}$$

on $\mathfrak{S}$. For an orbital surface $\mathbf{X}$ we call $\chi(\mathbf{X})$ the *orbital arithmetic genus* of $\mathbf{X}$ and $\mathbf{c}_1^2(\mathbf{X})$ the *orbital canonical self-intersection* . We can define in this manner invariants $\chi_P(\mathbf{X}'/\mathbf{X})$, $\chi(X)$, $\chi(\mathbf{C})$, $\mathbf{c}_1^2(\mathbf{C})$ and so on via linear combinations (3.4.27). The explicit descriptions are left to the reader. All results and proof methods for $\mathbf{e}$ and τ can be transferred to χ and $\mathbf{c}_1^2$. It should be mentioned that the explicit expressions for χ and $\mathbf{c}_1^2$ are more complicated than those of $\mathbf{e}$ and τ. We preferred the latter CHERN symbols in this monograph because they allow to give the most handy proofs. The reason will become clear later, when we show that $\mathbf{e}, \tau$ are the only "diagonalizing invariants" for the explicit HURWITZ type formulas for CHERN numbers of algebraic surfaces (see section 6.9). Moreover, for surfaces X with singularities, the invariant $\mathbf{c}_1^2(X)$ as defined above, seems not to have a very natural meaning like $\mathbf{e}$ and χ. We close this chapter with the following

Theorem 3.4.9 *For the category $\mathfrak{S}$ of regular orbital surfaces exists a two-dimensional $\mathbb{Q}$-vector space of orbital invariants $\mathfrak{S} \to \mathbb{Q}$ with rational values with respect to the class of* GALOIS*-finite coverings. This space is generated by two of the following orbital invariants: The* EULER *height $\mathbf{e} = \mathbf{c}_2$, the signature height τ, the orbital arithmetic genus χ and the orbital canonical self-intersection $\mathbf{c}_1^2$.*

3.5 Quasi-homogeneous Points, Quotient Points and Cusp Points

A two-dimensional *quasihomogeneous singularity* is a normal surface singularity (U, P), $P \in U$, such that the minimal resolution $\tilde{\rho} : \tilde{U} \to U$ of P is starlike; that means that $\tilde{\rho}^{-1}(P)$ is a configuration of curves described in Figure 3.5.1 where C is a smooth curve and the "rays" E are linear trees of projective lines; more precisely, E is a minimal resolution of a cyclic singularity. The curve C is called the *center* or *central curve* of $\tilde{\rho}^{-1}(P)$. It is uniquely defined only for non-cyclic singularities P. The cyclic ones are characterized by the conditions: C is rational and $\tilde{\rho}^{-1}(P)$ has at most

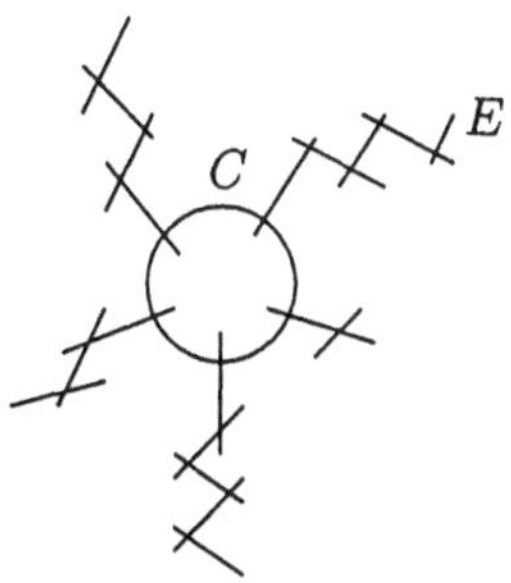

Figure 3.5.1

two "rays". Quasihomogeneous singularities can be defined in another way. For the equivalence of the definitions we refer the reader to the article [PIN] of PINKHAM. Let $X = Spec\ A$ be an affine algebraic variety over $\mathbb{C}$. There is a $\mathbb{C}^* = \mathbb{G}l_1(\mathbb{C})$-action on X if and only if A admits a graduation $A = \bigoplus_k A_k$, $A_k \cdot A_l \subseteq A_{k+l}$. We assume that $A_k = 0$ for $k < 0$ and $A_0 = \mathbb{C}$. Then the point P corresponding to the maximal ideal $\bigoplus_{k>0} A_k$ is the only fixed point of the $\mathbb{C}^* = A_0^*$-action. If additionally X is two-dimensional and normal, then P is the only singularity on X. By (second) definition, P is a quasihomogeneous surface singularity.

Coming back to the first definition we contract now the "rays" E of $\tilde{\rho}^{-1}(P)$ and obtain a reduced orbital curve $\mathbf{C} = (U_C, C)$ with some cyclic surface singularities on C. The resulting commutative diagram (3.5.1) fixes further notations. Tacitly we endowed the singular points P_i on C, U_C with

$$\begin{array}{ccc} \left(\tilde{U}, \tilde{\rho}^{-1}(P)\right) & \xrightarrow{\tilde{\rho}} & (U, P) \\ & \gamma \searrow \quad \nearrow \rho = \rho_P & \\ & (U_C, C) & \end{array} \tag{3.5.1}$$

singular coordinate systems, that means that P_i is understood as arranged point $(U_C, P_i; C, C_i)$. The images of the curve germs C_i on U are denoted by the same symbol. Let m be the number of them. Then we call a *reduced quasihomogeneous point.*

$(U,\ P;\ C_1,\ \ldots,\ C_m):$

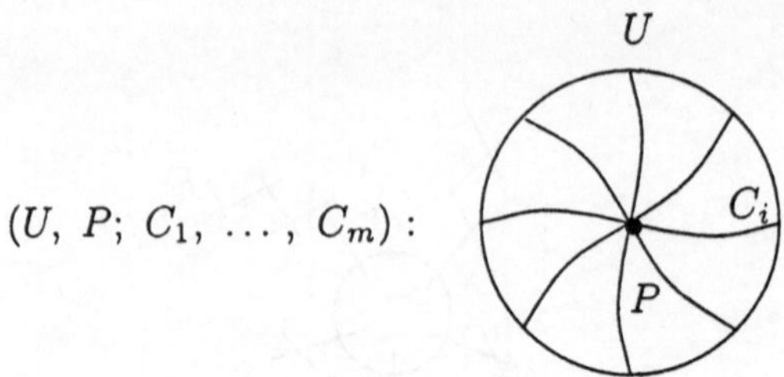

Definition 3.5.1 A *quasihomogeneous point* **P** is a tupel $(U, vP; v_1C_1, \ldots, v_mC_m)$, v, v_i natural numbers (later we allow also $v = \infty$) such that $(U, P; C_1, \ldots, C_m)$ is a reduced quasihomogeneous point. The corresponding orbital curve

$$\mathbf{C} = \mathbf{C}(\mathbf{P}) = (U_C, vC; \mathbf{P}_1, \ldots, \mathbf{P}_m)$$

and the morphism $\rho = \rho_{\mathbf{P}} : \mathbf{C} \longrightarrow \mathbf{P}$ supported by ρ_P are called central (small) resolutions of **P** or the *contraction* of **C**.

If there is no danger of misunderstanding, then we will denote a reduced quasihomogeneous point **P** simply by P. The *graph of a quasihomogeneous point* **P** is defined to be the atomic graph of **C(P)** (see picture (2.1.2)).

Proposition 3.5.2 *An orbital curve* **C** *is contractible to a quasihomogeneous point if and only if* $\tau(\mathbf{C}) < 0$.

Proof Looking at the graph of and the definition (2.4.3) of $\tau(\mathbf{C})$ one has

C:

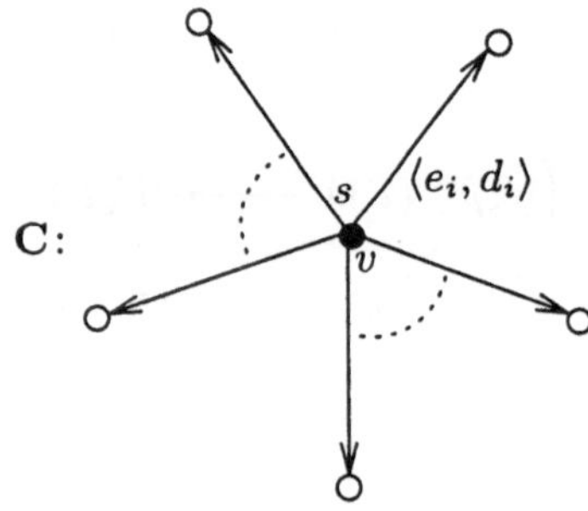

if and only if the minimal singularity resolution $E(C) \to C$ of the reduction C of $\mathbf{C}$ has a negative definit intersection matrix (MUMFORD'S criterion). The proof of this equivalence is a simple exercise of linear algebra. We refer to [HOI], Lemma 3.3.2. The statement is also a corollary of the next proposition. □

Definition 3.5.3 The quasihomogeneous point $\mathbf{P}$ or $\mathbf{C} = \mathbf{C}(\mathbf{P})$ is called *uniformizable* if there is a reduced orbital curve D (with $(D^2) < 0$) without surface singularities and a finite group G acting on (U_C, C) such that $\mathbf{C}(\mathbf{P}) = D/G$. In this situation we have a commutative diagram: [htp] The

$$\begin{array}{ccc} D & \xrightarrow{\rho_Q} & Q \\ \downarrow G & & \downarrow G \\ \mathbf{C}(\mathbf{P}) & \xrightarrow{\rho_P} & \mathbf{P} \end{array} \tag{3.5.3}$$

resolution ρ_Q is called a *uniformization* of $\rho_{\mathbf{P}}$.

Proposition 3.5.4 ([PIN], Theorem 3.2.3.1;[HOI]) *Let* $\mathbf{C}$ *be an orbital curve. We exclude the case where* $C = \mathbb{P}^1$ *and the point arrangement of* $\mathbf{C}$ *consists of exactly one non-trivial abelian point. Then there exists a uniformizable orbital curve* $\mathbf{C}'$ *which differs from* $\mathbf{C}$ *only by the central weight.*

PINKHAM proved the statement for reduced orbital curves. It has been extended to the general case by the author. The uniformizing curves have been constructed as zero sections of line bundles F' over these curves. The constructed curve $D \subset F'$, $\mathbf{C} = D/G$ is contractible if and only if C is contractible. One direction is clear. On the other hand, if C is contractible to P, then we get a commutative diagram where F is the normalization of

$$\begin{array}{ccc} F' & \xrightarrow{\psi} & F \\ g' \downarrow G & & g \downarrow G \\ F'/G = X' & \xrightarrow{\varphi} & X \\ \cup & & \Cup \\ C & \longrightarrow & P \end{array}$$

X in the function field $\mathbb{C}(F')$ of F'. The preimage $g^{-1}(P)$ is a finite point set because g is finite. But $\psi^{-1}(g^{-1}(P))$ is D. So $g^{-1}(P)$ consists of one point Q, and this is the contraction of D (along ψ).

Now we can give the

Second Proof of Proposition 3.5.2 Without loss of generality (Proposition 3.5.4) we can assume that $\mathbf{C} = D/G$. With the above notations it holds that

$$\frac{1}{3}(D^2) = \tau(D) = |G| \cdot \tau(\mathbf{C}) = \frac{|G|}{3v} \cdot \left(s + \sum e_i/d_i\right)$$

by the degree formula for GALOIS coverings of orbital curves. So we obtain the equivalent conditions

$$C \text{ contractible} \iff D \text{ contractible} \iff (D^2) < 0$$

$$\iff \tau(\mathbf{C}) < 0 \iff (3.5.2)$$

□

We see also that the condition $(D^2) < 0$ in definition 3.5.6 is superfluous because $\tau(\mathbf{C}) < 0$ implies $(D^2) < 0$ for any uniformization $D \longrightarrow \mathbf{C}$ by the degree formula.

Definition 3.5.5 For a quasihomogeneous point $\mathbf{P}$ with resolution $\mathbf{C} \to \mathbf{P}$ we set

$$\mathbf{e}(\mathbf{P}) = \mathbf{e}(\mathbf{C}), \ \tau(\mathbf{P}) = \tau(\mathbf{C})(< 0)$$

and call it the EULER *height* or the *signature height* of $\mathbf{P}$, respectively.

Definition 3.5.6 A quasihomogeneous point $\mathbf{P}$ is called a *cusp point*, iff $\mathbf{e}(\mathbf{P}) = 0$. It is called a *fraction point*, iff $\mathbf{e}(\mathbf{P}) > 0$ and $\mathbf{C}(\mathbf{P}) \neq (U, v\mathbb{P}^1, \mathbf{P}_1)$ with non-trivial $\mathbf{P}_1$. Notice that the definition does not depend on the central point weight.

The notations are justified by the following

Lemma 3.5.7 *(i) A quasihomogeneous point* $\mathbf{P}$ *is a fraction point if and only if there exists a uniformization* $O \longrightarrow \mathbf{P}'$;

(ii) it is a cusp point iff there exists a uniformization $E \longrightarrow \mathbf{C}(\mathbf{P}') = \mathbf{C}'$, *$E$ a reduced orbital elliptic curve with* $(E^2) < 0$.
Thereby $\mathbf{P}'$ *is a suitable point which differs from* $\mathbf{P}$ *only by the central point weight* .

Proof (i) We consider a quotient $\mathbf{P} = O/G$, G a finite group acting on $(\mathbb{C}^2, O)$. We blow up the point O and denote the exceptional line by L. Then $\mathbf{e}(L) = e(L) = e(\mathbb{P}^1) = 2 > 0$. Hence $\mathbf{e}(\mathbf{C}(\mathbf{P})) = \mathbf{e}(L/G) > 0$. Conversely, assume that $\mathbf{e}(\mathbf{P}) = \mathbf{e}(\mathbf{C}) > 0$. There exists a uniformization $D \underset{H}{\rightarrow} \mathbf{C}'$ of $\mathbf{C}'$ by Proposition 3.5.4. The EULER number $e(D) = \mathbf{e}(D)$ is greater than 0 because of the degree formula. Therefore D is the projective line on a surface (line bundle F over $D = \mathbb{P}^1$) with negative selfintersection number (D^2). Therefore it is contractible to a symmetric cyclic singularity Q and $\mathbf{P}' = Q/H$. Let V^* be the universal covering of $U_Q^* = U_Q \setminus \{Q\}$. Since $Q = O/\langle d, 1\rangle$ for a suitable d we may assume that $V^* = U_O^* \subseteq \mathbb{C}^2 \setminus \{O\}$. Now we lift the action of H from U_Q^* to U_O^* and obtain a finite group G acting on U_O^*, hence on U_O, which is connected with H and $\langle d, e\rangle$ by the exact sequence

$$1 \longrightarrow \langle d, e\rangle \longrightarrow G \longrightarrow H \longrightarrow 1.$$

Then $\mathbf{P}' = Q/H = (O/\langle d, e\rangle)/H = O/G$.

(ii) The vanishing property $\mathbf{e}(E) = e(E) = 0$, E a reduced orbital elliptic curve, is preserved for E/G. If $E \to Q$ is the contraction, then for $\mathbf{P} = Q/G$ it holds that $\mathbf{e}(\mathbf{P}) = 0$. Conversely, let $D \underset{G}{\rightarrow} \mathbf{C}'$ be a uniformization of $\mathbf{C}'$. From $\mathbf{e}(\mathbf{C}) = 0$ follows that $e(D) = \mathbf{e}(D) = 0$. Hence D is an elliptic curve. □

Now we classify all fraction and cusp points. We use the following

Convention 3.5.8 We identify the following graphs:

$$\underset{\langle d_i, e_i\rangle}{\longrightarrow}\!\!\circ_{v_i} \;=\; \underset{\langle v_i d_i, v_i e_i\rangle}{\longrightarrow}\!\!\circ$$

Corollary 3.5.9 *(i)* $\mathbf{P}$ *is a fraction point if and only if the central curve* C *of* $\mathbf{C}(\mathbf{P})$ *is rational and* $\mathbf{P}$ *has one of the graphs listed in Figure* 3.5.2

(ii) $\mathbf{P}$ *is a cusp point iff it has one of the graphical descriptions listed in Figure* 3.5.3.

Proof (i) The condition $\mathbf{e}(\mathbf{C}) > 0$ characterizing fraction points implies immediately $e(C) > 0$, hence $e(C) = 2$, $C \simeq \mathbb{P}^1$. With $e(C) = 2$ the condition $\mathbf{e}(\mathbf{C}) > 0$ can be written as

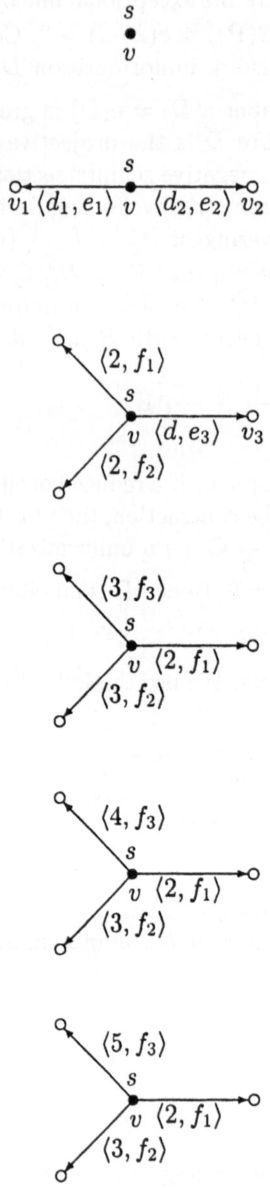

(0)	$s \leq -2$
(d_1, d_2)	$s \leq -2$
$(2, 2, d)$	$s \leq -2$ or: $s = -1,\ f_1 = f_2 = 0$ or: $s = -1,\ f_1 = 0,\ f_2 = 1,\ \frac{e_3}{d} < \frac{1}{2}$
$(2, 3, 3)$	$s \leq -2$ or: $s = -1,\ \frac{f_1}{2} + \frac{f_2}{3} + \frac{f_3}{3} < 1$
$(2, 3, 4)$	$s \leq -2$ or: $s = -1,\ \frac{f_1}{2} + \frac{f_2}{3} + \frac{f_3}{4} < 1$
$(2, 3, 5)$	$s \leq -2$ or: $s = -1,\ \frac{f_1}{2} + \frac{f_2}{2} + \frac{f_3}{5} < 1$

Figure 3.5.2

(0) $\mathbf{P} : \overset{s}{\underset{v}{\bullet}}, s \leq -1, C = C(\mathbf{P})$ is an elliptic curve.

(2, 2, 2, 2)

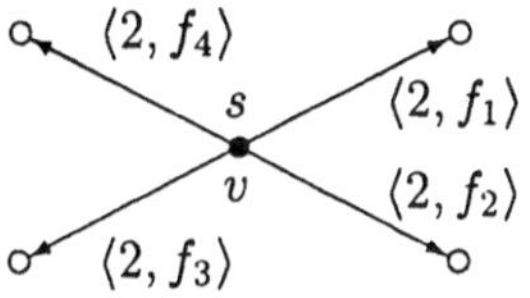

$\langle 2, f_4 \rangle$, $\langle 2, f_1 \rangle$, $\langle 2, f_2 \rangle$, $\langle 2, f_3 \rangle$; s, v

$s \leq -3$
or: $s = -2,\ f_1 = 0$
or: $s = -1,\ f_1 = f_2 = f_3 = 0$

(3, 3, 3)

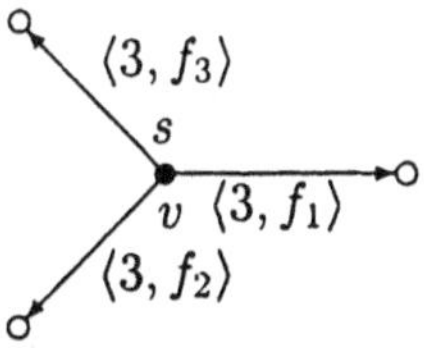

$\langle 3, f_3 \rangle$, $\langle 3, f_1 \rangle$, $\langle 3, f_2 \rangle$; s, v

$s \leq -3$
or: $s = -2,\ f_1 = 2$
or: $s = -1,\ \frac{f_1}{3} + \frac{f_2}{3} + \frac{f_3}{3} < 1$

(2, 4, 4)

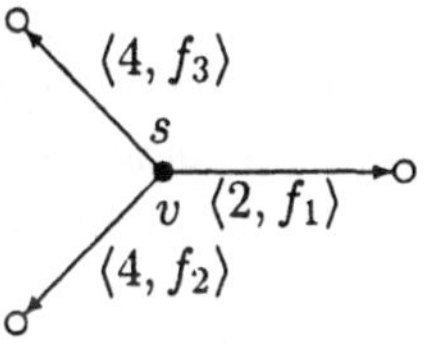

$\langle 4, f_3 \rangle$, $\langle 2, f_1 \rangle$, $\langle 4, f_2 \rangle$; s, v

$s \leq -3$
or: $s = -2,\ f_1 = 0$
or: $s = -2,\ f_2 \neq 3$
or: $s = -1,\ \frac{f_1}{2} + \frac{f_2}{4} + \frac{f_3}{4} < 1$

(2, 3, 6)

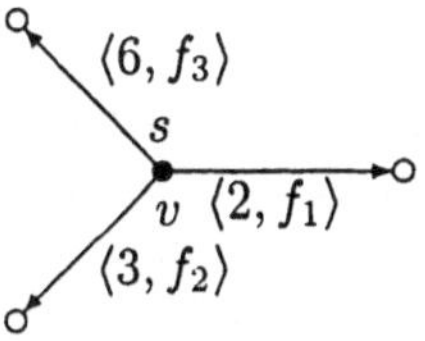

$\langle 6, f_3 \rangle$, $\langle 2, f_1 \rangle$, $\langle 3, f_2 \rangle$; s, v

$s \leq -3$
or: $s = -2,\ f_1 = 0$
or: $s = -2,\ f_2 \neq 2$
or: $s = -2,\ f_3 \neq 5$
or: $s = -1,\ \frac{f_1}{2} + \frac{f_2}{3} + \frac{f_3}{6} < 1$

Figure 3.5.3

$$2 - \sum_{i=1}^{k} \left(1 - \frac{1}{t_i}\right) > 0, \ f_i = e_i \cdot v_i, \ t_i = d_i v_i.$$

The summand $1 - \dfrac{1}{t_i}$ is not less than $\dfrac{1}{2}$, hence $k \leq 3$. A case by case checking yields the list of graphs. The restrictions for $s = (C^2)$ come from $\tau(\mathbf{C}) < 0$ or (3.5.2).

(ii) The graphs of cusp points can be received in the same manner with the equation

$$\mathbf{e}(\mathbf{C}) = 2 - \sum_{i=1}^{k} \left(1 - \frac{1}{t_i}\right) = 0. \tag{3.5.4}$$

□

Remark 3.5.10 The classification result (i) is well-known for graphs with reduced electronic parts

$$\underset{\langle t_i, f_i\rangle}{\longrightarrow\!\circ} = \underset{\langle t_i, f_i\rangle}{\longrightarrow\!\circ}\ _1 \ , \ \text{g.c.d.}\,(t_i, f_i) = 1$$

They correspond precisely to the quotients O/G of small subgroups $G \subset \mathbb{G}\mathrm{l}_2(\mathbb{C})$. Especially each fraction point $\mathbf{P}$ is supported by a quotient singularity P. Two small subgroups are conjugated in $\mathbb{G}\mathrm{l}_2(\mathbb{C})$ if and only if the graphs of the corresponding quotient points coincide. Equivalently, the quotient singularities have to be the same up to analytic isomorphy. The best reference is BRIESKORN's paper [BRI]. It should be of some interest to solve the following extended problem: Are the graphs of two quotient points $O/G, O/H$ equal if and only if the groups G, H are conjugated in $\mathbb{G}\mathrm{l}_2(\mathbb{C})$?

The graphs of type (d_1, d_2) correspond to weighted abelian points. From the "weighted" abelian point "$v \cdot \mathbf{P}$" itself the graph cannot be constructed because it is not clear which of the components of a resolution has to be weighted by v. In this case it is helpful to work with a third auxiliary curve branch through P (see Figure 3.5.4) indicating the weighted component. So in any case fraction points $\mathbf{P}$ can be understood as quotient singularities with three weighted curve branches on U_P through P as described in Figure 3.5.4. Weighted abelian points are useful in the following situation. It happens in general that a group G acting locally abelian on a surface G yields a stronger equivalence on a (smooth) curve $C \subset X$ than

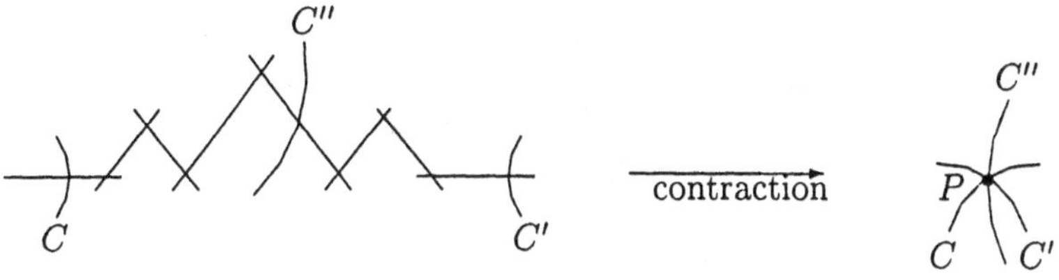

Figure 3.5.4

$N_G(C)$ does. In this case $\mathbf{D} = C/G \subset X/G$ has curve singularities and $D' = C/N_G(C) \longrightarrow C/G$ is the corresponding resolution. For example, C/G can have one double point Q. Then it can be resolved by means of σ-processes in the preimages $P \in X$ of Q. So we get commutative diagrams see 3.5.5:

$$\begin{array}{ccc} X' & \xrightarrow{\sigma} & X \\ \downarrow & & \downarrow \\ X'/G & \longrightarrow & X/G \end{array} \qquad \begin{array}{ccc} C' & \xrightarrow{\sim} & C \\ \downarrow N_G(C) & & \downarrow G - \text{equivalence} \\ \mathbf{D}' & \longrightarrow & \mathbf{D} \end{array} \tag{3.5.5}$$

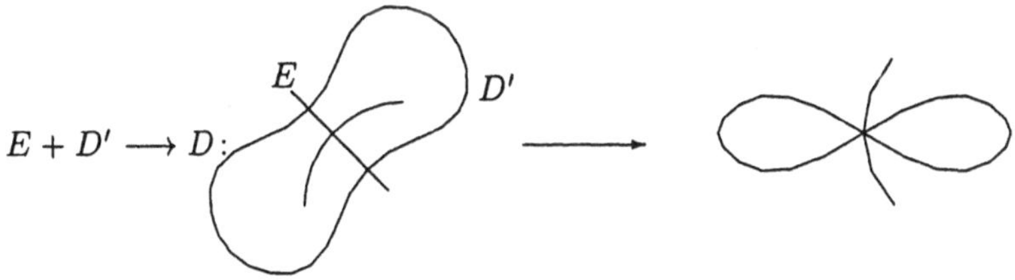

$E + D' \longrightarrow D$:

Figure 3.5.5

E is the quotient of the σ-exceptional lines. In this situation the use of graphs of weighted abelian points is useful. The abelian point $\mathbf{Q} \in \mathbf{D}$ is called a *double point* of $\mathbf{D}$. Arranging $\mathbf{Q}$, both $\mathbf{Q}$ and $\mathbf{Q}^{\perp}$ lie on $\mathbf{D}$.

At this place we also want to introduce *triple points*. They may appear in a situation as above described in diagram (3.5.5) in the case of non-abelian isotropy groups G_P. Then the tangent space $T_P(X)$ contains exactly three G_P-inequivalent eigenlines of non-trivial elements of G_P. The action of G may be transitive on the set of the $G_{P'}$-eigenlines of all points $P' \in GP$. Then P/G_P is a triple point on C/G: A mixed situation for non-abelian

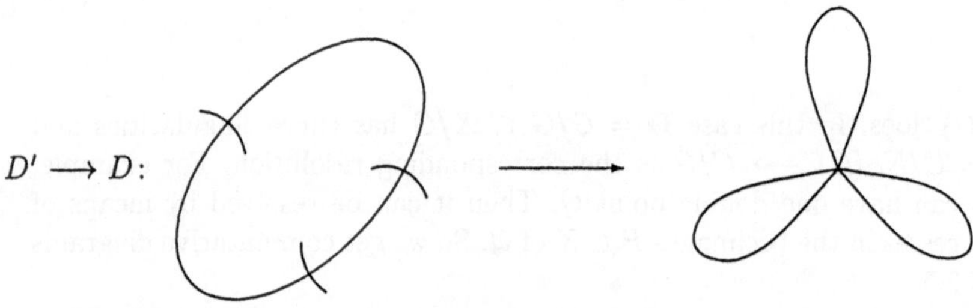

Figure 3.5.6

quotient points is described in the following Figure 3.5.7:

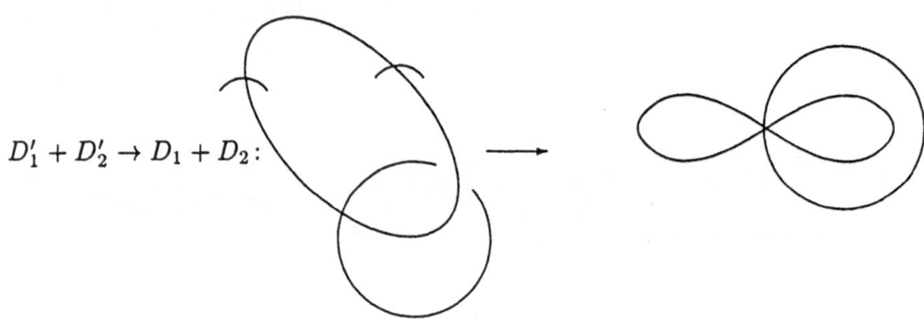

Figure 3.5.7

Here exactly two eigenlines are identified by the global action of G.

Remark 3.5.11 In this monograph we only consider ball cusp points. Therefore, we have the right to call them shortly cusp points omitting the word "ball". Cusp points come from quotients of groups Γ acting on the two-dimensional unit ball. If $\Gamma_{\tilde{\kappa}}$ is a cusp lattice, $\tilde{\kappa}$ a boundary point of the ball, then $\tilde{\kappa}/\Gamma_{\tilde{\kappa}}$ is a cusp point. In any case a cusp point is supported by a cusp singularity and any cusp singularity is a cusp point (with trivial weight). For details we refer to the next chapter 4 (section **4.5.**).

We close this section with the proof of useful lifting properties for the singularities considered above.

Proposition 3.5.12 *Let $f : X \to Y$ be a finite* GALOIS *covering of normal complex algebraic surfaces; $P \in X, Q = f(P)$, and $B_f(P) \subset U_Q$ the weighted branch curve germ of f_P defined as branch curve of a small finite analytic restriction $f_P : U_P \longrightarrow U_Q$ of f around P. If $\mathbf{Q}$, endowed with branch weights, is an abelian point, a quasi-homogeneous point, a cusp point or a quotient point, then P is a cyclic singularity, a quasi-homogeneous singularity, a cusp singularity or a quotient singularity, respectively.*

Proof The proof for abelian points has already been given in section 1.12, see Corollary 1.12.3. Now let $\mathbf{Q}$ be quasi-homogeneous, $\mathbf{Q} = (U_Q, vQ, \mathbf{B}_f)$, v a point weight which we will specify later. We consider the central resolution $\varphi_{\mathbf{Q}} : \mathbf{D} = \mathbf{D}(\mathbf{Q}) \to \mathbf{Q}$ and substitute on Y the neighbourhood U_Q by U_D. We obtain a global birational morphism $\varphi : Y' \to Y$ with restriction $\varphi_D : U_D \to U_Q$. Now let X' be the normalization of Y' in the function field $\mathbb{C}(X)$ of X (see [MU6]). Then we obtain a base change diagram: in

$$\begin{array}{ccc} X' & \xrightarrow{\psi} & X \\ \downarrow{\scriptstyle f'} & & \downarrow{\scriptstyle f} \\ Y' & \xrightarrow{\varphi} & Y \end{array} \tag{3.5.6}$$

the category of normal algebraic surfaces, ψ a birational morphism, f' a finite GALOIS covering. A component of $f^{-1}(D)$ through P is denoted by C. We restrict f' finitely to a small analytic neighbourhood U_C of C to get a finite covering $\mathbf{f'}_C : \mathbf{C} = (U_C, C) \to \mathbf{D}$. The point weight v can now be choosen as ramification index of f' at C. At each point of D we have by assumption an abelian branch situation. Therefore, all singularities of X' on D are cyclic. Consequently, $\mathbf{C} \to \mathbf{P}$ is the central resolution of a (reduced) quasi-homogeneous point. In the special situation of cusp points, or quotient points, we find $e(C) = 0$ or > 0, respectively, because $\mathbf{e}(\mathbf{D}) = 0$

or > 0. In the last conclusion one can use the degree formula (2.2.4). □

3.6 Quasi-smooth Orbital Surfaces

Let X be a quasi-smooth (complex compact normal algebraic) surface, that means that all singularities on X are quotient singularities. Allowing also mild singularities now on supporting curves we define *(quasi-smooth) orbital curves* on X as tupels

$$\mathbf{C} = (U_C, vC; \mathbf{P}_1 + \ldots + \mathbf{P}_a + \mathbf{R}_1 + \ldots + \mathbf{R}_b),$$

where $\mathbf{P}_1, ..., \mathbf{P}_a$ are abelian points arranged by C and $\mathbf{R}_1, ..., \mathbf{R}_b$ are weighted triple points (or double points) with graphs described in Figure (3.5.2), see also Figures (3.5.5), (3.5.6). The formal sum

$$\mathbf{A}(\mathbf{C}) = \mathbf{A}^o(\mathbf{C}) = \mathbf{P}_1 + \ldots + \mathbf{P}_a + \mathbf{R}_1 + \ldots + \mathbf{R}_b$$

is called the ***point arrangement*** of $\mathbf{C}$. We claim that all curve singularities and all non-abelian quotient singularities of X lying on C belong to $\{R_1, ..., R_b\}$ and that all cyclic singularities of X on C belong to $\{P_1, ..., P_a\}$ if they do not belong to the former set. Furthermore, we assume that $P_i \neq P_j \neq R_k \neq R_l$ for all possible $i, j, k, l, k \neq l$, $i \neq j$.

Definition 3.6.1 A *(quasi-smooth) arrangement* on X is a formal sum $\mathbf{A} = \mathbf{A}^1 + \mathbf{A}^0$,

$$\mathbf{A}^1 = \mathbf{C}_1 + \ldots + \mathbf{C}_k, \ \mathbf{A}^0 = \mathbf{P}_1 + \ldots + \mathbf{P}_k + \mathbf{R}_1 + \ldots + \mathbf{R}_m$$

such that $\mathbf{C}_i$ is a (quasi-smooth) orbital curve on X, $i = 1, \ldots, k$; $\mathbf{P}_j$ and $\mathbf{R}_l$ are abelian or triple points arranged by the curves $\mathbf{C}_i$ going through P_j or R_l, respectively. If there are less than two or three curves of the C_i's going through P_j or R_l, respectively, then auxiliary curve germs of weight 1 have to be used to arrange them. Furthermore, we claim that $P_i \neq P_j \neq R_l \neq R_n$ for $i \neq j, l \neq n$, and that as well all points of the point arrangements of the $\mathbf{C}_i$ belong to $\{\mathbf{P}_1, ..., \mathbf{R}_m\}$ as well as all surface singularities of X belong to $\operatorname{supp}(\mathbf{A}^0) = \{P_1, \ldots, R_m\}$.

A *(quasi-smooth) orbital surface* is a pair $\mathbf{X} = (X, \mathbf{A})$, A a quasi-smooth surface and $\mathbf{A}$ a quasi-smooth arrangement on X.

Since regular orbital surfaces are special cases of quasi-smooth orbital surfaces we can shortly speak about orbital surfaces until the definition has not been generalized furthermore. As for the regular case the orbital surfaces form a category. The transfer of a lot of definitions from the smaller category to the new one is obvious. So it is left to the reader when it is needed. The close relation between the two categories is contained in the following fact:

3.6.2 For each orbital surface $\mathbf{X}$ there is a central (partial) resolution $\varphi = \varphi_{\mathbf{X}} : \mathbf{X}' \to \mathbf{X}$, where $\mathbf{X}'$ is a regular orbital surface, $\varphi : X' \to X$ birational, such that the triple points $\mathbf{R} \in \mathbf{X}$ are centrally resolved substituting the weighted point $\mathbf{R}$ by the orbital curve $\mathbf{C}(\mathbf{R})$ see:

$$\begin{array}{ccc} \mathbf{X}' & \xrightarrow{\varphi} & \mathbf{X} \\ \cup & & \Cup \\ \mathbf{C}(\mathbf{R}) & \xrightarrow{\varphi_R} & \mathbf{R} \end{array} \tag{3.6.1}$$

Outside of the triple points φ restricts to an isomorphism. The weight v_R of the triple point $\mathbf{R}$ changes over to the weight $v_{C'}$ of $\boldsymbol{C} = \boldsymbol{C}(\boldsymbol{R})$. We call φ the *central (small) resolution* of $\mathbf{X}$.

We would like to uniformize orbital surfaces, if it is possible. For this purpose we study in more detail finite group actions on smooth surfaces.

Definition 3.6.3 Let the finite group G act effectively on the smooth surface X and let $C \subset X$ be a smooth (compact) curve on X. The embedded point $P \in C \subset X$ is called a *G-contact point* if the curve $GC = \sum_{g \in G} gC \subset X$ is not smooth at P. A G-contact point P of C is called a *G-cross point* if all branches of GC intersect each other transversally in P.

Example 3.6.4 If G_P is not abelian, $P \in C \subset X$, then P is a G-contact point of C.

Proof Since G_P is not abelian it cannot be diagonalized on the tangent space $T_P(X)$. Thus G_P does not act on C. Therefore P is a singularity of the curve $G_P \cdot C \subseteq GC$. □

Example 3.6.5 If

$$Z_G(C) = \{g \in G; g|C = id_C\} \neq 1,$$

or, equivalently, the quotient map $f : X \to X/G$ is ramified along C, then P is a smooth point of GC or a G-cross point of C.

Proof Let P be a G-contact point of C. Then there exists an element $g \in G$ such that P is singular on $C + gC$, that means C and gC intersect each other in P. The quotient map f is also ramified along gC. Therefore, there exists an element $\sigma \in Z_G(gC)$, $\sigma \neq 1$. From $P \in gC$ follows $\sigma \in G_P$. The eigenlines of elements of G_P in $T_P(X) \simeq \mathbb{C}^2$ cross each other. Therefore gC crosses C. □

Definition 3.6.6 The G-cross point $P \in C \subset X$ is called an *honest G-cross point* of C, if $(GC)_P > (G_P C)_P$, where $(GC)_P$ is the reduced divisor supported by all curves $gC, g \in G$, going through P.

Lemma 3.6.7 *Let $P \in C \subset X$ and G be as above. If $f : X \to X/G$ is ramified along C, then $D = C/G \subset X/G$ has at most double or triple points. The image point $f(P)$ is a singularity of D if and only if P is an honest G-cross point of C.*

Proof The curve C has at most G-cross points by 3.6.5. We blow up all honest G-cross points of C and get a surface X'. The proper preimage of C is denoted by C'.

The tangent space $T_P(X)$ has at most three G_P-inequivalent eigenlines. Therefore, there are at most three G_P-inequivalent components of GC through P. These components are projected all onto D. Their curve germs through P appear as points on the exceptional line $L'_P \subset X'$. The group G_P acts on this line and we have $G_P = N_G(L_P)$. So we get with

(1) three G_P-inequivalent GC-components through P, the following Figure 3.6.1 on $Y' = X'/G$:

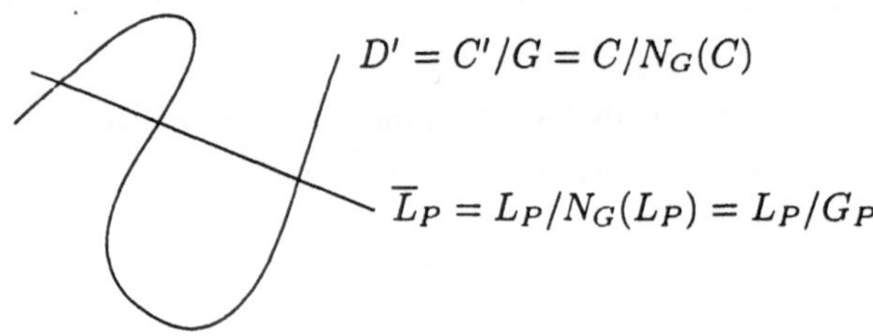

Figure 3.6.1

The quotient line $\overline{L}_P$ can be contracted. On this way we receive a triple point on $D = C/G$.

We omitted the following cases:

(2) G_P abelian or non-abelian, $(GC)_P$ has two G_P-inequivalent components:

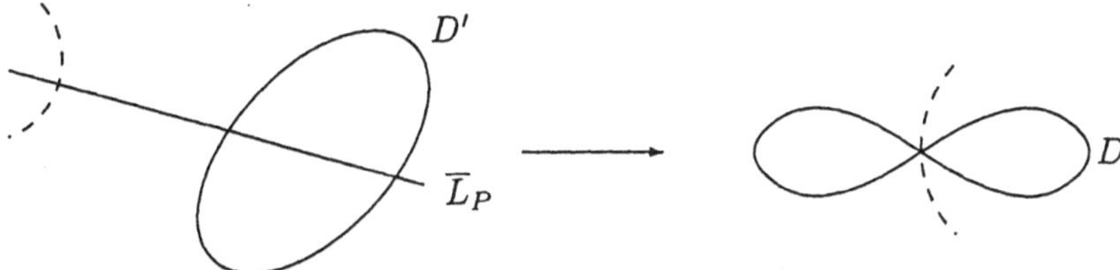

Figure 3.6.2

(3) $(GC)_P \subset GC$ has only one G_P-inequivalent component, see Figure 3.6.3

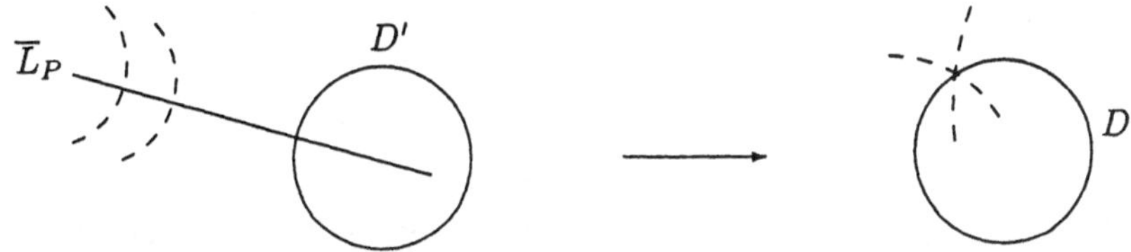

Figure 3.6.3

This is possible for all kinds of isotropy groups G_P. The case (2) is only possible for abelian G_P or for the types (2,2,d), (2,3,3) with graphs:

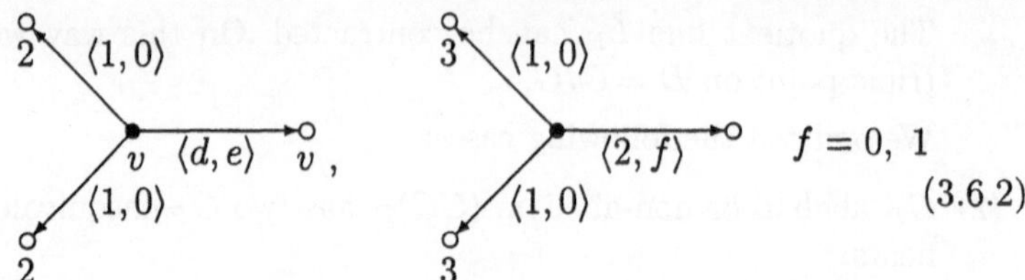

(3.6.2)

The case (1) is only possible for the type (2,2,2) with special graphs:

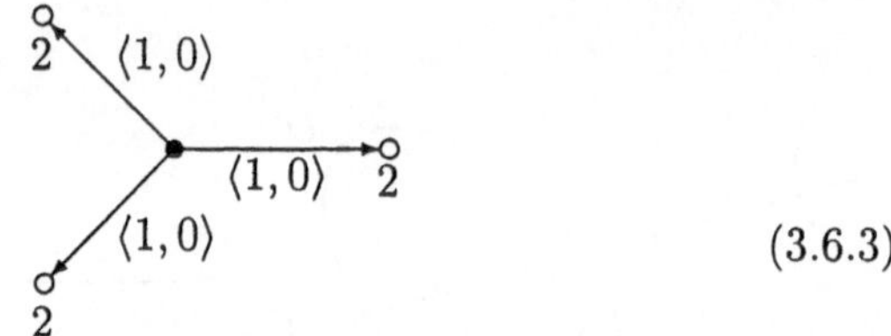

(3.6.3)

This is because $\overline{L}_P$ corresponds to the central point of the graph, and the electrons corresponding to the branches of D' through $\overline{L}_P$ must have the same weight v, which is the ramification index of $f : X \to X/G$ along C.

□

Now it is not difficult to define the quotient arrangement for X, G on $Y = X/G$, X smooth. We blow up all G-cross points of curves C on X with non-trivial decomposition group $Z_G(C)$. Then we blow up all remaining points P with non-abelian isotropy group G_P. So we get a commutative diagram:

$$\begin{array}{ccc} X' & \xrightarrow{\ \sigma\ } & X \\ \Big\downarrow{\scriptstyle f'} & & \Big\downarrow{\scriptstyle f} \\ Y' = X'/G & \xrightarrow{\ \varphi\ } & Y = X/G \end{array} \tag{3.6.4}$$

inducing $X'/G = \mathbf{Y}' \xrightarrow{\ \varphi\ } \mathbf{Y}$ in the following manner: The group action on X' is regular. So the quotient arrangement $\mathbf{A}'$ on Y' is well defined. Now we contract all images of the σ-exceptional lines on Y' coming back to Y. The branch weights of these curves are transferred to the contraction points. The weights of the other curves are not changed. On this way we obtain a contraction $\varphi_{\mathbf{A}} : \mathbf{A}' \to \mathbf{A}$ corresponding to central resolutions of orbital curves with triple points. This is the *quotient arrangement* on $Y = X/G$ we look for. The *orbital quotient surface* $\mathbf{Y} = (Y, \mathbf{A})$ will be denoted by

X/G. The morphism $\mathbf{f} : X \to X/G = \mathbf{Y}$ is called a *uniformization* if X is smooth. If X is not smooth, then we call $\mathbf{f} : X \to \mathbf{Y} = X/G$ a *quasi-uniformization* . The weights of all curves $\mathbf{C}$ in $\mathbf{A}(\mathbf{Y})$ are the ramification indices of $f : X \underset{G}{\to} Y = X/G$ along C and the weights of the triple points $\mathbf{P}$ are the ramification indices of $f' : X' \to \mathbf{Y}' = X'/G$ along the preimages of $\mathbf{C}(\mathbf{P})$ on X', where f' is constructed by small resolutions $\varphi_{\mathbf{P}}$ of all the weighted triple points of $\mathbf{A}^0$, its globalization $\varphi : \mathbf{Y}' \to \mathbf{Y}$ and base change diagram:

$$\begin{array}{ccc} X' & \xrightarrow{\psi} & X \\ f' \downarrow G & & G \downarrow f \\ Y' & \xrightarrow{\varphi} & Y \end{array} \tag{3.6.5}$$

in the category of normal surfaces, which defines X'. Observe that X' supports only cyclic singularities by Proposition 3.5.12 and that $P' \to \mathbf{Q}' = P'/G_P$, are (local) abelian quotient morphisms for all $P' \in X'$ by Corollary 1.12.3. So f' defines a quotient morphism $\mathbf{f}' : X' \to \mathbf{Y}' = X'/G$ in the category of regular orbital surfaces. Comparing the weights in $\mathbf{A}' = \mathbf{A}(\mathbf{Y}')$ and $\mathbf{A}$ we see that $\mathbf{A}$ is the contraction of $\mathbf{A}'$ along φ. Thus from (D.6.12) we can derive the "base change diagram":

$$\begin{array}{ccc} X' & \longrightarrow & X \\ \mathbf{f}' \downarrow G & & G \downarrow \mathbf{f} \\ \mathbf{Y}' & \longrightarrow & \mathbf{Y} \end{array} \tag{3.6.6}$$

in the category of orbital surfaces. Now we are able to define the EULER and signature heights $\mathbf{c}(\mathbf{Y})$. We set

$$\mathbf{c}_{\mathbf{Q}}(\mathbf{Y}'/\mathbf{Y}) = \begin{cases} 0; & \mathbf{Q} \text{ not a triple point} \\ \left(\frac{1}{t_1} + \frac{1}{t_2} + \frac{1}{t_3} - 1\right)/2v, \ \mathbf{c} = \mathbf{e}, & \mathbf{Q} \text{ a triple point} \\ -\left(\frac{1}{t_1} + \frac{1}{t_2} + \frac{1}{t_3} - 1\right)/2v, \ \mathbf{c} = \tau, & \mathbf{Q} \text{ a triple point} \end{cases} \tag{3.6.7}$$

Here we used a common form for the graphs of all the triple points listed in Figure (3.5.2), see 3.6.4

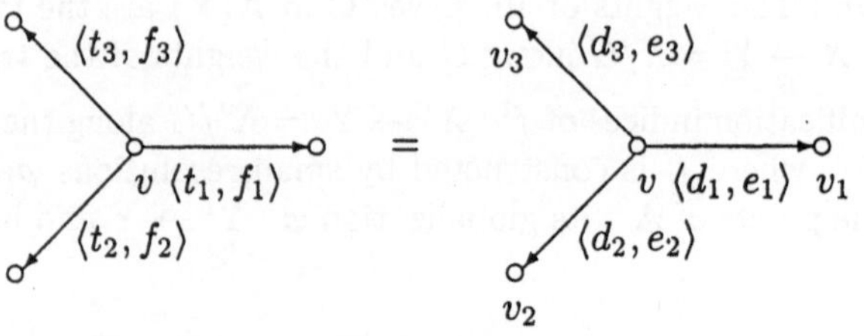

$$t_i = v_i d_i, \ f_i = v_i e_i$$

Figure 3.6.4

For the cases (0) and (d_1, d_2) we only added trivial "arrows" $\xrightarrow[\langle 1,0\rangle]{}\circ$

Definition 3.6.8 For an orbital surface $\mathbf{Y}$ with central resolution $\mathbf{Y}' \to \mathbf{Y}$ the EULER and *signature heights* of $\mathbf{Y}$ are defined by

$$\mathbf{c}(\mathbf{Y}') = \mathbf{c}(\mathbf{Y}) + \mathbf{c}(\mathbf{Y}'/\mathbf{Y}), \tag{3.6.8}$$

where

$$\mathbf{c}(\mathbf{Y}'/\mathbf{Y}) = \sum_{\mathbf{Q}} \mathbf{c}_{\mathbf{Q}}(\mathbf{Y}'/\mathbf{Y}).$$

Proposition 3.6.9 *Let* $\mathbf{f} : X \underset{G}{\to} \mathbf{Y}$ *be a uniformization of the quasi-smooth orbital surface* $\mathbf{Y}$. *Then it holds that*

$$\mathbf{c}(X) = \deg(\mathbf{f}) \cdot \mathbf{c}(\mathbf{Y}) = |G| \cdot \mathbf{c}(\mathbf{Y}). \tag{3.6.9}$$

Proof We first translate $\mathbf{c}_{\mathbf{Q}}(\mathbf{Y}'/\mathbf{Y})$ in terms of G_P-data, where $Q = f(P)$. For each triple point $\mathbf{Q}$ it holds that

$$\mathbf{c}_{\mathbf{Q}}(\mathbf{Y}'/\mathbf{Y}) = \begin{cases} \dfrac{1}{|G_P|}, & \mathbf{c} = \mathbf{e} \\ -\dfrac{1}{|G_P|}, & \mathbf{c} = \tau \end{cases} \tag{3.6.10}$$

We have only to uniformize the small resolution $\varphi_{\mathbf{Q}} : \mathbf{E} \to \mathbf{Q}$ of $\mathbf{Q}$ as described in the following diagram:

$$\begin{array}{ccc} & X' & & X \\ & \cup & & \Cup \\ \mathbb{P}^1 \simeq & L & \longrightarrow & P \\ & \mathbf{f}'_L \Big\downarrow G_P & & \Big\downarrow G_P \\ \mathbf{Y}' \supset & \mathbf{E} & \longrightarrow & \mathbf{Q} \in \mathbf{Y} \end{array} \tag{3.6.11}$$

The degree formula for $\mathbf{e}$ applied to $\mathbf{f'}_L$ yields

$$\begin{aligned} 2 = e(\mathbb{P}^1) &= e(L) = \mathbf{e}(L) = \frac{|G_P|}{v_L}\mathbf{e}(\mathbf{E}) \\ &= \frac{|G_P|}{v}\left(2 - \sum_{i=1}^{3}\left(1 - \frac{1}{t_i}\right)\right) \\ &= \frac{|G_P|}{v}\left(\frac{1}{t_1} + \frac{1}{t_2} + \frac{1}{t_3} - 1\right). \end{aligned} \tag{3.6.12}$$

This proves (3.6.10) for $\mathbf{c} = \mathbf{e}$. With this relation we obtain

$$\begin{aligned} |G|\mathbf{e}(\mathbf{Y}'/\mathbf{Y}) &= |G| \sum_{\mathbf{Q}} \mathbf{e}_{\mathbf{Q}}(\mathbf{Y}'/\mathbf{Y}) \\ &= \sum_{\substack{\mathbf{Q}\text{ triple} \\ \text{point}}} |G| \cdot \frac{1}{|G_P|} = \sum_{\mathbf{Q}} [G : G_P] \cdot 1 \\ &= \sum_{\mathbf{Q}} \sum_{P \to \mathbf{Q}} 1 = |\{G - \text{cross points of X }\}| \\ &= e(X') - e(X) = e(X'/X), \end{aligned} \tag{3.6.13}$$

where P is called a *G-cross point of X*, if there exists a curve C on X with $Z_G(C) \neq 1$ and P is a G-cross point of C, or G_P is not abelian. Finally, we get

$$|G| \cdot \mathbf{e}(\mathbf{Y}) = |G| \cdot (\mathbf{e}(\mathbf{Y}') - \mathbf{e}(\mathbf{Y}'/\mathbf{Y})) = e(X') - e(X'/X) = e(X).$$

The proof for $\mathbf{c} = \boldsymbol{\tau}$ is the same. □

Corollary 3.6.10 *The degree formula (3.6.9) holds for quasi-uniformizations and quasi-uniformiizable morphisms* $\mathbf{f} : X \to \mathbf{Y}$ *in the category of quasi-smooth orbital surfaces.*

Proof First observe that

$$\mathbf{c_Q}(\mathbf{Y}'/\mathbf{Y}) = \begin{cases} \mathbf{e}(\mathbf{E_Q})/2v, & \mathbf{c} = \mathbf{e} \\ -\mathbf{e}(\mathbf{E_Q})/2v, & \mathbf{c} = \boldsymbol{\tau} \end{cases} \qquad (v = v(\mathbf{E_Q})) \tag{3.6.14}$$

by (3.6.7) and (3.6.12). It suffices to assume that $\mathbf{f} : X \underset{G}{\to} \mathbf{Y}$ is a quasi-uniformization. With $\mathbf{Q} = \mathbf{f}(P)$ and obvious notations we get

$$\mathbf{c}_P(X'/X) = \pm\mathbf{e}(E_P)/2 = \pm\frac{|G_P|}{2v}\mathbf{e}(\mathbf{E_Q}) = |G_P|\mathbf{c_Q}(\mathbf{Y}'/\mathbf{Y})$$

by the degree formula applied to $E_P \to \mathbf{E_Q}$ sitting in the morphism $X' \underset{G}{\to} \mathbf{Y}'$ of regular orbital surfaces and by (3.6.14). Hence $\mathbf{c_Q}(\mathbf{Y}'/\mathbf{Y}) = \frac{1}{|G_P|}\mathbf{c}_P(X'/X)$. In analogy with (3.6.13) it follows that $|G|\mathbf{c}(\mathbf{Y}'/\mathbf{Y}) = \mathbf{c}(X'/X)$ and finally $|G|\mathbf{c}(\mathbf{Y}) = \mathbf{c}(X)$. □

Theorem 3.6.11 *On the category of quasi-smooth orbital surfaces there exists at least a two-dimensional $\mathbb{Q}$-space of functions with rational values which are orbital with respect to the class of all* GALOIS*-finite (= quasi-uniformizable) morphisms.*

Corollary 3.6.12 *For* $\mathbf{c} = \chi$, τ, e, c_1^2 *and* GALOIS *coverings* $X \to \mathbf{Y} = X/G$, X, $\mathbf{Y}$ *quasi-smooth, the degree formulas*

$$c(X) = |G|\mathbf{c}(\mathbf{Y}) = |G| \cdot \mathbf{c}(X/G) \tag{3.6.15}$$

are valid. They connect explicitly the CHERN *numbers of* X *and* Y.

The relations (3.6.15) are the *explicit two-dimensional* HURWITZ *formulas* for GALOIS coverings and CHERN numbers.
Proof It is only necessary to check that a quotient morphism

$$\mathbf{f} : X \to X/G = \mathbf{Y}$$

belongs to our category, that means that the branch locus on $\mathbf{Y}$ has at most triple points. We can locally uniformize our map. Let $P \in X$ be a quotient singularity. With the canonical resolution of P we get the commutative diagrams:

$$\begin{array}{ccc} E_P & \longrightarrow & P \\ \downarrow G_P & & \mathbf{f}_P \downarrow G_P \\ \mathbf{E_Q} & \longrightarrow & \mathbf{Q}\,, \end{array} \qquad \begin{array}{ccccc} U_{E_P} & \longrightarrow & U_P & \supset & U_P^* \\ \downarrow G_P & & \downarrow G_P & & \downarrow G_P \\ U_{E_Q} & \longrightarrow & U_Q & \supset & U_Q^*\,, \end{array} \qquad \begin{array}{ccc} X' & \longrightarrow & X \\ \mathbf{f}' \downarrow G & & \mathbf{f} \downarrow G \\ \mathbf{Y}' & \longrightarrow & \mathbf{Y} \end{array} \tag{3.6.16}$$

with $U_P^* = U_P \setminus \{O\}$. We let H_0 be the uniquely determined small subgroup of $\mathbb{Gl}_2(\mathbb{C})$ such that $P = O/H_0$. Since $U_O^* = U_O \setminus \{O\} \subset \mathbb{C}^2$ is simply-connected and H_O is small, the quotient morphism $U_O^* \underset{H_0}{\to} U_P^*$ is the universal covering of U_P^*. Therefore, the action of G_P can be lifted from U_P to U_O. We get on this way a group extension

$$1 \to H_O \to F_O \to G_P \to 1.$$

The action of F_O on U_O^* can be extended to U_O by HARTOG'S theorem. Then $\mathbf{Q}$ is the quotient of $O \in U_O$ by F_O and $U_Q = U_O/F_O$. We uniformized on this way $\mathbf{f_P}$. The branch locus of f_P around Q is the same as that of $U_O \to U_O/F_O$ around Q. So there can be at most three G_P-inequivalent points R on E_P with ramification curve C_R crossing E_P at R. The stationary group G_P acts locally abelian on (U_{E_P}, E_P). Thus C_R and E_P are the only curve germs through R allowing a non-trivial ramification index. Both are obviously not G-equivalent. Hence R is not a G-cross point of Y'. Therefore $\mathbf{f}'$ is a morphism of regular orbital surfaces. It is the central resolution of $\mathbf{f}$. The quotient morphism $\mathbf{f}$ belongs to our category of (quasi-smooth) orbital surfaces. □

3.7 Open Orbital Surfaces

We introduce in the same style as we have done for quasi-smooth surfaces locally quasi-homogeneous orbital surfaces. A locally quasi-homogeneous surface is a normal (compact, complex, algebraic) surface X which supports only quasi-homogeneous singularities. A quasi-homogeneous arrangement on X admits quasi-homogeneous points.

Definitions 3.7.1 A *quasi-homogeneous orbital surface* is a pair $\widehat{\mathbf{X}}=(\widehat{X},\widehat{\mathbf{A}})$ where $\hat{\mathbf{A}}$ is a *quasi-homogeneous arrangement* on $\hat{X}$. The latter is defined as image of a contraction $\mathbf{A}' \to \hat{\mathbf{A}}$ of a regular arrangement $\mathbf{A}'$ on a surface X' with at most cyclic singularities along a birational morphism $\varphi : X' \to \hat{X}$ such that locally at each weighted quasi-homogeneous point $\mathbf{P} \in \mathbf{A}$ the global morphism

$$\varphi : \mathbf{X}' = (X', \mathbf{A}') \to \widehat{\mathbf{X}}$$

restricts to the central resolution $\varphi_{\mathbf{P}} : \mathbf{C}(\mathbf{P}) \to \mathbf{P}$. The global morphism φ is called the central resolution of $\widehat{\mathbf{X}}$. It is assumed to be isomorphic outside of the weighted quasi-homogeneous points. If the quasi-homogeneous orbital surface has at most quotient points and cusp points, then we call it an *orbital surface with cusp points.* We consider now a partial resolution of an orbital surface with cusp points. We only take central resolutions of all the cusp points. The resulting resolution

$$\mathbf{X} \to \widehat{\mathbf{X}},\ \mathbf{E}_\kappa \longrightarrow \kappa \text{ for all cusp points } \kappa$$

is called the *central resolution of cusp points* of $\widehat{\mathbf{X}}$. The resolving surface $\mathbf{X}$ is a (quasi-smooth) orbital surface. Its central resolution is $\mathbf{X}'$.

Definitions 3.7.2 Let $\widehat{\mathbf{X}}$ be an orbital surface with cusp points. The surface $X_f = \hat{X} \setminus \{$ support of cusp points$\}$ is called an *open surface;* the index f comes from "finite part". Let $\widehat{\mathbf{A}} = \widehat{\mathbf{A}}^1 + \widehat{\mathbf{A}}^0$ be the arrangement of $\mathbf{X}$,

$$\widehat{\mathbf{A}}^0 = \mathbf{P}_0 + \ldots + \mathbf{P}_l + \mathbf{R}_1 + \ldots + \mathbf{R}_m + \kappa_1 + \ldots + \kappa_n$$
$$\widehat{\mathbf{A}}^1 = \widehat{\mathbf{C}}^1 + \ldots + \widehat{\mathbf{C}}^k,$$

$\widehat{C}$ a (compact) curve on $\hat{X}$. The curve $C_f = \widehat{C}_f = \widehat{C} \setminus \{\kappa_1, \ldots, \kappa_n\}$ on X_f is called an *open curve* on $\hat{X}$. Omitting cusp points we get *open orbital curves* $\mathbf{C}^i_f$ supported by $C^i_f \subseteq \widehat{C}^i$. The formal sum

$$\mathbf{A}_f = \mathbf{A}^1_f + \mathbf{A}^0_f = (\mathbf{C}^1_f + \ldots + \mathbf{C}^k_f) + (\mathbf{P}_1 + \ldots + \mathbf{P}_l + \mathbf{R}_1 + \ldots + \mathbf{R}_m)$$

is called the *open arrangement* on X_f *induced* by $\hat{\mathbf{A}}$. Generally, an (open) arrangement on X_f (on $\hat{X}$) is defined to be an induced open arrangement. An *open orbital surface* $\mathbf{X}_f$ is a pair $(\mathbf{X}_f, \mathbf{A}_f)$, $\mathbf{A}_f$ an (open) arrangement on X_f (or $\hat{X}$).

Further Notations 3.7.3 We denote by $\mathbf{X}_f = (X_f, \mathbf{A}_f)$ an open orbital surface. It is induced by an orbital surface $\hat{\mathbf{X}}$ with cusp singularities, which is called the *cuspidal compactification* of $\mathbf{X}_f$. Its central resolution of cusp points is denoted by $\mathbf{X}$ and is called the *toroidal compactification* of $\mathbf{X}_f$. The corresponding arrangements on $\mathbf{X}_f$, $\hat{\mathbf{X}}$, $\mathbf{X}$ are denoted by $\mathbf{A}_f = \mathbf{A}^1_f + \mathbf{A}^0_f$, $\widehat{\mathbf{A}} = \widehat{\mathbf{A}}^1 + \widehat{\mathbf{A}}^0$ or $\mathbf{A} = \mathbf{A}^1 + \mathbf{A}^0$, respectively. The central resolution of (all quasi-homogeneous points) of $\hat{\mathbf{X}}$ is denoted by $\mathbf{X}'$. So we dispose on a commutative diagram of *birational orbital morphism* :

$$\begin{array}{ccccc} \widehat{\mathbf{X}}_f = \mathbf{X}_f & \longrightarrow & \mathbf{X} & \longrightarrow & \widehat{\mathbf{X}} \\ \uparrow & & \uparrow & & \\ \mathbf{X}'_f & \longrightarrow & \mathbf{X}' & & \end{array} \tag{3.7.1}$$

If $\mathbf{C}_f$ is an (open) orbital curve on $\mathbf{X}_f$, then we use similar notations

$$\widehat{\mathbf{C}} \subset \widehat{\mathbf{X}},\ \mathbf{C} \subset \mathbf{X},\ \mathbf{C}' \subset \mathbf{X}'$$

for the corresponding "compactifications". Compactifications become unique, if we introduce the weight ∞ at cusp point or compactifying curves, see below.

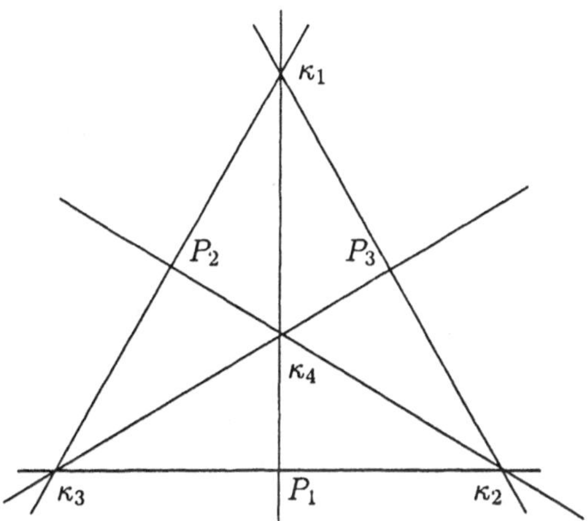

Figure 3.7.1

Example 3.7.4 Let κ_1, κ_2, κ_3, κ_4 be four points in general position on $\mathbb{P}^2 = \hat{X}$. The projective line through κ_i, κ_j, $i \neq j$, is denoted by $\hat{L}_{ij} = \hat{L}_{ji}$. The six lines $\hat{L}_{ij}$ are drawn in Figure 3.7.1.

We endow all lines with weight 3. Then

$$\widehat{\mathbf{A}} = \sum_{i<j} \widehat{\mathbf{L}}_{ij} + \mathbf{P}_1 + \mathbf{P}_2 + \mathbf{P}_3 + \kappa_1 + \kappa_2 + \kappa_3 + \kappa_4 \tag{3.7.2}$$

is an orbital arrangement on $\hat{X}$, $\hat{\mathbf{X}} = (\hat{X}, \hat{\mathbf{A}})$ an orbital surface with four cusp points. The cusp points and orbital lines have the graphs as shown in Figure 3.7.2 or Figure 3.7.3, respectively.

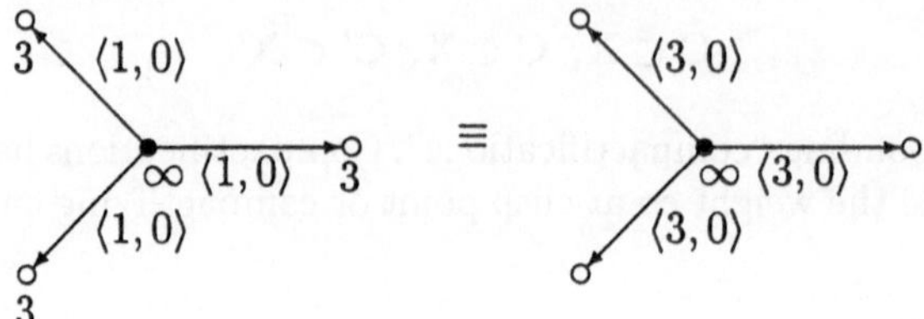

Figure 3.7.2

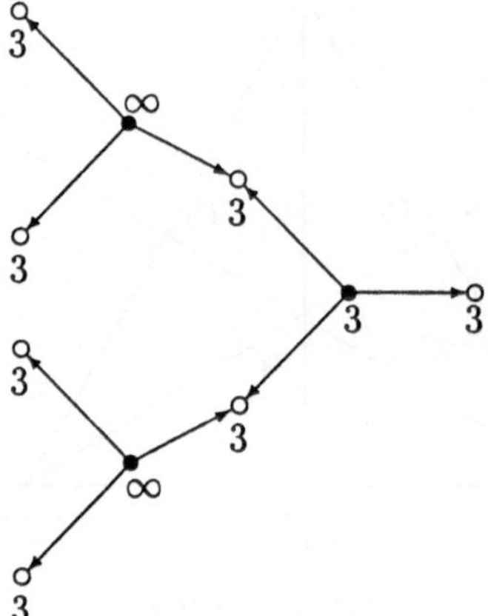

Figure 3.7.3

The molecular graph (Figure 3.7.4) of $\hat{\mathbf{A}}$ can be drawn by joining special points on a tetrahedron (see Figure 3.7.4). We introduce the weight ∞ for the cusp points and omitted the trivial coordinates $\langle 1, 0\rangle$ and arrows. Furthermore, we have

$$X_f = \mathbb{P}^2 \setminus \{\kappa_1, \kappa_2, \kappa_3, \kappa_4\}.$$

The orbital surface $\mathbf{X}' = \mathbf{X}$ is received by blowing up the four cusp points. The exceptional curves on X correspond to the graphs Figure 3.7.5. The configuration of $\mathbf{A}$ on $\mathbf{X}$ looks like Figure 3.7.6. The graph of $\mathbf{A}$ lies on a tetrahedron again, see Figure 3.7.7. The *graph of the open arrangement* $\mathbf{A}_f$ is defined to be the same as that of $\mathbf{A}$.

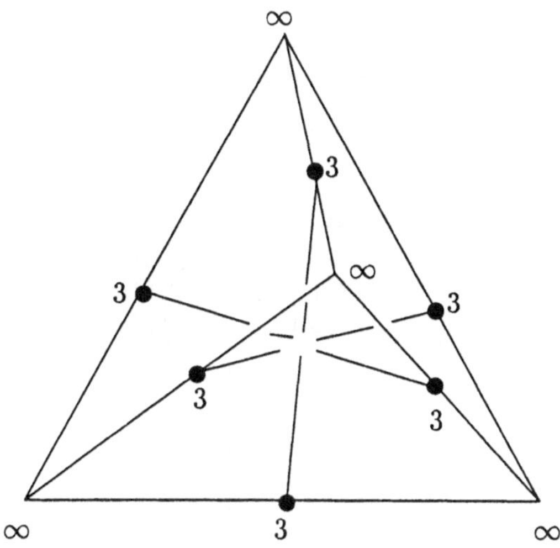

Figure 3.7.4

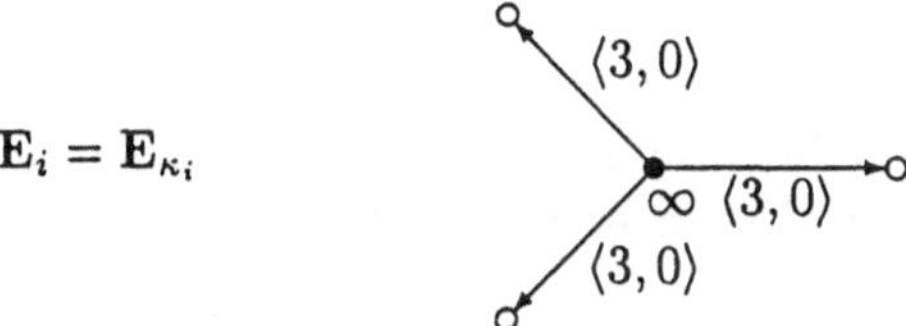

Figure 3.7.5

Definition 3.7.5 Let $\mathbf{X}_f$ be an open orbital surface and $c \in \mathbb{Q}e + \mathbb{Q}\tau$ a CHERN symbol. The orbital CHERN height $\mathbf{c}(\mathbf{X}_f)$ is defined to be

$$\mathbf{c}(\mathbf{X}_f) \underset{Df}{=} \mathbf{c}(\mathbf{X}) + \sum_{\kappa} \mathbf{c}_\kappa(\mathbf{X}), \mathbf{c}_\kappa(\mathbf{X}) = \mathbf{c}_\kappa(\mathbf{X}_f) \underset{Df}{=} -\frac{1}{v_\kappa}\mathbf{c}(\mathbf{E}_\kappa). \tag{3.7.3}$$

The sum runs over all cusp points, $\mathbf{E}_\kappa$ is the central resolution of κ and $v_\kappa = v_{\mathbf{E}_\kappa}$.

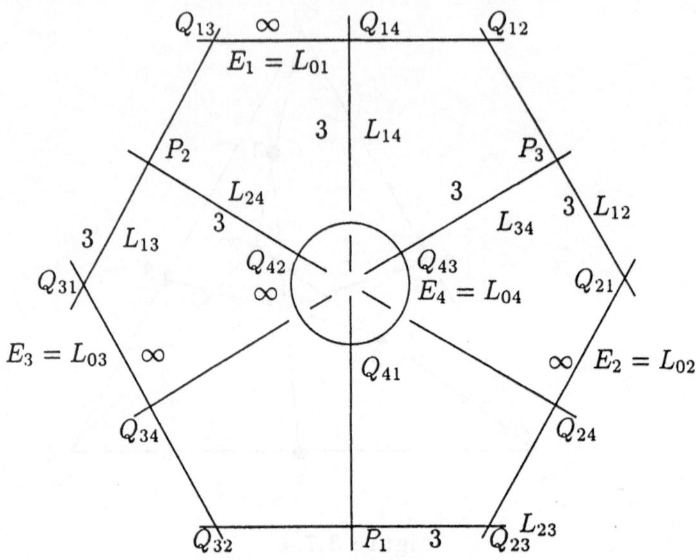

Figure 3.7.6

Proposition 3.7.6 *The orbital* CHERN *heights* $\mathbf{c}(\mathbf{X}_f)$ *are orbital invariants in the category of open orbital surfaces with respect to the class of* GALOIS *finite coverings.*

Proof Let $\mathbf{f}^0 : X_f \underset{G}{\rightarrow} \mathbf{Y}_f = X_f/G$ be a quotient morphism of open orbital surfaces, X_f reduced. The action on X_f extends to $\hat{X}$ and f^0 extends to the quotient morphism $\hat{f} : \hat{X} \rightarrow \hat{Y}$. A local uniformization argument (see Proposition 3.5.4) shows that the action of G around a cusp $\kappa \in \hat{X} \setminus X_f$ can be extended to an action around $E_\kappa \subset X$. So $\hat{f}$ can be lifted to $f : X \underset{G}{\rightarrow} \mathbf{Y} = X/G$. It suffices to work with $c = e, \tau$. In both cases it holds that

$$\sum_{\kappa \rightarrow \lambda} \mathbf{c}_\kappa(X) = |G| \cdot \mathbf{c}_\lambda(\mathbf{Y}), \ \lambda = f(\kappa). \tag{3.7.4}$$

Namely, for $c = e$ we have

$$\mathbf{e}_\lambda(\mathbf{Y}) = \mathbf{e}(\mathbf{E}_\lambda) = e(E_\lambda) - \sum \left(1 - \frac{1}{v_i d_i}\right) = 0$$

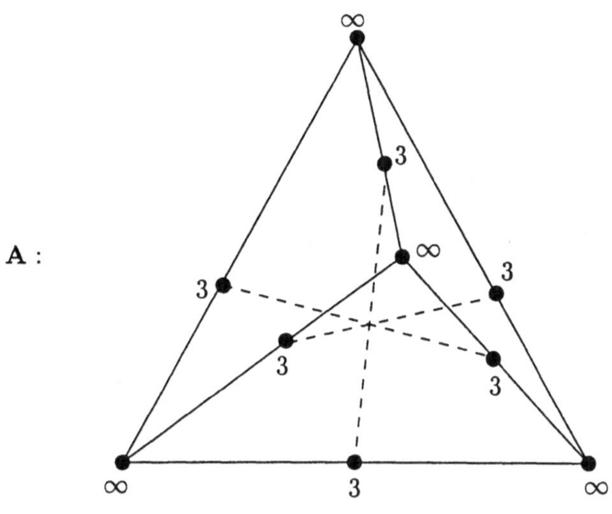

Figure 3.7.7

because of (3.5.4). So the relation (3.7.4) is trivial. For $c = \tau$ we use the degree formula for signature heights of orbital curves, the definition (2.4.3) of these numbers and receive:

$$\begin{aligned}\sum_{\kappa\to\lambda} \tau_\kappa(X) &= \sum_{\kappa\to\lambda} \tau\,(E_\kappa) = [G:G_\kappa]\,\tau\,(E_\kappa)\\ &= [G:G_\kappa]\cdot|N_G\,(E_\kappa)|\cdot\tau\,(\mathbf{E}_\lambda)\,/v_\kappa\\ &= [G:G_\kappa]\cdot|G_\kappa|\cdot\tau\,(\mathbf{E}_\lambda)\,/v_\kappa\\ &= |G|\cdot\tau_\lambda(\mathbf{Y}).\end{aligned}$$

With the help of (3.7.4), definition (3.7.3) and the degree formula (3.6.15) for quasi-smooth orbital surfaces applied to $\mathbf{f} : X \to \mathbf{Y} = X/G$ we obtain

$$\begin{aligned}|G|\cdot\mathbf{c}\,(\boldsymbol{Y}_f) &= |G|\cdot\left(\mathbf{c}(\boldsymbol{Y}) + \sum_\lambda \mathbf{c}_\lambda\,(\boldsymbol{Y})\right)\\ &= |G|\cdot\mathbf{c}(\mathbf{Y}) + \sum_\lambda |G|\cdot\mathbf{c}_\lambda(\mathbf{Y})\\ &= \mathbf{c}(X) + \sum_\lambda\sum_{\kappa\to\lambda}\mathbf{c}_\kappa(X)\end{aligned}$$

$$= \mathbf{c}(X) + \sum_{\kappa} \mathbf{c}_{\kappa}(X) = \mathbf{c}(X_f)$$

The degree formula for quasi-uniformizations of open orbital surfaces is proved. Its extension to quasi-uniformizable morphisms is trivial. □

Example 3.7.7 We calculate the orbital heights of $\mathbf{X}_f$ for $\widehat{X} = \mathbb{P}^2$, $X = \tilde{\mathbb{P}}^2$ and $\mathbf{X}_f$ described in Example 3.7.4, especially by Figure (3.7.6). By definition and the orbital decomposition (3.3.8), we have

$$\begin{aligned} \mathbf{e}\,(\mathbf{X}_f) &= \mathbf{e}(\mathbf{X}) \\ &= \mathbf{e}\left(\tilde{\mathbb{P}}^2\right) - \sum_{1 \le i < j \le 4} \left(1 - \frac{1}{3}\right) \mathbf{e}\,(\mathbf{L}_{ij}) - \sum_{l=1}^{3} \mathbf{e}\,(\mathbf{P}_i) - \sum_{i \ne j} \mathbf{e}\,(\mathbf{Q}_{ij}) \end{aligned}$$

For the calculation of the summands we use Table **3**.B at the end of the chapter.

$$\begin{aligned} \mathbf{e}\,(P_i) &= \left(1 - \frac{1}{3}\right)\left(1 - \frac{1}{3}\right), \\ \mathbf{e}\,(Q_{ij}) &= \left(1 - \frac{1}{v_i}\right)\left(1 - \frac{1}{3}\right), \quad v_i = v_{E_i} \\ \lambda_e(3) \cdot \mathbf{e}\,(\mathbf{L}_{ij}) &= \left(1 - \frac{1}{3}\right)\left[e\,(L_{ij}) - \left(1 - \frac{1}{3}\right) - \left(1 - \frac{1}{v_i}\right) - \left(1 - \frac{1}{v_j}\right)\right], \\ \mathbf{e}\left(\tilde{\mathbb{P}}^2\right) &= e\left(\tilde{\mathbb{P}}^2\right) = e(\mathbb{P}^2) + 4 = 7; \end{aligned}$$

hence

$$\mathbf{e}(\mathbf{X}) = 7 + 3\left(1 - \frac{1}{3}\right)\left(1 - \frac{1}{3}\right) - 6\left(1 - \frac{1}{3}\right) \cdot 2 = \frac{1}{3}. \tag{3.7.5}$$

Observe that the weights v_i of the cusp resolving curves E_i have been disappeared. The same happens for the signature because

$$\begin{aligned} \lambda_\tau(E_i)\tau(\mathbf{E}_i) + \tau_{\kappa_i}(\mathbf{X}) &= \left(v_i - \frac{1}{v_i}\right) \cdot \tau(\mathbf{E}_i) + \frac{1}{v_i}\tau(\mathbf{E}_i) \\ &= v_i\tau(\mathbf{E}_i) = v_i \cdot \tau(E_i)/v_i = \tau(E_i) \end{aligned} \tag{3.7.6}$$

We calculate the summands of the orbital decomposition

$$\begin{aligned}
\tau(\mathbf{X}_f) &= \tau(X) - \sum_{1\le i<j\le 4} \lambda_\tau(3)\cdot\tau(\mathbf{L}_{ij}) - \sum_{i=1}^{4}\tau(E_i)\\
\tau\left(\tilde{\mathbb{P}}^2\right) &= \tau\left(\tilde{\mathbb{P}}^2\right) = \tau(\mathbb{P}^2) - 4 = 1-4 = -3\\
\lambda_\tau(3)\cdot\tau(\mathbf{L}_{ij}) &= \left(3-\frac{1}{3}\right)[(-1)+0]/3\cdot 3 = -8/27 \qquad (3.7.7)\\
\tau(E_i) &= \frac{1}{3}\left[(E_i^2)+0\right] = -\frac{1}{3},
\end{aligned}$$

hence

$$\tau(\mathbf{X}_f) = -3 + 6\cdot 8/27 + 4/3 = \frac{1}{9} \qquad (3.7.8)$$

Example 3.7.8 Let $\mathbf{X}_f$ be a reduced open orbital surface with smooth open surface X_f as support. Then it holds that

$$\begin{aligned}
\tau(X_f) &= \tau(\mathbf{X}_f) = \tau(X) - \sum_\kappa \frac{1}{3}(E_\kappa^2),\\
\mathbf{e}(X_f) &= \mathbf{e}(\mathbf{X}_f) = e(X),\\
\chi(X_f) &= \chi(\mathbf{X}_f) = \chi(X) - \frac{1}{12}\sum_\kappa (E_\kappa^2)\\
\mathbf{c}_1^2(X_f) &= \mathbf{c}_1^2(\mathbf{X}_f) = c_1^2(X) - \sum_\kappa (E_\kappa^2)
\end{aligned}$$

So, notice that the orbital CHERN heights of a reduced smooth open surface do not coincide in general with the CHERN numbers of their toroidal smooth compactifications. This is only true for the EULER numbers. It turns out that the above orbital CHERN heights are the so-called *logarithmic* CHERN *numbers* (see [BHH])

3.8 Orbital Decompositions

We decompose compactified arrangements of open surfaces into a finite and an infinite part. Let Y_f be an open surface, $\hat{Y}, Y$ their compactifications and Y' the central resolution of $\hat{Y}$. The corresponding compactifications (and resolution) of an open arrangement $\mathbf{A}_f$ on Y_f we denote again by $\hat{\mathbf{A}}$, $\mathbf{A}$ or $\mathbf{A}'$, respectively. The sum of all orbital points and curves of $\mathbf{A}$, whose support lie on $Y_\infty = Y \setminus Y_f$ is denoted by $\mathbf{A}_\infty$. So we have

$$\mathbf{A} = \mathbf{A}_f + \mathbf{A}_\infty = (\mathbf{A}_f^1 + \mathbf{A}_f^0) + (\mathbf{A}_\infty^1 + \mathbf{A}_\infty^0) \text{ on } Y.$$

Contracting cusp curves on Y' or resolving orbital triple points on Y we get decompositions

$$\widehat{\mathbf{A}} = \widehat{\mathbf{A}}_f + \widehat{\mathbf{A}}_\infty = \mathbf{A}_f + \widehat{\mathbf{A}}_\infty = (\mathbf{A}_f^1 + \mathbf{A}_f^0) + \widehat{\mathbf{A}}_\infty \text{ on } \widehat{Y}$$
$$\mathbf{A}' = \mathbf{A}_f' + \mathbf{A}_\infty' = \mathbf{A}_f' + \mathbf{A}_\infty = \left(\mathbf{A}_f'^1 + \mathbf{A}_f'^0\right) + (\mathbf{A}_\infty^1 + \mathbf{A}_\infty^0) \text{ on } Y'.$$

The arrangements $\mathbf{A}_f = \hat{\mathbf{A}}_f, \mathbf{A'}_f$ are correctly defined on the open surfaces Y_f or Y_f', respectively. If $\mathbf{D}_f$ is a component of $\mathbf{A}_f^1$, then we introduce in obvious manner further notations by the commutative diagrams:

$$\begin{array}{ccccc} \mathbf{D}_f' & \longrightarrow & \mathbf{D}_f & \xrightarrow{=} & \mathbf{D}_f \\ \cap & & \cap & & \cap \\ \mathbf{D}' & \longrightarrow & \mathbf{D} & \longrightarrow & \widehat{\mathbf{D}} \end{array} \quad \text{on} \quad \begin{array}{ccccc} Y_f' & \longrightarrow & Y_f & \xrightarrow{=} & \widehat{Y}_f \\ \cap & & \cap & & \cap \\ Y' & \longrightarrow & Y & \longrightarrow & \widehat{Y} \end{array} \qquad (3.8.1)$$

First we define stepwise:

$$\mathbf{c}(\mathbf{D}) = \mathbf{c}\left(\mathbf{D}'\right) - \mathbf{c}\left(\mathbf{D}'/\mathbf{D}\right),$$
$$\mathbf{c}\left(\mathbf{D}'/\mathbf{D}\right) = \sum_{\mathbf{Q}\in\mathbf{D}} \mathbf{c}_Q\left(\mathbf{D}'/\mathbf{D}\right), \qquad (3.8.2)$$
$$\mathbf{c}_Q\left(\mathbf{D}'/\mathbf{D}\right) = \begin{cases} -\sum\limits_{\mathbf{D}'\ni\mathbf{Q}'\to\mathbf{Q}} \dfrac{1}{3v_\mathbf{Q}\cdot d_{Q'}}\,; & \mathbf{c} = \boldsymbol{\tau};\mathbf{Q} \text{ triple point} \\ 0; & \mathbf{c} = \mathbf{e} \text{ or } \boldsymbol{\tau} \\ & \text{otherwise} \end{cases}$$

Now let $\mathbf{f} : X \underset{G}{\to} \mathbf{Y}$ be a uniformization and $\mathbf{f}_C : C \to \mathbf{D}$ the restriction to an irreducible smooth preimage curve C of D. With the central resolution diagram (3.8.3) we fix further notations

$$\begin{array}{ccc} C' & \longrightarrow & C \\ \downarrow{\scriptstyle \mathbf{f}_{C'}'} & & \downarrow{\scriptstyle \mathbf{f}_C} \\ \mathbf{D}' & \longrightarrow & \mathbf{D} \end{array} \qquad \begin{array}{ccc} X' & \longrightarrow & X \\ {\scriptstyle \mathbf{f}'}\downarrow{\scriptstyle G} & & {\scriptstyle G}\downarrow{\scriptstyle \mathbf{f}} \\ \mathbf{Y}' & \longrightarrow & \mathbf{Y} \end{array} \qquad (3.8.3)$$

3.8.1 *With the above notations it holds that*

$$\mathbf{c}(C) = |f_C| \cdot \mathbf{c}(\mathbf{D}) = \deg(f_C) \cdot \mathbf{c}(\mathbf{D})$$

Proof For $\mathbf{c} = \mathbf{e}$ we get

$$\mathbf{e}(C) = e(C) = e(C') = \mathbf{e}(C') = |f_{C'}| \cdot \mathbf{e}(\mathbf{D}') = |f_C| \cdot \mathbf{e}(\mathbf{D}).$$

Now let c be the signature symbol τ. We denote by $\mathbf{A}_3^0 = \mathbf{A}_3^0(\mathbf{D}) \subseteq \mathbf{A}^0(\mathbf{D})$ the subarrangement of weighted points because these are the triple points. Then

$$\begin{aligned} 3\tau(C') = \left(C'^2\right) &= \sum_{\mathbf{Q}\in\mathbf{A}_3^0} \sum_{C\ni P\to\mathbf{Q}} (-1) \\ &= \sum_{\mathbf{Q}\in\mathbf{A}_3^0} \sum_{\mathbf{D}'\ni\mathbf{Q}'\to\mathbf{Q}} \sum_{C'\ni P'\to\mathbf{Q}'} (-1) + (C^2) \end{aligned} \tag{3.8.4}$$

because we have to blow up each preimage point of $\mathbf{Q} \in \mathbf{A}_3^0$ to get the central resolution described in diagram (3.8.3). We calculate

$$\begin{aligned} \sum_{\mathbf{D}'\ni\mathbf{Q}'\to\mathbf{Q}} \sum_{P'\to Q'} (-1) &= \sum_{\mathbf{D}'\ni\mathbf{Q}'\to\mathbf{Q}} [N_G(C') : N_G(C')_{P'}] \cdot (-1) \\ &= -\frac{N_G(C')}{v_{\mathbf{D}'}} \sum_{\mathbf{D}'\ni\mathbf{Q}'\to\mathbf{Q}} \frac{1}{v_{\mathbf{Q}} \cdot d_{Q'}} \\ &= +3|f_C|\tau_Q(\mathbf{D}'/\mathbf{D}), \end{aligned} \tag{3.8.5}$$

where $v = v_{\mathbf{D}} = v_{\mathbf{D}'}$ is the weight of $\mathbf{D}$ and $v_{\mathbf{Q}} = v_{\mathbf{E}}$ the weight of $\mathbf{Q}$ and $\mathbf{E} = \mathbf{E}_{\mathbf{Q}}$. From (3.8.4) and (3.8.5) it follows that

$$\begin{aligned} \tau(C') - \tau(C) = \tau(C'/C) &= \sum_{\mathbf{Q}} |f_C| \cdot \tau_Q(\mathbf{D}'/\mathbf{D}) \\ &= |f_C| \cdot \tau(\mathbf{D}'/\mathbf{D}) \\ &= |f_C| \cdot [\tau(\mathbf{D}') - \tau(\mathbf{D})]. \end{aligned}$$

The summands $\tau(\mathbf{C}')$ and $|f_C| \cdot \tau(\mathbf{D}')$ are equal by the degree formula for regular orbital curves. This proves the statement of 3.8.1. □

The orbital decompositions we look for look like

$$\mathbf{c}(\mathbf{Y}_f) = \mathbf{c}(Y_f) - \mathbf{c}(\mathbf{A}_f^1) - \mathbf{c}(\mathbf{A}_f^0) - \mathbf{c}(\mathbf{A}_\infty). \tag{3.8.6}$$

It has been found already for orbital heights of regular orbital surfaces. We try to extend it to quasi smooth open orbital surfaces. We know already $\mathbf{c}(\mathbf{Y}_f)$, hence also $\mathbf{c}(Y_f)$, and we set

$$\mathbf{c}(\mathbf{A}_f^1) = \sum_{\mathbf{D}_f \subset \mathbf{A}_f^1} \lambda_c(\mathbf{D}) \cdot \mathbf{c}(\mathbf{D}). \tag{3.8.7}$$

it remains to define $\mathbf{c}(\mathbf{A}_\infty)$ and

$$\mathbf{c}(\mathbf{A}_f^0) = \sum_{\mathbf{Q}\in\mathbf{A}_f^0} \mathbf{c}(\mathbf{Q}). \tag{3.8.8}$$

For abelian points $\mathbf{Q}$ the orbital invariants $\mathbf{c}(\mathbf{Q})$ are known. So we have to fill one gap, namely to define $\mathbf{c}(\mathbf{Q})$ for weighted triple points (fraction points) $\mathbf{Q} \in \mathbf{A}_f^0$.

From the global point of view there are three types of triple points $\mathbf{Q} \in \mathbf{Y}$. Through $\mathbf{Q}$ can go one, two or three different global curves. The corresponding pictures, together with the small resolutions, are described in Figure (3.8.1)

The weighted graphs of these configurations are described in Figure 3.8.2.

We introduced triangles for the central part of the graph of a triple point. Now we set

$$\begin{aligned}\mathbf{c}(\mathbf{Q}) = &- \mathbf{c}_Q(Y'/Y) + \mathbf{c}_Q(\mathbf{Y}'/\mathbf{Y}) + \lambda_c(\mathbf{E})\mathbf{c}(\mathbf{E}) \\ &+ \sum_{\mathbf{D}\ni\mathbf{Q}} \lambda_c(\mathbf{D})\mathbf{c}_{\mathbf{Q}}(\mathbf{D}'/\mathbf{D}) + \sum_{\mathbf{i}=\mathbf{1}}^{\mathbf{3}} \mathbf{c}\left(\mathbf{Q}_i'\right)\end{aligned} \tag{3.8.9}$$

The motivation is given in the proof of Proposition 3.8.4, below. First we calculate these numbers for $\mathbf{c} = \mathbf{e}, \boldsymbol{\tau}$, that means we express $\mathbf{c}(\mathbf{Q})$ directly by data of the above graphs. One finds

$$\begin{aligned}\mathbf{e}(\mathbf{Q}) &= -\,\mathbf{e}(E)/2 + \mathbf{e}(\mathbf{E})/2v + \left(1 - \frac{1}{v}\right)\mathbf{e}(\mathbf{E}) + \sum_{i=1}^{3} e\left(\mathbf{Q}_i'\right) \\ &= -\frac{1}{2}\left(\frac{1}{d_1} + \frac{1}{d_2} + \frac{1}{d_3} - 1\right) \\ &\quad + \left(1 - \frac{1}{2v}\right)\left(\frac{1}{v_1 d_1} + \frac{1}{v_2 d_2} + \frac{1}{v_3 d_3} - 1\right) \\ &\quad + \sum_{i=1}^{3} \frac{1}{d_i}\left(1 - \frac{1}{v}\right)\left(1 - \frac{1}{v_i}\right),\end{aligned}$$

hence

$$\begin{aligned}\mathbf{e}(\mathbf{Q}) &= \left(\frac{1}{2} - \frac{1}{v}\right) \cdot \mathbf{e}(E) + \frac{1}{2v}\mathbf{e}(\mathbf{E}) \\ &= \left(\frac{1}{2} - \frac{1}{v}\right)\left(\sum_i \frac{1}{d_i} - 1\right) + \frac{1}{2v}\left(\sum_i \frac{1}{v_i d_i} - 1\right)\end{aligned} \tag{3.8.10}$$

We calculate $\boldsymbol{\tau}(\mathbf{Q})$ for quotient points.

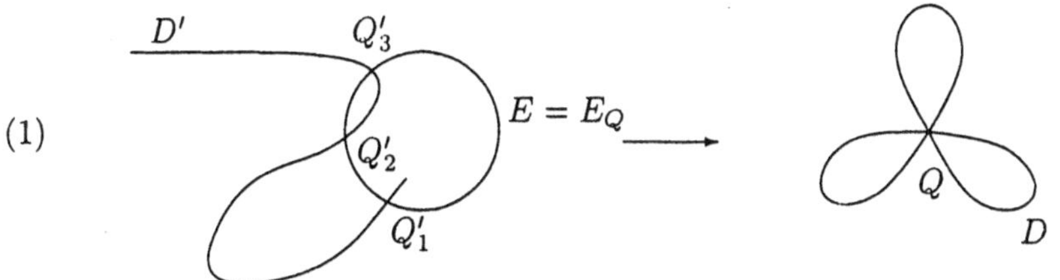

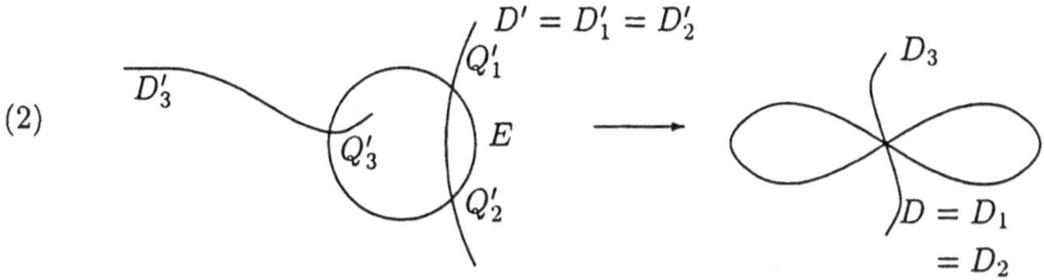

(3)

D'_3 Q'_3 Q'_1 $D' = D'_1$

E

Q'_2 D'_2

$D_1 = D$

D_2

D_3

Figure 3.8.1

Definition 3.8.2 A *quotient point* is a fraction point $\mathbf{Q}$ satisfying the following relation:

$$\mathbf{e}(\mathbf{E}_Q) = -6\boldsymbol{\tau}(\mathbf{E}_Q) \text{ (relative proportionality),} \tag{3.8.11}$$

where $\mathbf{E}_Q \to \mathbf{Q}$ is the central resolution of $\mathbf{Q}$.

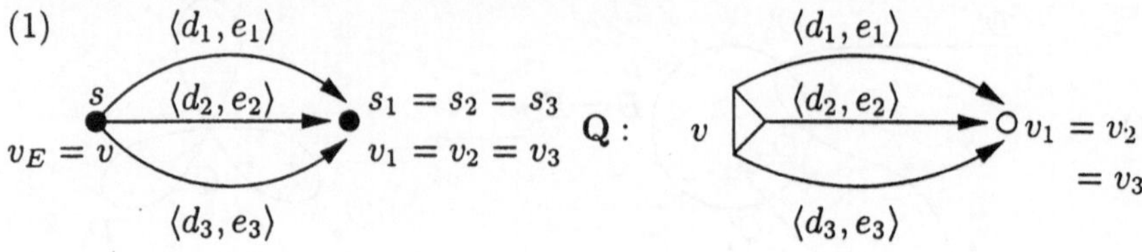

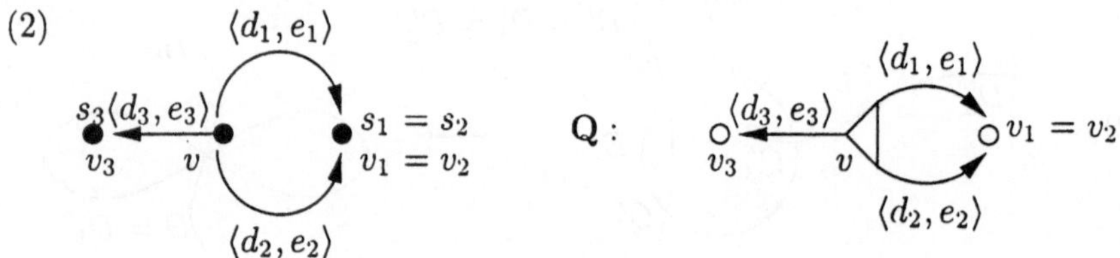

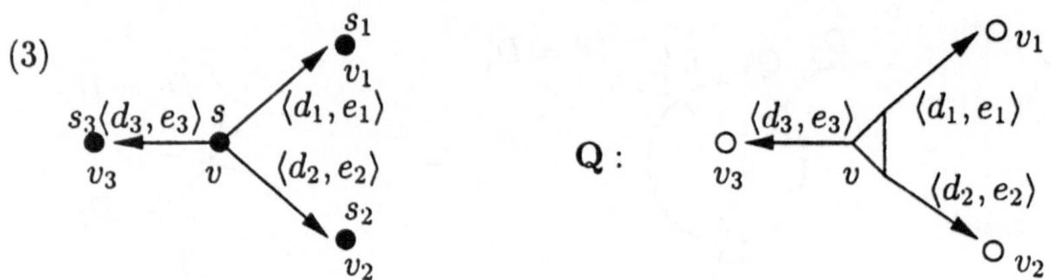

Figure 3.8.2

Example 3.8.3 A quotient point $\mathbf{Q} = O/G, G \subset \mathbb{Gl}_2(\mathbb{C})$ a finite group: This comes from the uniformization $O \underset{G}{\to} \mathbf{Q}$, its central resolution (σ-process at O) $L = E_O \underset{G}{\to} \mathbf{E}_Q = E_O/G$,

$$\mathbf{e}(L) = e(L) = 2 = -2(L^2) = -6\tau(L) \tag{3.8.12}$$

and the degree formulas for $\mathbf{e}, \tau$ in the category of orbital curves.

Now one finds

$$\tau(\mathbf{Q}) = \mathbf{e}(E)/2 - \mathbf{e}(\mathbf{E})/2v - \frac{1}{6}\left(v - \frac{1}{v}\right)\cdot \mathbf{e}(\mathbf{E}) - \frac{1}{3v}\sum_{i=1}^{3}\frac{1}{d_i}\left(v_i - \frac{1}{v_i}\right)$$

$$= \frac{1}{2}\left(\frac{1}{d_1} + \frac{1}{d_2} + \frac{1}{d_3} - 1\right) - \frac{1}{6}\left(v + \frac{2}{v}\right)\mathbf{e}(\mathbf{E}) - \frac{1}{3}\sum_{i=1}^{3}\frac{1}{vd_i}\left(v_i - \frac{1}{v_i}\right)$$

$$= \frac{1}{2}\left(\frac{1}{d_1} + \frac{1}{d_2} + \frac{1}{d_3} - 1\right) - \frac{1}{6}\left(v + \frac{2}{v}\right)\left(\frac{1}{v_1 d_1} + \frac{1}{v_2 d_2} + \frac{1}{v_3 d_3} - 1\right)$$

$$= +\frac{1}{3v}\left(\left(\sum_i \frac{1}{v_i d_i}\right) - 1\right) - \frac{1}{3v}\sum_i \frac{v_i}{d_i}$$

and finally

$$\tau(\mathbf{Q}) = -\frac{1}{2}\left(1 - \sum_{i=1}^{3}\frac{1}{d_i}\right) + \frac{1}{6v}\left(1 - \sum_{i=1}^{3}\frac{1}{v_i d_i}\right) + \frac{1}{3v}\left(1 - \sum_{i=1}^{3}\frac{v_i}{d_i}\right)$$

(3.8.13)

or

$$\tau(\mathbf{Q}) = 3\tau(E) - \frac{1}{v}\tau(\mathbf{E}) + \frac{1}{3v}\left(1 - \sum_{i=1}^{3}\frac{v_i}{d_i}\right).$$

Observe that $\mathbf{e}(Q) = \tau(Q) = 0$.

Proposition 3.8.4 *Let* $\mathbf{Y}$ *be the toroidal compactification of an open (quasi-smooth) orbital surface* $\mathbf{Y}_f$.

For $\mathbf{c} = \mathbf{e}, \tau$ *there exist orbital decompositions*

$$\mathbf{c}(\mathbf{Y}_f) = \mathbf{c}(Y_f) - \sum_{\mathbf{D}_f \subset \mathbf{A}_f^1} \lambda_c(v_{\mathbf{D}})\cdot \mathbf{c}(\mathbf{D}) - \sum_{\mathbf{Q}\in \mathbf{A}^0} \mathbf{c}(\mathbf{Q}) \qquad (3.8.14)$$

Proof According to (3.7.3), (3.6.8), (3.8.2) and (3.8.9) we have

$$\mathbf{c}(\mathbf{Y}_f) = \mathbf{c}(\mathbf{Y}) + \sum_{\kappa\in\widehat{\mathbf{A}}_\infty} \mathbf{c}_\kappa(\mathbf{Y}) = \mathbf{c}(\mathbf{Y}') + \sum_\kappa \mathbf{c}_\kappa(\mathbf{Y}) - \mathbf{c}(\mathbf{Y}'/\mathbf{Y})$$

$$= \mathbf{c}(\mathbf{Y}') - \sum_\kappa \frac{1}{v_\kappa}\mathbf{c}(\mathbf{E}_\kappa) - \sum_{\mathbf{Q}\in\mathbf{A}_3^0} \mathbf{c}_Q(\mathbf{Y}'/\mathbf{Y})$$

$$= \mathbf{c}(Y') - \sum_\kappa \frac{1}{v_\kappa}\mathbf{c}(\mathbf{E}_\kappa) - \sum_{\mathbf{Q}} \lambda_c(v_\kappa)\mathbf{c}(\mathbf{E}_\kappa)$$

$$
\begin{aligned}
&- \sum_{\mathbf{D}_f' \subset \mathbf{A}_f'^1} \lambda_c(\mathbf{D}') \cdot \mathbf{c}(\mathbf{D}') - \sum_{\mathbf{Q}} \lambda_c(\mathbf{E}_Q) \mathbf{c}(\mathbf{E}_Q) \\
&- \sum_{\mathbf{Q}} \mathbf{c}_Q(\mathbf{Y}'/\mathbf{Y}) - \sum_{\mathbf{Q}} \sum_{\mathbf{Q}' \to \mathbf{Q}} \mathbf{c}(\mathbf{Q}') - \sum_{\mathbf{P} \in \mathbf{A}_2'^0} \mathbf{c}(\mathbf{P}) \\
= &\left(\mathbf{c}(Y') - \sum_{\kappa} v_\kappa \cdot \mathbf{c}(\mathbf{E}_\kappa) - \mathbf{c}(Y'/Y) - \sum_{\mathbf{Q}} \mathbf{c}(Q) \right) \\
&+ \sum_{\mathbf{Q}} \mathbf{c}(Q) + \sum_{Q} \mathbf{c}_Q(Y'/Y) - \sum_{\mathbf{C}} \lambda_c(\mathbf{C}) \cdot \mathbf{c}_{\mathbf{Q}}(\mathbf{C}'/\mathbf{C}) \\
&- \sum_{\mathbf{Q} \in \mathbf{D}} \lambda_c(\mathbf{E}_{\mathbf{Q}}) \cdot \mathbf{c}(\mathbf{E}_Q) - \sum_{\mathbf{Q}} \mathbf{c}_Q(\mathbf{Y}'/\mathbf{Y}) - \sum_{\mathbf{Q}'} \mathbf{c}(\mathbf{Q}').
\end{aligned}
$$

Since $\mathbf{c}(Q) = 0$ by (3.8.10), (3.8.13) we obtain with Definition (3.8.9):

$$
\mathbf{c}(\mathbf{Y}_f) = \mathbf{c}(Y_f) - \sum_{\mathbf{D}_f \subset \mathbf{A}_f^1} \lambda_c(\mathbf{D}) \cdot \mathbf{c}(\mathbf{D}) - \sum_{\mathbf{Q} \in \mathbf{A}^0} \mathbf{c}(\mathbf{Q}) \tag{3.8.15}
$$

Surely, one has first to define

$$
\begin{aligned}
\mathbf{c}(Y_f) &= \mathbf{c}(Y') - \sum_{\kappa} v_\kappa \cdot \mathbf{c}(E_\kappa) - \mathbf{c}(Y'/Y) \\
&= \mathbf{c}(Y) - \sum_{\kappa} v_\kappa \cdot \mathbf{c}(E_\kappa),
\end{aligned}
$$

$$
\mathbf{c}(Y_f) = \begin{cases} \mathbf{e}(Y), & \mathbf{c} = \mathbf{e} \\ \tau(Y) - \sum_{\kappa} \tau(E_\kappa), & \mathbf{c} = \tau \end{cases} \tag{3.8.16}
$$

and then to check that (3.8.15) applied to Y_f preserves this definition. This is true because $\mathbf{c}(Q) = 0$ and $\lambda_c(D) = 0$. □

Corollary 3.8.5 *For* $\mathbf{c} = \mathbf{e}, \tau$ *there are orbital decompositions*

$$
\mathbf{c}(\mathbf{Y}_f) = \mathbf{c}(Y_f) - \mathbf{c}(\mathbf{A}_f^1) - \mathbf{c}(\mathbf{A}_f^0) - \sum_{\kappa \in \widehat{\mathbf{Y}}_\infty} \mathbf{c}(\kappa) \tag{3.8.17}
$$

with

$$
\mathbf{c}(\mathbf{A}_f^1) = \sum_{\mathbf{D}_f \subset \mathbf{A}_f^1} \lambda_c(\mathbf{D}) \cdot \mathbf{c}(\mathbf{D}_f), \mathbf{c}(\mathbf{A}_f^0) = \sum_{\mathbf{Q} \in \mathbf{A}_f^0} \mathbf{c}(\mathbf{Q}), \tag{3.8.18}
$$

$$\mathbf{c}(\mathbf{D}_f) = \mathbf{c}(\mathbf{D}) + \sum_{\mathbf{Q}\in\mathbf{A}^0_\infty(\mathbf{C})} \mathbf{c}_{\mathbf{Q}}(\mathbf{D}) \qquad (3.8.19)$$
$$= \mathbf{c}(\mathbf{D}') - \mathbf{c}(\mathbf{D}'/\mathbf{D}) + \sum_{\mathbf{Q}\in\mathbf{A}^0(\mathbf{D}')} \mathbf{c}_Q(\mathbf{D}')$$

$$\mathbf{c}(\mathbf{D}_f) = \begin{cases} e(D') - \sum\limits_{\mathbf{Q}'\in\mathbf{A}^0_f(\mathbf{D}')} \left(1 - \frac{1}{v'_{Q'} d_{Q'}}\right), & \mathbf{c} = \mathbf{e} \\ \frac{1}{3v}\Big[\left(D'^2\right) + \sum\limits_{\mathbf{Q}'\in\mathbf{A}^1_f(\mathbf{D}')} e_{Q'}/d_{Q'} & \mathbf{c} = \tau \\ \quad + \frac{1}{3} \sum\limits_{\mathbf{Q}\in\mathbf{A}^1_3(\mathbf{D})} \sum\limits_{\mathbf{D}'\ni\mathbf{Q}'\to\mathbf{Q}} \frac{1}{v_{\mathbf{Q}}\cdot d_{Q'}}\Big], \end{cases} \qquad (3.8.20)$$

with $\mathbf{Q}' : \underset{v = v_{\mathbf{Q}}}{\circ} \xrightarrow{\langle d_{\mathbf{Q}'}, e_{\mathbf{Q}'}\rangle} \underset{v'_{\mathbf{Q}'}}{\circ}$.

Remark 3.8.6 For practical calculations this is the most important orbital decomposition for orbital heights of open orbital surfaces because one has only to know the orbital curves on the open surface itself. Only in $\mathbf{c}(Y_f)$ appear the compactification curves E_κ. Observe that according to (3.8.16), (3.8.17) and (3.8.18) the weights of the orbital compactification curves do not appear. That is the reason to set them equal to ∞ as we have done in example 3.7.4, see Figure 3.7.6. More precisely, the contribution $\lambda_c(\mathbf{E}_\kappa)\cdot\mathbf{c}(\mathbf{E}_\kappa)$ reduces to

$$\left(v - \frac{1}{\infty}\right)\cdot\mathbf{c}(\mathbf{E}_\kappa) = v\cdot\mathbf{c}(\mathbf{E}_\kappa) = \begin{cases} 0, & \mathbf{c} = \mathbf{e} \\ \tau(E_\kappa), & \mathbf{c} = \tau \end{cases}$$

which has been shifted to $\mathbf{c}(Y_f)$.

Proof of Corollary 3.8.5. We consider cusps as *quadruple points* described in Figure (3.8.3) together with the canonical resolution If κ is of type (3,3,3), (2,4,4) or (2,3,6) (see Figure (3.5.3)), then we added a curve germ D_4 through κ which intersects E_κ transversally in a non-singular surface point Q_4. The corresponding graphs are described in Figure (3.8.4): With the notations of the picture and graphs we calculate

$$\mathbf{c}(\kappa) = \sum_{i=1}^{4}\mathbf{c}(\mathbf{Q}_i) - \sum_{i=1}^{4}\lambda_c(\mathbf{D}_i)\cdot\mathbf{c}_{Q_i}(\mathbf{D}_i) \qquad (3.8.21)$$

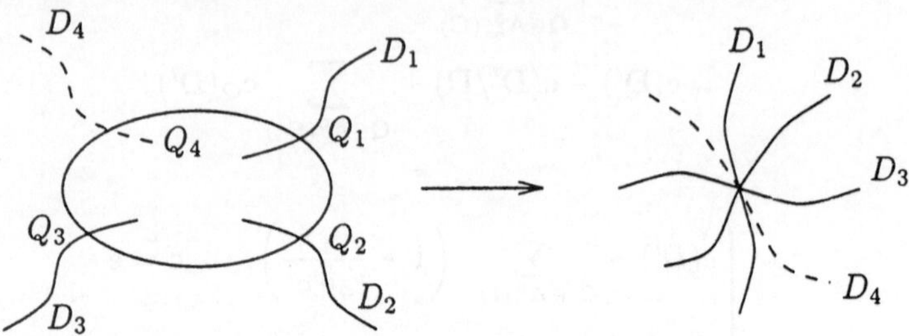

Figure 3.8.3

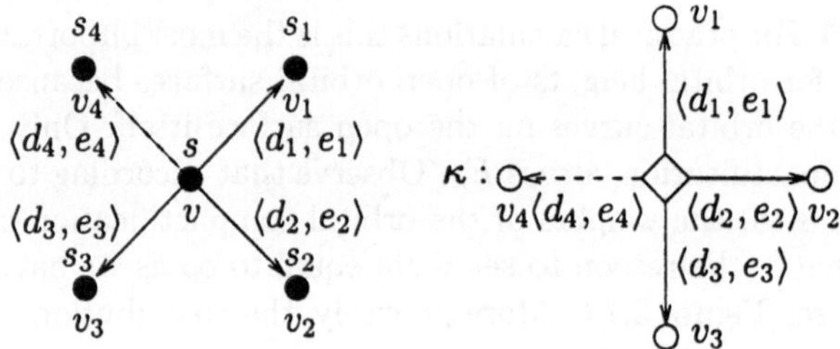

Figure 3.8.4

for $\mathbf{c} = \mathbf{e}, \boldsymbol{\tau}$. This expressions are to read as

$$\begin{aligned}
\mathbf{e}(\boldsymbol{\kappa}) &= \sum_{i=1}^{4} \frac{1}{d_i}\Big(1-\frac{1}{v}\Big)\Big(1-\frac{1}{v_i}\Big) - \sum_{i=1}^{4}\Big(1-\frac{1}{v_i}\Big)\Big(1-\frac{1}{vd_i}\Big) \\
&= \sum_i \frac{1}{d_i}\Big(1-\frac{1}{v_i}\Big) - \sum_i \Big(1-\frac{1}{v_i}\Big),
\end{aligned}$$

$$\mathbf{e}(\boldsymbol{\kappa}) = -\sum_{i=1}^{4}\Big(1-\frac{1}{v_i}\Big)\Big(1-\frac{1}{d_i}\Big), \tag{3.8.22}$$

and

$$\tau(\kappa) = 0 + \sum_i \left(v_i - \frac{1}{v_i}\right)\frac{e_i'}{d_i v_i}, \quad \left(e_i' = e_i \text{ at cusps}\right)$$

$$\tau(\kappa) = \sum_{i=1}^{4}\left(1 - \frac{1}{v_i^2}\right)\frac{e_i'}{d_i}. \tag{3.8.23}$$

Comparing (3.8.14) and (3.8.17) we have to prove that

$$\sum_{\mathbf{D}_f \subset \mathbf{A}_f^1} \lambda_c(v_{\mathbf{D}}) \cdot \mathbf{c}(\mathbf{D}) + \sum_{\mathbf{Q}\in\mathbf{A}^0} \mathbf{c}(\mathbf{Q}) = \mathbf{c}(\mathbf{A}_f^1) + \mathbf{c}(\mathbf{A}_f^0) + \sum_{\kappa\in\widehat{\mathbf{Y}}_\infty} \mathbf{c}(\kappa) \tag{3.8.24}$$

The sum $\sum_{\mathbf{Q}\in\mathbf{A}^0}$ splits into $\sum_{\mathbf{Q}\in\mathbf{A}_f^0} + \sum_{\kappa\in\mathbf{A}_\infty^0}$. The other splitting we need has been described in (3.8.19). With Definition (3.8.21) we see that the identity (3.8.24) holds. The corollary is proved.

□

Example 3.8.7 We turn back to our example 3.7.4. The information given in the tetrahedral graph and the CHERN numbers of the basic surface $\mathbb{P}^2$ are sufficient for the calculation of the orbital heights of the open orbital surface $\tilde{\mathbf{P}}_f^2 = \mathbf{P}_f^2$. According to (3.8.17) we get

$$\tau(\mathbf{P}_f^2) = \left[-3 - 4\cdot\frac{1}{3}(-1)\right] - 6\left[\left(3 - \frac{1}{3}\right)\frac{1}{3\cdot 3}(-1)\right] - 0 - 0 = \frac{1}{9}$$

$$\mathbf{e}(\mathbf{P}_f^2) = [7 - 0] - 6\left[\left(1 - \frac{1}{3}\right)\left(2 - \left(1 - \frac{1}{3}\right)\right)\right] - 3\left[\left(1 - \frac{1}{3}\right)\left(1 - \frac{1}{3}\right)\right] - 0 = \frac{1}{3}.$$

Tables 3.A Invariants of orbital curves with (at most) abelian (singular) points

3.A.0 Local invariants

c^1	$c^1(\mathbf{Q})$	$c^1(Q)$	$\mathbf{c}^1(\mathbf{Q}) = c^1(\mathbf{Q}) - c^1(Q)$
e^1	$1 - \frac{1}{v'd}$	$1 - \frac{1}{d}$	$\frac{1}{d}\left(1 - \frac{1}{v'}\right)$
τ^1	$-\frac{f}{3vd}$	$-\frac{f}{3d}$	$\frac{f}{3d}\left(1 - \frac{1}{v}\right)$

$\mathbf{Q}_i = \mathbf{Q} : \underset{v\ \langle d,f\rangle\ v'}{\circ\!\!\longrightarrow\!\!\circ}$ abelian point on orbital curve

$\mathbf{C} = (U_C, vC; \mathbf{Q}_1 + \ldots \mathbf{Q}_k)$

3.A.1 Global invariants

c	$c(\mathbf{C})$	$h_c(\mathbf{C}) = \mathbf{c}(\mathbf{C})$	$h_c(C) = \mathbf{c}(C)$
e	$e(C)$	$e(C) - \sum_{i=1}^{k}\left(1 - \frac{1}{v_i d_i}\right)$	$e(C) - \sum_{i=1}^{k}\left(1 - \frac{1}{d_i}\right)$
τ	$\frac{1}{3v}(C^2)$	$\frac{1}{3v}\left((C^2) + \sum_{i=1}^{k}\frac{f_i}{d_i}\right)$	$\frac{1}{3}\left((C^2) + \sum_{i=1}^{k}\frac{f_i}{d_i}\right)$

$\mathbf{Q}_i : \underset{v\ \langle d_i,f_i\rangle\ v'}{\circ\!\!\longrightarrow\!\!\circ}$, (C^2) selfintersecting index on minimal singularity resolution (of U_C)

Table 3.B Invariants of abelian points on orbital surfaces

c	$c(\mathbf{P})$	$\mathrm{c}(\mathbf{P})$
e	$-1+\frac{1}{v}+\frac{1}{v'}-\frac{1}{vv'd}$	$\frac{1}{d}\left(1-\frac{1}{v}\right)\left(1-\frac{1}{v'}\right)$
τ	$\frac{1}{3}\left(3l_P+Tr(P)-\frac{f}{v^2d}-\frac{f'}{v'^2d}\right)$ $\Vert$ $-4S(P)+\frac{f}{3d}\left(1-\frac{1}{v^2}\right)+\frac{f'}{3d}\left(1-\frac{1}{v'^2}\right)$	0

$\mathbf{P}=(U_P,P;vC,v'C')$: $\underset{v\ \langle d,f\rangle\ v'}{\circ\!\!\longrightarrow\!\!\circ}$

l_P = number of components of the minimal singularity resolution of P

$Tr(P)$ = sum of selfintersections of these components

$S(P) = -\frac{1}{12}\left(3l_p+Tr(P)-\frac{f}{d}-\frac{f'}{d}\right)$
(Eisenstein-Dedekind-Hirzebruch sum)

4 Ball Quotient Surfaces

4.1 Ball Lattices

We denote by

$$\mathbb{B} = \left\{(z_1, z_2) \in \mathbb{C}^2; |z_1|^2 + |z_2|^2 < 1\right\} \tag{4.1.1}$$

the two-dimensional complex unit ball. Up to biholomorphic equivalance it is the only irreducible symmetric domain of complex dimension 2. Its group of biholomorphic automorphisms is the projective group $\mathbb{P}U((2,1),\mathbb{C}) = \mathbb{P}SU((2,1),\mathbb{C})$ acting on $\mathbb{B}$ by fractional linear transformations. With obvious notations the corresponding (special) unitary group is defined by

$$\begin{aligned} \mathbb{U}((4.2.1),\mathbb{C}) &:= \{A \in \mathbb{G}\mathrm{l}_3(\mathbb{C});{}^t\overline{A} \cdot \mathrm{diag}(1,1,-1) \cdot A = \mathrm{diag}(1,1,-1)\}, \\ \mathbb{S}U(2,1),\mathbb{C} &:= \mathbb{U}((2,1),\mathbb{C}) \cap \mathbb{S}\mathrm{l}_3(\mathbb{C}). \end{aligned} \tag{4.1.2}$$

All these real Lie groups $\mathbb{U}((2,1),\mathbb{C}), \mathbb{S}U((2,1),\mathbb{C}), \mathbb{P}SU((2,1),\mathbb{C}) = \mathbb{P}U((2,1),\mathbb{C})$ act transitively on $\mathbb{B}$. We prefer mainly to work with $\mathbf{G} = \mathbf{G}_{\mathbb{R}} = \mathbb{S}U(2,1)$. Then $\mathbf{G}(\mathbb{R}) = \mathbb{S}U((2,1),\mathbb{C})$ is a simple simply-connected Lie group, see [HEL]. Basic definitions and results can also be transferred to the other Lie groups in (4.1.2). The ball $\mathbb{B}$ can be identified with the space of cosets $\mathbf{G}(\mathbb{R})/\mathbf{K}$, where $\mathbf{K} = \mathbb{S}(\mathbb{U}(2) \times \mathbb{U}(1))$ is the maximal subgroup of $\mathbf{G}(\mathbb{R})$ stabilizing the zero point $O = (0,0) \in \mathbb{B}$.

Remark 4.1.1 In order to be more flexible in applications we remark that the notation $\mathbb{U}((2,1),\mathbb{C})$ can be more generally used for the invariant (unitary) group of a hermitian vector space $(V, \langle\ ,\ \rangle)$ with $\dim_{\mathbb{C}}(V) = 3$ and a hermitian form $\langle\ ,\ \rangle$ of signature (2,1). The ball $\mathbb{B}$ appears as subspace

$$\mathbb{B} = \mathbb{P}\{\mathbb{C}v \subset V; \langle v,v\rangle < 0\} \subset \mathbb{P}V \cong \mathbb{P}^2(\mathbb{C})$$

of all complex lines in V generated by a "negative" vector v. The group $\mathbb{U}((2,1),\mathbb{C})$ acts on $\mathbb{B}$ via the natural composition

$$\begin{aligned} \mathbb{U}((2,1),\mathbb{C}) \subset \mathbb{G}\mathrm{l}_3(V) \longrightarrow \mathbb{P}G\mathrm{l}(V) = \mathrm{Aut}_{\mathrm{hol}}(\mathbb{P}V) \cong \\ \mathbb{P}G\mathrm{l}_3(\mathbb{C}) \cong \mathrm{Aut}_{\mathrm{hol}}\left(\mathbb{P}^2(\mathbb{C})\right). \end{aligned}$$

Let Γ be a discrete subgroup of $\mathbf{G}(\mathbb{R})$. It is equivalent to say that Γ acts *properly discontinuously* on $\mathbb{B}$, this means that the set $\{\gamma\in\Gamma;\gamma(M)\cap L\neq\emptyset\}$ is finite for all compact subsets L, M of $\mathbb{B}$. Especially, it holds that

4.1.2 *for each $P \in \mathbb{B}$ the stationary groups (isotropy groups)*

$$\Gamma_P := \{\gamma \in \Gamma; \gamma P = P\}$$

are finite.

Furthermore,

4.1.3 *there exists a small analytic neighbourhood U of P in $\mathbb{B}$ such that*

$$\{\gamma \in \Gamma; \gamma(U) \cap U \neq \emptyset\} = \Gamma_P.$$

Moreover, for any $Q \in \Gamma P \backslash \{P\}$ there exist analytic neighbourhoods U, V of P or Q, respectively, in $\mathbb{B}$ such that $\gamma(U) \cap V = \emptyset$ for all $\gamma \in \Gamma$. For the proof of these properties we refer to [SH4],I, § 1.1.

Now it is easy to see that the (smooth) complex analytic structure of the ball $\mathbb{B}$ goes down to a normal complex analytic structure on the quotient space $\Gamma\backslash\mathbb{B}$ with at most quotient singularities. The discrete subgroup Γ of $\mathbf{G}(\mathbb{R})$ is called *cocompact* , if the quotient space $\Gamma\backslash\mathbb{B}$ is compact. By a theorem of Borel [BO1], see 4.1.8 below, there exists a *cofinite* normal subgroup Γ' of Γ (that means Γ/Γ' is finite) acting *freely* (that means fixed point free) on $\mathbb{B}$. By means of automorphic forms the smooth compact complex surface $\Gamma'\backslash\mathbb{B}$ can be embedded into a complex projective space $\mathbb{P}^n(\mathbb{C})$, see [SC2] IX.3.2. Then also $\Gamma\backslash\mathbb{B}$ is a (normal compact complex) algebraic surface because it is the finite quotient of the smooth algebraic surface $\Gamma\backslash\mathbb{B}$ by the finite group $G := \Gamma/\Gamma'$, see e.g. [MU4]. On this place we change over to write $\mathbb{B}/\Gamma$ instead of the left quotient $\Gamma\backslash\mathbb{B}$ because finite quotients of surfaces and their branch loci play the main role in our monograph and we decided to write them as right quotients in the previous and following chapters. So we can write $\mathbb{B}/\Gamma = (\mathbb{B}/\Gamma)/G$ instead of $(\Gamma \setminus \mathbb{B})/G$. More confusions will be avoided on this way, if one considers towers of finite quotients.

Definition 4.1.4 We call a discrete subgroup Γ of $\mathbf{G}(\mathbb{R})$ a *ball lattice* , if the following properties are satisfied:

(1) the volume of a Γ-fundamental domain with respect to the $\mathbf{G}(\mathbb{R})$ -invariant hermitian (Bergmann) metric on $\mathbb{B}$, which is uniquely determined up to a nontrivial constant factor, is finite;

(2) the quotient surface $\mathbb{B}/\Gamma$ can be compactified by means of finitely man cusp singularities to a (normal complex compact) algebraic surface $\widehat{\mathbb{B}/\Gamma}$.

For the construction of BERGMANN metrics and "nice" fundamental domains we refer to [HEM], [P-S] or [RAG].

In general one defines a *lattice* Γ of a local compact group G as a discrete subgroup of G such that on the coset space G/Γ exists a finite G-invariant measure (see [RAG], 1.8). Since the invariant HAAR measure is unique up to a constant factor and $\mathbb{B}$ differs from our $G = \mathbf{G}(\mathbb{R}) = \mathbb{S}U((2,1),\mathbb{C})$ only by a compact subgroup, the condition (1) is the general lattice condition. Sublattices of $\mathbf{G}(\mathbb{R})$ in this general sense will be called $\mathbf{G}$-lattices.

The k-rank of an algebraic group defined over the field k is the dimension of a maximal k-defined subtorus (abelian LIE subgroup). Our group $\mathbf{G} = \mathbb{S}U(2,1)$ is simple of $\mathbb{R}$ -rank 1. Therefore, each lattice Γ in $\mathbf{G}(\mathbb{R})$ is irreducible, hence a *discrete subgroup of rank 1* in the sense of RAGHUNATHAN. For the definitions and result we refer to RAGUNATHAN'S book [RAG], especially to the definitions 13.2, 13.7, 13.11 and Theorem 13.19. Lattices Γ of rank 1 of connected simple linear groups, especially ball lattices, have the following properties:

(3) Γ is finitely generated.

(4) Γ contains a torsion free normal subgroup of finite index.

(5) Up to Γ-conjugation the lattice Γ has only finitely many maximal unipotent subgroups.

(6) If U_Γ is a maximal unipotent subgroup of Γ and U its ZARISKI-closure, then $U/U \cap \Gamma$ is compact.

For the proofs we refer to [RAG], Cor. 13.15, Cor. 13.16 (better 6.11), Theorem 13.3 and Theorem 2.1, respectively.

We are mainly interested in the following

Examples 4.1.5 Each cocompact discrete subgroup of $\mathbf{G}(\mathbb{R})$ is a ball lattice. Also arithmetic subgroups of $\mathbf{G}(\mathbb{R})$ are ball lattices. For a general definition and proof we refer to [RAG], 10.12. We concentrate our attention to the following principal case. Let $\mathbf{G}_\mathbb{Q}$ be a $\mathbb{Q}$ -form of $\mathbf{G}$, this means that $\mathbf{G}_\mathbb{Q}$ is defined over $\mathbb{Q}$ and after the extension $\mathbb{R}/\mathbb{Q}$ of definition fields in the sense of algebraic geometry one gets an isomorphism $\mathbb{R} \times \mathbf{G}_\mathbb{Q} \cong \mathbf{G} = \mathbb{S}U((2,1),\mathbb{C})$. Then the group $\mathbf{G}_\mathbb{Q}(\mathbb{Z})$ is an *arithmetic group* but also the subgroups Γ of $\mathbb{S}U((2,1),\mathbb{R})$ *commensurable* with $\mathbf{G}_\mathbb{Q}(\mathbb{Z})$, which means that the intersection $\Gamma \cap \mathbf{G}_{\mathbb{Q}}(\mathbb{Z})$ is a cofinite subgroup as well in Γ as in $\mathbf{G}_{\mathbb{Q}}(\mathbb{Z})$. Infinitely many typical examples one gets on the following way: Let $K = \mathbb{Q}(\sqrt{-d})$ be an imaginary quadratic number field, d a square-free positive integer. Then there exists a $\mathbb{Q}$ -defined algebraic group $\mathbf{G}_{\mathbb{Q}}$ such that $\mathbf{G}_{\mathbb{Q}}(\mathbb{Q}) = \mathbb{S}U((2,1),K)$. The ball lattices $\mathbb{S}U((2,1),\mathfrak{O}_K), \mathfrak{O}_K$ the ring of integers in K, are called (special) PICARD *modular groups*. Their quotient surfaces — the (special) PICARD *modular surfaces* — will be investigated in more detail in chapter 5. The PICARD modular groups are not cocompact.

Remark. By a result of MOSTOW [D-M] there exist non-arithmetic ball lattices not only of cocompact type but also of non-compact type.

Definition 4.1.6 A ball lattice $\Gamma \subset \mathbb{U}((2,1),\mathbb{C})$ is called *neat,* if and only if the eigenvalues of each element $\gamma \in \Gamma$ generate a torsionfree subgroup of $\mathbb{C}^*$. If Γ is a neat ball lattice, then the surface $\mathbb{B}/\Gamma$ and also its compactification $\widehat{\mathbb{B}/\Gamma}$ is called a *neat ball quotient surface.*

4.1.7 *Each neat ball lattice Γ acts freely (without fixed points) on $\mathbb{B}$, and the corresponding neat (in general non-compact) ball quotient surface $\mathbb{B}/\Gamma$ is smooth.*

Proof Assume that there is a fixed point $P \in \mathbb{B}$ of an element $1 \neq \gamma \in \Gamma$. The order of γ is finite by 4.1.2. Therefore, the eigenvalues of γ are unit roots. One of them is not trivial. This is a torsion element of $\mathbb{C}^*$, hence Γ is not neat in contradiction to our assumption. □

From 4.1.3 it follows that $\mathbb{B}/\Gamma$ is locally isomorphic to $\mathbb{B}$. Indeed, $\mathbb{B}$ is the universal covering of $\mathbb{B}/\Gamma$. Therefore the quotient surface is smooth in the neat case.

A basic role for the calculation of CHERN invariants of ball quotient surfaces plays the following

Proposition 4.1.8 *(*BOREL, *[BO1]) Each ball lattice Γ contains a neat normal sublattice Γ' of finite index.*

Proof Since Γ is finitely generated by (3) and a subgroup of $\mathbb{G}\mathrm{l}_3(\mathbb{C})$, this follows from [RAG], Theorem 6.11. □

Corollary 4.1.9 *The ball quotient surface $\mathbb{B}/\Gamma$ is quasismooth.*

Proof With the above notations the surface $\mathbb{B}/\Gamma'$ is a smooth finite covering of $\mathbb{B}/\Gamma$ with GALOIS group $G := \Gamma/\Gamma'$. therefore $\mathbb{B}/\Gamma = (\mathbb{B}/\Gamma')/G$ supports at most quotient singularities, but this means that it is quasismooth. □

4.2 Neat Ball Cusp Lattices

If Γ is a ball lattice of non-compact type, then we look for uniformizations of small open analytic punctured neighbourhoods $\widehat{U} \setminus \{\widehat{\kappa}\}$ around cusp singularities $\widehat{\kappa}$ (see 4.1.4,(2)) on $\widehat{\mathbb{B}/\Gamma}$. As first step we look for the universal covering and the fundamental group $\pi_1\left(\widehat{U} \setminus \{\widehat{\kappa}\}\right)$ in the case of a neat lattice Γ. It turns out that this can be done by the study of neighbourhoods of special boundary points of the ball $\mathbb{B}$. From classifying numerical invariants of these local fundamental groups we will derive cusp contributions for the calculation of CHERN invariants of the minimal smooth compactification $\overline{\mathbb{B}/\Gamma}$ of $\mathbb{B}/\Gamma$.

The set of boundary points of the ball $\mathbb{B}$ is denoted by $\partial\mathbb{B}$. There are biunivoque correspondences with the set $\mathcal{P} = \mathcal{P}_{\mathbb{R}}(\mathbf{G})$ of minimal parabolic $\mathbb{R}$ -defined subgroups of $\mathbf{G}$ (BOREL subgroups) and with the set $\mathcal{U} = \mathcal{U}_{\mathbb{R}}(\mathbf{G})$ of maximal unipotent $\mathbb{R}$ -defined subgroups of $\mathbf{G}$. The LIE group $\mathbf{G}(\mathbb{R})$ acts transitively on $\mathcal{P}$ and $\mathcal{U}$ via conjugation by their elements. For these statements we refer to the review chapter XII of [RAG]. More precisely we dispose on the following realizations of the bijective correspondences:

$$\begin{array}{ccccc} \mathcal{U} & \longleftrightarrow & \mathcal{P} & \longleftrightarrow & \partial\mathbb{B} \\ \mathbf{U} & \longleftarrow\!\!\!\!\mapsto & \mathbf{P} & \longleftarrow\!\!\!\!\mapsto & \kappa \end{array}$$

$$\begin{aligned} \mathbf{P} &= N_G(\mathbf{U}) = \text{ normalizer of } \mathbf{U} \text{ in } \mathbf{G} : \\ \mathbf{P}(\mathbb{R}) &= \{g \in \mathbf{G}(\mathbb{R});\quad g\mathbf{U}(\mathbb{R})g^{-1} = \mathbf{U}(\mathbb{R})\} ; \\ \mathbf{U} &= \mathbf{U}(\mathbf{P}) = \text{ unipotent radical of } \mathbf{P} : \\ \mathbf{U}(\mathbb{R}) &= \{u \in \mathbf{P}(\mathbb{R});\ u \text{ unipotent } \} ; \\ \kappa &= \kappa(\mathbf{P}) \\ &= \kappa(\mathbf{U}) : \text{ fixed point of all elements of } \mathbf{P}(\mathbb{R}) \text{ (or } \mathbf{U}(\mathbb{R})); \\ \mathbf{P}_\kappa &= \text{ stabilizer of } \kappa : \\ \mathbf{P}_\kappa(\mathbb{R}) &= \{g \in \mathbf{G}(\mathbb{R});\ g\kappa = \kappa\};\quad \mathbf{U}_\kappa = \mathbf{U}(\mathbf{P}_\kappa).4.2.1 \end{aligned} \tag{4.2.1}$$

Two maximal unipotent subgroups cannot have the same boundary fixed point because of the following

Lemma 4.2.1 (RAGUNATHAN, [RAG], 12.15.) *If* $\mathbf{G}$ *is a connected semi-simple algebraic group of k-rang 1 over a field k and* $\mathbf{U}$, $\mathbf{U}'$ *are two maximal unipotent k-subgroups of* $\mathbf{G}$, *then* $\mathbf{U} = \mathbf{U}'$ *or* $\mathbf{U} \cap \mathbf{U}' = \mathbf{1}$, *where* $\mathbf{1}$ *is the trivial group.*

□

The surjective property in (4.2.1) comes from the transitive action of $\mathbf{G}(\mathbb{R})$ on $\partial\mathbb{B}$, $\mathcal{U}$ or $\mathcal{P}$, respectively, and from the study of the situation at a special boundary point. For this purpose we use coordinates and move the ball $\mathbb{B}$ in $\mathbb{P}^2(\mathbb{C})$ to the unbounded SIEGEL domain

$$\mathbb{V} := \left\{(z,u) \in \mathbb{C}^2;\ 2 \cdot \operatorname{Im} z - |u|^2 > 0\right\}, \tag{4.2.2}$$

for example by the linear transformation

$$g : \begin{pmatrix} z_1 \\ z_2 \\ 1 \end{pmatrix} \mapsto \frac{1}{\sqrt{2}} \begin{pmatrix} 1 & 0 & 1 \\ 0 & \sqrt{-2} & 0 \\ i & 0 & -i \end{pmatrix} \begin{pmatrix} z_1 \\ z_2 \\ 1 \end{pmatrix} =: \begin{pmatrix} z \\ u \\ 1 \end{pmatrix}$$

sending the $\mathbb{B}$ -boundary point $\kappa_0 := (1,0) = \mathbb{P}^t(1,0,0)$ to the $\mathbb{V}$ -boundary point $\infty := {}^t(1,0,0)$. The above statements are correct also for $\mathbb{V}$ and its boudary $\partial\mathbb{V}$ instead of $\mathbb{B}, \partial\mathbb{B}$. The group $g\mathbf{G}g^{-1}(\mathbb{R})$ acting on $\mathbb{V}$ is denoted by G.

The parabolic subgroup $P_\infty = \mathbf{P}_\infty(\mathbb{R})$ of G has the

4.2.2 LANGLANDS *decomposition* $P_\infty = A_\infty \cdot M_\infty \cdot U_\infty$, where

$$A_\infty = \left\{ \begin{pmatrix} r & 0 & 0 \\ 0 & 1 & 0 \\ 0 & 0 & r^{-1} \end{pmatrix}; r \in \mathbb{R}_+ \right\},$$

$$M_\infty = \left\{ \begin{pmatrix} \tau & 0 & 0 \\ 0 & \tau^{-2} & 0 \\ 0 & 0 & \tau \end{pmatrix}; \tau \in \mathbb{C}, |\tau| = 1 \right\},$$

$$U_\infty = \left\{ \begin{pmatrix} 1 & i\bar{a} & i|a|^2/2 + r \\ 0 & 1 & a \\ 0 & 0 & 1 \end{pmatrix} =: [a,r]; a \in \mathbb{C}, r \in \mathbb{R} \right\}.$$

Especially, we see that all our unitary groups U have real dimension 3, and we have exact sequences

$$1 \longrightarrow \Delta(U) := [U,U] \cong \mathbb{R} \longrightarrow U \longrightarrow \mathbb{C} \cong U/[U,U] \longrightarrow 1. \tag{4.2.3}$$

For $U = U_\infty$ we get more explicitly

$$\begin{array}{ccccccc} 1 \longrightarrow \Delta_\infty = \{[0,r];\ r \in \mathbb{R}\} \longrightarrow U_\infty & \longrightarrow & \mathbb{C} & \longrightarrow & 1 \\ [a,r] & \mapsto & a & & \end{array} \qquad (4.2.4)$$

and the rule

$$\begin{aligned} [a,r]\cdot[b,s] &= \left[a+b, r+s+i(\bar{a}b-\bar{b}a)/2\right]4.2 \qquad (4.2.5) \\ &= \left[a+b, r+s-\operatorname{Im}\,\bar{a}b\right] \end{aligned}$$

from which follows that

4.2.3 *$\Delta(U)$ is the center of U.*

Now we change over to ball lattices Γ and their maximal unipotent subgroups.

Definition 4.2.4 Let $U = U_\kappa$ be the (maximal) unipotent subgroup of G associated to the ball cusp κ. The subgroup Γ'_κ of U is called a *neat ball cusp lattice* , if Γ'_κ is a sublattice of U or, equivalently, if U/Γ'_κ is compact. The discrete subgroup Γ_κ of $P = P_\kappa$ is a *ball cusp lattice* , if $\Gamma'_\kappa := U \cap \Gamma_\kappa$ is a neat ball cusp lattice.

Definition 4.2.5 Let Γ be a lattice in $\mathbf{G}(\mathbb{R})$. Then $\kappa \in \partial\mathbb{B}$ is a Γ-*cusp*, if $\Gamma_\kappa := P_\kappa \cap \Gamma$ is a ball cusp lattice.

Notation 4.2.6 The set of Γ-cusps is denoted by $\partial_\Gamma\mathbb{B}$. By section 1,(5) the Γ-conjugation class $\partial_\Gamma\mathbb{B}/\Gamma$ is a finite set called the *cusp number* $h(\Gamma)$ of Γ.

4.2.7 *Non-trivial maximal unipotent subgroups of an arithmetic ball lattice $\Gamma \subset \mathbf{G}_\mathbb{Q}(\mathbb{Q})$ are neat ball cusp lattices.*

Proof The Baily-Borel compactification $\widehat{\mathbb{B}/\Gamma}$ is realized by finitely many cusp singularities coming from ball cusp lattices $\Gamma_\kappa = \Gamma \cap \mathbf{P}(\mathbb{Q})$ of $\mathbb{Q}$-defined minimal parabolic (Borel) subgroups $\mathbf{P} = \mathbf{P}_\kappa$ of $\mathbf{G}_\mathbb{Q}$ (see [B-B]). It is easy to see that Γ_κ is commensurable with the sublattice $\mathbf{U}_\kappa(\mathbb{Z})$ of the $\mathbb{Q}$-defined (maximal) unipotent subgroup $\mathbf{U}_\kappa$ of $\mathbf{P}_\kappa$ (see [RAG], 13.1). □

Remark 4.2.8 In the case of arithmetic ball lattices as above Γ-cusps are also called *rational boundary points* of the $\mathbb{Q}$-defined ball group $\mathbf{G}_{\mathbb{Q}}$ because their set only depends on $\mathbf{G}_{\mathbb{Q}}$. If $\mathbf{G}_{\mathbb{Q}}$ is fixed, then this set is sometimes denoted by $\partial_{\mathbb{Q}}\mathbb{B}$. Denoting the set of $\mathbb{Q}$-defined BOREL subgroups or maximal unipotent subgroups by $\mathcal{P}_{\mathbb{Q}}$ or $\mathcal{U}_{\mathbb{Q}}$, respectively, we can extend the correspondences (4.2.1) to a commutative diagram (4.2.6) in the category of sets.

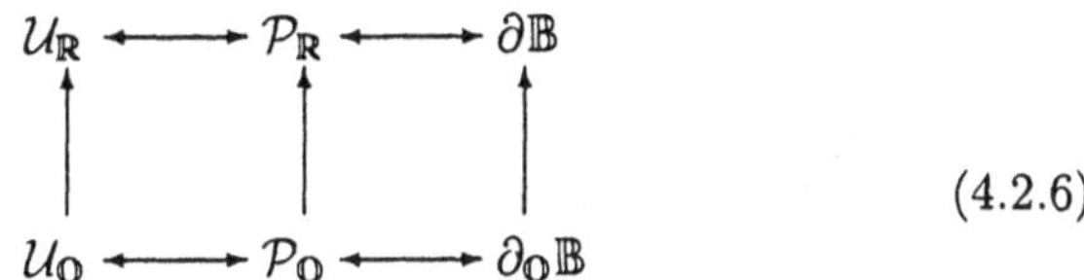

(4.2.6)

Now let Γ_κ be a neat ball lattice. Then the sequence (4.2.3) extends to the commutative diagram (4.2.7) of group homomorphisms.

$$\begin{array}{ccccccccc} 0 & \longrightarrow & \mathbb{R} & \longrightarrow & U_\kappa & \longrightarrow & \mathbb{C} & \longrightarrow & 0 \\ & & \uparrow & & \uparrow & & \uparrow & & \\ 0 & \longrightarrow & \Delta_\kappa & \longrightarrow & \Gamma_\kappa & \longrightarrow & \Lambda_\kappa & \longrightarrow & 0 \end{array} \tag{4.2.7}$$

where $\Delta_\kappa \cong \mathbb{Z}$ and $\Lambda_\kappa \cong \mathbb{Z}^2$ are lattices in the additive groups of $\mathbb{R}$ or $\mathbb{C}$, respectively.

We restrict our attention now to $\kappa = \infty$. Coming from a ball lattice, Γ_κ acts on the unbounded domain $\mathbb{V} \subset \mathbb{C}^2$ defined in (4.2.2). By 4.2.2 the action extends linearly to $\mathbb{C}^2$. In two steps we factorize $\mathbb{C}^2$ first by Δ_κ and then the quotient by Λ_κ. The group Δ_κ acts on the first factor of $\mathbb{C} \times \mathbb{C}$ by translations. Therefore $\mathbb{C}^2/\Delta_\kappa \cong \mathbb{C}^* \times \mathbb{C} \subset \mathbb{C} \times \mathbb{C}$. We will see that the action of Λ_κ on $\mathbb{C}^* \times \mathbb{C}$ extends to $\mathbb{C} \times \mathbb{C}$. On the second factor it acts by translation (see 4.2.2). Therefore $F_\kappa = F_\kappa(\Gamma) = F(\Gamma_\kappa) := \mathbb{C}^2/\Lambda_\kappa$ is a line bundle F_κ/T_κ over an elliptic curve $T_\kappa = T_\kappa(\Gamma) = T(\Gamma_\kappa) := \mathbb{C}/\Lambda_\kappa$. The situation is described in diagram:

$$\begin{array}{ccccc} & & \mathbb{C} \times \mathbb{C} & \longrightarrow & F_\kappa = \mathbb{C} \times \mathbb{C}/\Lambda_\kappa \\ & & \uparrow & & \uparrow \\ \mathbb{C} \times \mathbb{C} & \longrightarrow & \mathbb{C}^* \times \mathbb{C} & & \overline{\mathbb{V}/\Gamma_\kappa} \\ \uparrow & & \uparrow & & \uparrow \\ \mathbb{V} & \longrightarrow & \mathbb{V}/\Delta_\kappa & \longrightarrow & \mathbb{V}/\Gamma_\kappa \end{array} \tag{4.2.8}$$

On this way we can find explicitly the (local) toroidal compactification of $\mathbb{V}/\Gamma_\kappa$ corresponding to the cusp κ, filling in the elliptic curve T_κ as image of $(0 \times \mathbb{C})$ along the Λ_κ-quotient morphism. The quotient $\mathbb{V}/\Gamma_\kappa$ appears as bundle of punctured discs in F_κ around (over) the zero section T_κ. The (Baily-Borel) point compactification $\widehat{\mathbb{V}/\Gamma_\kappa} = \mathbb{V}/\Gamma_\kappa \cup \{\widehat{\kappa}\}$ is received by contracting T_κ to the cusp point $\widehat{\kappa}$. In order to be more precise we define Δ_κ-invariant neighbourhoods

$$\mathbb{V}_C = \left\{(z,u) \in \mathbb{C}^2;\ 2\,\mathrm{Im}\,z - |u|^2 > C\right\}.$$

For $C \gg 0$ the subgroup of a ball lattice Γ acting on $\mathbb{V}_C$ is nothing else but Γ_κ; for elements $1 \neq \gamma \in \Gamma$ it holds that $\gamma\mathbb{V}_C \cap \mathbb{V}_C = \emptyset$. Therefore $\mathbb{V}_C/\Gamma_\kappa$ is an open analytic neighbourhood of the cusp singularity $\widehat{\kappa} \in \widehat{\mathbb{V}/\Gamma}$. The diagram (4.2.8) can be extended to:

$$\begin{array}{ccccc}
F_\kappa & & & & \\
\uparrow & & & & \\
\overline{\mathbb{V}/\Gamma_\kappa} & \longleftarrow & \overline{\mathbb{V}_C/\Gamma_\kappa} & \longrightarrow & \overline{\mathbb{V}/\Gamma} \\
\uparrow & & \uparrow & & \uparrow \\
\mathbb{V}/\Gamma_\kappa & \longleftarrow & \mathbb{V}_C/\Gamma_\kappa & \longrightarrow & \mathbb{V}/\Gamma
\end{array} \qquad (4.2.9)$$

For the calculation of Chern invariants (heights) of ball quotient surfaces we need their cusp contributions. One of them is the Euler number of T_κ which is zero. the other is the selfintersection T_κ^2 of T_κ in $\overline{\mathbb{V}/\Gamma}$. We will learn how to calculate it explicitly from euclidean volumes of fundamental domains of Δ_κ in $\mathbb{R}$ and Λ_κ in $\mathbb{C}$, see (4.2.7). This is also a starting point for the fine classification of the cusp line bundles, hence of (neat) ball cusp singularities at all.

Omitting the index $\kappa = \infty$ we start with the classification of neat ball cusp groups $\Gamma = \Gamma_\kappa$ as abstract groups. All of them act effectively on $\mathbb{V} = \mathbb{V}_0$. This will be investigated more carefully. With the help of the rule (4.2) we first observe that Γ is *not abelian.*

Definition 4.2.9 (see [RAG], II) A *nilpotent group of rank* n is a finitely generated nilpotent group Γ with a filtration

$$\Gamma = \Gamma_0 \supset \Gamma_1 \supset \ldots \supset \Gamma_k = 1$$

such that

- Γ_i is a normal subgroup of Γ_{i-1} for $i = 1, \ldots k$;

- Γ_{i-1}/Γ_i is an abelian group for $i = 1, ..., k$;
- $n = \sum_{i=1}^{k} \operatorname{rank}(\Gamma_{i-1}/\Gamma_i)$.

Independently of the chosen normal filtration with abelian factors the ranks of them are finite and their sum does not change ([RAG], Prop. 2.8).

Theorem 4.2.10 ([RAG], 2.10, 2.18) *The group Γ is isomorphic to a lattice of a simply-connected nilpotent* LIE *group N if and only if it is a torsion free nilpotent group. In this case the rank of Γ is equal to the dimension of the* ZARISKI *closure U of the lattice.*

□

Corollary 4.2.11 *A group Γ is isomorphic to a neat ball cusp lattice if and only if it is a non-abelian torsion free nilpotent group of rank 3.*

Proof By the classification of LIE algebras there exist up to isomorphy only two simply-connected LIE groups of dimension 3, namely the commutative LIE group $\mathbb{R}^3$ and the non-commutative one $U = U_\infty$ considered above. □

From (4.2.7) we know a canonical filtration of Γ presenting Γ as central group extension

$$1 \to \mathbb{Z} \to \Gamma \underset{\pi}{\to} \mathbb{Z}^2 \to 1.$$

The cohomology group $H^2(\mathbb{Z}^2, \mathbb{Z})$ classifies congruence classes of such group extensions (with trivial action of $\mathbb{Z}^2$ on $\mathbb{Z}$), see e.g. [MAL], IV.4.7. It holds that $H^2(\mathbb{Z}^2, \mathbb{Z}) \cong \mathbb{Z}$ (see e.g. [MU3], I, §2, Lemma 1). Therefore, the torsion free nilpotent groups of rank 3 are classified on this way.

More explicitly, we let δ, β, γ be three generators of the non-abelian nilpotent group Γ of rank 3 such that $\pi(\beta), \pi(\gamma)$ generate $\mathbb{Z}^2$ and δ the center $\Delta \cong \mathbb{Z}$ of Γ. Each element $g \in \Gamma$ can be written as $\beta^b\gamma^c\delta^d$, where $b, c, d \in \mathbb{Z}$ are uniquely determined by g. Therefore,

Proposition 4.2.12 *The abstract group structure of a neat ball cusp lattice Γ is uniquely determined by the negative integer t satisfying*

$$\gamma^{-1}\beta^{-1}\gamma\beta = \delta^t \tag{4.2.10}$$

with generators β, γ, δ of Γ as described above.

Proof The invariant $t = 0$ corresponds to the commutative case. If t is positive, then we change from δ to δ^{-1}. Base changes do not change $|t|$ because $\pi(\beta), \pi(\gamma) \in \mathbb{Z}^2$ have to be changed by a $\mathbb{Gl}_2(\mathbb{Z})$ -action (determinant ± 1). The rest is clear. □

The correspondence $(\pi(\beta), \pi(\gamma)) \mapsto t$ in (4.2.10) extends to a non-degenerate scew-symmetric form λ on $\pi(\Gamma)$ on $\mathbb{Z}^2$. For a lattice $\Gamma \subset U = U_\infty$ it extends from $\Lambda = \Lambda_\kappa \subset \mathbb{C}$ to $\mathbb{C}$ by $\mathbb{R}$-linear extension, see (4.2.7). Working with a positive orientied basis of Λ we see that $\lambda = t \cdot \det / f$, where f is the (positive) volume of a fundamental domain of Λ in $\mathbb{C}$, and det is the normalized determinant form on $\mathbb{C} = \mathbb{R} + \mathbb{R}i$ sending $(1, i)$ to 1. With the notations of (4.2.2) and (4.2.4) we take an *oriented basis* $\beta = [b, r], \gamma = [c, s], \delta = [0, q]$ satisfying (4.2.10). By (4.2) we can calculate t in these terms:

$$\begin{aligned}[0, tq] = \delta^t &= [-b, -r][-c, -s][b, r][c, s] \\ &= [-b - c, -r - s - \operatorname{Im} \bar{b}c] = [0, -2 \operatorname{Im} \bar{b}c],\end{aligned}$$

hence

$$t = (-2 \operatorname{Im} \bar{b}c)/q = -2 \det(b, c)/q = -2 \operatorname{vol}(\Lambda)/\operatorname{vol}(\Delta), \tag{4.2.11}$$

where vol means the (oriented, positive) volume of a fundamental domain of the corresponding lattices in $\mathbb{C}$ or $\mathbb{R}$, respectively. Since t is negative, we see that q is positive, indeed.

Isomorphy classes of line bundles over the elliptic curve $T = \mathbb{C}/\Lambda$ are classified by hermitian forms h on $\mathbb{C}$ such that the imaginary part Im h has integral values on Λ, together with quasi-characters $\alpha : \Lambda \longrightarrow S^1 := \{c \in \mathbb{C}; |c| = 1\}$ connected with h by the relations

$$\alpha(\lambda_1 + \lambda_2) = \alpha(\lambda_1)\alpha(\lambda_2) \cdot \exp\{\pi i \cdot \operatorname{Im} h(\lambda_1, \lambda_2)\} \quad \text{for all } \lambda_1, \lambda_2 \in \Lambda. \tag{4.2.12}$$

More precisely, the above pairs (h, α) form a group $AH(\Lambda)$ via componentwise addition or multiplication, respectively, with natural projection onto

$$H(\Lambda) = \{h : \mathbb{C}^2 \longrightarrow \mathbb{C} \text{ hermitian; } \operatorname{Im}\ h(\Lambda \times \Lambda) \subseteq \mathbb{Z}\}.$$

We denote by Pic T the group of isomorphy classes of line bundles over T and by $\operatorname{Pic}^0 T$ its subgroup of bundles (invertible $\mathcal{O}_T$-sheaves) of degree 0.

Theorem of APPELL-HUMBERT **4.2.13** (see e.g. [MU3], I, §2) *With the above notations there is a commutative diagram (4.2.13) of canonical group homomorphisms with exact rows and vertical isomorphisms*

$$\begin{array}{ccccccccc} 1 & \longrightarrow & Hom(\Lambda, S^1) & \longrightarrow & AH(\Lambda) & \longrightarrow & H(\Lambda) & \longrightarrow & 1 \\ & & \downarrow & & \downarrow & & \downarrow & & \\ 1 & \longrightarrow & Pic^0 T & \longrightarrow & Pic T & \rightarrow & Ker[H^2(T,\mathbb{Z}) \to H^2(T,\mathcal{O}_T)] & \longrightarrow & 1 \end{array}$$

(4.2.13)

□

One knows that $H(\Lambda)$ classifies the topological isomorphy classes of complex line bundles over T.

Now the ball cusp lattices $\Gamma = \Gamma_\infty$ with generators $\beta = [b,r]$, $\gamma = [c,s]$, $\delta = [0,q]$ and $\Lambda = \mathbb{Z}a + \mathbb{Z}b$ as above fit into this theorem by the following correspondence:

$$\Gamma \mapsto (\Lambda, h, \alpha),\ h(w,u) := -2\overline{w}u/q,$$
$$\alpha : b \mapsto e^{2\pi i r/q}, c \mapsto e^{2\pi i s/q} \text{ extended by (4.2.12) to } \Lambda.$$

Then Im h is integral on Λ because Im $h(b,c) = t \in \mathbb{Z}$ by (4.2.11) and 4.2.12.

Explicitly, the Theorem of APPELL-HUMBERT corresponds to the triple (Λ, h, α) the T-line bundle $F(\Lambda, h, \alpha) = \mathbb{C} \times \mathbb{C}/\Lambda$, where Λ acts by the following automorphisms

$$\varphi_\lambda : \mathbb{C} \times \mathbb{C} \longrightarrow \mathbb{C} \times \mathbb{C}, \tag{4.2.14}$$
$$(w,u) \mapsto (w \cdot \alpha(\lambda) \cdot \exp\{\pi h(u,\lambda) + \pi h(\lambda,\lambda)/2\}, u + \lambda),$$

on $\mathbb{C} \times \mathbb{C}$. But this is precisely the action of Λ on $\mathbb{C}^* \times \mathbb{C}$ in diagram (4.2.8). Namely, $\mathbb{C}^* = \mathbb{C}/\Delta$ and Δ is generated by $[0,q]$; therefore the coordinates w on $\mathbb{C}^*$ and z on the uniformizing $\mathbb{C}$ are connected by $w = e^{2\pi i z/q}$. For example the action of $[b,r]$ on $\mathbb{V} \subset \mathbb{C} \times \mathbb{C}$,

$$[b,r](z,u) = \left(z + i\bar{b}u + i|b|^2/2 + r, u + b\right), \tag{4.2.14'}$$

see 4.2.2, goes modulo Δ down to

$$(w,u) \mapsto e^{2\pi iz/q} \cdot e^{2\pi ir/q} \cdot e^{-\pi(2\bar{b}u+\bar{b}b)/q}$$
$$= w \cdot \alpha(b) \cdot e^{\pi(h(u,b)+h(b,b)/2)}$$

Comparing with (4.2.14) and the definition of the cusp bundle $F_\kappa(\Gamma)$ in (4.2.8) we see that

4.2.14 *the line bundles $F(\Lambda, h, \alpha)$ and $F_\kappa(\Gamma)$ over the elliptic curve $T = \mathbb{C}/\Lambda$ coincide.*

Denoting this bundle by F we calculate now the selfintersection number $(T^2) = (T^2)_F$ of the zero section T. For this purpose we consider the dual bundle $F^* = F(-h, -\alpha^{-1})$ of F on T. The hermitian form $-h$ is positive definite, hence

$$(\dim H^0(T, F^*))^2 = \det(-\operatorname{Im}\ h) = |t|^2$$

by a classical formula, see e.g. [MU3], I, §3. From the RIEMANN-ROCH Theorem for line bundles on ellipitic curves it follows that

$$\deg F^* = \dim H^0(T, F^*) = |t|,$$

hence

$$(T^2)_F = \deg F = -\deg F^* = t \tag{4.2.15}$$

because the selfintersection number of a (smooth) curve on a (smooth) surface coincides with the degree of its normal bundle.

We call a line bundle L on an elliptic curve T *negative*, if the selfintersection number of its zero section is negative. The set of all negative line bundles over T is denoted by $\mathrm{Pic}^-(T)$. We close this section with the following transferred restriction of the Theorem of APPELL-HUMBERT.

Proposition 4.2.15 *Let $\mathbf{U} = \mathbf{U}_\kappa$ be a maximal unipotent subgroup of the ball group $\mathbf{G} = \mathbb{S}U(2,1)$ and $U = \mathbf{U}(\mathbb{R})$. There is a bijective correspondence*

$$\{\textit{neat ball cusp lattices in } U\} \longleftrightarrow \bigcup_{\Lambda \textit{ lattice in } \mathbb{C}} \mathrm{Pic}^-(\mathbb{C}/\Lambda)$$
$$\Gamma \mapsto F_\kappa(\Gamma)$$

Proof We can assume that $\kappa = \infty$. Since the union is understood to be disjoint we can also fix a lattice Λ of $\mathbb{C}$. Let $F = F(\Lambda, h, \alpha)$ be a line bundle over $T = \mathbb{C}/\Lambda$ whose isomorphy class belongs to $\mathrm{Pic}^- T$. As a consequence of APPELL-HUMBERT'S theorem it is uniquely determined by h and α. The negative definite hermitian form h can be written as $h(w,u) = -2\overline{w}u/q, 0 < q \in \mathbb{R}$. It defines the sublattice Δ of $[U,U]$ generated by $[0,q]$. Choose two generators b, c of $\Lambda \subset \mathbb{C}$ with positive orientation. Up to $\mathbb{Z}$-multiples the quasi-character α determines uniquely real numbers r, s by

$$\alpha(b) = e^{2\pi i r/q}, \alpha(c) = e^{2\pi i s/q}.$$

The neat ball cusp lattices we look for is generated by $\Delta, [b,r], [c,s]$.

The inverse construction has already been done. □

4.3 Invariants of Neat Ball Quotient Surfaces

Following the basic ideas of HIRZEBRUCH [HI2] and a generalization of MUMFORD [MU4] to non-compact algebraic quotient varieties of symmetric domains by neat lattices we explain the proportionality principle relating different CHERN numbers of the toroidal compactification $\overline{\mathbb{B}/\Gamma}$ of ball quotient surfaces. We restrict our attention to the ball $\mathbb{B} = K\backslash G, K = \mathrm{Stab}_G O,\ O \in \mathbb{B}$, of complex dimension $n = 2$, but partly we work in a more general situation.

The dual symmetric space of the ball is $\mathbb{B}^\vee = \mathbb{P}^2 (= \mathbb{P}^2(\mathbb{C})$ more precisely). The compact form $G^c = \mathbb{S}U(3,\mathbb{R})$ of $G = \mathbb{S}U((2,1),\mathbb{C})$ sits in the complexification $G^{\mathbb{C}} = \mathbb{S}\mathrm{l}_3(\mathbb{C})$ of G acting on $\mathbb{B}^\vee = K\backslash G^c = NK^{\mathbb{C}}$, where N is the unipotent radical of the stabilizer of $G^{\mathbb{C}}$ at O. In terms of LIE groups the canonical embedding of $\mathbb{B}$ into $\mathbb{B}^\vee$ is described by

$$\mathbb{B} = K\backslash G \longrightarrow NK^{\mathbb{C}} = \mathbb{B}^\vee.$$

Now we consider pairs (E', h') of holomorphic G-vector bundles E' on $\mathbb{B}$ with invariant hermitian metric h'. It can be canonically extended to a metrized holomorphic $G^{\mathbb{C}}$ -vector bundle $(E^\vee, h^\vee)$ on $\mathbb{B}^\vee$ in the following manner: Let E_O be the fibre of E' at O. Then K acts on E_O and we have $E' = E_O \times_K G$. We define $E^\vee := E_O \times_{NK^{\mathbb{C}}} G^{\mathbb{C}}$, where the action of N on E_O is defined to be trivial. In analogous manner one extends h' to a $G^{\mathbb{C}}$ -invariant hermitian metric $h^\vee$ on $\mathbb{B}^\vee$.

On the other hand, since Γ acts freely on $\mathbb{B}$, the Γ-equivariant pair (E', h') descends to a metrized holomorphic vector bundle (E, h) on the quotient variety $\mathbb{B}/\Gamma$. By a Theorem of MUMFORD [MU4] this pair can be extended in a unique manner to a pair $(\overline{E}, \overline{h})$ on the toroidal compactification $\overline{\mathbb{B}/\Gamma}$. The crucial point is to prove that

4.3.1 *the poles of h along the compactification divisor $T = \left(\overline{\mathbb{B}/\Gamma}\right)\backslash(\mathbb{B}/\Gamma)$ are logarithmically restricted in the following sense:*

Definition 4.3.2 Let $\overline{X}$ be a smooth complex compact variety, X an open subvariety such that $D = \overline{X} \backslash X$ is a divisor with normal crossings, $\overline{E}$ a holomorphic vector bundle on $\overline{X}$ with restriction E to X, h a hermitian metric on E. The metric is called *logarithmically restricted* (along D), if for each point $P \in D$, any polycylindric neighbourhood $\overline{U}$ of P such that $U = \overline{U} \cap X \cong \Delta^{*a} \times \Delta^b, \Delta^*$ the punctured unique disc $\Delta \backslash \{0\}$, and each holomorphic C^∞-base field $e_1, ..., e_r$ of $\overline{E}$ over $\overline{U}$ it holds that

$$|h(e_i, e_j)|, |\det(h(e_i, e_j))^{-1}| < C \cdot \sum_{i=1}^{a} \log |z_i|^{2N}$$

on $\overline{U}$, where $z_1, ..., z_a$ are coordinates on Δ^a, and $C > 0$, $N \in \mathbb{N}_+$ are suitable constants only depending on U and E.

The connection of the MUMFORD-extension $\overline{E}$ with the equivalent bundles we started with is illustrated in diagram:

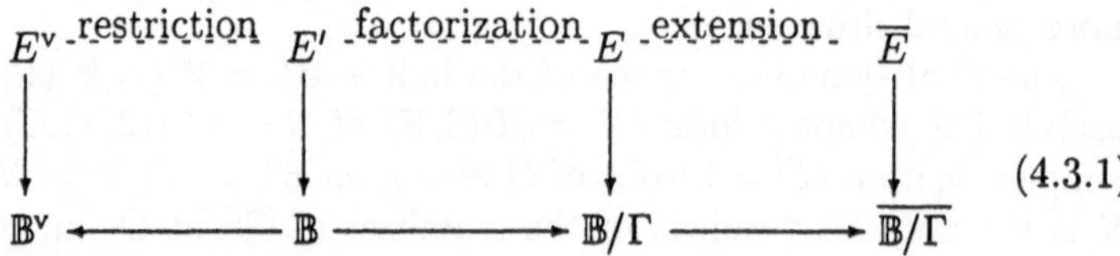

(4.3.1)

Main Example 4.3.3 ([MU4]) For the G-equivariant cotangent bundle $E' = \mathcal{T}_{\mathbb{B}}^*$ one gets the cotangent bundles $E^\vee = \mathcal{T}_{\mathbb{P}^2}^*$ on $\mathbb{P}^2$ or $E = \mathcal{T}_{\mathbb{B}/\Gamma}$ on $\mathbb{B}/\Gamma$, respectively. If one starts with the canonical bundle $K' = \mathcal{T}_{\mathbb{B}}^* \wedge \mathcal{T}_{\mathbb{B}}^*$ then $K^\vee$ and $K = \mathcal{T}_{\mathbb{B}/\Gamma}^* \wedge \mathcal{T}_{\mathbb{B}/\Gamma}^*$ are the canonical bundles on $\mathbb{P}^2$ or $\mathbb{B}/\Gamma$, respectively. The corresponding locally free sheaves $\mathcal{O}(\overline{E})$ or $\mathcal{O}(\overline{K})$ of MUMFORD'S extensions $\overline{E}$ or $\overline{K}$ are the $\mathcal{O}_{\overline{\mathbb{B}/\Gamma}}$ -sheaves $\Omega^i(\log T)$ of germes of *logarithmic i-forms* along the compactification divisor T for $i = 1$ or 2, respectively. $\Omega^i(\log T)$ is defined by the following properties:

(i) $\Omega^i(\log T)$ is a subsheave of the sheaf of meromorphic i-forms on $\overline{\mathbb{B}/\Gamma}$;

(ii) $\Omega^i(\log T)|\mathbb{B}/\Gamma$ is the sheaf of germs of holomorphic i-forms on $\mathbb{B}/\Gamma$;

(iii) For any $t \in \operatorname{supp} T$ let (w, u) be local parameters on $\overline{\mathbb{B}/\Gamma}$ around t such that T is defined by $w = 0$. Then

$$\Omega^1(\log T)(U) = \left\{a(w,u)dw/w + b(w,u)du; a, b \in \mathcal{O}_{\overline{\mathbb{B}/\Gamma}}(U)\right\},$$

$$\Omega^2(\log T)(U) = \left\{a(w,u)(dw \wedge du)/w; a \in \mathcal{O}_{\overline{\mathbb{B}/\Gamma}}(U)\right\}$$

for small open analytic neighbourhoods U of t.

Let X be a smooth compact complex manifold of dimension n and F a holomorphic vector bundle of rank r on X. In the cohomology groups on X with constant real (or complex) coefficients we consider the CHERN *classes* $c_i(F) \in H^{2i}(X, \mathbb{R})$, see e.g. [WEL]. Together they form the *total* CHERN *class*

$$1 + c_1(F) + c_2(F) + \ldots + c_r(F) \in H^{even}(X, \mathbb{C}). \tag{4.3.2}$$

the CHERN classes $c_i(F)$ are represented by closed real $2i$-forms $\gamma_i(F)$ on X. Each form

$$\gamma_a(F) := \gamma_{a_1}(F) \wedge \ldots \wedge \gamma_{a_k}(F), \ a = (a_1, \ldots, a_k), a_i \in \mathbb{N}_+, \tag{4.3.3}$$

of highest degree $\sum_{i=1}^{k} a_i = n$ is called a CHERN *form* of F, especially $\gamma_n(F)$ is the EULER-CHERN *form* of F on X. The corresponding volumes

$$c_a(F)[X] := \int_X \gamma_a(F) \tag{4.3.4}$$

are called the CHERN *numbers* of the bundle F. If $F = \mathcal{T}_X$ is the tangent bundle on X then one writes shortly $c_a(X)$ instead of $c_a(\mathcal{T})$ and calls them the CHERN *numbers* of X. Especially $c_n(X)$ is called the EULER *number* of X.

With the additional notations of diagram (4.3.1) there is a remarkable connection between the CHERN numbers of $E^\vee$ and $\overline{E}$, for neat arithmetic groups Γ and bounded symmetric domains in general.

Theorem 4.3.4 (HIRZEBRUCH-MUMFORD's Proportionality Principle [MU4]) *For all* CHERN *numbers of $E^\vee$ and $\overline{E}$ it holds that*

$$c_a(\overline{E}) = V \cdot c_a(E^\vee)$$

with a constant V only depending on Γ (assumed to be neat).

A sketch of proof, especially for ball lattices, is given at the end of this section.

Corollary 4.3.5 *For two-dimensional neat ball quotient surfaces it holds that*

$$c_1^2(\overline{E}) = c_1^2(E^\vee) \cdot c_2(\mathbb{B}/\Gamma)/c_2(\mathbb{P}^2),$$
$$c_2(\overline{E}) = c_2(E^\vee) \cdot c_2(\mathbb{B}/\Gamma)/c_2(\mathbb{P}^2),$$

where $c_2(\mathbb{B}/\Gamma) := \int_{\mathcal{F}(\Gamma)} \gamma_2$ *denotes the* EULER *volume of a* Γ*-fundamental domain* $\mathcal{F}(\Gamma)$ *on* $\mathbb{B}$ *defined by the* EULER-CHERN *form* γ_2 *of the* BERGMANN *metric on* $\mathbb{B}$.

Proof First we derive 4.3.5 from the theorem. In MUMFORD'S proof of 4.3.4 the constant V is given as volume $\mathrm{vol}_\gamma(\mathbb{B}/\Gamma) = \int_{\mathcal{F}(\Gamma)} \gamma$ of a Γ-fundamental domain with respect to a volume form γ on $\mathbb{B}$ not depending on Γ. By HIRZEBRUCH'S proof in the cocompact case the same result holds with the same volume form γ. BOREL proved in [BO1] the existence of a cocompact neat lattice Γ. Working with the tangent bundle $E = \overline{E}$ on $\mathbb{B}/\Gamma = \overline{\mathbb{B}/\Gamma}$ and the EULER-CHERN class c_n the identity of 4.3.4 is

$$c_n\left(\overline{\mathbb{B}/\Gamma}\right) = \mathrm{vol}_\gamma(\mathbb{B}/\Gamma) \cdot c_n(\mathbb{B}^\vee). \tag{4.3.5}$$

On the other hand using (4.3.4) we have $c_n\left(\overline{\mathbb{B}/\Gamma}\right) = c_n(\mathbb{B}/\Gamma) = \int_{\mathbb{B}/\Gamma} \gamma_n = \mathrm{vol}_{\gamma_n}(\mathbb{B}/\Gamma)$. A comparison with (4.3.5) gives $c_n(\mathbb{B}^\vee)\cdot\gamma = \gamma_n$. For $n = 2$ the correct substitution is $V = c_2(\mathbb{B}/\Gamma)/c_2(\mathbb{P}^2)$. □

Proposition 4.3.6 *Let* Γ *be a neat ball lattice,* $T = \sum_{\kappa \bmod \Gamma} T_\kappa$ *the compactification divisor of* $\mathbb{B}/\Gamma \subset \overline{\mathbb{B}/\Gamma}$. *The the* CHERN *numbers of* $\overline{\mathbb{B}/\Gamma}$ *are related with the* EULER *volume* $c_2(\mathbb{B}/\Gamma)$ *of a* Γ*-fundamental domain on* $\mathbb{B}$ *by*

$$c_2\left(\overline{\mathbb{B}/\Gamma}\right) = c_2(\mathbb{B}/\Gamma) \quad , \quad c_1^2\left(\overline{\mathbb{B}/\Gamma}\right) = 3c_2(\mathbb{B}/\Gamma) + (T^2); \tag{4.3.6}$$

for the arithmetic genus and the signature of $\overline{\mathbb{B}/\Gamma}$ *it holds that*

$$\chi\left(\overline{\mathbb{B}/\Gamma}\right) = c_2(\mathbb{B}/\Gamma)/3 + (T^2)/12 \quad , \quad \tau\left(\overline{\mathbb{B}/\Gamma}\right) = c_2(\mathbb{B}/\Gamma)/3 + (T^2)/3. \tag{4.3.7}$$

Proof We need the CHERN homomorphism *ch* of the GROTHENDIECK ring $\mathcal{K}(X)$ from a smooth complex compact algebraic variety X into its cohomology ring (see [HI3])

$$ch : \mathcal{K}(X) \longrightarrow H^*(X, \mathbb{C}). \tag{4.3.8}$$

If $\xi \in \mathcal{K}(X)$ is the class of a vector bundle on X of rank r, then

$$ch(\xi) = \sum_{i=1}^{r} \exp(\gamma_i), \text{ where } \sum_{j=0}^{r} c_j(\xi) X^j = \prod_{i=1}^{r} (1 + \gamma_j X) \tag{4.3.9}$$

is the formal product decomposition. We have

$$\begin{aligned} ch(\xi) &= 1 + c_1(\xi) + c_1^2(\xi)/2 + \ldots \text{(terms of higher degree), if } r = 1; \\ ch(\xi) &= 2 + c_1(\xi) + \left(c_1^2(\xi) - c_2(\xi)\right)/2 + \ldots \\ &\quad \text{(terms of higher degree), if } r = 2; \end{aligned} \tag{4.3.10}$$

The CHERN homomorphism can be transferred and extended to the category of coherent $\mathcal{O} = \mathcal{O}_X$-sheaves. The locally free sheaf corresponding to a vector bundle F will be denoted by $\mathcal{O}(F)$ and the vector bundle corresponding to a locally free sheaf $\mathcal{M}$ is denoted by $\{\mathcal{M}\}$. If D is a divisor on X, then we use the notations $\mathcal{O}(D)$ or $\{D\}$ for the corresponding sheaf or line bundle, respectively.

We apply the above formulas of the HIRZEBRUCH-GROTHENDIECK-RIEMANN-ROCH Theory now to our main example described in 4.3.3. With $X = \overline{\mathbb{B}/\Gamma}$, the notations and statements of 4.3.3 we get two exact sequences of coherent $\mathcal{O}$-modules

$$\begin{aligned} 0 &\longrightarrow \Omega^2_{\overline{\mathbb{B}/\Gamma}} \longrightarrow \mathcal{O}(\overline{K}) = \Omega^2(\log T) \longrightarrow \mathcal{O}(\{T\}|T) \longrightarrow 0, \\ 0 &\longrightarrow \Omega^1_{\overline{\mathbb{B}/\Gamma}} \longrightarrow \mathcal{O}(\overline{E}) = \Omega^1(\log T) \longrightarrow \mathcal{O}(\{T\}|T) \longrightarrow 0. \end{aligned}$$

where $\mathcal{O}(\{T\}|T)$ is considered as $\mathcal{O}$-module which is zero outside of T. The additivity of the CHERN homomoprhisms applied to both sequences yields

$$ch(\overline{E}) - ch\left(\Omega^1_{\overline{\mathbb{B}/\Gamma}}\right) = ch(\mathcal{O}(\{T\}|T)) = ch(\overline{K}) - ch\left(\Omega^2_{\overline{\mathbb{B}/\Gamma}}\right).$$

According to (4.3.10) we compare the quadratic parts of the left and right side of this identity. One obtains

$$\begin{aligned} &\left(c_1^2(\overline{E}) - 2c_2(\overline{E})\right)/2 - \left(c_1^2\left(\overline{\mathbb{B}/\Gamma}\right) - 2c_2\left(\overline{\mathbb{B}/\Gamma}\right)\right)/2 \\ &= c_1^2(\overline{K})/2 - c_1^2\left(\overline{\mathbb{B}/\Gamma}\right). \end{aligned}$$

Since $\overline{K} = \overline{E} \wedge \overline{E}$ by 4.3.3 (iii) we know that $c_1(\overline{E}) = c_1(\overline{K})$. Therefore all c_1^2-members in the identity can be omitted, thus

$$c_2(\overline{E}) = c_2\left(\overline{\mathbb{B}/\Gamma}\right).$$

Together with the second proportionality formula of Corollary 4.3.5 one gets

$$c_2\left(\overline{\mathbb{B}/\Gamma}\right) = c_2(\overline{E}) = c_2(E^\vee)\cdot c_2(\mathbb{B}/\Gamma)/c_2(\mathbb{P}^2) = c_2(\mathbb{B}/\Gamma)$$

because $E^\vee$ is the cotangent bundle of $\mathbb{P}^2$, hence $c_2(E^\vee) = c_2(\mathbb{P}^2)$. We proved the first part of (4.3.6).

The first proportionality formula of 4.3.5 yields for the special case $\overline{K}$

$$c_1^2(\overline{K}) = c_1^2(K^\vee)\cdot c_2(\mathbb{B}/\Gamma)/c_2(\mathbb{P}^2) = 3c_2(\mathbb{B}/\Gamma)$$

because $c_1^2/c_2 = 9/3 = 3$ on $\mathbb{P}^2$. On the other hand the second formula of 4.3.3 (iii) can be read as

$$\mathcal{O}(\overline{K}) = \Omega^2(\log T) = \Omega^2_{\overline{\mathbb{B}/\Gamma}} \otimes \mathcal{O}(T). \tag{4.3.11}$$

Using the same notation K for the canonical bundles and a canonical divisor on $\overline{\mathbb{B}/\Gamma}$, and regarding c_1^2 as the selfintersection product of corresponding divisors we receive

$$(K+T)^2 = c_1^2(\overline{K}) = 3c_2(\mathbb{B}/\Gamma). \tag{4.3.11'}$$

Now we use the *adjunction formula*

$$2(g(E)-1) = -e(D) = ((K+D)\cdot D), (g \text{ genus }, e \text{ Euler number}) \tag{4.3.12}$$

for smooth curves D on smooth (compact complex) surfaces X. Since the components T_κ of T are elliptic curves, the EULER numbers of T_κ are 0, hence $((K+T_\kappa)\cdot T_\kappa) = 0$. Coming from cusp singularity resolutions of $\widehat{\mathbb{B}/\Gamma}$ the sum T of all T_κ's is disjoint, therefore $((K+T)\cdot T) = 0$. Together with (4.3.11) one gets

$$\begin{aligned}(K^2) - (T^2) &= (K^2) + (K\cdot T) = ((K+T)\cdot K)\\ &= ((K+T)(K+T)) = 3c_2(\mathbb{B}/\Gamma).\end{aligned}$$

The second formula of the proposition is proved. The formulas of (4.3.7) follow from those of (4.3.6) by NOETHER'S formula $12\chi = c_1^2 + c_2$ (see (3.2.3)) or the Index Theorem $4\chi = c_2 + \tau$ (see (3.2.4)), respectively.

□

Corollary 4.3.7 *In terms of global and local volumes the signature of the toroidal compactified neat ball quotient surface $\overline{\mathbb{B}/\Gamma}$ is expressed by*

$$\tau\left(\overline{\mathbb{B}/\Gamma}\right) = c_2(\mathbb{B}/\Gamma)/3 - 2 \sum_{\kappa \bmod \Gamma} |\Lambda_\kappa|/3|\Delta_\kappa|,$$

where $\mathbb{C} \supset \Lambda_\kappa = \Gamma_\kappa/\Delta_\kappa, \Gamma_\kappa$ the neat ball cusp subgroup of Γ at the cusp $\kappa, \mathbb{R} \supset \Delta_\kappa$ is the center of Γ_κ, see diagram (4.2.7), and $|\Lambda_\kappa|$, $|\Delta_\kappa|$ are the euclidean volumes of fundamental domains of the lattices $\Lambda_\kappa, \Delta_\kappa$ in $\mathbb{C}$ or $\mathbb{R}$, respectively.

Proof We have only to express the last summand in the second formula of (4.3.7) in local volume terms. The compactification divisor and its self-intersection split into sums

$$T = \sum_{\kappa \bmod \Gamma} T_\kappa \quad , \quad (T^2) = \sum_{\kappa \bmod \Gamma} (T_\kappa^2).$$

The selfintersection $t_\kappa := (T_\kappa^2)$ coincide with that of the zero section of T_κ in the cusp bundle $F_\kappa(\Gamma)$, see (4.2.9). Now we only have to compare the formulas (4.2.15) and (4.2.11) for t_κ. We obtain

$$(T_\kappa^2) = t_\kappa = -2\,\mathrm{vol}\,(\Lambda_\kappa)/\,\mathrm{vol}\,(\Delta_\kappa) = -2|\Lambda_\kappa|/|\Delta_\kappa|,$$

which was to be proved. □

Proof of Theorem 4.3.4 (sketch). We use the notations of the beginning of this section until diagram (4.3.1), especially K is now a maximal compact subgroup of G again. The Lie algebras of $G, K, G^c, G^{\mathbb{C}}$ are denoted by $\mathfrak{g}, \mathfrak{k}, \mathfrak{g}^c$ or $\mathfrak{g}^{\mathbb{C}}$, respectively. We dispose on (Cartan) decompositions

$$\mathfrak{g} = \mathrm{Lie}\, G = \mathfrak{k} \oplus \mathfrak{p} \quad , \quad \mathfrak{g}^c = \mathrm{Lie}\, G^c = \mathfrak{k} \oplus i\mathfrak{p} \text{ in } \mathfrak{g}^{\mathbb{C}}$$

and exponential maps

$$\exp : \mathfrak{k} + \mathfrak{p} \longrightarrow G \quad , \quad \exp : \mathfrak{k} + i\mathfrak{p} \longrightarrow G^c$$

restricting both the exponential map of $\mathfrak{g}^{\mathbb{C}}$. Restricting derivatives to the complex tangent space of the symmetric domains at O we get isomorphisms

$$d\exp : \mathfrak{p} \xrightarrow{\sim} \mathcal{T}_O(\mathbb{B}) \quad , \quad d\exp : i\mathfrak{p} \xrightarrow{\sim} \mathcal{T}_O(\mathbb{B}^\vee)$$

Now we use a basis $a_1, \cdots, a_n$ of $\mathfrak{p}$ and corresponding coordinates $\xi_1, \ldots, \xi_n$. Multiplying with i we get a basis $ia_1, \ldots, ia_n$ of $i\mathfrak{p}$ and corresponding coordinates $\xi_1' = i\xi_1, \ldots, \xi_n' = i\xi_n$ (for the two-ball we have $n = 2$). In these coordinates we dispose on matrices (θ_{ij}), $(\theta_{ij}^\vee)$ of real two-forms corresponding to the curvature tensors of the metrized bundles (E', h') and $(E^\vee, h^\vee)$ on our symmetric spaces $\mathbb{B}$ or $\mathbb{B}^\vee$ (around O), respectively.

The connection with the total sum of CHERN forms at O is given by

$$1 + \gamma_1 + \ldots + \gamma_n = \det(I_n - \theta_{ij}/2\pi i),$$
$$1 + \gamma_n^\vee + \ldots + \gamma_n^\vee = \det\left(I_n - \theta_{ij}^\vee/2\pi i\right),$$

where I_n denotes the unit matrix with n rows.

We remember that $(E', h'), (E^\vee, h^\vee)$ are determined by the fibre $E_O = E'_O = E_O^\vee$ because $E' = G \times_K G_O, E^\vee = G^c \times_K E_O$. The extensions of the hermitian metric h_O on E_O to h' or $h^\vee$ around O are managed by the exponential map via the relations

$$h^\vee((\exp ia, e), (\exp ia, f)) = h'((\exp a, e), (\exp a, f)),$$
$$e, f \in E_O, a \in \mathfrak{p}.$$

The coordinate calculations of curvatures at O in tangent spaces at O of $\mathbb{B}$ or $\mathbb{B}^\vee$, respectively, are the same up to i-multiplication. Since

$$d\xi'_k \wedge d\xi'_l = id\xi_k \wedge id\xi_l = -d\xi_k \wedge d\xi_l$$

we get $\theta_{ij}^\vee = -\theta_{ij}$, hence $\theta^\vee = -\theta$ for the curvature tensor and

$$\gamma_k(E^\vee, h^\vee)(O) = (-1)^k \gamma_k(E', h')(O)$$

for the CHERN forms at O.

Assume that the EULER-CHERN form γ_n is not 0. The quotients of two equivariant CHERN forms γ_α, $\gamma_\alpha^\vee$ of highest degree, $\alpha = (\alpha_1, \ldots, \alpha_r)$, $\alpha_1 + \ldots + \alpha_r = n$, are constants. Especially we obtain the relations

$$\gamma_\alpha/\gamma_n = (-1)^n \gamma_\alpha^\vee/(-1)^n \gamma_n^\vee = \gamma_\alpha^\vee/\gamma_n^\vee.$$

These relations descend to the quotient surface $\mathbb{B}/\Gamma$. The CHERN forms extend simultaneously to sections of $\overline{E}$ over $\overline{\mathbb{B}/\Gamma}$ preserving the constant relations. For the two-ball we extended γ_1^2 in the logarithmic and canonical sheaf, see 4.3.3. Taking integrals we get the relations for the CHERN numbers we look for.

□

4.4 Γ-Rational Discs

A *linear (complete)* subdisc $\mathbb{D}$ of $\mathbb{B} \subset \mathbb{P}^2$ is the intersection of a projective line $L \subset \mathbb{P}^2$ and $\mathbb{B}$ in $\mathbb{P}^2$ (if it is not void). Each line L through $\mathbb{B}$ is the projection $\mathbb{P}(X^{\perp}) \subset \mathbb{P}\mathbb{C}^3 = \mathbb{P}^2$, where $X^{\perp}$ is the orthogonal complement of a *positive vector* $X \in \mathbb{C}^3$, this means that $\langle X, X \rangle > 0$. We work with the hermitian (2,1)-metric $\langle\ ,\ \rangle$ on $\mathbb{C}^3$ defining $\mathbb{B}$ as space of projections of all negative lines in $\mathbb{C}^3$. Since each positive vector can be extended to an orthogonal basis of $\mathbb{C}^3$, our ball group $G = \mathbb{S}U((2,1),\mathbb{C})$ acts transitively on the set of all linear subdiscs of $\mathbb{B}$.

Definition 4.4.1 Let Γ be a ball lattice and $\mathbb{D}$ a linear subdisc of $\mathbb{B}$. The disc $\mathbb{D}$ is called a *Γ-rational disc* or shortly a *Γ-disc* , if the (normalizing) *decomposition group of* Γ at $\mathbb{D}$

$$N_{\Gamma}(\mathbb{D}) := \{\gamma \in \Gamma; \gamma\mathbb{D} \subseteq \mathbb{D}\} \tag{4.4.1}$$

is a $\mathbb{D}$-lattice, this means that there is a $N_{\Gamma}(\mathbb{D})$-fundamental domain of finite volume with respect to the POINCARÉ metric on $\mathbb{D}$.

The ineffective kernel of the action of $N_{\Gamma}(\mathbb{D})$ on $\mathbb{D}$ is the (centralizing) *inertia group of* Γ *at* $\mathbb{D}$

$$Z_{\Gamma}(\mathbb{D}) := \{\gamma \in \Gamma; \gamma|\mathbb{D} = id_{\mathbb{D}}\} \tag{4.4.2}$$

This is a finite group by 4.1.2 because it fixes (all) points of $\mathbb{D}$. The extension of the action to $\mathbb{P}^2 \supset \mathbb{B}$ fixes pointwise the projective line L through $\mathbb{D}$. By the unitary trick for finite groups we recognize that the action of $Z_{\Gamma}(\mathbb{D})$ on the tangent space of $\mathbb{P}^2$ at P diagonalizes, and we get a faithful representation of this group on a line M crossing L at P with fixed point P. Therefore,

4.4.2 *The centralizing group $Z_{\Gamma}(\mathbb{D})$ is a finite cyclic normal subgroup of $N_{\Gamma}(\mathbb{D})$.*

The exact sequence

$$1 \longrightarrow Z_{\Gamma}(\mathbb{D}) \longrightarrow N_{\Gamma}(\mathbb{D}) \longrightarrow \Gamma_{\mathbb{D}} \longrightarrow 1 \tag{4.4.3}$$

defines $\Gamma_{\mathbb{D}}$ acting effectively on $\mathbb{D}$; the quotient curves $\mathbb{D}/N_{\Gamma}(\mathbb{D})$ and $\mathbb{D}/\Gamma_{\mathbb{D}}$ coincide.

Proposition 4.4.3 *Let Γ be a ball lattice, $p : \mathbb{B} \longrightarrow \mathbb{B}/\Gamma$ the natural projection to the ball quotient surface $\mathbb{B}/\Gamma$ and $\mathbb{D} \subset \mathbb{B}$ a Γ-rational disc. Then $\mathbb{D}/\Gamma := p(\mathbb{D}) \subset \mathbb{B}/\Gamma$ is an (in general non-compact) algebraic curve.*

Proof We work with a cofinite neat normal subgroup Γ' of Γ, see 4.1.8, and set up the following commutative diagram (4.4.4) of maps

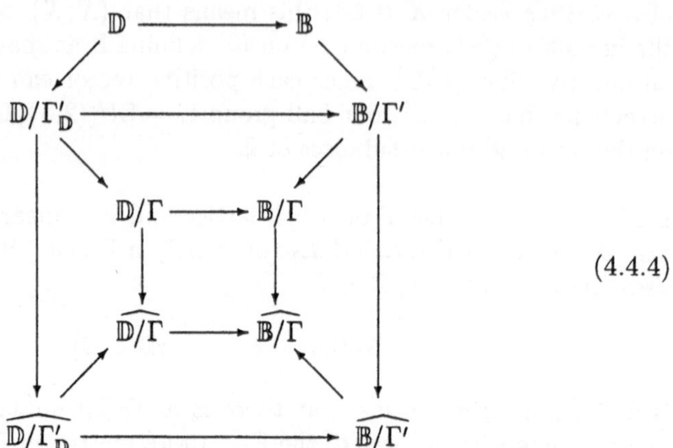

(4.4.4)

where the last three hats denote compactifications by cusp points as usual, and $\widehat{\mathbb{D}/\Gamma}$ is the closure of $\mathbb{D}/\Gamma$ on $\widehat{\mathbb{B}/\Gamma}$. These are compact complex algebraic curves or surfaces. Up to (possibly) $\widehat{\mathbb{D}/\Gamma}$ they are normal, especially $\widehat{\mathbb{D}/\Gamma'_{\mathbb{D}}}$ is smooth. Now we investigate the maps in more detail.

The GALOIS covering $(\mathbb{B}/\Gamma')/(\mathbb{B}/\Gamma)$ extends to a quotient map $\widehat{\mathbb{B}/\Gamma'} \longrightarrow \widehat{\mathbb{B}/\Gamma}$ by HARTOG's Theorem (for normal complex singularities, see e.g. [BAS]) applied to the extensions of the action of elements $g \in G := \Gamma/\Gamma'$ around the cusp singularities.

Observe that $Z_{\Gamma'}(\mathbb{D}) = 1$ because the neat group Γ' has only trivial finite subgroups and by 4.4.2. Therefore $N_{\Gamma'}(\mathbb{D}) = \Gamma'_{\mathbb{D}}$. The holomorphic embedding $\mathbb{D} \longrightarrow \mathbb{B}$ descends to a holomorphic map $\mathbb{D}/N_{\Gamma'}(\mathbb{D}) \longrightarrow \mathbb{B}/\Gamma'$ because the covering maps $\mathbb{B} \longrightarrow \mathbb{B}/\Gamma'$, $\mathbb{D} \longrightarrow \mathbb{D}/\Gamma'_{\mathbb{D}}$ are local isomorphisms. By an extension theorem of BOREL in [BO4] each holomorphic map of the punctured disc Δ^* into $\widehat{\mathbb{B}/\Gamma'}$ extends to a holomorphic map $\Delta \longrightarrow \widehat{\mathbb{B}/\Gamma'}$. This is the local situation we need for extending $\mathbb{D}/\Gamma'_{\mathbb{D}} \longrightarrow \widehat{\mathbb{B}/\Gamma'}$ holomorphically to $\widehat{\mathbb{D}/\Gamma'_{\mathbb{D}}}$ as described in the last row of the diagram.

Obviously, $\widehat{\mathbb{D}/\Gamma}$ is the image of $\widehat{\mathbb{D}/\Gamma'}$ along the GALOIS covering $\widehat{\mathbb{B}/\Gamma'} \longrightarrow \widehat{\mathbb{B}/\Gamma}$. By a theorem of CHOW (see e.g. [MU3], I, §3) holomorphic maps of algebraic varieties are algebraic. Therefore $\widehat{\mathbb{D}/\Gamma}$ is a complete algebraic curve on $\widehat{\mathbb{B}/\Gamma}$ with ZARISKI open part $\mathbb{D}/\Gamma$. □

In the situation above the curves $\mathbb{D}/\Gamma'$ and $\mathbb{D}/\Gamma$ are not smooth in general. We want to characterize those points $P \in \mathbb{B}$ whose images are singularities. For this purpose we introduce the following notations:

$$\begin{aligned} \Gamma\mathbb{D} &:= \{\gamma\mathbb{D};\ \gamma \in \Gamma\}, \\ \{\Gamma\mathbb{D}\}_P &:= \{\text{discs of } \Gamma\mathbb{D} \text{ going through } P \in \mathbb{B}\}, \\ \{\Gamma P\}_\mathbb{D} &:= \{\text{ points of the orbit } \Gamma P \text{ lying on } \mathbb{D}\}. \end{aligned} \tag{4.4.5}$$

Lemma 4.4.4 *For $P \in \mathbb{D}$ the correspondence $\gamma P \mapsto \gamma^{-1}\mathbb{D}$ defines a bijective map*

$$\{\Gamma P\}_\mathbb{D}/\Gamma_\mathbb{D} \mapsto \{\Gamma\mathbb{D}\}_P/\Gamma_P.$$

Proof The map is correctly defined. Namely, let $\gamma P \sim \beta P \bmod N_\Gamma(\mathbb{D})$, this means $\gamma P = \delta\beta P$ with $\delta \in N_\Gamma(\mathbb{D})$; with $\sigma := \gamma^{-1}\delta\beta \in \Gamma_P$ one gets $\gamma^{-1}\mathbb{D} = \sigma\beta^{-1}\mathbb{D}$, that means $\gamma^{-1}\mathbb{D} \sim \beta^{-1}\mathbb{D} \bmod \Gamma_P$.

Conversely, γ, β, σ determine $\delta := \gamma\sigma\beta^{-1} \in N_\Gamma(\mathbb{D})$ such that $\gamma P = \delta\beta P$, hence $\gamma P \sim \beta P \bmod N_\Gamma(\mathbb{D})$. □

Definition 4.4.5 With the notations of (4.4.5) we call $P \in \mathbb{D}$ an *honest Γ-cross point* of $\mathbb{D}$, if $\{\Gamma\mathbb{D}\}_P \neq \Gamma_P\mathbb{D}$. By Lemma 4.4.4 it is equivalent to say that $\{\Gamma P\}_\mathbb{D} \neq \Gamma_\mathbb{D}P$. A *Γ-cross point* of $\mathbb{D}$ is a point $P \in \mathbb{D}$ such that $\{\Gamma\mathbb{D}\}_P \neq \{\mathbb{D}\}$.

Proposition 4.4.6 *Let Γ be a ball lattice, $p : \mathbb{B} \longrightarrow \mathbb{B}/\Gamma$ the quotient map and $\mathbb{D}$ a Γ-rational disc. Then it holds that*

(i) For $P \in \mathbb{D}$ the image point $\overline{P} := p(P)$ is a regular point of the curve $\mathbb{D}/\Gamma \subset \mathbb{B}/\Gamma$ if and only if P is not an honest Γ-cross point of $\mathbb{D}$;

(ii) the set of honest Γ-cross point of $\mathbb{D}$ is the $\Gamma_\mathbb{D}$ -orbit of a finite set of points $P_1, \ldots, P_k \in \mathbb{D}$,

(iii) the set of honest Γ-cross points of $\mathbb{D}$ is a discrete subset of $\mathbb{D}$;

(iv) $\{\Gamma\mathbb{D}\}_P$ is a finite set for all $P \in \mathbb{D}$;

(v) the canonical morphism $\mathbb{D}/\Gamma_\mathbb{D} \longrightarrow \mathbb{D}/\Gamma$ is birational, hence $\mathbb{D}/\Gamma_\mathbb{D}$ is the normalization of $\mathbb{D}/\Gamma$;

(vi) the canonical morphism $\widehat{\mathbb{D}/\Gamma_\mathbb{D}} \longrightarrow \widehat{\mathbb{D}/\Gamma}$ is birational, hence $\widehat{\mathbb{D}/\Gamma_\mathbb{D}}$ is the normalization of $\widehat{\mathbb{D}/\Gamma}$.

Proof (i) The quotient curve $\mathbb{D}/\Gamma_\mathbb{D}$ is smooth. Therefore, the point $\overline{P}$ is regular if and only if Γ and $\Gamma_\mathbb{D}$ yield the same equivalence on $\{\Gamma P\}_\mathbb{D}$. Now the statement follows by the definition of honest Γ-cross points.

(ii),(iii) The algebraic curve $\mathbb{D}/\Gamma$ has only finitely many singularities. Therefore the preimages of. Sing $(\mathbb{D}/\Gamma)$ on $\mathbb{B}$ along p and on $\mathbb{D}$ along $\mathbb{D} \longrightarrow \mathbb{D}/\Gamma$ are discrete sets on $\mathbb{B}$ or $\mathbb{D}$, respectively. By (i) these preimages on $\mathbb{D}$ are the honest Γ-cross points on $\mathbb{D}$. Up to Γ-equivalence their number is finite. Also their preimage along $\mathbb{D}/\Gamma_{\mathbb{D}} \longrightarrow \mathbb{D}/\Gamma$ is a finite set $\{P'_1, ..., P'_k\}$. Now (ii) follows immediately, and (iii) too.

(iv) Assume that $\{\Gamma\mathbb{D}\}_P$ is not finite. Since Γ_P is finite it follows that P is an honest Γ-cross point of $\mathbb{D}$ by Definition 4.4.5. Moreover, by Lemma 4.4.4 also $\{\Gamma P\}_{\mathbb{D}}/\Gamma_{\mathbb{D}}$ is an infinite set. This contradicts with (ii).

(v),(vi) Let $\mathbb{D}^*$ be the (open analytic) complement in $\mathbb{D}$ of the set of honest Γ-cross points. By (ii) the quotient curve $\mathbb{D}^*/\Gamma_{\mathbb{D}}$ is a ZARISKI open set on $\mathbb{D}/\Gamma_{\mathbb{D}}$. On $\mathbb{D}^*$ the Γ- and $\Gamma_{\mathbb{D}}$ -equivalence coincide by definition of honest Γ-cross points, hence $\mathbb{D}^*/\Gamma_{\mathbb{D}} = \mathbb{D}^*/\Gamma$.

□

Definition 4.4.7 Let Γ be a ball lattice and $\mathbb{D}$ a Γ-rational subdisc of $\mathbb{B}$. A cofinite subgroup Γ' of Γ is called $\mathbb{D}$ *-neat*, if it is a neat ball lattice and $\mathbb{D}/\Gamma' \subset \mathbb{B}/\Gamma'$ is smooth. By 4.4.6(i) the latter condition is equivalent to the absence of honest Γ'-cross points on $\mathbb{D}$.

We want to study now what happens at the cusp points. For this purpose we change over again via projective transformation from $\mathbb{B}$ to the SIEGEL realization $\mathbb{V}$ of the ball defined in (4.2.2). Then linear subdiscs of $\mathbb{B}$ are transformed to linear subdiscs of $\mathbb{V}$. As we mentioned already at the beginning of the section the ball group G acts transitively on the set of linear subdiscs. Therefore, we can use a special one going through $\kappa = \infty$. Working with coordinates (z, u) on $\mathbb{V}$ we choose $\mathbb{D}$ defined by $u = 0$. Following the maps of diagram (4.2.8) the image of $\mathbb{D}$ on $\mathbb{C}^* \times \mathbb{C}$ has coordinates $(w, 0)$. Therefore, $\mathbb{D}/\Gamma_\kappa$ lives on one fibre of the cusp bundle $F_\kappa(\Gamma)$ through the zero section T_κ. By diagram (4.2.9) this is precisely the local situation around the compactifying elliptic curve T_κ of the ball quotient. Under global Γ-identifications it may happen that the quotient curve $\mathbb{D}/\Gamma$ appears on several fibres through T_κ. In any case we get the following result:

Lemma 4.4.8 *Let Γ be a neat ball lattice, $\mathbb{D} \subset \mathbb{B}$ a Γ-rational disc, $\overline{\mathbb{B}/\Gamma}$ the toroidal compactification of $\mathbb{B}/\Gamma$ and $\overline{\mathbb{D}/\Gamma}$ the closure of $\mathbb{D}/\Gamma$ on $\overline{\mathbb{B}/\Gamma}$. The curve $\overline{\mathbb{D}/\Gamma}$ intersects the compactification curve $T = \sum_{\kappa \bmod \Gamma} T_\kappa$ transversally at each common point. Especially $\overline{\mathbb{D}/\Gamma}$ is smooth at these points.*

In order to establish a relative proportionality principle for quotient curves on ball quotient surfaces we restrict ourselves now to neat ball lattices Γ and linear subdiscs $\mathbb{D}$ of the two-ball such that Γ is also $\mathbb{D}$ -neat. We have a commutative diagram:

$$\begin{array}{ccccccc} \mathbb{P}^2 = \mathbb{B}^\vee & \longleftarrow & \mathbb{B} & \longrightarrow & \mathbb{B}/\Gamma & \longrightarrow & \overline{\mathbb{B}/\Gamma} \\ \uparrow & & \uparrow & & \uparrow & & \uparrow \\ \mathbb{P}^1 = \mathbb{D}^\vee & \longleftarrow & \mathbb{D} & \longrightarrow & \mathbb{D}/\Gamma & \longrightarrow & \overline{\mathbb{D}/\Gamma} \end{array} \tag{4.4.6}$$

with smooth analytic embeddings at the vertical arrows. Without loss of generality we let now $\mathbb{D} \subset \mathbb{B}$ be the special disc $\mathbb{D} := (0 \times \mathbb{C}) \cap \mathbb{B}$. Working again with equivalent hermitian vector bundles (E', h') over $\mathbb{B}$, their extensions and quotients as described in Diagram (4.3.1), we look now at their restrictions according to Diagram (4.4.6). Starting with $f' := E'|\mathbb{D}$ together with the $\mathbb{D}$-restriction of the hermitian product h' we get by the HIRZEBRUCH-MUMFORD construction a commutative diagram:

$$\begin{array}{ccccccc} f^\vee & \overset{\text{restriction}}{\dashleftarrow} & f' & \overset{\text{factorisation}}{\dashrightarrow} & f & \overset{\text{extension}}{\dashrightarrow} & \overline{f} \\ \downarrow & & \downarrow & & \downarrow & & \downarrow \\ \mathbb{D}^\vee & \longleftarrow & \mathbb{D} & \longrightarrow & \mathbb{D}/\Gamma_{\mathbb{D}} & \longrightarrow & \overline{\mathbb{D}/\Gamma_{\mathbb{D}}} \end{array} \tag{4.4.7}$$

We check that the restriction of the bundles of (4.3.1) over surfaces lead to the same bundles over curves as in the construction just described. This becomes clear, if we consider the LIE groups taking part. We work with

$$G = \mathbb{S}U((2,1),\mathbb{C}), G^{\mathbb{C}} = \mathbb{S}l_3(\mathbb{C}), K = \mathbb{S}(\mathbb{U}(2) \times \mathbb{U}(1)), K^{\mathbb{C}} = \mathbb{S}(\mathbb{S}l_2(\mathbb{C}) \times \mathbb{C}^*),$$
$$H = \mathbb{S}U((1,1),\mathbb{C}), H^{\mathbb{C}} = \mathbb{S}l_2(\mathbb{C}), k = \mathbb{S}(\mathbb{U}(1) \times \mathbb{U}(1)), k^{\mathbb{C}} = \mathbb{S}(\mathbb{C}^* \times \mathbb{C}^*)$$

and the restriction diagrams

$$\begin{array}{ccc} G^{\mathbb{C}} & \longleftarrow & G \\ \uparrow & & \uparrow \\ H^{\mathbb{C}} & \longleftarrow & H \end{array} \qquad \begin{array}{ccc} \mathbb{P}^2 = NK^{\mathbb{C}}\backslash G^{\mathbb{C}} & \longleftarrow & K\backslash G = \mathbb{B} \\ \uparrow & & \uparrow \\ \mathbb{P}^1 = nk^{\mathbb{C}}\backslash H^{\mathbb{C}} & \longleftarrow & k\backslash H = \mathbb{D} \end{array}$$

where

$$N = \left\{ \begin{pmatrix} 1 & 0 & 0 \\ 0 & 1 & 0 \\ a & b & 1 \end{pmatrix}; a, b \in \mathbb{C} \right\}, \quad n = \left\{ \begin{pmatrix} 1 & 0 \\ c & 1 \end{pmatrix}; c \in \mathbb{C} \right\}$$

The embeddings are induced by

$$H^{\mathbb{C}} \longrightarrow G^{\mathbb{C}}, \; X \longmapsto \left(\begin{array}{c|cc} 1 & 0 & 0 \\ \hline 0 & & \\ 0 & & X \end{array}\right)$$

Now it is clear that

$$E^{\vee}|\mathbb{D}^{\vee} = E' \times_{NK^{\mathbb{C}}} G^{\mathbb{C}}|\mathbb{D}^{\vee} = f' \times_{nk^{\mathbb{C}}} H^{\mathbb{C}} = f^{\vee}.$$

Furthermore, the quotient equivalence of the categories of holomorphic Γ-vector bundles on $\mathbb{B}$ and holomorphic vector bundles on $\mathbb{B}/\Gamma$ fits into the equivalence of the categories of holomorphic $\Gamma_{\mathbb{D}}$-vector bundles on $\mathbb{D}$ and holomorphic vector bundles on $\mathbb{D}/\Gamma_{\mathbb{D}}$ by restrictions. Therefore, $f = E|\left(\mathbb{D}/\Gamma_{\mathbb{D}}\right)$.

The G-invariant metric h' on E' restricts to an H-invariant metric on f'. Descending to $\mathbb{D}/\Gamma \subset \mathbb{B}/\Gamma$ we can use the metric $h|f$ for the construction of $\overline{e}$. By MUMFORD's construction the holomorphic bundle $\overline{E}$ is uniquely determined extending the logarithmically restricted metric h of E over $\mathbb{B}/\Gamma$ to a hermitian metric $\overline{h}$ of $\overline{E}$ over $\overline{\mathbb{B}/\Gamma}$. The same is true for the restriction $(f, h|(\mathbb{D}/\Gamma_{\mathbb{D}}))$. Therefore $\overline{f}$ and $\overline{E}|\left(\overline{\mathbb{D}/\Gamma_{\mathbb{D}}}\right)$ coincide.

Theorem 4.4.9 (Relative Proportionality) *With the notations of the bundle diagram* (4.4.7) *it holds that*

$$c_1\left(\overline{E}|\left(\overline{\mathbb{D}/\Gamma}\right)\right) = c_1(E^{\vee}|\mathbb{D}^{\vee}) \cdot c_1(\mathbb{D}/\Gamma_{\mathbb{D}})/c_1(\mathbb{D}^{\vee}), \tag{4.4.8}$$

where $c_1(\mathbb{D}/\Gamma_{\mathbb{D}})$ is the volume of a $\Gamma_{\mathbb{D}}$-fundamental domain on $\mathbb{D}$ with respect to the POINCARÉ *metric on $\mathbb{D}$ and $c_1\left(\mathbb{D}^{\vee}\right) = c_1\left(\mathbb{P}^1\right) = 2$ is the* EULER *number of $\mathbb{P}^1$.*

Proof We apply the HIRZEBRUCH-MUMFORD Proportionality Theorem 4.3.4 to the restricted bundle diagram (4.4.7) and receive

$$c_1(\overline{f}) = v \cdot c_1(f^{\vee}) \tag{4.4.9}$$

with a constant v only depending on Γ. As in the proof of Corollary 4.3.5 it is easy to see that $v = c_1(\mathbb{D}/\Gamma_{\mathbb{D}})/c_1(\mathbb{D}^{\vee})$. By the considerations above the formula (4.4.8) is nothing else but (4.4.9) now.

□

As an immediate consequence we obtain:

Proposition 4.4.10 *Let* $\mathbb{D}$ *be a linear subdisc of the two-ball* $\mathbb{B}$, Γ *a* $\mathbb{D}$-*neat ball lattice and* $T = \left(\overline{\mathbb{B}/\Gamma}\right) \setminus (\mathbb{B}/\Gamma)$ *the compactification divisor. Then the* EULER *number of* $\overline{\mathbb{D}/\Gamma} \subset \overline{\mathbb{B}/\Gamma}$, *its selfintersection index and its intersection number with a canonical divisor* K *of* $\overline{\mathbb{B}/\Gamma}$ *are related with the* EULER-POINCARE *volume* $c_1(\mathbb{D}/\Gamma_{\mathbb{D}})$ *of a* $\Gamma_{\mathbb{D}}$-*fundamental domain on* $\mathbb{D}$ *by the formulas*

$$e\left(\overline{\mathbb{D}/\Gamma}\right) = c_1(\mathbb{D}/\Gamma_{\mathbb{D}}) + \left(T \cdot \overline{\mathbb{D}/\Gamma}\right); \tag{a}$$

$$\left(K \cdot \overline{\mathbb{D}/\Gamma}\right) = -\left(T \cdot \overline{\mathbb{D}/\Gamma}\right) - 3c_1(\mathbb{D}/\Gamma_{\mathbb{D}}); \tag{b}$$

$$\left(\overline{\mathbb{D}/\Gamma}^2\right) = c_1(\mathbb{D}/\Gamma_{\mathbb{D}}). \tag{c}$$

Proof (a) Let $t = \left(\overline{\mathbb{D}/\Gamma}\right) \setminus (\mathbb{D}/\Gamma)$ be the compactification divisor of $\mathbb{D}/\Gamma$ and $|t| = \deg t$ its number of points. By a classical formula (see e.g. [SH4], Theorem 4.2.11) one knows that

$$e\left(\overline{\mathbb{D}/\Gamma}\right) = c_1\left(\mathbb{D}/\Gamma_{\mathbb{D}}\right) + |t|.$$

On the other hand it holds that $|t| = \left(T \cdot \overline{\mathbb{D}/\Gamma}\right)$ because $\overline{\mathbb{D}/\Gamma}$ intersects T transversally at each common point by Lemma 4.4.8. Now (a) follows immediately.

(b) We apply the relative proportionality formula (4.4.8) to the canonical bundle $E' = \mathcal{J}_{\mathbb{B}}^* \wedge \mathcal{J}_{\mathbb{B}}^*$ of $\mathbb{B}$. By (4.3.11) $\overline{E}$ corresponds to the logarithmic canonical divisor $K + T$ on $\overline{\mathbb{B}/\Gamma}$ and $E^\vee$ to the canonical divisor $K_{\mathbb{P}^2}$ on $\mathbb{P}^2$, see 4.3.3. The classical formula for the intersection product of curves as first CHERN number of the restriction of the normal bundle, the relative proportionality formula (4.4.8) and the choice $K_{\mathbb{P}^2} = -3H, H$ a projective line on $\mathbb{P}^2$, yield

$$\begin{aligned}\left((K+T) \cdot \overline{\mathbb{D}/\Gamma}\right) &= c_1\left(\overline{E}|\overline{\mathbb{D}/\Gamma}\right) = c_1(E^\vee|\mathbb{D}^\vee) \cdot c_1(\mathbb{D}/\Gamma_{\mathbb{D}})/c_1(\mathbb{D}^\vee)\\ &= (H \cdot K_{\mathbb{P}^2}) \cdot c_1(\mathbb{D}/\Gamma_{\mathbb{D}})/2\\ &= (-3H \cdot H) \cdot c_1(\mathbb{D}/\Gamma_{\mathbb{D}})/2\\ &= -3 \cdot c_1(\mathbb{D}/\Gamma_{\mathbb{D}})/2.\end{aligned}$$

This proves (b).

(c) is a consequence of (a), (b) and the adjunction formula (4.3.12) applied to the curve $\overline{\mathbb{D}/\Gamma}$:

$$-e\left(\overline{\mathbb{D}/\Gamma}\right) = \left(\overline{\mathbb{D}/\Gamma}^2\right) + \left(K \cdot \overline{\mathbb{D}/\Gamma}\right).$$

□

At the end of this section we prove a general existence theorem for $\mathbb{D}$-neat lattices for given linear subdiscs of $\mathbb{B}$ in the arithmetic case.

Definition 4.4.11 For an arithmetic ball group Γ commensurable with $\mathbf{G}_{\mathbb{Q}}(\mathbb{Z})$ the linear subdisc $\mathbb{D}$ of $\mathbb{B}$ is called *arithmetic* (with respect to Γ or $\mathbf{G}/\mathbb{Z}$), if the $\mathbb{D}$-normalizing LIE group $H = N_G(\mathbb{D})$ is defined over $\mathbb{Q}$.

In this case $\mathbb{D}$ is a Γ-rational disc because $N_\Gamma(\mathbb{D})$ is obviously commensurable with the arithmetic $\mathbb{D}$-lattice $\mathbf{H}_{\mathbb{Q}}(\mathbb{Z})$.

Proposition 4.4.12 *Let Γ be an arithmetic ball group and $\mathbb{D}$ a Γ-rational disc. Then there exists a cofinite $\mathbb{D}$ -neat normal subgroup Γ' of Γ.*

Proof First we deal with the principal case $\Gamma = \mathbf{G}_{\mathbb{Q}}(\mathbb{Z})$, see 4.1.5. By an argument of BOREL in [BO3] there are only finitely many positive integers a such that the congruence sub-group $\Gamma(a)$, defined by the exact sequence

$$1 \longrightarrow \Gamma(a) \longrightarrow \Gamma \longrightarrow \mathbf{G}_{\mathbb{Q}}(\mathbb{Z}/\mathbb{Z}a),$$

is not neat. We prove, moreover, that for given $P \in \mathbb{D}$.

4.4.13 *There exist only finitely many principal congruence subgroups $\Gamma(a)$ such that $\Gamma P \cap \mathbb{D}$ contains an honest $\Gamma(a)$-cross point.*

This is sufficient. Namely, the set of Γ-orbits of honest Γ-cross points of $\mathbb{D}$ is finite by Proposition 4.4.6 (i) because the algebraic curve has only finitely many singularities.

Let $P_1, ..., P_k$ be a complete set of representants of $\{\Gamma P\}_{\mathbb{D}}$ modulo $\Gamma_{\mathbb{D}}$. We define a finite representative set $\{\gamma_{ij}\} \subset \Gamma$ by

$$\{\Gamma\mathbb{D}\}_{P_i} \setminus \{\mathbb{D}\} =: \{\mathbb{D}_{ij} := \gamma_{ij}\mathbb{D}; j = 1, \ldots, n_j\},\ P_j = \gamma_{ij}P. \tag{4.4.10}$$

The normalizer $\mathbf{N}$ of $\mathbf{H} = \mathbf{H}(\mathbb{D})$ in $\mathbf{G}$, see 4.4.11, is also a $\mathbb{Q}$-subgroup of $\mathbf{G}$ by [BO3], I.1.7. By Theorem 4.2.3.1 of the same monograph there is a faithful representation $\alpha : G \longrightarrow \mathbb{G}\mathrm{l}(E)$, E a finite-dimensional vector space, all defined over $\mathbb{Q}$, and one-dimensional $\mathbb{Q}$-subspace $L \subset E$ such that

$$N = \mathbf{N}(\mathbb{R}) = \{g \in G;\ \alpha(g)L = L\}.$$

On this way α defines by restriction a character

$$\rho : \mathbf{N}(\mathbb{Q}) \longrightarrow \mathbb{Q}^*, \quad \alpha(g)X = \rho(g)X \text{ for (a fixed) } X \in L(\mathbb{Z}),\ g \in \mathbf{N}(\mathbb{Q}).$$

It restricts to the identity on $\mathbf{H}(\mathbb{Q})$ because $H \cong \mathbb{S}\mathrm{l}_2(\mathbb{R})$ is simple. Since $N_\Gamma(\mathbb{D})$ is commensurable with $\mathbf{H}(\mathbb{Z})$ we have $\rho(N_\Gamma(\mathbb{D})) \subset \{\pm 1\}$.

Observe that $X_{ij} := \gamma_{ij}X \neq \pm X$. Otherwise we could conclude that $\gamma_{ij} \in N_\Gamma(\mathbb{D})$ in contradiction to the choice of γ_{ij} in (4.4.10). Namely, assume that $\gamma_{ij}X = \pm X$. Then $\gamma_{ij}N\gamma_{ij}^{-1}L = L$, hence $\gamma_{ij}N\gamma_{ij}^{-1} = N$ and $\gamma_{ij} \in N \cap \Gamma = N_\Gamma(\mathbb{D})$. Therefore, for almost all integers a it holds that

$$X \equiv \pm X_{ij} \bmod a \text{ for } i = 1, \ldots, n,\ j = 1, \ldots, n_j.$$

For these numbers we can show that the points of $\{\Gamma P\}_\mathbb{D}$ loose the property to be honest cross points with respect to the congruence subgroups $\Gamma(a)$.

Assume for example, that P is an honest $\Gamma(a)$-cross point. Then there are elements $\gamma' \in \Gamma(a)$, $\delta \in N_\Gamma(\mathbb{D})$ and a pair i, j such that $\gamma'\mathbb{D} = \delta\gamma_{ij}\mathbb{D}$. Therefore, $\gamma_{ij}^{-1}\delta\gamma' \in N_\Gamma(\mathbb{D})$, hence $\gamma_{ij}^{-1}\delta\gamma'X = \pm X$, thus $\gamma''X = \pm X_{ij}$ for $\gamma'' := \delta^{-1}\gamma'\delta \in \Gamma'$. But then we come to the contradiction

$$\pm X \equiv \gamma''X = \gamma''\delta^{-1}X = \gamma_{ij}(\gamma_{ij}^{-1}\delta^{-1}\gamma')X = \pm\gamma_{ij}X = \pm X_{ij} \bmod a.$$

If Γ' is a sublattice of Γ, then there are less honest Γ'-cross points on $\mathbb{D}$ than Γ-cross points because $\{\Gamma'\mathbb{D}\} \subset \{\Gamma\mathbb{D}\}$. Therefore, $\Gamma'(a) := \Gamma(a) \cap \Gamma'$ kills the crossing property of a point $P \in \mathbb{D}$, if $\Gamma(a)$ does.
If $\Gamma \not\subset \mathbf{G}(\mathbb{Z})$, then we find a quotient ring $\mathbb{Z}_d := \bigcup_{n \geq 0} \mathbb{Z}/d^n$ such that $\Gamma \subset \mathbf{G}(\mathbb{Z}_d)$, see Borel [BO3], proof of 17.4. Then the same arguments work for $\mathbb{Z}_d$ instead of $\mathbb{Z}$.

□

4.5 Cusp Singularities, Reflections and Elliptic Points

We already studied neat ball cusp singularities coming from unipotent ball cusp lattices. Let Γ be an arbitrary ball lattice and $\kappa \in \partial_\Gamma\mathbb{B}$ a Γ-cusp. As in 4.2.5 we denote the corresponding ball cusp lattice $\Gamma \cap P_\kappa$ by Γ_κ and the intersection with U_κ by $\Gamma_{\kappa u}$. Then $\Gamma_{\kappa u}$ is a neat ball cusp lattice and we have a normal subgroup filtration

$$1 \subset \Delta_\kappa \subset \Gamma_{\kappa u} \subseteq \Gamma_\kappa. \tag{4.5.1}$$

Omitting the index κ we remember that it sits in the LANGLAND's decomposition of the parabolic (BOREL) subgroup $\mathbf{T} \cdot \mathbf{M} \cdot \mathbf{U}, \mathbf{T}$ a subtorus of $\mathbf{G}$ of $\mathbb{R}$ -rank 1, $\mathbf{T} \cdot \mathbf{M} = Z(\mathbf{T})$ the centralizer of $\mathbf{T}$ (see [RAG], 12.4). For the real points we have

$$P = \mathbf{T}(\mathbb{R}) \cdot M \cdot U \cong \mathbb{R}^* \cdot S^1 \cdot U = \mathbb{R}^*_+ \cdot S^1 \cdot U,$$

compare with the decomposition 4.2.2 at $\kappa = \infty$. The elements of $\mathbb{R}^*$ do not centralize U. With the notation of 4.2.2 the adjoint action of $t \in \mathbb{R}^*_+$ is explicitly given by $[a, r] \mapsto [ta, t^2 r]$. Therefore, it holds that $\Gamma_\kappa \cap \mathbf{T}(\mathbb{R})_+ = 1$. Otherwise $\Gamma_{\kappa u}$ could not be a sublattice of U because $\Lambda_\kappa = \Gamma_{\kappa u}/\Delta_\kappa$ is not in this case, in contradiction to the defining cusp condition. Furthermore, Γ_κ can only be a discrete subgroup of $S^1 \cdot U$, if its projection on S^1 is a finite group (consisting of unit roots). Therefore,

4.5.1 $G_\kappa := \Gamma_\kappa/\Gamma_{\kappa u}$ *is a finite cyclic group.*

With the notation of 4.2.2 again we set for abbreviation $\theta := \operatorname{diag}(\theta, \theta^{-2}, \theta)$. The adjoint action of the unit root θ on U and the action on $\mathbb{V}$ is described by

$$\theta[a, r]\theta^{-1} = \left[\overline{\theta}^3 a, r\right] \text{ or } \theta(z, u) = \left(z, \overline{\theta}^3 u\right), \tag{4.5.2}$$

respectively. Thus the affine action (4.2.14) of $\Gamma_{\kappa u}$ on $\mathbb{V}$ extends to an affine action of Γ_κ on $\mathbb{V} \subset \mathbb{C}^2$. Moreover, the latter descends to a Γ_κ-action on $\mathbb{V}/\Delta \subset \mathbb{C}^* \times \mathbb{C}$ also extending linearly to $\mathbb{C} \times \mathbb{C}$ as $\Lambda_\kappa = \Gamma_{\kappa u}/\Delta$ does, see (4.2). Furthermore, the Γ_κ-action descends to $\mathbb{V}/\Gamma_{\kappa u} \subset \mathbb{C}^2/\Lambda_\kappa = F_\kappa(\Gamma)$ (look at the cusp bundle diagram (4.2.8)). We have to check that the implication

$$\begin{aligned} &(z', u') \sim (z, u) \bmod \Gamma_{\kappa u} \\ \Rightarrow\ &\gamma(z', u') \sim \gamma(z, u) \bmod \Gamma_{\kappa u} (\text{ for all } \gamma \in \Gamma_{\kappa u}) \end{aligned}$$

holds. Indeed, if $(z', u') = \lambda(z, u)$, $\lambda \in \Gamma_{\kappa u}$, then $\gamma(z', u') = \lambda' \gamma(z, u)$ for $\lambda' = \gamma\lambda\gamma^{-1} \in \Gamma_{\kappa u}$. Altogether

4.5.2 *For each Γ-cusp $\kappa \in \partial\mathbb{B}$ the cusp bundle diagram (4.2.8) extends to a commutative diagram*

$$\begin{array}{ccccc} F_\kappa & \longrightarrow & F'_\kappa = F_\kappa/G_\kappa & \longrightarrow & \widehat{F_\kappa} \\ \uparrow & & \uparrow & & \uparrow \\ \overline{\mathbb{V}/\Gamma_{\kappa u}} & \longrightarrow & (\mathbb{V}/G_\kappa)' & \longrightarrow & \widehat{\mathbb{V}/G_\kappa} \\ \uparrow & & \uparrow & & \uparrow \\ \mathbb{V}/\Gamma_{\kappa u} & \longrightarrow & \mathbb{V}/\Gamma_\kappa & \xrightarrow{=} & \mathbb{V}/\Gamma_\kappa \end{array} \tag{4.5.3}$$

The action of the cyclic group Γ_κ is compatible with the bundle morphism $F_\kappa \longrightarrow T_\kappa = T(\Gamma_{\kappa u})$. Especially G_κ acts on T_κ understood as zero section of F_κ.

The last two statements follow from the second identity of (4.5.2). The right lower square describes locally the situation at any cusp point of the BAILY-BOREL compactification of a ball quotient surface together with the small (toroidal) resolution. For a ball lattice Γ we get globally a commutative diagram:

$$\begin{array}{ccc} (\mathbb{B}/\Gamma)' & \longrightarrow & \widehat{\mathbb{B}/\Gamma} \\ \uparrow\!\cup & & \uparrow\!\cup \\ (\mathbb{B}/\Gamma)'_f & \longrightarrow & \mathbb{B}/\Gamma \end{array} \qquad (4.5.4)$$

using notations of section **3**.7, compare with (3.7.1).

Proposition 4.5.3 *Let Γ be a ball lattice. Then the cusp singularities $\widehat{\kappa}$ of $\widehat{\mathbb{B}/\Gamma}$ are finite quotients of neat cusp singularities. The small resolution curve of $\widehat{\kappa}$ is the quotient T_κ/G_κ of the elliptic curve $T = T_\kappa \subset F_\kappa$ by the cyclic group G_κ as described in diagram (4.5.3). The group $\overline{G}_\kappa := G_\kappa/Z_{G_\kappa}(T_\kappa)$ acting effectively on T_κ is cyclic of order 2, 3, 4 or 6, if G_κ acts not trivially on F_κ. At least in the latter three cases $\widehat{\kappa}$ is a quotient singularity.*

Proof If G_κ acts trivially, then we are in the situation of section 2, diagram (4.2.8). Then the small resolution of κ is an elliptic curve.

Assume now that G_κ acts not trivially. Then by (4.5.2) G_κ has at least one fixed point on F_κ, namely the image point O of $(w,u) = (0,0)$ along $\mathbb{C}^2 \longrightarrow \mathbb{C}^2/\Lambda_\kappa = F_\kappa$. Therefore, the quotient curve T_κ/G_κ is isomorphic to $\mathbb{P}^1$. By Lemma 3.5.7 the ball cusp singularity $\widehat{\kappa}$ is a (reduced) cusp point in the sense of Definition 3.5.6. The resolution graphs have been classified in 3.5.9(ii). Since G_κ fixes $O \in T_\kappa$, the group order $|\overline{G}_\kappa|$ appears there as highest number in one of the classifying tupels (2,2,2,2), (3,3,3), (2,4,4), (2,3,6) (according to HURWITZ' genus formula 2.2.5 applied to the covering $T_\kappa \longrightarrow \mathbb{P}^1$). The last statement comes from comparison with the list of quotient singularities Figure 3.5.2.

□

Now, we come to the study of branch curves of the (locally finite) quotient morphism $p : \mathbb{B} \longrightarrow \mathbb{B}/\Gamma$, Γ a ball lattice.

Definition 4.5.4 An element $\sigma \in \mathbb{U}((2,1),\mathbb{C})$ is called a *(ball) reflection* if it has finite order and $\dim_{\mathbb{C}} \mathrm{Fix}_{\mathbb{B}}\, \sigma = 1$. If the reflection σ belongs to the ball lattice Γ, then it is called a Γ*-reflection.* In this case we call the linear subdisc $\mathrm{Fix}_{\mathbb{B}}\, \sigma$ a Γ*-reflection disc.* It will be denoted by $\mathbb{D}_\sigma$.

Indeed, if σ fixes a one-dimensional subset $\mathbb{D} \subset \mathbb{B} \subset \mathbb{P}^2$, then it fixes pointwise a projective line going through $\mathbb{B}$ because of the linear action. Therefore, $\mathrm{Fix}_{\mathbb{B}}\, \sigma$ is the intersection of that line with $\mathbb{B}$, hence a (complete) linear subdisc of $\mathbb{B}$ by the definition at the beginning of section 4.

If σ is a Γ-reflection disc and Γ' a normal sublattice of Γ, then σ *normalizes* Γ', this means that $\sigma\Gamma'\sigma^{-1} \subseteq \Gamma'$. If this property is satisfied for any given ball lattice Γ' and a reflection σ, then we call σ a Γ'*-normal reflection* and $\mathbb{D}_\sigma$ a Γ'*-normal reflection disc.*

Lemma 4.5.5 *Let Γ' be a neat ball lattice and σ a Γ'-normal reflection. Then Γ' is a $\mathbb{D}_\sigma$-neat ball lattice.*

Proof According to Definition 4.4.7 we have to show that there is no Γ'-cross point on $\mathbb{D}_\sigma$. Let $P, Q \in \mathbb{D}_\sigma$ be two Γ'-equivalent points. We must prove that P and Q are $N_{\Gamma'}(\mathbb{D}_\sigma)$-equivalent. By assumption it holds that $Q = \gamma P$ for a suitable $\gamma \in \Gamma'$, hence

$$Q \in \gamma\mathbb{D}_\sigma \cap \mathbb{D}_\sigma = \mathbb{D}_{\gamma\sigma\gamma^{-1}} \cap \mathbb{D}_\sigma .$$

Since σ normalizes Γ' one finds $\gamma' \in \Gamma'$ such that $\sigma\gamma^{-1} = \gamma'\sigma$; the element $\gamma\gamma' = (\gamma\sigma\gamma^{-1})\sigma^{-1} \in \Gamma'$ has Q as fixed point. But Γ' acts as a neat group freely on $\mathbb{B}$, hence $\gamma\sigma\gamma^{-1} = \sigma$ and $\gamma\mathbb{D}_\sigma = \mathbb{D}_{\gamma\sigma\gamma^{-1}} = \mathbb{D}_\sigma$. Therefore γ is an element of $N_{\Gamma'}(\mathbb{D}_\sigma)$, thus the points P and Q are $N_{\Gamma'}(\mathbb{D}_\sigma)$-equivalent. □

Corollary 4.5.6 *Let Γ be a ball lattice, σ a Γ-reflection. Then there is a commutative diagram (4.5.5) with uniformizations on the left (horizontal) side, finite* GALOIS *coverings on the right (horizontal) side and smooth embeddings at the left vertical arrows. Assume that Γ acts effectively on $\mathbb{B}$. Then for $C := \mathbb{D}/\Gamma'_{\mathbb{D}}$ the ramification index of the finite surface covering at C is equal to the order of $Z_\Gamma(\mathbb{D})$.*

$$\begin{array}{ccccc} \mathbb{B} & \longrightarrow & \mathbb{B}/\Gamma' & \longrightarrow & \mathbb{B}/\Gamma \\ \uparrow & & \uparrow & & \uparrow \\ \mathbb{D} & \longrightarrow & \mathbb{D}/\Gamma'_{\mathbb{D}} & \longrightarrow & \mathbb{D}/\Gamma_{\mathbb{D}} \end{array} \qquad (4.5.5)$$

Proof By Proposition 4.1.8 there exists a cofinite neat normal subgroup Γ' of Γ. This is a $\mathbb{D}_\sigma$-neat ball lattice by Lemma 4.5.5 and the remark before. From Proposition 4.4.6 it follows that $\mathbb{D}/\Gamma'_{\mathbb{D}} = \mathbb{D}/\Gamma'$ is a smooth curve on the smooth surface $\mathbb{B}/\Gamma'$, see also Definition 4.4.7. Furthermore $\mathbb{B}/\Gamma$ is the GALOIS quotient surface of $\mathbb{B}/\Gamma'$ by $G = \Gamma/\Gamma'$.

The ramification index at C is the order of the cyclic inertia group $Z_G(C)$. Because of local isomorphy of the uniformization $p' : \mathbb{B} \longrightarrow \mathbb{B}/\Gamma'$ around an arbitrary point $P \in \mathbb{D} = \mathbb{D}_\sigma$ and its image $P' := p'(P)$ the (faithful) representations of Γ_P and $G_{P'}$ on the tangent spaces of $\mathbb{B}$ at P or of $\mathbb{B}/\Gamma'$ at P', respectively, are equivalent. If we choose P' general on C, then $G_{P'}$ coincides with $Z_G(C)$ because there are only finitely many possibilities of non-cyclic subgroups of G and they can only appear as stationary groups of finitely many points of the quasiprojective surface $\mathbb{B}/\Gamma'$. So, for a general choice of $P \in \mathbb{D}$ we get

$$|Z_\Gamma(\mathbb{D})| = |\Gamma_P| = |G_{P'}| = |Z_G(C)|. \tag{4.5.6}$$

□

Definition 4.5.7 Let Γ be a ball lattice acting effectively on $\mathbb{B}$. A point $P \in \mathbb{B}$ is called a Γ-*elliptic point* , if the isotropy group Γ_P is not trivial and not generated by a Γ-reflection.

Example 4.5.8 If P is the interesection point of two different Γ–reflection discs, then P is a Γ-elliptic point.

Lemma 4.5.9 *The point $P \in \mathbb{B}$ is a Γ-elliptic point if and only if its image point $\overline{P}$ on $\mathbb{B}/\Gamma$ is a surface singularity or Γ_P is generated by Γ-reflections but not by only one of them.*

Proof Choose a neat normal sublattice Γ' of Γ. The image of P on $\mathbb{B}/\Gamma'$ or on $\mathbb{B}/\Gamma$ is denoted by P' or $\overline{P}$, respectively. The canonical map $\mathbb{B}/\Gamma' \longrightarrow \mathbb{B}/\Gamma$ is a finite GALOIS covering with GALOIS group $G := \Gamma/\Gamma'$. By local isomorphy the isotropy groups Γ_P and $G_{P'}$ are isomorphic, and $\overline{\mathbf{P}}$ is the fraction point $P/\Gamma_P \cong P'/G_{P'}$ in the sense of section **3**.5, see Lemma 3.5.7, with finite uniformization $P' \longrightarrow \overline{\mathbf{P}}$. By CHEVALLEY'S Lemma 1.1.1 $\overline{P}$ is not a surface singularity if and only if $G_{P'}$ is generated by reflections. Because of the equivalent representations of Γ_P and $G_{P'}$ on tangent spaces, see 4.6.2, below, it is the same to say that Γ_P is generated by reflections. Exclude now the case when Γ_P is trivial or generated by one reflection. This means precisely on the one hand that P is Γ-elliptic and on the other hand that $\overline{P}$ is a singularity or Γ_P is generated by reflections but not by only one.

□

Corollary 4.5.10 *The set $Ell(\Gamma)$ of Γ-elliptic points is a discrete subset of $\mathbb{B}$. It is the Γ-orbit of finitely many ball points.*

Proof If P is an elliptic point, then also γP for each $\gamma \in P$ is because $\Gamma_{\gamma P} = \gamma \Gamma_P \gamma^{-1}$. Therefore $Ell(\Gamma)$ consists of complete Γ-orbits of ball points. Furthermore, there is a disjoint union

$$Ell(\Gamma) = Ell_{sing}(\Gamma) \cup Ell_{reg}(\Gamma) \tag{4.5.7}$$

corresponding to elliptic points P with singular image point $\overline{P} \in \mathbb{B}/\Gamma$ or with non-singular ones, respectively. Since there are only finitely many singularities on any normal algebraic surface, the set $Ell_{sing}(\Gamma)$ consists of finitely many Γ-orbits. By the Lemma and Definition 4.4.5 the other subset $Ell_{reg}(\Gamma)$ consists of Γ-cross points of suitable Γ-reflection discs. By local isomorphy the images of them appear on $\mathbb{B}/\Gamma'$ as intersection points of 1–dimensional components of the ramification locus of the finite GALOIS covering $\mathbb{B}/\Gamma' \longrightarrow \mathbb{B}/\Gamma$, see the proof of 4.5.9. Therefore $Ell_{reg}(\Gamma)$ consists of finitely many Γ'-orbits; the number of Γ-orbits is not greater.

□

In Chapter 5, see proof of Proposition 5.4.5 we will see that there exist only finitely many isomorphy classes of isotropy groups Γ_P simultaneously for all PICARD modular groups Γ.

4.6 Orbital Ball Quotient Surfaces and Molecular Graphs

We are now able to extend the calculations of invariants of disc quotients by $\mathbb{D}$-neat lattices (see Prop. 4.4.10) to the general case. Let Γ be a ball lattice, $\mathbb{D} = \mathbb{D}_\sigma$ a Γ-reflection disc and Γ_1 a neat normal subgroup of finite index of Γ with factor group $G = \Gamma/\Gamma_1$. We consider the following commutative diagram (4.6.1) with lower part in the category of orbital surfaces with BAILY-BOREL compactifications $\widehat{\mathbb{B}/\Gamma_1}$, $\widehat{\mathbb{B}/\Gamma}$ of $\mathbb{B}/\Gamma_1$ or $\mathbb{B}/\Gamma$, respectively. The morphism $\overline{\mathbb{B}/\Gamma_1} \to \widehat{\mathbb{B}/\Gamma_1}$ resolves the (neat) cusp singularities; $\widehat{\mathbb{B}/\Gamma}$and $\overline{\mathbb{B}/\Gamma}$ are compact orbital quotient surfaces. Since $\overline{\mathbb{B}/\Gamma_1}$ is smooth,

$$
\begin{array}{ccccccccc}
 & & & & & & \mathbb{B} & & \\
 & & & & & & \downarrow \Gamma_1 & & \\
\overline{\mathbb{B}/\Gamma_1}' & \longrightarrow & \overline{\mathbb{B}/\Gamma_1} & \longrightarrow & \widehat{\mathbb{B}/\Gamma_1} & \supset & \mathbb{B}/\Gamma_1 & & \Gamma \\
\downarrow G & & \downarrow G & & \downarrow G & & \downarrow G & & \\
\overline{\mathbb{B}/\Gamma}' & \longrightarrow & \overline{\mathbb{B}/\Gamma} & \longrightarrow & \widehat{\mathbb{B}/\Gamma} & \supset & \mathbb{B}/\Gamma & &
\end{array}
\tag{4.6.1}
$$

the orbital surface $\overline{\mathbb{B}/\Gamma}$ is quasismooth and $\overline{\mathbb{B}/\Gamma}'$ is the canonical resolution of quotient points (triple points) in the sence of 3.6.2. So $\overline{\mathbb{B}/\Gamma}'$ is a locally abelian orbital surface. $\mathbb{B}/\Gamma$ is a (quasismooth) open (orbital) surface in the sense of definition 3.7.2. the notations in diagram (4.6.1) correspond to the notations of diagram (3.7.1).

The *ball quotient arrangement* of $\mathbb{B}/\Gamma$ is defined to be the quotient arrangement of the quotient map $\mathbb{B}/\Gamma_1 \underset{G}{\rightarrow} \mathbb{B}/\Gamma = (\mathbb{B}/\Gamma_1)/G$. It is easy to see that it does not depend on the choice of $\Gamma_1 \subset \Gamma$. Namely, the one-dimensional part of the branch locus comes from the Γ-reflection discs.

Lemma 4.6.1 *Let g be an element of G and C a one-dimensional irreducible component of the fixed point set $Fix_{\overline{\mathbb{B}/\Gamma_1}}(g)$ not lying on the compactification divisor $T = \overline{\mathbb{B}/\Gamma_1} \setminus (\mathbb{B}/\Gamma_1)$. Then there is a reflection $\sigma \in \Gamma$ such that $C = \overline{\mathbb{D}_\sigma/\Gamma_1}$. Moreover, there are natural bijections*

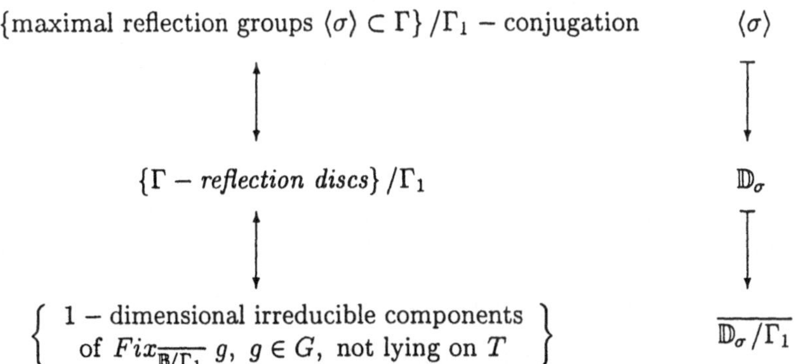

Proof The second map is correctly defined because $\overline{\mathbb{D}_\sigma/\Gamma_1}$ is a component of $\operatorname{Fix}\overline{\sigma}$, where $\overline{\sigma}$ denotes the image of σ in $G = \Gamma/\Gamma_1$. This map is injective.

Namely, if $\overline{\mathbb{D}_\sigma/\Gamma_1} = \overline{\mathbb{D}_\tau/\Gamma_1}$ for two Γ-refelctions σ, τ, then outside of T we get $\mathbb{D}_\sigma/\Gamma_1 = \mathbb{D}_\tau/\Gamma_1$. We choose a general point $P \in \mathbb{D}_\sigma$ such that the image point $\overline{P} \in \mathbb{B}/\Gamma_1$ is not an intersection point of two different quotients $\mathbb{D}/\Gamma_1$ of Γ-reflection discs on $\mathbb{B}/\Gamma_1$. For a suitable $\gamma_1 \in \Gamma_1$ it holds that

$$P \in \gamma_1\mathbb{D}_\tau = \mathbb{D}_{\gamma_1\tau\gamma_1^{-1}}, P \in \mathbb{D}_\sigma \cap \mathbb{D}_{\gamma_1\tau\gamma_1^{-1}}$$

because $\overline{P} \in \mathbb{D}_\sigma/\Gamma_1 = \mathbb{D}_\tau/\Gamma_1$. From our choice of P it follows that $\mathbb{D}_\sigma/\Gamma_1 = \mathbb{D}_{\gamma_1\tau\gamma_1^{-1}}/\Gamma_1$, hence $\mathbb{D}_{\gamma_1\tau\gamma_1^{-1}} = \gamma'\mathbb{D}_\sigma$ for a suitable $\gamma' \in \Gamma_1$. Therefore,

$$\mathbb{D}_\sigma = \gamma'^{-1}\mathbb{D}_{\gamma_1\tau\gamma_1^{-1}} = \left(\gamma'^{-1}\gamma_1\right)\mathbb{D}_\tau = \gamma_2\mathbb{D}_\tau, \gamma_2 \in \Gamma_1.$$

Thus, the second map is injective.

The bijectivity of the first map is obvious. So it remains to prove the first part of the lemma. We need the following fact:

4.6.2 *The canonical homomorphism $\varphi : \Gamma \to G = \Gamma/\Gamma_1$ induces for each point $P \in \mathbb{B}$ an isomorphism of isotropy groups $\varphi_P : \Gamma_P \to G_{\overline{P}}, \overline{P} \in \mathbb{B}/\Gamma_1$ image of P, such that the natural representations of Γ_P and $G_{\overline{P}}$ in the tangent spaces $T_P(\mathbb{B})$ or $T_{\overline{P}}(\mathbb{B}/\Gamma_1)$ are equivalent.*

Proof If $\varphi_P(\gamma) = 1$, $\gamma \in \Gamma_P$, then $\gamma \in \Gamma_1 \cap \Gamma_P = \Gamma_{1P} = 1$, hence $\operatorname{Ker}\varphi_P = 1$. Now let g be an element of $G_{\overline{P}}$, $g = \gamma\Gamma_1$, $\gamma \in \Gamma$. Therefore $\gamma\Gamma_1 \cdot \Gamma_1 P = g\overline{P} = \overline{P} = \Gamma_1 P$, hence $\gamma P = \gamma_1 P$ for a suitable $\gamma_1 \in \Gamma_1$, $\gamma' = \gamma^{-1}\gamma_1 \in \Gamma_P$, $g^{-1} = \gamma'\Gamma_1 \in \varphi_P(\Gamma_P)$. Therefore g lies in the image $\varphi_P(\Gamma_P)$, so φ_P is surjective, hence bijective. For the second part of 4.6.2 we take $\gamma \in \Gamma_P$, $g = \varphi_P(\gamma)$ and remark that $p : \mathbb{B} \to \mathbb{B}/\Gamma_1$ is a universal covering, hence locally isomorphic. So we get commutative diagrams:

$$\begin{array}{ccc} \mathbb{B} & \xrightarrow{p} & \mathbb{B}/\Gamma_1 \\ \gamma\downarrow\wr & & \wr\downarrow g \\ \mathbb{B} & \xrightarrow{p} & \mathbb{B}/\Gamma_1 \end{array} \qquad \begin{array}{ccc} T_P(\mathbb{B}) & \xrightarrow{\sim} & T_{\overline{P}}(\mathbb{B}/\Gamma') \\ (d\gamma)_P\downarrow\wr & & (dg)_{\overline{P}}\downarrow\wr \\ T_P(\mathbb{B}) & \xrightarrow{\sim} & T_{\overline{P}}(\mathbb{B}/\Gamma') \end{array} \tag{4.6.2}$$

□

Now we can finish the proof of the lemma. Let C be a component of $\operatorname{Fix}_{\overline{\mathbb{B}/\Gamma_1}}(g), P \in \mathbb{B}$ such that the image point $\overline{P}$ lies in C. Since g acts trivially on C, its linearization $(dg)_{\overline{P}}$ at $\overline{P}$ has diagonal form $\begin{pmatrix} 1 & 0 \\ 0 & \zeta \end{pmatrix}$. The same is true for $(d\gamma)_P$ by 4.6.2 or diagrams (4.6.2), where $\gamma = \varphi_{\overline{P}}^{-1}(g)$ is the preimage of g along φ_P. Therefore, γ is a Γ-reflection and $p(\mathbb{D}_\gamma) = C$. The lemma is proved.

□

The ramification curves on $\mathbb{B}/\Gamma_1$ of $q : \mathbb{B}/\Gamma_1 \to \mathbb{B}/\Gamma$ are images of Γ-reflection discs. The branch curves on $\mathbb{B}/\Gamma$ of the same map are the images of these ramification curves. Therefore,

Corollary 4.6.3 *The branch curves of q are the images of the Γ-reflection discs. The ramification index of a branch curve $\mathbb{D}/\Gamma = \mathbb{D}_\sigma/\Gamma$ is the order of the cyclic (effective) Γ-reflection group $Z_\Gamma(\mathbb{D})$.*

For the last statement we have only to check that

$$Z_G(C) \simeq Z_\Gamma(\mathbb{D}),\ C = \mathbb{D}/\Gamma_1$$

by means of diagrams (4.6.2) at a general point $P \in \mathbb{D}$. If Γ does not act effectively, one has to work with the corresponding projective group $\mathbb{P}\Gamma \subset \mathbb{P}\mathbb{U}((2,1),\mathbb{C})$.

The ramification indices appear as weights $v_{\mathbb{D}/\Gamma} = |Z_\Gamma(\mathbb{D})|$ of the open orbital curves $\mathbb{D}/\Gamma = C/G$. The diagram (4.6.1) restricts to quotient curves of Γ-reflection discs $\mathbb{D}$ and the related curves in the following manner, see Diagram:

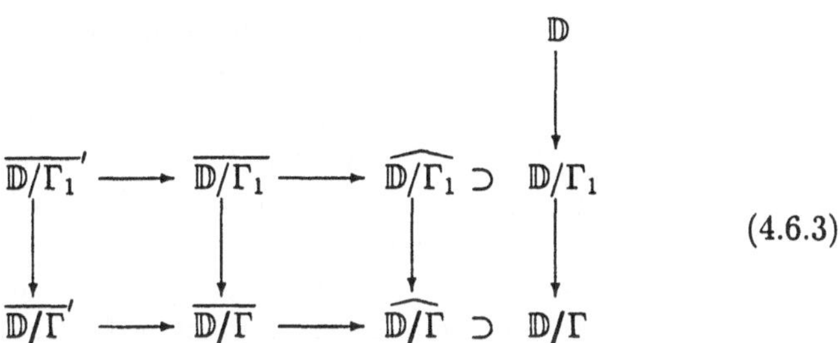

$$\begin{array}{ccccccc} & & & & & & \mathbb{D} \\ & & & & & & \downarrow \\ \overline{\mathbb{D}/\Gamma_1}' & \longrightarrow & \overline{\mathbb{D}/\Gamma_1} & \longrightarrow & \widehat{\mathbb{D}/\Gamma_1} & \supset & \mathbb{D}/\Gamma_1 \\ \downarrow & & \downarrow & & \downarrow & & \downarrow \\ \overline{\mathbb{D}/\Gamma}' & \longrightarrow & \overline{\mathbb{D}/\Gamma} & \longrightarrow & \widehat{\mathbb{D}/\Gamma} & \supset & \mathbb{D}/\Gamma \end{array} \qquad (4.6.3)$$

$\overline{\mathbb{D}/\Gamma_1}$ is smooth because Γ_1 is a $\mathbb{D}$-neat ball lattice, see Lemma 4.4.8. The orbital quotient curves $\overline{D/\Gamma}$ (compact) and D/Γ (open) have at most triple points by Lemma 3.6.7. They are simultaneously resolved for all quotients of Γ-reflection discs $\mathbb{D}/\Gamma$ by $\overline{\mathbb{B}/\Gamma}' \to \overline{\mathbb{B}/\Gamma}$ because the latter morphism resolves all triple points. We want to recognize the preimages of triple points of B/Γ on $\mathbb{B}$.

Lemma 4.6.4 *The point $P \in \mathbb{B}$ is the preimage of a (weighted) triple point of $\mathbb{B}/\Gamma$ if and only if either*

(i) Γ_P is a non-abelian (finite) group, or

(ii) Γ_P is abelian and P is an honest Γ-cross point of a Γ-reflection disc $\mathbb{D} \subset \mathbb{B}$.

In any of these cases P is a Γ-elliptic point.

Proof The types of the weighted triple points have been described in Figure 3.8.1 from a global point of view. Since $G_{\overline{P}} \simeq \Gamma_P$, $\overline{P}$ the image of P on $\mathbb{B}/\Gamma_1$, by 4.6.2 it suffices to check only the case of weighted abelian points. The problem is reduced to prove that P is an honest Γ-cross point of the Γ-reflection $\mathbb{D} = \mathbb{D}_\sigma$ if and only if $\overline{P}$ is an honest G-cross point of the image curve $C = \mathbb{D}/\Gamma'$. We prove first the implication:

$$P \in \mathbb{D} \text{ not honest} \Rightarrow \overline{P} \in C \text{ not honest} .$$

Let $\overline{P}$ be an intersection point of C and gC, $g \in G$. We have to show that $gC = g_{\overline{P}}C$ for a suitable $g_{\overline{P}} \in G_{\overline{P}}$. We lift the situation to $\mathbb{D}$ setting $g = \gamma\Gamma_1$. Without loss of generality we can assume that $\gamma\mathbb{D}$ goes through P. Namely, since $\overline{P}$ lies on gC there exists $\gamma_1 \in \Gamma_1$ such that $\gamma_1 P \in \gamma\mathbb{D}$. Instead of γ we can work with $\gamma_1^{-1}\gamma$. Obviously, P is the intersection point of $\mathbb{D}$ and $\gamma\mathbb{D}$. Since P is not an honest Γ-cross point there exists $\gamma_P \in \Gamma_P$ such that $\gamma\mathbb{D} = \gamma_P\mathbb{D}$. It follows that $gC = g_{\overline{P}}C$, $g_{\overline{P}} = \varphi_P(\gamma_P) \in G_{\overline{P}}$.

$$\overline{P} \in C \text{ not honest } \Rightarrow P \in \mathbb{D} \text{ not honest} .$$

In order to prove this implication we assume that $\{P\} = \mathbb{D} \cap \gamma\mathbb{D}$, $\gamma \in \Gamma$. Then $\overline{P}$ is an intersection point of C and gC, $g = \gamma\Gamma_1$. By the assumption of our implication there exists $g_{\overline{P}} \in G_{\overline{P}}$ such that $gC = g_{\overline{P}}C$. According to 4.6.2 we set $\gamma_P = \varphi_P^{-1}(g_{\overline{P}}) \in \Gamma_P$. Since $g_{\overline{P}}^{-1}gC = C$ we find $\gamma_1 \in \Gamma_1$ via the second bijection of Lemma 4.6.1 such that $\gamma_P^{-1}\gamma\mathbb{D} = \gamma_1\mathbb{D}$. We show that $\gamma_1\mathbb{D} = \mathbb{D}$. Then $\gamma\mathbb{D} = \gamma_P\mathbb{D}$ which has to be proved. Assume that $\gamma_1\mathbb{D} \neq \mathbb{D}$. Then P is the intersection point of $\mathbb{D}$ and $\gamma_1\mathbb{D}$ because $P \in \gamma\mathbb{D}$, hence $P \in \gamma_P^{-1}\gamma\mathbb{D} = \gamma_1\mathbb{D}$. So P is a Γ_1-cross point of $\mathbb{D}$. But $\mathbb{D}$ is Γ_1-neat by Lemma 4.5.5. From the definition 4.4.7 it follows that P is not an honest Γ_1-cross point of $\mathbb{D}$. By definition 4.4.5 there exists $\gamma_1' \in \Gamma_{1P}$ such that $\gamma_1\mathbb{D} = \gamma_1'\mathbb{D}$. Since Γ_1 is neat it acts freely on $\mathbb{B}$, hence $\Gamma_{1P} = 1$ and $\gamma_1' = 1$, thus $\gamma_1\mathbb{D} = \mathbb{D}$.

Now we prove the last statement of the lemma. If Γ_P is not abelian, then we take two elements $\gamma, \delta \in \Gamma_P$ not commuting with each other. The commutator $\gamma\delta\gamma^{-1}\delta^{-1}$ is not trivial. The determinant of its representation in the tangent space $T_P(\mathbb{B})$ is equal to 1. Therefore $\gamma\delta\gamma^{-1}\delta^{-1}$ is not a reflection. Hence it is an elliptic element with isolated fixed point P. This means that P is a Γ-elliptic point. The same holds in case (ii). This has been already proved, see Prop. 4.4.6. □

For the calculation of invariants of ball quotient surfaces it is convenient to work with *fixed point formations* on the ball. It consists of a complete system of representations of Γ-elliptic points on $\mathbb{B}$ modulo Γ and a complete system of representatives of Γ-reflection discs modulo Γ.

Example 4.6.5 [HO4]. Let Γ be the congruence subgroup of $\mathbb{U}((2,1),\mathcal{O})$, $\mathcal{O} = \mathbb{Z} + \mathbb{Z}\omega$ the ring of integral EISENSTEIN numbers, $\omega = e^{2\pi i/3}$ a primitive third unit root, with respect to the ideal $\mathcal{O}(1-\omega)$ of $\mathcal{O}$, defined by the exact sequence

$$1 \longrightarrow \Gamma \longrightarrow \mathbb{U}((2,1),\mathcal{O}) \longrightarrow \mathbb{U}((2,1),\mathcal{O}/(1-\omega)\mathcal{O}) \longrightarrow 1$$

A fixed point formation of Γ on $\mathbb{B}$ looks like Figure (4.6.1) It consists of

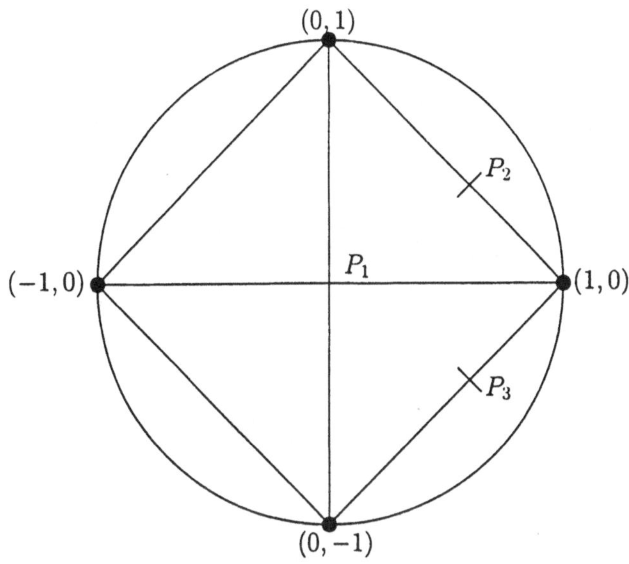

Figure 4.6.1

6 discs "going through" the pairs of four Γ-cusps and three Γ-inequivalent elliptic points, which are intersection points of two Γ-inequivalent discs. The arrangement of $\widehat{\mathbb{B}/\Gamma}$ has been drawn in Figure 3.7.1. The three elliptic points go down to $\mathbf{P}_1$, $\mathbf{P}_2$, $\mathbf{P}_3$ and κ_1, κ_2, κ_3, κ_4 corresponds to the four cusps (see [HO4]).

The orbital points $\mathbf{P}_1$, $\mathbf{P}_2$, $\mathbf{P}_3$ are abelian points. The preimages of these points in $\mathbb{B}$ are described by two crossing disc germs. These points are not blown up along $\overline{\mathbb{B}/\Gamma'} \longrightarrow \overline{\mathbb{B}/\Gamma}$ in diagram (4.6.1). The blown up points are the weighted triple points. We indicate their preimages on a Γ-fixed point formation by a thick point on the boundary.

Example 4.6.6 [SV1], see [HO2] . Let $\Gamma = S\mathbb{U}((2,1),\mathcal{O})$ be the (full) GAUSS lattice of $\mathbb{B}, \mathcal{O} = \mathbb{Z} + \mathbb{Z}i$ the ring of integral GAUSS numbers. A Γ-fixed point formation is described in Figure (4.6.2): It consists of three

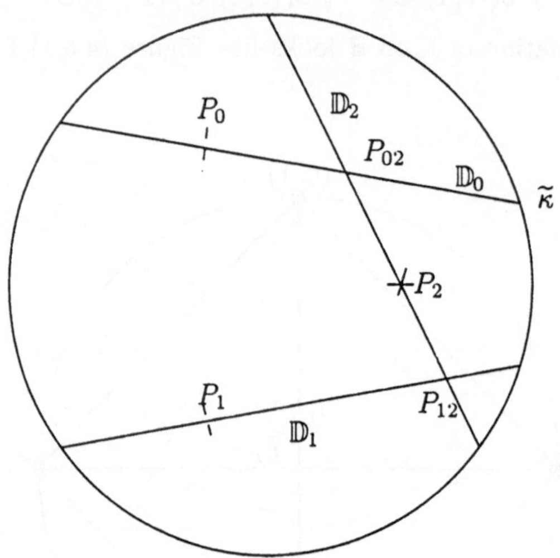

Figure 4.6.2

discs, three points with abelian isotropy groups and two cross points going down to triple point on $\mathbb{B}/\Gamma$, see Table 4.6.1.

The rough picture of a fixed point formation is not sufficient for the calculation of invariants of $\mathbb{B}/\Gamma$ or $\mathbb{D}/\Gamma \subset \mathbb{B}/\Gamma$. One needs essentially the structures of the stationary groups Γ_P, more precisely their representation in the tangent spaces $T_P(\mathbb{B})$. It turns out that the graph of the corresponding orbital point $\mathbf{P} = P/\Gamma_P \in \mathbb{B}/\Gamma$ is sufficient for the calculation of the contribution of $\mathbf{P}$ to the invariants. The possible graphs of these quotient points have been described in Corollary 3.5.9(i).

Very important are the atomic graphs of the quotients $\mathbb{D}/\Gamma$ of Γ-reflection discs $\mathbb{D}$. The center • has to be connnected with the centers □ of cusp points (weighted quadruple points), the centers △ of weighted triple points and with electrons ∘ corresponding to abelian points on D/Γ. It looks like a star as described in Figure (4.6.3): We omitted the abelian coordinates $\langle d, e\rangle$ at the arrows and the weights of the centers △ and the electrons ∘ (up to

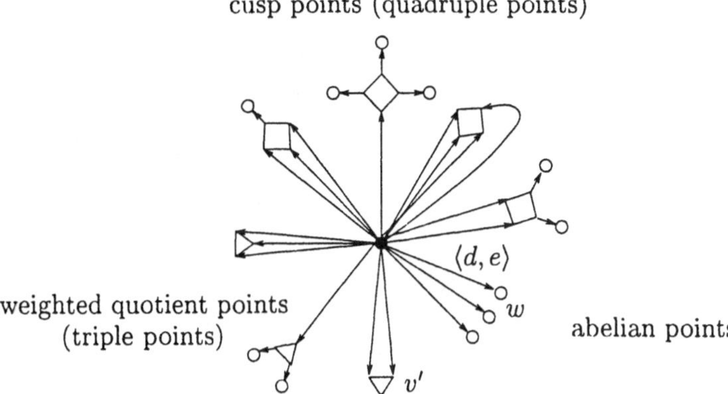

Figure 4.6.3

one case). On the left side we listed all the possibilities of connecting the center with a triple point. These possibilities have been described in Figure 3.8.1 and Figure 3.8.2. Also all possibilities of connections with a cusp point graph have been described in Figure (4.6.3). We do not distinguish between graphs of $\mathbb{D}/\Gamma, \widehat{\mathbb{D}/\Gamma}$.

Example 4.6.7 In the case of the congruence EISENSTEIN lattice introduced in 4.6.5 the quotient curve $\mathbb{D}/\Gamma$ has the graph described in Figure (4.6.4) The graph can be read off from the configuration Figure 3.7.1.

Example 4.6.8 [HO2], (3.5.5). If Γ is the full GAUSS lattice of $\mathbb{B}$ introduced in 4.6.6 then we have three orbital curves $\mathbb{D}_0/\Gamma$, $\mathbb{D}_1/\Gamma$, $\mathbb{D}_2/\Gamma$ coming from the three Γ-inequivalent reflection discs. They are drawn in the Figures (4.6.5), (4.6.6) or (4.6.7), respectively. For the selfintersection numbers we refer to 4.7.8, (4.7.10).

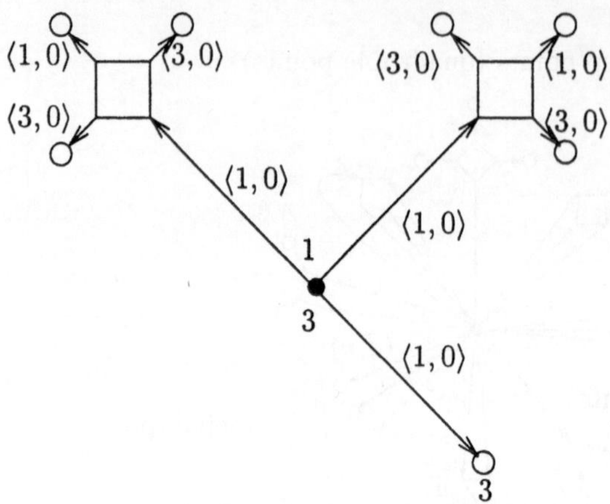

Figure 4.6.4

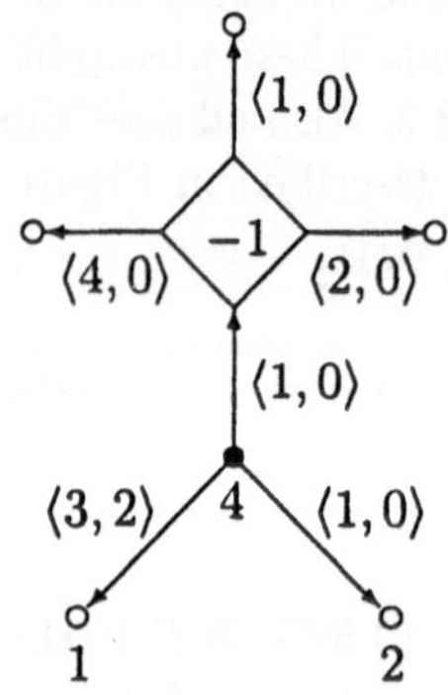

Figure 4.6.6

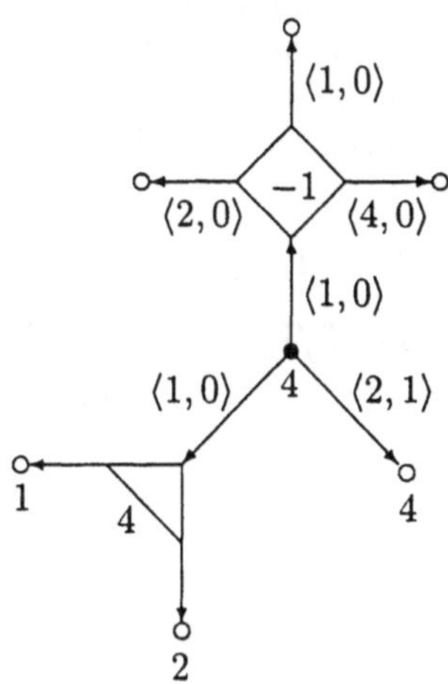

Figure 4.6.5

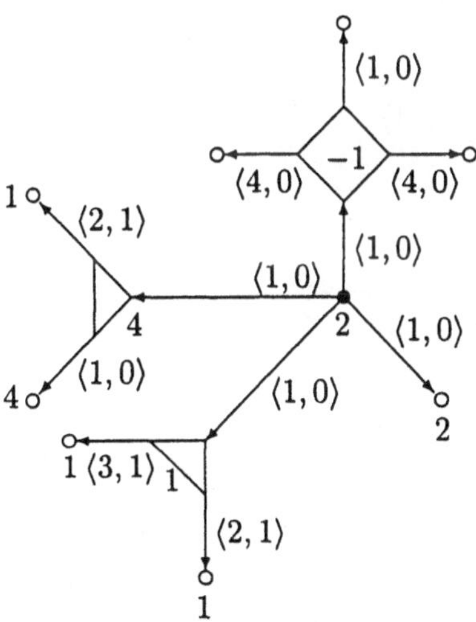

Figure 4.6.7

In Table 4.6.1 are listed the representations of the isotropy groups Γ_P, $P = P_0, P_1, P_2, P_{02}, P_{12}$ with the notations of Figure (4.6.2), in the tangent space $T_P(\mathbb{B}) \simeq \mathbb{C}^2$. The table is adapted from Table I.4.7 in [HO2], where SHVARTSMAN's classification of isotropy groups of Γ has been used [SV1]. The number -1 in $\Diamond$ is the selfintersection number (E_κ^2) of the canonical resolution of the cusp point.

Table 4.6.1

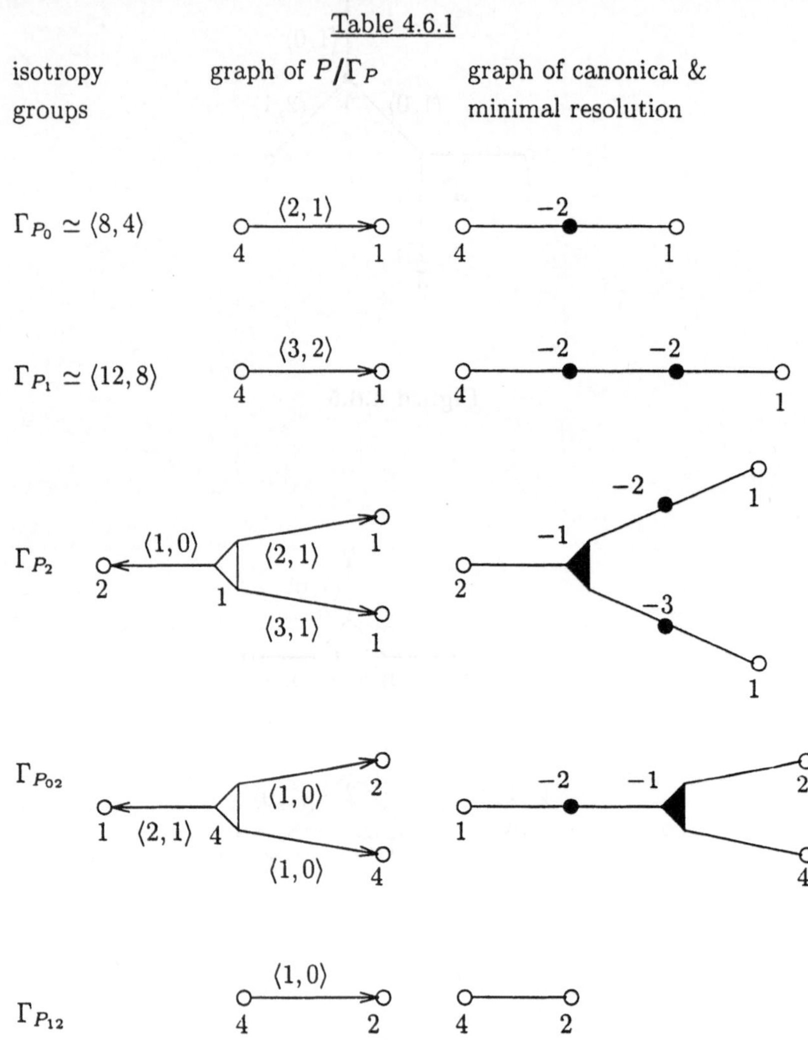

For the representation in the tangent space we used the following representative generators for the above isotrope groups:

$\Gamma_{P_0} \simeq \langle 8,4\rangle : \begin{pmatrix} j & 0 \\ 0 & -1 \end{pmatrix}, j = e^{\pi i/4};$

$\Gamma_{P_0} \simeq \langle 12,8\rangle : \begin{pmatrix} i\omega & 0 \\ 0 & \omega^2 \end{pmatrix}, \omega = e^{2\pi i/3};\ \Gamma_{P_2} : \begin{pmatrix} \omega & 0 \\ 0 & \omega^2 \end{pmatrix}, \begin{pmatrix} 0 & 1 \\ 1 & 0 \end{pmatrix};$

$\Gamma_{P_{02}} : \begin{pmatrix} i & 0 \\ 0 & 1 \end{pmatrix}, \begin{pmatrix} 1 & 0 \\ 0 & i \end{pmatrix}, \begin{pmatrix} 0 & -1 \\ 1 & 0 \end{pmatrix};\ \Gamma_{P_{12}} : \begin{pmatrix} i & 0 \\ 0 & 1 \end{pmatrix}, \begin{pmatrix} 1 & 0 \\ 0 & -1 \end{pmatrix}.$

The Γ-fixed point configuration Figure (4.6.2) teaches us how to connect the atomic graphs to the molecular graph of the arrangement of $\mathbb{B}/\Gamma$. It is described in Figure (4.6.8):

Molecular graph of $\mathbb{B}/\Gamma$
($\Gamma$the GAUSS lattice of $\mathbb{B}$)

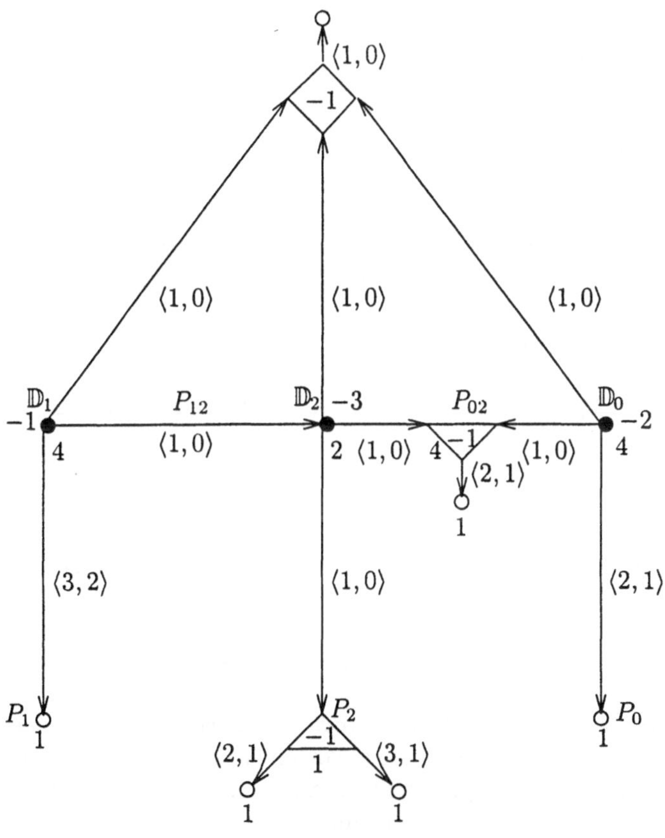

Figure 4.6.8

The molecular graph of the arrangement of $\mathbb{B}/\Gamma$ of the congruence EISENSTEIN lattice Γ of $\mathbb{B}$ introduced in 4.6.5 can be read off from the configuration 3.7.1 or better from 3.7.6. It is drawn in Figure (4.6.9) as tetrahedron. We omitted trivial coordinates $\langle 1, 0 \rangle$ of abelian points, the direction of the arrows and trivial arrows $\circ\!\xrightarrow[\langle 1,0\rangle]{}$

Molecular graph of $\mathbb{B}/\Gamma$
(Γ the congruence Eisenstein lattice of $\mathbb{B}$)

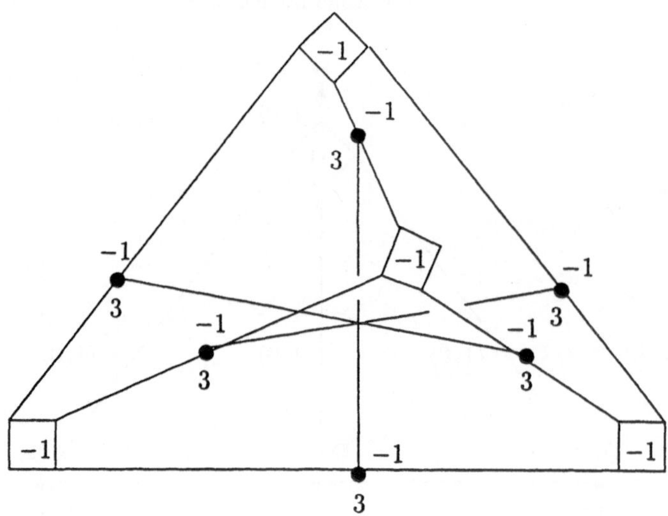

Figure 4.6.9

The graph is a version of 3.7.7 distinguishing the cusp curves in a more visible manner.

Now it should be clear what one has to understand under the atomic graph of a quotient $\mathbb{D}/\Gamma$ of a Γ-reflection curve and how to construct it.

Definition 4.6.9 The *molecular graph* of an orbital ball quotient surface $\mathbb{B}/\Gamma$ is the molecular graph of the arrangement of $\mathbb{B}/\Gamma$. It is constructed by means of the atomic graphs of the orbital curves $\mathbb{D}/\Gamma$ on $\mathbb{B}/\Gamma$ connecting intersection points of the curves supporting the derived arrangement on $\overline{\mathbb{B}/\Gamma}'$ and adding isolated abelian points of the arrangement coming from small cyclic isotropy groups $\Gamma_P \subset \Gamma$.

4.7 Invariants of Disc Quotient Curves

The invariants $e\left(\overline{\mathbb{D}/\Gamma}'\right)$ and $\left(\overline{\mathbb{D}/\Gamma}'^2\right)$ of the smooth quotient curves of Γ-reflection discs $\mathbb{D}$ on the locally abelian surface $\overline{\mathbb{B}/\Gamma}'$ can be expressed by the volumes $c_1(\mathbb{D}/\Gamma_{\mathbb{D}})$ of fundamental domains and the numerical data of the atomic graph of $\mathbb{D}/\Gamma$ (without selfintersection weights).

Theorem 4.7.1 *Let Γ be a ball lattice and $\mathbb{D}$ a Γ-reflection disc, $\overline{\mathbb{B}/\Gamma}'$ the canonical resolution of $\overline{\mathbb{B}/\Gamma}$. Then for the smooth curve $\overline{\mathbb{D}/\Gamma}' \subset \overline{\mathbb{B}/\Gamma}'$ it holds that*

$$e\left(\overline{\mathbb{D}/\Gamma}'\right) = e\left(\overline{\mathbb{D}/\Gamma_{\mathbb{D}}}\right) = c_1(\mathbb{D}/\Gamma_{\mathbb{D}}) + \sum_{i\in Q}\left(1-\frac{1}{v_i d_i}\right) + |K|, \qquad \text{(eqn:4.i)}$$

where K denotes the set of cusp arrows connecting the center of the graph of $\mathbb{D}/\Gamma$ with one of the cusp point symbols $\diamond$ (see Figure (4.6.3)) and Q is the set of quotient arrows of the graph connecting the center with a triple point symbol $\triangle$ or with an abelian point symbol $\circ$; $\langle d_i, e_i\rangle$ is the singularity type corresponding to the arrow i.

$$\left(\overline{\mathbb{D}/\Gamma}'^2\right) = \frac{1}{2}v_{\mathbb{D}}\cdot c_1(\mathbb{D}/\Gamma_{\mathbb{D}}) - \sum_{i\in Q}\frac{v_{\mathbb{D}}\cdot\delta_i + v_i e_i}{v_i d_i} - \sum_{j\in K}\frac{e_j}{v_j d_j}, \qquad \text{(eqn:4.ii)}$$

where $v_{\mathbb{D}} = |Z_\Gamma(\mathbb{D})/\text{ineffective kernel}\,|$ is the weight of the orbital curve $\mathbb{D}/\Gamma$, and

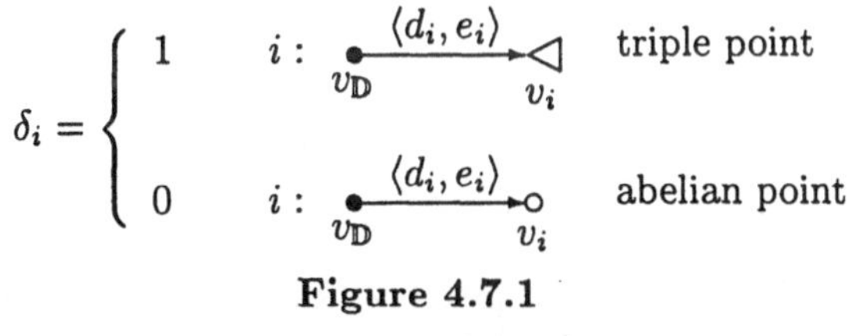

$$\delta_i = \begin{cases} 1 & i: \underset{v_{\mathbb{D}}}{\bullet}\xrightarrow{\langle d_i, e_i\rangle}\underset{v_i}{\triangleleft} \quad \text{triple point} \\[2ex] 0 & i: \underset{v_{\mathbb{D}}}{\bullet}\xrightarrow{\langle d_i, e_i\rangle}\underset{v_i}{\circ} \quad \text{abelian point} \end{cases}$$

Figure 4.7.1

Proof Let Γ_1 be a neat sublattice of Γ. Γ_1 is $\mathbb{D}$ -neat by Lemma 4.5.5. We use the notations of the diagram (4.6.3) being the restriction of diagram (4.6.1). The group $N_G\left(\overline{\mathbb{D}/\Gamma_1}'\right), G = \Gamma/\Gamma_1$, acts locally abelian along $\overline{\mathbb{D}/\Gamma_1}'$. We consider the quotient morphism of orbital curves $f : \overline{\mathbb{D}/\Gamma_1} \longrightarrow \overline{\mathbb{D}/\Gamma}$ admitting quotient points.

First we prove the identity (ii) combining Proposition 4.4.10(c) with the orbital property of signatures τ of orbital curves with quotient points, see 3.8.1. One finds that

$$\frac{1}{6}c_1(\mathbb{D}/\Gamma_{1\mathbb{D}}) = \tau\left(\overline{\mathbb{D}/\Gamma_1}\right) = |f| \cdot \tau\left(\overline{\mathbb{D}/\Gamma}\right).$$

Since the volumes of fundamental domains behave multiplicatively it holds that $c_1(\mathbb{D}/\Gamma_{1\mathbb{D}}) = |f| \cdot c_1(\mathbb{D}/\Gamma_{\mathbb{D}})$. Consequently, we get

$$\frac{1}{6}c_1(\mathbb{D}/\Gamma_{\mathbb{D}}) = \tau\left(\overline{\mathbb{D}/\Gamma}\right) = \tau\left(\overline{\mathbb{D}/\Gamma'}\right) - \tau\left(\overline{\mathbb{D}/\Gamma}'/\overline{\mathbb{D}/\Gamma}\right). \tag{4.7.1}$$

The orbital curve $\overline{\mathbb{D}/\Gamma}'$ is locally abelian. Hence

$$\tau\left(\overline{\mathbb{D}/\Gamma}'\right) = \left[\left(\overline{\mathbb{D}/\Gamma}'^2\right) + \sum_Q \frac{e_i}{d_i} + \sum_K \frac{e_j}{d_j}\right] / 3v_{\mathbb{D}} \tag{4.7.2}$$

by definition (2.4.3). The relative orbital signature $\tau\left(\overline{\mathbb{D}/\Gamma}'/\overline{\mathbb{D}/\Gamma}\right)$ has been defined in (3.8.2). With our notations it can be expressed in the following manner:

$$\tau\left(\overline{\mathbb{D}/\Gamma}'/\overline{\mathbb{D}/\Gamma}\right) = -\sum_Q \frac{\delta_i}{3v_i d_i}$$

Together with (4.7.1) and (4.7.2) we obtain

$$\frac{1}{2}v_{\mathbb{D}}c_1(\mathbb{D}/\Gamma_{\mathbb{D}}) = \left(\overline{\mathbb{D}/\Gamma}'^2\right) + \sum_Q \frac{v_{\mathbb{D}}\delta_i + v_i e_i}{v_i d_i} + \sum_K \frac{e_j}{d_j}$$

More directly we can prove the identity (i). It is a version of the classical formula

$$e\left(\overline{\mathbb{D}/\Gamma_{\mathbb{D}}}\right) = c_1(\mathbb{D}/\Gamma_{\mathbb{D}}) + \sum_{\mathbb{D}\ni P \bmod \Gamma_{\mathbb{D}}} \frac{|\Gamma_{\mathbb{D},P}| - 1}{|\Gamma_{\mathbb{D},P}|} + |\partial_{\Gamma_{\mathbb{D}}}\mathbb{D}/\Gamma_{\mathbb{D}}| \tag{4.7.3}$$

We refer the reader to SHIMURA [SH1], 2., § 2.5, where he can find a proof of (4.7.3). We have only to translate this formula. The number $|\partial_{\Gamma_{\mathbb{D}}}\mathbb{D}/\Gamma_{\mathbb{D}}|$ is nothing else but the number of cusp points of the smooth curve $\overline{\mathbb{D}/\Gamma_{\mathbb{D}}}$. This curve coincides with $\overline{\mathbb{D}/\Gamma}'$. The cusp points of $\overline{\mathbb{D}/\Gamma}'$ are the intersection points of this curve with the compactification divisor $E = \Sigma E_\kappa = \overline{\mathbb{B}/\Gamma} \setminus (\mathbb{B}/\Gamma)$. These intersection points appear in the atomic graph of $\mathbb{D}/\Gamma$ as arrows connecting the center with one of the cusp point centers $\diamond$. So we have biunivoque correspondences

$$\partial_{\Gamma_{\mathbb{D}}}\mathbb{D}/\Gamma_{\mathbb{D}} \Longleftrightarrow \{\text{cusp points of } \overline{\mathbb{D}/\Gamma_{\mathbb{D}}}\} \Longleftrightarrow K \tag{4.7.4}$$

The group $\Gamma_{\mathbb{D}}$ is defined to be $N_\Gamma(\mathbb{D})/Z_\Gamma(\mathbb{D})$. Since $\overline{\mathbb{D}/\Gamma}'$ is smooth, there is a biunivoque correspondence

$$\begin{array}{ccc} \mathbb{D}/\Gamma_{\mathbb{D}} & \Longleftrightarrow & (\mathbb{D}/\Gamma)' \subset \overline{\mathbb{D}/\Gamma}', \\ P & \longleftrightarrow & \overline{P} \end{array}$$

where $(\mathbb{D}/\Gamma)'$ is the open part $\overline{\mathbb{D}/\Gamma}' \setminus E$ of $\overline{\mathbb{D}/\Gamma}'$ defining E. We blow up the ball $\mathbb{B}$ at all preimages of triple points of $\mathbb{B}/\Gamma$. We get a modification $\mathbb{B}' \to \mathbb{B}$, $(\mathbb{D}/\Gamma)'$ lies on $\mathbb{B}'/\Gamma$ and all isotropy groups $\Gamma_{P'}, P' \in \mathbb{B}'$ are abelian. The isotropy group of $P' \in \mathbb{B}$ is $N_\Gamma(\mathbb{D})_{P'}$. We divide it by the $\mathbb{D}$ -ineffective kernel $Z_\Gamma(\mathbb{D})$ and receive

$$|\Gamma_{\mathbb{D},P'}| = |N_\Gamma(\mathbb{D})_{P'}|/|Z_\Gamma(\mathbb{D})| = v_{\mathbb{D}} \cdot v_i \cdot d_i/v_{\mathbb{D}}. \tag{4.7.5}$$

Let $\overline{P}_i$ be the image point of P' on $(\mathbb{D}/\Gamma)'$. The group $\Gamma_{\mathbb{D},P'}$ is non-trivial if and only if $\overline{\mathbf{P}}_i : \underset{v_{\mathbb{D}}}{\circ}\!\xrightarrow[\langle d_i, e_i\rangle]{}\!\underset{v_i}{\circ}$ belongs to the point arrangement of $(\mathbb{D}/\Gamma)'$. With (4.7.3), (4.7.4) and (4.7.5) we see that the formula (i) for the EULER number of $\overline{\mathbb{D}/\Gamma}'$ is correct.

□

Corollary 4.7.2 *Let Γ be an arithmetic ball lattice and $\mathbb{D}$ be a Γ-arithmetic disc in the sense of definition 4.4.11. Then the formulas (i), (ii) of Theorem 4.7.1 are valid for $\overline{\mathbb{D}/\Gamma}' \subset \overline{\mathbb{B}/\Gamma}'$, where $\overline{\mathbb{B}/\Gamma}' \longrightarrow \overline{\mathbb{B}/\Gamma}$ is the canonical resolution of all triple points of $\overline{\mathbb{D}/\Gamma} \subset \overline{\mathbb{B}/\Gamma}$.*

Proof The proof is the same as that of the theorem. We have only to remark that Γ contains a $\mathbb{D}$ -neat normal sublattice Γ_1 by Proposition 4.4.12. Then we can establish a diagram (4.6.3) as restriction of a diagram of type (4.6.1). □

It is quite natural to conjecture that the formulas (i), (ii) hold for any ball lattice Γ and any Γ-rational disc $\mathbb{D}$. Formula (i) should be clear because we did not need the intermediate group Γ_1 for the proof.

We want to deduce now a necessary criterion for an open orbital curve with non-trivial weight to be a quotient of a Γ-reflection disc, Γ a ball lattice. Let $\widehat{\mathbf{C}}$ be an orbital curve with (at most) quotient and cusp points, $\mathbf{C}$ its resolution of cusp points, $\mathbf{C}_f$ the open part and $\mathbf{C}'$ the canonical (locally abelian) resolution (see (4.7.6), (3.8.1)):

$$\begin{array}{ccccc} \mathbf{C}' & \longrightarrow & \mathbf{C} & \longrightarrow & \widehat{\mathbf{C}} \\ \cup & & \cup & & \cup \\ \mathbf{C}'_f & \longrightarrow & \mathbf{C}_f & \xrightarrow{=} & \mathbf{C}_f \end{array} \tag{4.7.6}$$

For all these orbital curves we have one and the same atomic graph of type Figure (4.6.3). Let Q be the set of quotient point arrows and K the set of cusp point arrows in the graph of $\mathbf{C}$. We set

Definition 4.7.3

$$\mathbf{e}_f(\mathbf{C}) = e(C') - \sum_{i\in Q}\left(1 - \frac{1}{v_i d_i}\right) - |K| \tag{e}$$

$$\tau_f(\mathbf{C}) = \frac{1}{3v}\left[\left(C'^2\right) + \sum_{i\in Q}\left(\frac{v\delta_i + v_i e_i}{v_i d_i}\right) + \sum_{j\in K}\frac{e_j}{d_j}\right] \tag{τ}$$

where $v = v_{\mathbf{C}}$ is the weight of $\mathbf{C}$.

Proposition 4.7.4 *(Relative proportionality). The open orbital curve $\mathbf{C}_f$ can only be a quotient curve of a Γ-reflection disc $\mathbb{D}$ of a ball lattice Γ, if*

$$v = v_{\mathbf{C}} > 1 \text{ and} \tag{i}$$

$$6\tau_f(\mathbf{C}) = \mathbf{e}_f(\mathbf{C}) < 0. \tag{ii}$$

If $\mathbf{C}_f = \mathbb{D}/\Gamma$ for a Γ-reflection disc $\mathbb{D}$, the conditions (i) and (ii) are satisfied. The condition (ii) is also satisfied for any quotient curve $\mathbb{D}/\Gamma$ of a Γ-rational disc $\mathbb{D}$ of an $\mathbb{Q}$ -arithmetic ball group Γ. So it is a necessary criteroin for $\mathbf{C}_f$ to be of such quotient type $\mathbb{D}/\Gamma$.

Proof If $\mathbf{C}_f = \mathbb{D}/\Gamma$, then it holds that

$$\mathbf{e}_f(\mathbf{C}) = c_1(\mathbb{D}/\Gamma_{\mathbf{D}}), \tau_f(\mathbf{C}) = \frac{1}{6}c_1(\mathbb{D}/\Gamma_{\mathbf{D}}) \tag{4.7.7}$$

by Definition 4.7.3, Theorem 4.7.1 (i), (ii) and Corollary 4.7.2. Hence condition (ii) is satisfied. If $\mathbb{D}$ is of reflection type, then also (i) is satisfied.

□

Corollary 4.7.5 *Let $\mathbb{D}/\Gamma$ be an open orbital curve, $\mathbb{D}$ of arithmetic or reflection type. Assume that the atomic graph of $\mathbb{D}/\Gamma$ (without selfintersection numbers) is known. then the* EULER *number of $\overline{\mathbb{D}/\Gamma}'$ determines the selfintersection number $\left(\overline{\mathbb{D}/\Gamma}'^2\right)$ and vice versa.*

Proof If the EULER number is known then we can calculate $\mathbf{e}_f(\mathbb{D}/\Gamma)$ by 4.7.3,(e). We find then $\tau_f(\mathbb{D}/\Gamma)$ by relative proportionality 4.7.4,(ii) and finally $\left(\overline{\mathbb{D}/\Gamma}'^2\right)$ by 4.7.3,(τ). The steps can be inversed.

□

Example 4.7.6 (Reflection discs of the congruence EISENSTEIN lattice of $\mathbb{B}$). Let $\mathbf{C}_f$ be an open orbital curve with graph

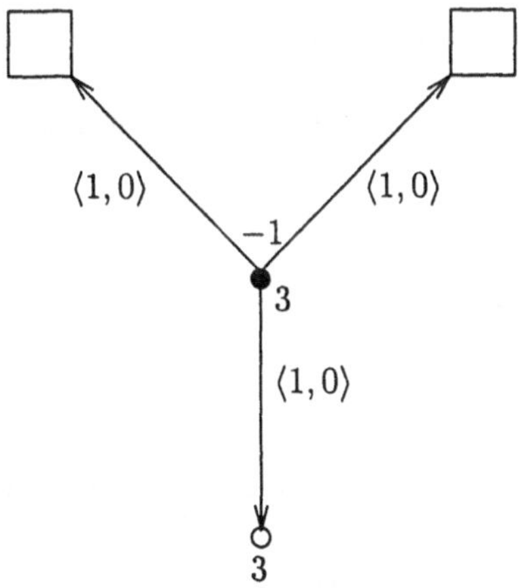

Figure 4.7.2

and $e = e(C) = 2$ that means $C \simeq \mathbb{P}^1, C_f \simeq \mathbb{P}^1 \setminus \{2 \text{ points }\}$. Then

$$\mathbf{e}_f(\mathbf{C}) = 2 - \left(1 - \frac{1}{3}\right) - 2 = -\frac{2}{3}$$
$$\tau_f(\mathbf{C}) = \frac{1}{3 \cdot 3}(-1) + 0 + 0 = -\frac{1}{9}.$$

The relations (i), (ii) of Proposition 4.7.4 are satisfied. $\mathbf{C}_f$ is really a quotient curve $\mathbb{D}/\Gamma$ of a Γ-reflection disc. Indeed, the graph (4.7.2) is part of the molecular graph (4.6.9) of the congruence EISENSTEIN lattice Γ of $\mathbb{B}$, see also (4.6.4). the corresponding geometric configuration $\underline{\text{on}}$ $\mathbb{P}^2 = \widehat{\mathbb{B}/\Gamma}$ has been described in Figure 3.7.1; see also Figure 3.7.6 for $\overline{\mathbb{B}/\Gamma}$.

Remark 4.7.7 There is a classical geometric method for finding volumes of fundamental domains of disc lattices. This method due to FORD is carefully described in [LEH]. Let $\Gamma_\mathbb{D}$ be a disc lattice acting on $\mathbb{D} \subset \mathbb{C} \subset \mathbb{P}^1$. If $\gamma \in \Gamma_\mathbb{D}$ is not a rotation around 0 or a reflection with respect to the euclidean metric of $\mathbb{C}$, then one corresponds to γ the isometry circle $K(\gamma) \subset \mathbb{P}^1$ of γ consisting of all points P where $(d\gamma)_P$ is an euclidean isometry. These circles $K(\gamma)$ cut out a domain in $\mathbb{D}$ with finite non-euclidean (EULER-POINCARÉ) volume. The euclidean motions of $\Gamma_\mathbb{D}$ subdivide this domain into $\Gamma_\mathbb{D}$-fundamental domains. Only finitely many isometry circles $K(\gamma_1), \ldots, K(\gamma_n)$ are sufficient for this construction. They can be found explicitly by means of an algorithm. On this way we can calculate $c_1(\mathbb{D}/\Gamma_\mathbb{D})$ and discuss the next

Examples 4.7.8 (Reflection discs of the full GAUSS lattice of $\mathbb{B}$). We continue with the discussions of the examples 4.6.8 calculating the orbital heights $\mathbf{c}(\mathbb{D}_i/\Gamma_{\mathbb{D}_i})$, the EULER numbers $e\left(\overline{\mathbb{D}_i/\Gamma_{\mathbb{D}_i}}'\right)$ and the selfintersection numbers $\left(\overline{\mathbb{D}_i/\Gamma_{\mathbb{D}_i}}'^2\right)$, $i = 0, 1, 2$. By the isometry circle method described above one finds that

$$c_1(\mathbb{D}_0/\Gamma_{\mathbb{D}_0}) = -\frac{1}{4}, \; c_1(\mathbb{D}_1/\Gamma_{\mathbb{D}_1}) = -\frac{1}{6}, \; c_1(\mathbb{D}_2/\Gamma_{\mathbb{D}_2}) = -\frac{1}{2}. \tag{4.7.8}$$

From (4.7.7) it follows that

$$\mathbf{e}_f(\mathbb{D}_0/\Gamma) = -\frac{1}{4}, \; \mathbf{e}_f(\mathbb{D}_1/\Gamma) = -\frac{1}{6}, \; \mathbf{e}_f(\mathbb{D}_2/\Gamma) = -\frac{1}{2},$$
$$\tau_f(\mathbb{D}_0/\Gamma) = -\frac{1}{24}, \; \tau_f(\mathbb{D}_1/\Gamma) = -\frac{1}{36}, \; \tau_f(\mathbb{D}_2/\Gamma) = -\frac{1}{12}.$$

Now, we use the graphs (4.6.5), (4.6.6) and (4.6.7) for calculating the integral invariants. For $e\left(\overline{\mathbb{D}_i/\Gamma}'\right)$ we use formula 4.7.3,(e).

$$
\begin{aligned}
e\left(\overline{\mathbb{D}_0/\Gamma}'\right) &= -\frac{1}{4} + \left(1 - \frac{1}{2}\right) + \left(1 - \frac{1}{4\cdot 1}\right) + 1 = 2 \\
e\left(\overline{\mathbb{D}_1/\Gamma}'\right) &= -\frac{1}{6} + \left(1 - \frac{1}{2\cdot 1}\right) + \left(1 - \frac{1}{1\cdot 3}\right) + 1 = 2 \\
e\left(\overline{\mathbb{D}_2/\Gamma}'\right) &= -\frac{1}{2} + \left(1 - \frac{1}{4\cdot 1}\right) + \left(1 - \frac{1}{1\cdot 1}\right) + \left(1 - \frac{1}{4\cdot 1}\right) + 1 \\
&= 2
\end{aligned} \tag{4.7.9}
$$

For the selfintersection numbers we use formula 4.7.3,(τ).

$$
\begin{aligned}
\left(\overline{\mathbb{D}_0/\Gamma}'^2\right) &= 3\cdot 4\left(-\frac{1}{24}\right) - \frac{1}{2} - \left(\frac{4\cdot 1}{4\cdot 1} + \frac{0}{1}\right) - \frac{0}{1} = -2 \\
\left(\overline{\mathbb{D}_1/\Gamma}'^2\right) &= 3\cdot 4\left(-\frac{1}{36}\right) - \frac{0}{2\cdot 1} - \frac{2}{1\cdot 3} - \frac{0}{1} = -1 \\
\left(\overline{\mathbb{D}_2/\Gamma}'^2\right) &= 3\cdot 2\left(-\frac{1}{12}\right) - \frac{2\cdot 0 + 4\cdot 0}{4\cdot 1} - \frac{2\cdot 1 + 0\cdot 1}{1\cdot 1} \\
&\quad - \frac{2\cdot 1 + 4\cdot 0}{4\cdot 1} - \frac{0}{1} = -3
\end{aligned} \tag{4.7.10}
$$

4.7.9 The curves $\overline{\mathbb{D}_i/\Gamma}'$, $i = 0, 1, 2$, are smooth and rational, hence isomorphic to $\mathbb{P}^1$, with selfintersection numbers -2, -1 or -3, respectively on $\overline{\mathbb{B}/\Gamma}'$.

□

4.8 Invariants of Ball Quotient Surfaces

In a similar manner as we have done for open orbital quotient curves of ball discs we establish formulas for the integral invariants (signature, arithmetic genus, EULER number, selfintersection number of the canonical class). The central role play the orbital invariants $\mathbf{e}(\mathbb{B}/\Gamma), \tau(\mathbb{B}/\Gamma)$ of the open orbital surfaces $\mathbb{B}/\Gamma, \Gamma$ a ball lattice. It turns out that these numbers are non-euclidean volumes of a Γ-fundamental domain with respect to the EINSTEIN-KÄHLER metric with negative RICCI curvature. More precisely, we prove the following

Theorem 4.8.1 *Let Γ be a ball lattice and $\mathbb{B}/\Gamma$ the open orbital quotient surface. Then*

$$\mathbf{e}(\mathbb{B}/\Gamma) = \mathbf{e}_f\left(\overline{\mathbb{B}/\Gamma}\right) = c_2(\mathbb{B}/\Gamma)$$
$$\tau(\mathbb{B}/\Gamma) = \tau_f\left(\overline{\mathbb{B}/\Gamma}\right) = \frac{1}{3}c_2(\mathbb{B}/\Gamma).$$

Proof By BOREL'S theorem [BO1] there exists a normal neat sublattice Γ_1 of Γ. The finite quotient group Γ/Γ_1 is denoted by G. We consider $\mathbb{B}/\Gamma_1$ as (open) orbital surface with trivial branch weights or, in other words, as reduced orbital surface. The lattice Γ_1 is neat, hence it acts freely on $\mathbb{B}$ and all cusp singularities are neat, i.e. that their minimal resolutions are elliptic curves. The orbital heights of $\mathbb{B}/\Gamma_1$ have been defined in section **3**.7, see 3.7.8, by means of the CHERN invariants:

$$\begin{aligned}\mathbf{e}(\mathbb{B}/\Gamma_1) &= \mathbf{e}\left(\overline{\mathbb{B}/\Gamma_1}\right) = e\left(\overline{\mathbb{B}/\Gamma_1}\right)\\ \tau(\mathbb{B}/\Gamma_1) &= \tau\left(\overline{\mathbb{B}/\Gamma_1}\right) - \sum_{\kappa\in K_1}\tau(T_\kappa)\\ &= \tau\left(\overline{\mathbb{B}/\Gamma_1}\right) - \sum_{\kappa\in K_1}\frac{1}{3}(T_\kappa^2),\end{aligned} \tag{4.8.1}$$

where K_1 is the set of cusp points of $\widehat{\mathbb{B}/\Gamma}_1$ and T_κ the canonical resolution of $\kappa \in K_1$. From HIRZEBRUCH-MUMFORD'S proportionality theorem and RIEMANN-ROCH theory it follows that

$$\begin{aligned}\mathbf{e}(\mathbb{B}/\Gamma_1) &= c_2(\mathbb{B}/\Gamma_1) = \int_{\mathcal{F}(\Gamma_1)}\gamma_2\\ \tau(\mathbb{B}/\Gamma_1) &= \frac{1}{3}c_2(\mathbb{B}/\Gamma_1).\end{aligned} \tag{4.8.2}$$

This follows directly from (4.8.1) and the formulas of Proposition 4.3.6.

Now we have only to apply the degree formulas for orbital heights of open orbital surfaces, see Proposition 3.7.6. Using (4.8.2) we get for $c = e, \tau$

$$\begin{aligned}\mathbf{c}(\mathbb{B}/\Gamma) &= \mathbf{c}((\mathbb{B}/\Gamma_1)/\mathbf{G}) = \frac{1}{|G|}\mathbf{c}(\mathbb{B}/\Gamma_1)\\ &= \frac{1}{|G|}\delta_c \cdot c_2(\mathbb{B}/\Gamma_1) = \delta_c \cdot c_2(\mathbb{B}/\Gamma),\\ \delta_c &= \begin{cases}1 & c = e\\ 1/3 & c = \tau\end{cases}\end{aligned}$$

because a Γ_1-fundamental domain of $\mathbb{B}$ can be received by patching together $|G|$ suitable examples of a Γ-fundamental domain of $\mathbb{B}$ moving one of them by non-euclidean (metric invariant) transformations. This proves our theorem.

□

Remark 4.8.2 We assumed that Γ acts effectively on $\mathbb{B}$. If it does not act effectively, then one has to work with the projective group $\mathbb{P}\Gamma$ and with $G = \mathbb{P}\Gamma/\Gamma'$.

The connection between the CHERN invariants of $\overline{\mathbb{B}/\Gamma}$ and the orbital heights of $\mathbb{B}/\Gamma$ can be explicitly described by means of the molecular graph (of the arrangement) of $\mathbb{B}/\Gamma$. We learned how to calculate the selfintersection numbers of the 1–dimensional irreducible components $\mathbb{D}/\Gamma$, $\mathbb{D} \subset \mathbb{B}$ a Γ-reflection disc, in the last section. So, we assume that the molecular graph of $\mathbb{B}/\Gamma$ is endowed with these numbers. Now one has only to combine Theorem 4.8.1 with the definition of orbital heights of open orbital surfaces. For the cusp part we need Definition 3.7.5, for the triple point part the Definition 3.6.8 together with (3.6.6), and for abelian quotient points the basic definitions (3.3.5) and (3.4.7).

Before we begin with the corresponding substitutions we summarize the connection between the EINSTEIN-KÄHLER metric of the ball and the CHERN invariants of ball quotient surfaces in the following scheme:

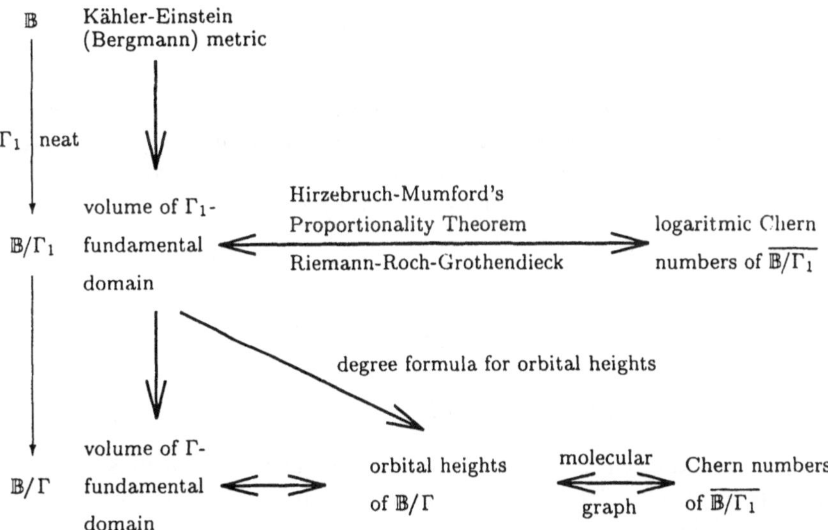

(4.8.3)

This scheme makes visible that the heights of orbital surfaces imitate volumes of fundamental domains with respect to non-euclidean metrics and the degree formula is the translation of multiplicative behaviour of the volumes.

Theorem 4.8.3 *Let Γ be a ball lattice and $\mathbb{B}/\Gamma$ the corresponding open orbital surface,*

$$\begin{array}{ccccc} \overline{\mathbb{B}/\Gamma}' & \longrightarrow & \overline{\mathbb{B}/\Gamma} & \longrightarrow & \widehat{\mathbb{B}/\Gamma} \\ \| & & \| & & \| \\ \mathbf{Y}' & \longrightarrow & \mathbf{Y} & \longrightarrow & \widehat{\mathbf{Y}} \end{array}$$

(partial) small resolutions as in diagram (4.6.1). Then

$$e\left(\overline{\mathbb{B}/\Gamma}'\right) = c_2(\mathbb{B}/\Gamma) + \sum_{\mathbf{D} \bmod \Gamma} \left(1 - \frac{1}{|\mathbb{D}|_\Gamma}\right) c_1(\mathbb{D}/\Gamma_{\mathbf{D}}) \qquad \text{(e)}$$

$$+ \sum_{\mathbf{P} \in \mathbf{Q}} \mathbf{e}_f(\mathbf{P}) + 2|\mathbf{K}_{rat}|$$

$$\tau\left(\overline{\mathbb{B}/\Gamma}'\right) = \frac{1}{3} c_2(\mathbb{B}/\Gamma) + \frac{1}{6} \sum_{\mathbf{D} \bmod \Gamma} \left(|\mathbb{D}|_\Gamma - \frac{1}{|\mathbb{D}|_\Gamma}\right) c_1(\mathbb{D}/\Gamma_{\mathbf{D}}) \qquad (\tau)$$

$$+ \sum_{\mathbf{P} \in \mathbf{Q}} \tau_f(\mathbf{P}) + \sum_{\kappa \in \mathbf{K}} (\tau_f(\kappa)),$$

where $\mathbf{Q}$, $\mathbf{K}$, $\mathbf{K}_{\mathrm{rat}}$ *denote the set of orbital quotient points, cusp points or rational cusp points, respectively,* $|\mathbb{D}|_\Gamma$ *is the order of the effective order of* $Z_\Gamma(\mathbb{D})$ *and* $c_2(\mathbb{B}/\Gamma)$, $c_1(\mathbb{D}/\Gamma_{\mathbf{D}})$ *are the non-euclidean volumes of fundamental domains. The values for* $\mathbf{c}_f(\mathbf{P})$ *and* $\mathbf{c}_f$ *are listed in table 4.8.1:*

where

$$\mathbf{e}(\mathbf{E_Q}) = \frac{1}{v_1 d_1} + \frac{1}{v_2 d_2} + \frac{1}{v_3 d_3} - 1 = \frac{2v}{|\Gamma_{\tilde{P}}|}, \ \mathbb{B} \ni \tilde{Q} \to Q,$$

$|\Gamma_{\tilde{Q}}|$ the effective order of $\Gamma_{\tilde{Q}}$; (4.8.4)

l_p, l_i, l_j are the lengths of the minimal singularity resolutions of P, P_i' or S_j', respectively; $Tr(P)$, $Tr(P_i')$, $Tr(\kappa)$ are the traces of the intersection matrices of the minimal singularity resolutions of P, P_j' or $\mathbf{E}_\kappa$, respectively; δ_i has been defined in 4.7.1(δ); there is an euclidean volume formula for the central summand (E_κ^2) of E_κ:

$$(E_\kappa^2) = -\left(\frac{2|\Lambda_\kappa|}{|\Delta_\kappa|} + \sum_{j=1}^{4} \frac{e_j}{d_j}\right) \tag{4.8.5}$$

with the notations of 4.3.7.

Proof We denote the branch arrangement of $\overline{\mathbb{B}/\Gamma}$ by $\mathbf{B} = \mathbf{B^1} + \mathbf{B^0}$ and the related arrangement of $(\mathbb{B}/\Gamma)' = \mathbf{Y}'$, coming from canonical resolutions of the triple points, by $\mathbf{B}' = \mathbf{B'^1} + \mathbf{B'^0}$. We use the orbital decomposition

$$\mathbf{c}(Y') = \mathbf{c}(\mathbf{Y}') + \mathbf{c}\left(\mathbf{B'^1}\right) + \mathbf{c}\left(\mathbf{B'^0}\right) \tag{4.8.6}$$

for $c = e$ or τ, see Lemma 3.3.3 or Lemma 3.4.2, respectively. Because of the definitions 3.7.5 and 3.6.8 of $\mathbf{c}_f(\mathbf{Y}')$ or $\mathbf{c}_f(\mathbf{Y})$ respectively, we can write

$$\mathbf{c}(\mathbf{Y}') = \mathbf{c}_f(\mathbf{Y}) + \sum_{P \in \mathbf{Q}_3} \mathbf{c}_{\mathbf{P}}(\mathbf{Y}'/\mathbf{Y}) + \sum_{\kappa \in K} \frac{1}{v_\kappa} \mathbf{c}(\mathbf{E}_\kappa) \tag{4.8.7}$$

Now we introduce in (4.8.8) for a quotient $\mathbb{D}/\Gamma$ of a Γ-reflection disc $\mathbb{D} \subset \mathbb{B}$ the following notations according to diagram (4.6.3).

$$\begin{array}{ccccc} \overline{\mathbb{D}/\Gamma}' & \longrightarrow & \overline{\mathbb{D}/\Gamma} & \longrightarrow & \widehat{\mathbb{D}/\Gamma} \\ \| & & \| & & \| \\ \mathbf{C}' & \longrightarrow & \mathbf{C} & \longrightarrow & \widehat{\mathbf{C}} \end{array} \tag{4.8.8}$$

With the Definitions 4.7.3, (e) and (τ), we can write

$$\mathbf{e}(\mathbf{C}') = \mathbf{e}_f(\mathbf{C}) + \sum_{j \in K(\mathbf{C}')} \frac{1}{v_{\kappa_j} d_j}, \tag{4.8.9}$$

$$\tau(\mathbf{C}') = \tau_f(\mathbf{C}) - \frac{1}{3} \sum_{i \in Q(\mathbf{C}')} \delta_i / v_i d_i, \tag{4.8.10}$$

where $K(\mathbf{C}') \subset K$ demotes the set of cusp points of $\mathbf{C}'$ represented by an arrow $j : \underset{v_{\mathbf{C}}}{\bullet} \!\longrightarrow\! \Diamond$ to the symbol $\Diamond$ of the cusp $\kappa = \kappa_j$, and $Q(\mathbf{C}')$ denotes the set of abelian points of $\mathbf{C}'$ represented by arrows $i : \underset{v_{\mathbf{C}}}{\bullet} \!\longrightarrow\! \circ$ or $i : \underset{v_{\mathbf{C}}}{\bullet} \!\longrightarrow\! \triangleleft$ in the molecular graph of $\mathbb{B}/\Gamma$. From (4.8.6), (4.8.7), (4.8.9), (4.8.10) and the definition of $\mathbf{c}(\mathbf{B'^1})$ and $\mathbf{c}(\mathbf{B'^0})$ in 3.3.3 and 3.4.2 it follows that

$$c(Y') - c\left(B'^0\right) = \mathbf{c}_f(\mathbf{Y}) + \sum_{\mathbf{P}} \mathbf{c}_{\mathbf{P}}(\mathbf{Y}'/\mathbf{Y}) + \sum_{\mathbf{K}} \frac{1}{v_\kappa}\mathbf{c}(\mathbf{E}_\kappa)$$
$$+ \sum_{\mathbf{C}} \lambda_c(\mathbf{C}) \cdot \left\{ \begin{matrix} \mathbf{e}_f(\mathbf{C}) + \sum\limits_{j \in \mathbf{K}(\mathbf{C}')} 1/v_\kappa d_j, & c = e \\ \tau_f(\mathbf{C}) - \frac{1}{3} \sum\limits_{i \in \mathbf{Q}(\mathbf{C}')} \delta_i/v_i d_i, & c = \tau \end{matrix} \right\}$$
$$+ \sum_{\kappa} \lambda_c(\mathbf{E}_\kappa) \cdot \mathbf{c}(\mathbf{E}_\kappa) + \sum_{\mathbf{P}} \lambda_c(\mathbf{E}_{\mathbf{P}}) \cdot \mathbf{c}(\mathbf{E}_{\mathbf{P}})$$
$$+ \sum_{\mathbf{P}'} \mathbf{c}(\mathbf{P}') + \sum_{j \in K} \mathbf{c}\left(\mathbf{S}'_j\right) \tag{4.8.11}$$

where $\mathbf{P}$ runs through all triple points of $\mathbb{B}/\Gamma$ and $\mathbf{P}'$ through all abelian points of $(\mathbb{B}/\Gamma)'$, the open part of $\overline{\mathbb{B}/\Gamma}'$. We decompose also $c(B'^0)$ in finite and infinite part:

$$c\left(B'^0\right) = \sum_{P'} c(P') + \sum_{j \in K} c(S'_j),$$

where $\mathbf{S}'_j \to \kappa_j$ is the abelian point on $\mathbf{E}_\kappa$, $\kappa = \kappa_j$, corresponding to the arrow $j : \longrightarrow\Diamond$ or $\Diamond\!\longrightarrow\!\circ$ in the molecular graph of $\mathbb{B}/\Gamma$. Now we can write (4.8.11) as

$$c(Y') = \mathbf{c}_f(\mathbf{Y}) + \sum_{\mathbf{C}} \lambda_c(\mathbf{C}) \cdot \mathbf{c}_f(\mathbf{C}) + \sum_{\mathbf{P}} \mathbf{c}_f(\mathbf{P}) + \sum_{\kappa} \mathbf{c}_f(\kappa) \tag{4.8.12}$$

with

$$\begin{aligned} \mathbf{e}_f(\mathbf{P}) &= e(P) + \mathbf{e}(\mathbf{P}) \\ \tau_f(\mathbf{P}) &= \tau(P) + \tau(\mathbf{P}) \end{aligned} \tag{4.8.13}$$

$$\mathbf{P}' \overset{\sim}{\to} \mathbf{P} \text{ an abelian point on } \mathbb{B}/\Gamma, \tag{4.8.14}$$

$$\mathbf{e}_f(\mathbf{P}) = \mathbf{e}_{\mathbf{P}}(\mathbf{Y}'/\mathbf{Y}) + \left(1 - \frac{1}{v_{\mathbf{P}}}\right)\mathbf{e}(\mathbf{E}_{\mathbf{P}}) + \sum_{i=1}^{3} \left(e\left(P'_i\right) + \mathbf{e}\left(\mathbf{P}'_i\right)\right), \tag{4.8.15}$$

$$\tau_f(\mathbf{P}) = \tau_{\mathbf{P}}(\mathbf{Y}'/\mathbf{Y}) + \left(v_{\mathbf{P}} - \frac{1}{v_{\mathbf{P}}}\right)\tau(\mathbf{E}_{\mathbf{P}}) + \sum_{i=1}^{3} \left(\tau\left(P'_i\right) + \tau(\mathbf{P}')\right)$$
$$- \frac{1}{3} \sum_{i=1}^{3} \left(v_i - \frac{1}{v_i}\right) \delta_i/v d_i \tag{4.8.16}$$

for triple points $\mathbf{P}$,

$$\mathbf{e}_f(\kappa) = \sum_{j=1}^{4}\left(e\left(S_j'\right) + \mathbf{e}\left(\mathbf{S}'_j\right)\right) + \sum_{j=1}^{4}\left(1 - \frac{1}{v_j}\right)\frac{1}{v_\kappa d_j} + \frac{1}{v_\kappa}\mathbf{e}(\mathbf{E}_\kappa) + \left(1 - \frac{1}{v_\kappa}\right)\mathbf{e}(\mathbf{E}_\kappa) \tag{4.8.17}$$

$$\tau_f(\kappa) = \sum_{j=1}^{4}\left(\tau\left(S_j'\right) + \tau\left(\mathbf{S}_j'\right)\right) + \frac{1}{v_\kappa}\tau(\mathbf{E}_\kappa) + \left(v_\kappa - \frac{1}{v_\kappa}\right)\tau(\mathbf{E}_\kappa) \tag{4.8.18}$$

for orbital cusp points $\kappa \in \mathbf{K}$.

We have only to check now that these are the local invariants listed in Table 4.8.1 because we know already that the first two expressions of (4.8.12) coincide with the corresponding first parts of 4.8.3, (e),(τ) by Theorem 4.8.1 and (4.7.7).

With (3.3), (3.3.12), (4.8.13) and the notations of the table we get

$$\mathbf{e}_f(\mathbf{P}) = 1 - \frac{1}{vd} - \frac{1}{v'd} + \frac{1}{vv'd}.$$

The signature height $\tau(\mathbf{P})$ has been defined to be 0 in Lemma 3.4.2 and $\tau(P)$ is defined in (3.4.2). We obtain

$$3\tau_f(\mathbf{P}) = 3\tau(P) = 3l_P + Tr(P) - \frac{e}{d} - \frac{e'}{d}.$$

We use these calculations tacitly also for the next local invariants.

Let $\mathbf{P}$ be a triple point. In (3.6.7), (3.6.2) and (3.6.14) one finds that

$$\mathbf{e}_\mathbf{P}(\mathbf{Y}'/\mathbf{Y}) = \left(\frac{1}{v_1 d_1} + \frac{1}{v_2 d_2} + \frac{1}{v_3 d_3} - 1\right)/2v = \frac{1}{|\Gamma_{\tilde{P}}|} = \mathbf{e}(\mathbf{E}_\mathbf{P})/2v$$

with the notations of Table 4.8.1. Substitution into (4.8.15) yields

$$\begin{aligned}
\mathbf{e}_f(\mathbf{P}) &= \mathbf{e}(\mathbf{E}_\mathbf{P})/2v_\mathbf{P} + \left(1 - \frac{1}{v_\mathbf{P}}\right)\mathbf{e}(\mathbf{E}_\mathbf{P}) + \sum_{i=1}^{3}\left(e\left(P_i'\right) + \mathbf{e}\left(\mathbf{P}_i'\right)\right) \\
&= \left(1 - \frac{1}{2v}\right)\mathbf{e}(\mathbf{E}_\mathbf{P}) + \sum_{i=1}^{3}\left(1 - \frac{1}{vd_i} - \frac{1}{v_i d_i} + \frac{1}{vv_i d_i}\right) \\
&= \left(1 - \frac{1}{2v}\right)\mathbf{e}(\mathbf{E}_\mathbf{P}) + \sum_i\left(1 - \frac{1}{vd_i}\right) - \mathbf{e}(\mathbf{E}_\mathbf{P}) + 1 + \frac{1}{v}\mathbf{e}(\mathbf{E}_\mathbf{P}) + \frac{1}{v} \\
&= \frac{1}{2v}\mathbf{e}(\mathbf{E}_\mathbf{P}) + \left(2 + \frac{1}{v}\right) - \frac{1}{v}\sum_i\frac{1}{d_i}.
\end{aligned}$$

In order to calculate $\tau_f(\mathbf{P})$ we remember to the relative proportionality (3.8.11) and to the definition of $\tau_{\mathbf{P}}(\mathbf{Y}'/\mathbf{Y})$ in (3.6.7):

$$\tau(\mathbf{E_P}) = -\frac{1}{6}\mathbf{e}(\mathbf{E_P}), \tau_{\mathbf{P}}(\mathbf{Y}'/\mathbf{Y}) = -\mathbf{e_P}(\mathbf{Y}'/\mathbf{Y}) = -\mathbf{e}(\mathbf{E_P})/2v. \tag{4.8.19}$$

Now (4.8.16) can be transferred to

$$\begin{aligned}\tau_f(\mathbf{P}) = & -\mathbf{e}(\mathbf{E_P})/2v - \frac{1}{6}\left(v - \frac{1}{v}\right)\mathbf{e}(\mathbf{E_P}) \\ & + \sum_{i=1}^{3}\left(\tau\left(P_i'\right) - \frac{1}{3}\left(v_i - \frac{1}{v_i}\right)/vd_i\right) \\ = & -\frac{1}{6}\left(v + \frac{2}{v}\right)\mathbf{e}(\mathbf{E_P}) + \sum_{i=1}^{3}\left(\tau\left(P_i'\right) - \frac{1}{3}\left(v_i - \frac{1}{v_i}\right)/vd_i\right)\end{aligned}$$

$$\begin{aligned}3\tau_f(\mathbf{P}) = & \sum_{i=1}^{3}\left(3l_i - \left(v_i - \frac{1}{v_i}\right)/vd_i + Tr\left(P_i'\right) - \frac{e_i}{d_i} - \frac{e_i'}{d_i}\right) \\ & - \left(\frac{1}{2}v + \frac{1}{v}\right)\mathbf{e}(\mathbf{E_P}).\end{aligned}$$

In (4.8.17) the last two summands vanish because of the definition of an orbital cusp point, see 3.5.6. Therefore, we receive

$$\begin{aligned}\mathbf{e}_f(\kappa) &= \sum_{j=1}^{4}\left(e\left(S_j'\right) + \mathbf{e}\left(\mathbf{S}_j'\right)\right) + \sum_{j=1}^{4}\left(1 - \frac{1}{v_j}\right)\frac{1}{vd_j} \\ &= \sum_{j=1}^{4}\left(1 - \frac{1}{vd_j} - \frac{1}{v_jd_j} + \frac{1}{vv_jd_j}\right) + \sum_{j=1}^{4}\left(1 - \frac{1}{v_j}\right)\frac{1}{vd_j} \\ &= \begin{cases} 0, & \kappa \text{ neat} \\ 4 - \sum\limits_{j=1}^{4}\dfrac{1}{v_jd_j} = 4 - 2 = 2, & \kappa \text{ not neat}. \end{cases}\end{aligned}$$

We used the relation (3.5.4).

For the last identity we remark that the abelian points $\mathbf{S}_j'$ are quotients of O by a cyclic group of order 2, 3, 4 or 6, see 3.5.9(ii). It follows that S_j' is a symmetric point in any case, that means that $e_j' = e_j$. So (4.8.18) changes to

$$\begin{aligned} 3\tau_f(\kappa) &= 3\sum_{j=1}^{4}\tau\left(S_j'\right) + 3v\tau(\mathbf{E}_\kappa) \\ &= \sum_{j=1}^{4}\left(3l_j + Tr\left(S_j'\right) - \frac{e_j}{d_j} - \frac{e_j}{d_j}\right) + \left(E_\kappa^2 + \Sigma\frac{e_j}{d_j}\right) \\ &= (E_\kappa^2) + \sum_{j=1}^{4} Tr\left(S_j'\right) + \sum_{j=1}^{4}\left(3l_j - \frac{e_j}{d_j}\right). \end{aligned}$$

The Theorem 4.8.3 is proved.

□

Remark 4.8.4 Let **P** be a triple point. In Table 4.8.1 we can write

$$\mathbf{e}_f(\mathbf{P}) = e(E_\mathbf{P}) + \frac{1}{v}\left[\frac{1}{2}\mathbf{e}(\mathbf{E}_\mathbf{P}) + \left(1 - \frac{1}{d_1} - \frac{1}{d_2} - \frac{1}{d_3}\right)\right]$$

$$\begin{aligned} 3\tau_f(\mathbf{P}) &= \left(1 + \frac{2}{v^2}\right)\left((E_\mathbf{P}^2) + \sum_i \frac{e_i}{d_i}\right) \\ &\quad + \sum_i\left(3l_i - \delta_i + Tr\left(P_i'\right) - \frac{e_i}{d_i} - \frac{e_i'}{d_i}\right) \end{aligned}$$

We vary the central weight $v = v_\mathbf{P} = v_{\mathbf{E}_\mathbf{P}}$ and denote the corresponding triple points by $\mathbf{P}(v)$. Now we let v tend to ∞. Then we get

$$\lim_{v\to\infty} \mathbf{e}_f(\mathbf{P}(v)) = e(E_\mathbf{P})$$

$$\lim_{v\to\infty} 3\tau_f(\mathbf{P}(v)) = Tr(\mathbf{P}) + \sum_i\left(3l_i - \delta_i - \frac{e_i'}{d_i}\right)$$

Comparing with $\mathbf{e}_f(\kappa)$ and $3\tau_f(\kappa)$ we can interpret a cusp point as an orbital point with weight infinity, $\delta_j = 0$ for $j = 1,2,3,4$. Only in the proof of the formulas we used the central weight v_κ depending on the neat subgroup $\Gamma_1 \subset \Gamma$. For different such subgroups we get different arbitrarily high weights. That is the reason why it has to disappear in the end formulas.

Example 4.8.5 (Full Gauss lattice of the ball). We continue to discuss the orbital ball quotient surface $\mathbb{B}/\Gamma$, $\Gamma = \mathbb{U}((2,1),\ \mathcal{O}_K)$, $K = \mathbb{Q}(i)$, $\mathcal{O}_K = \mathbb{Z} + \mathbb{Z}i$ the ring of integral Gauss numbers. The molecular graph of $\mathbb{B}/\Gamma$ has been drawn in Figure (4.6.8) starting from the fixed point configuration

Figure (4.6.2). There is exactly one cusp point κ. The volume formula (4.8.5) allows to calculate $(E_\kappa^2) = -1$ by means of an elementary analysis of the cusp group $\Gamma_{\tilde{\kappa}}$. The euclidean volumes $|\Lambda_\kappa|$ and $|\Delta_\kappa|$ can be read off from a special basis of $\Gamma_{\tilde{\kappa},u}$, see (4.2.10), (4.2.11). Now it is easy to calculate

$$\mathbf{e}_f(\kappa) = 2, \tau_f(\kappa) = (E_\kappa^2) = -1. \tag{4.8.20}$$

For the quotient points we use Table 4.6.1:

$$\begin{aligned} \mathbf{e}_f(\mathbf{P}_0) &= 1 - \frac{1}{4\cdot 2} - \frac{1}{1\cdot 2} + \frac{1}{4\cdot 1\cdot 2} = \frac{1}{2}, \\ \mathbf{e}_f(\mathbf{P}_1) &= 1 - \frac{1}{4\cdot 3} - \frac{1}{1\cdot 3} + \frac{1}{4\cdot 1\cdot 3} = \frac{2}{3}, \\ \mathbf{e}_f(\mathbf{P}_{12}) &= 1 - \frac{1}{4\cdot 1} - \frac{1}{2\cdot 1} + \frac{1}{4\cdot 2\cdot 1} = \frac{3}{8}. \end{aligned} \tag{4.8.21}$$

For $\mathbf{P} = \mathbf{P}_2$, $\mathbf{P}_{02}$ we first calculate

$$\begin{aligned} &\mathbf{e}(\mathbf{E}_{\mathbf{P}_2}) = \frac{2v}{|\Gamma_{\tilde{P}_2}|} = \frac{2\cdot 1}{6} = \frac{1}{3}, \ \mathbf{e}(\mathbf{E}_{\mathbf{P}_{02}}) = \frac{2\cdot 4}{32} = \frac{1}{4}, \\ &\tau(\mathbf{E}_{\mathbf{P}_2}) = -\frac{1}{6}\mathbf{e}(\mathbf{E}_{\mathbf{P}_2}) = -\frac{1}{18}, \ \tau(\mathbf{E}_{\mathbf{P}_{02}}) = -\frac{1}{24}, \\ &(E_{P_2}^2) = 3v\cdot\tau(\mathbf{E}_{\mathbf{P}_2}) - \sum_{i=1}^{3}\frac{e_i}{d_i} = 3\cdot 1\left(-\frac{1}{18}\right) - \frac{1}{3} - \frac{1}{2} - \frac{0}{1} = -1, \\ &(E_{P_{02}}^2) = 3\cdot 4\left(-\frac{1}{24}\right) - \frac{1}{2} - \frac{0}{1} - \frac{0}{1} = -1. \end{aligned} \tag{4.8.22}$$

These selfintersection numbers on $\overline{\mathbb{B}/\Gamma}'$ have already been filled into the canonical resolution graphs in Table 4.6.1. With (4.8.22) we can calculate

$$\mathbf{e}_f(\mathbf{P}_2) = 2 + \frac{1}{1}\left[\frac{1}{2}\cdot\frac{1}{3} + \left(1 - \frac{1}{1} - \frac{1}{2} - \frac{1}{3}\right)\right] = 2 - \frac{2}{3} = \frac{4}{3}$$

$$\mathbf{e}_f(\mathbf{P}_{02}) = 2 + \frac{1}{4}\left[\frac{1}{2}\cdot\frac{1}{4} + \left(1 - \frac{1}{1} - \frac{1}{1} - \frac{1}{2}\right)\right] = 2 - \frac{11}{32}$$

The local signatures are

$$3\tau_f(\mathbf{P}_0) = \left[3\cdot 1 - 2 - \frac{1}{2} - \frac{1}{2}\right] = 0$$

$$3\tau_f(\mathbf{P}_1) = \left[3\cdot 2 - 4 - \frac{2}{3} - \frac{2}{3}\right] = \frac{2}{3}$$

$$3\tau_f(\mathbf{P}_{12}) = \left[3\cdot 0 + 0 - \frac{0}{1} - \frac{0}{1}\right] = 0$$

$$\begin{aligned}3\tau_f(\mathbf{P}_2) = & -\left(\frac{1}{2}\cdot 1+\frac{1}{1}\right)\frac{1}{3}+\left[\left(3\cdot 1-0-2-\frac{1}{2}-\frac{1}{2}\right)\right.\\ & +\left(3\cdot 1-0-3-\frac{1}{3}-\frac{1}{3}\right)\\ & \left.+\left(3\cdot 0\left(-2+\frac{1}{2}\right)/1\cdot 1+0-\frac{0}{1}-\frac{0}{1}\right)\right]\\ = & -\frac{1}{2}+\left[0-\frac{2}{3}-\frac{3}{2}\right]=-2-\frac{2}{3}\end{aligned}$$

$$\begin{aligned}3\tau_f(\mathbf{P}_{02}) = & -\left(\frac{1}{2}\cdot 4+\frac{1}{4}\right)\frac{1}{4}+[\left(3\cdot 0\left(-2+\frac{1}{2}\right)/4+0-\frac{0}{1}-\frac{0}{1}\right)\\ & +\left(3\cdot 0\left(-4+\frac{1}{4}\right)/4+0\right)+\left(3\cdot 1-0-2-\frac{1}{2}-\frac{1}{2}\right)]\\ = & -\frac{9}{16}+\left[-\frac{3}{8}-\frac{15}{16}-0\right]=-\frac{30}{16}=-\frac{15}{8}\end{aligned}$$

$$\begin{aligned}e\left(\overline{\mathbb{B}/\Gamma}'\right) & = \frac{1}{32}+\left[\frac{3}{4}\left(-\frac{1}{4}\right)+\frac{3}{4}\left(-\frac{1}{6}\right)+\frac{1}{2}\left(-\frac{1}{2}\right)\right]\\ & +\left[\frac{1}{2}+\frac{2}{3}+\frac{3}{8}+\frac{4}{3}+2-\frac{11}{32}\right]+2=6\\ 3\tau\left(\overline{\mathbb{B}/\Gamma}'\right) & = \frac{1}{32}+\frac{1}{2}\left[\frac{15}{4}\left(-\frac{1}{4}\right)+\frac{15}{4}\left(-\frac{1}{6}\right)+\frac{3}{2}\left(-\frac{1}{2}\right)\right]\\ & +\left[0+\frac{2}{3}+0\left(-2-\frac{2}{3}\right)-\frac{15}{8}\right]-1=-6\end{aligned} \tag{4.8.23}$$

The volume $c_2(\mathbb{B}/\Gamma) = 3\tau(\mathbb{B}/\Gamma) = \frac{1}{32}$ will be calculated in the next chapter, see Chapter **5**, Appendix. The arithmetic genus of any quasismooth model of $\widehat{\mathbb{B}/\Gamma}$ is

$$\chi\left(\widehat{\mathbb{B}/\Gamma}\right)=\chi\left(\overline{\mathbb{B}/\Gamma}'\right)=\frac{1}{4}\left(e\left(\overline{\mathbb{B}/\Gamma}'\right)+\tau\left(\overline{\mathbb{B}/\Gamma}'\right)\right)=\frac{1}{4}(6-2)=1. \tag{4.8.24}$$

We will see that $\widehat{\mathbb{B}/\Gamma}$ is a rational surface and present its fine classification together with the general concept of classification of PICARD modular surfaces in Chapter **5**.

Table 4.8.1

kind of point *atomic graph* *local picture*

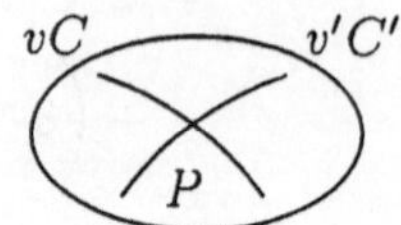

P abelian point:

$v \quad \langle d,e\rangle \quad v'$

$\mathbf{e}_f(\mathbf{P}) = 1 - 1/vd - 1/v'd + 1/v'vd,$

$3\tau_f(\mathbf{P}) = 3l_P + \mathrm{Tr}(P) - e/d - e'/d$

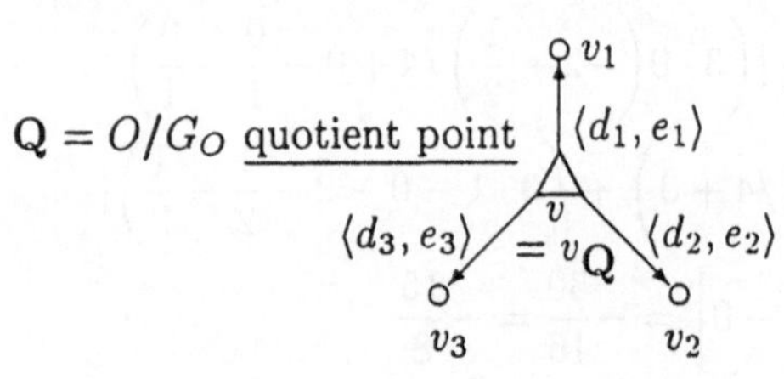

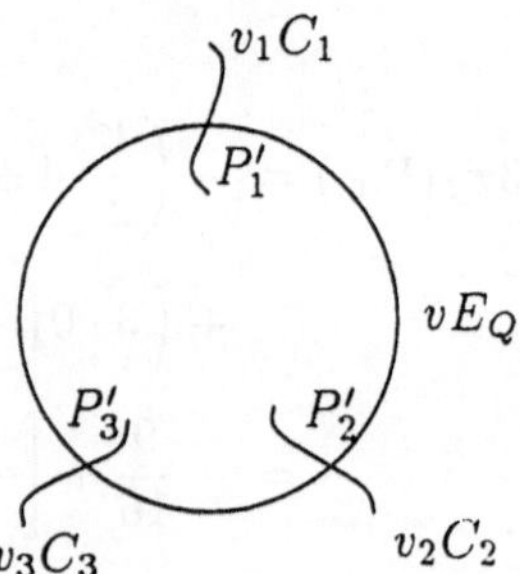

$$\begin{aligned}\mathbf{e}_f(\mathbf{Q}) &= 2 + \tfrac{1}{v}\left[\tfrac{1}{2}\mathbf{e}\left(\mathbf{E}_Q\right) + 1 - 1/d_1 - 1/d_2 - 1/d_3\right] \\ &= e\left(E_Q\right) = \tfrac{1}{v}\left[\tfrac{1}{2}\mathbf{e}\left(\mathbf{E}_Q\right) - \mathbf{e}\left(E_Q\right)\right] = 1 + \left|G_0^{\mathrm{is}}\right| / \left|G_0\right|\end{aligned}$$

$$\begin{aligned}3\tau_f(\mathbf{Q}) &= -\left(v/2 + 1/v\right)\mathbf{e}\left(\mathbf{E}_Q\right) \\ &\quad + \sum_{i=1}^{3}\left[3l_i - \left(v_i - 1/v_i\right)vd_i + \mathrm{Tr}\left(P_i'\right) - e_i/d_i - e_i'/d_i\right] \\ &= \left(1 + 2/v^2\right)\left[\left(E_Q^2\right) + \sum_{i=1}^{3} e_i/d_i\right] \\ &\quad + \sum_{i=1}^{3}\left[3l_i - \left(v_i - 1/v_i\right)vd_i + \mathrm{Tr}\left(P_i'\right) - e_i/d_i - e_i'/d_i\right]\end{aligned}$$

κ cusp point

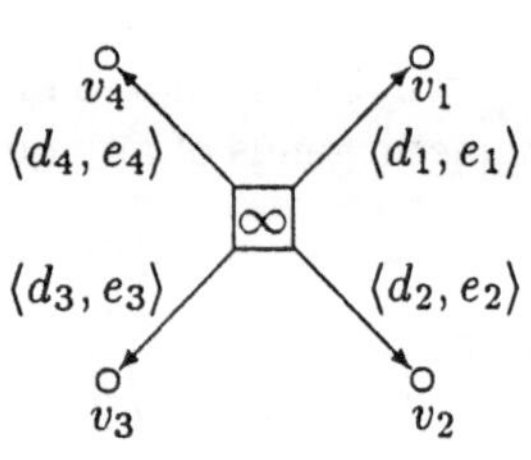

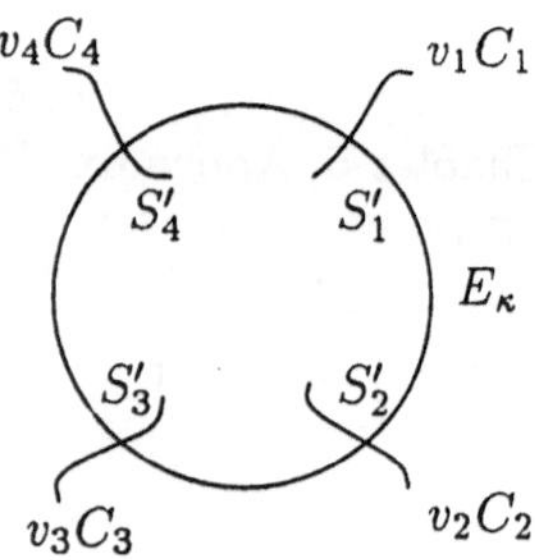

$$\mathbf{e}_f\left(\kappa\right) = e\left(E_\kappa\right) = \begin{cases} 0, & \text{elliptic} \\ 2, & \text{rational} \end{cases},$$

$$3\tau_f(\kappa) = \mathrm{Tr}(\kappa) + \sum_{j=1}^{4}\left(3l_j - e_j/d_j\right)$$

4.9 Global Proportionality

Theorem 4.9.1 (Proportionality) *The open orbital surface* $\mathbf{Y}_f$ *can only be an orbital ball quotient surface* $\mathbb{B}/\Gamma$, *if*

$$3\tau_f(\mathbf{Y}) = \mathbf{e}_f(\mathbf{Y}) > 0, \tag{4.9.1}$$

with

$$\mathbf{c}_f(\mathbf{Y}) = c(Y') - \sum_{\mathbf{C}\subset\mathbf{Y}_f} \lambda_c(\mathbf{C})\mathbf{c}_f(\mathbf{C}) - \sum_{\mathbf{P}\in\mathbf{Y}_f} \mathbf{c}_f(\mathbf{P}) - \sum_{\kappa} \mathbf{c}_f(\kappa) \tag{4.9.2}$$

adapted from (4.8.12). If $\mathbf{Y}_f = \mathbb{B}/\Gamma$ *is an orbital ball quotient surface, then the proportionality condition (4.9.1) is satisfied. Therefore, it is a necessary condition for* $\mathbf{Y}_f$ *to be a ball quotient.*

Proof Let Γ be a ball lattice, $\mathbf{Y}_f = \mathbb{B}/\Gamma$ and $\mathbf{Y}' \to \mathbf{Y} \to \hat{\mathbf{Y}} \supset \mathbf{Y}_f$ the lower row of (4.6.1). From Theorem 4.8.1 (or Theorem 4.8.3) it follows that

$$\mathbf{e}_f(\mathbf{Y}) = c_2(\mathbb{B}/\Gamma) = 3\tau_f(\mathbf{Y}).$$

The EINSTEIN-KÄHLER metric on $\mathbb{B}$ is positive in the sense that the volume $c_2(\mathbb{B}/\Gamma)$ of a Γ-fundamental domain is a positive number.

□

We get finer ball quotient conditions if we add to (4.9.1) the conditions of relative proportionality 4.7.4,(ii) and (3.8.11).

Theorem 4.9.2 (Local-Global Proportionality) *Let* $\hat{\mathbf{Y}}$ *be an orbital surface with (at most) quasihomogeneous points and canonical resolution* $\mathbf{Y}' \to \mathbf{Y} \to \hat{\mathbf{Y}}$. *The set of weighted points is denoted by* $\mathbf{W} = \mathbf{W}(\hat{\mathbf{Y}})$. *Furthermore, let* $\mathbf{W}_f(\hat{\mathbf{Y}}) = \mathbf{W}_f(\mathbf{Y})$ *be a subset of* $\mathbf{W}$ *and* $\mathbf{Y}_f$ *the (non-compact) orbital surface supported by* $Y_f = \hat{Y} \setminus W_f$. *The following conditions are necessary for* $\mathbf{Y}_f$ *to be an orbital ball quotient surface:*

$$\mathbf{e}(\mathbf{E}_\kappa) = 0, \tau(\mathbf{E}_\kappa) < 0 \tag{Prop ∞}$$

for all $\kappa \in \mathbf{W}_\infty = \mathbf{W} \setminus \mathbf{W}_f$

$$-\mathbf{e}(\mathbf{E}_P) = 6\tau(\mathbf{E}_\mathbf{P}) < 0 \tag{Prop 0}$$

for all $\mathbf{P} \in \mathbf{W}_f$, $\mathbf{E}_\mathbf{P} \to \mathbf{P}$ *the canonical resolution of* $\mathbf{P}$, $\mathbf{E}_\mathbf{P} \subset \mathbf{Y}'$.

$$\mathbf{e}_f(\mathbf{C}) = 6\tau(\mathbf{C}) < 0 \qquad \text{(Prop 1)}$$

for all components $\hat{\mathbf{C}}$ *of* $\hat{\mathbf{B}}^1$ *of the one-dimensional part of the arrangement* $\widehat{\mathbf{B}} = \widehat{\mathbf{B}^1} + \widehat{\mathbf{B}^0}$ *of* $\widehat{\mathbf{Y}}, \mathbf{C} \to \widehat{\mathbf{C}}$ *the canonical resolution of* $\widehat{\mathbf{C}}$ *on* $\mathbf{Y}$.

$$3\tau_f(\mathbf{Y}) = \mathbf{e}_f(\mathbf{Y}) > 0. \qquad \text{(Prop 2)}$$

Remark 4.9.3 If one knows the geometric data

$$(E_Q^2),\ e(E_Q),\ Q \in W,\ E_Q \to Q \text{ the canonical resolution}$$
$$\left(C'^2\right),\ e(C'),\ C \leftarrow C' \subset Y',$$
$$\text{and } c(Y') \text{ for two of the Chern symbols } e,\ \tau,\ c_1^2,\ \chi,$$

then all the conditions can be checked by means of the molecular graph of $\mathbf{B}$. The invariants (E_Q^2) and (C'^2) should be appear in the graph.

Remark 4.9.4 In the graph of $\mathbf{B}$ cannot appear stars of quasihomogeneous points with more than four arrows at the center. Only triple and quadruple points can occur because of (Prop ∞), (Prop 0), the definition of quotient points ($\mathbf{e}(\mathbf{E_P}) > 0$), the definition of cusp points ($\mathbf{e}(\mathbf{E}_\kappa) = \mathbf{0}$) and the classification of these points in Corollary 3.5.9. The conditions $\tau(\mathbf{E_Q}) < 0$, $\mathbf{Q} \in \mathbf{W}$ are superfluous because we assumed that all orbital points are quasihomogeneous, and this property is characterized by the above condition in 3.5.2. If we endow the cusp points with weight ∞, then (Prop 0) and (Prop ∞) can be uniquely written as

$$-\mathbf{e}(\mathbf{E_P}) = 6\tau(\mathbf{E_P}) = \frac{2}{v}\left((E_P^2) + \sum_i \frac{e_i}{d_i}\right) \qquad \text{(Prop *)}$$

with the new definition $\tau(\mathbf{E}_\kappa) = 0$ for $\kappa \in \mathbf{W}_\infty$ with weight ∞.

Proof of Theorem 4.9.2. Condition (Prop 2) comes from Theorem 4.9.1 and (Prop 1) from Proposition 4.7.4. For *honest quotient points* $\mathbf{P} = O/G, O \in \mathbb{C}^2, G$ a finite subgroup of $Gl_2(\mathbb{C})$ we know that the equality of (Prop 0) holds, see (3.8.11). The inequality comes from the proof step (3.8.12) or from the definition 3.5.6 of quotient points supported by quotient singularities. It should be remarked that not any quotient point is an honest one. This depends essentially on the central weight $v = v_{\mathbf{E_P}}$. If $\mathbf{P}$ lies on an open ball quotient surface $\mathbb{B}/\Gamma$, then we choose a neat normal sublattice Γ_1 of Γ and a point $O \in \mathbb{B}/\Gamma_1$ lying over $\mathbf{P}$. Obviously $\mathbf{P} = O/G_O$ for $G = \Gamma/\Gamma_1$. So $\mathbf{P}$ is an honest quotient point and (Prop 0) is satisfied at $\mathbf{P}$. The condition (Prop ∞) is the definition 3.5.6 of cusp points.

□

4.10 Orbital Decompositions and the Finiteness Theorem

We decompose the local invariants of Table 4.8.1 in order to change in the decomposition (4.9.2) of $\mathbf{c}_f(\mathbf{Y})$, the invariants $c(Y')$ by a new one $H_c(\mathbf{Y}_f)$, $c = e, \tau$. The points $\mathbf{P}$, κ in the table live on $\widehat{Y}$. We divide them into three classes:

$\mathbf{A}$: abelian points, $\mathbf{W}_f$: triple points, $\mathbf{W}_\infty$: cusp points, $\mathbf{W} = \mathbf{W}_f \cup \mathbf{W}_\infty$ weighted points.

For $\mathbf{S} \in \mathbf{A} \cup \mathbf{W}$ we define *local heights at points* by

$$H_c(\mathbf{P}) = \mathbf{c}_f(\mathbf{P}) - \mathbf{c}_f(P). \tag{4.10.1}$$

The old invariants can be read off from Table 4.8.1. The new ones can be calculated by (4.10.1). We list them together with the "very old" invariants $c_f(S)$ in Table 4.10.1. The notations of weights are taken from the graphs in Table 4.8.1.

Definition 4.10.1

$$H_c(Y_f) = \mathbf{c}_f(Y) \mp |\mathbf{W}_f| = c(Y) - \sum_{S \in W} c_f(S)$$

$$H_c(\mathbf{C}_f) = \mathbf{c}_f(\mathbf{C}), \text{ (see 7.9)}$$

$$H_c(\mathbf{Y}_f) = \mathbf{c}_f(\mathbf{Y}), \text{ (see 9.3)}.$$

From (4.9.2) we receive the

Proposition 4.10.2 *(Orbital decomposition of heights)*

$$H_c(\mathbf{Y}_f) = H_c(Y_f) - \sum_{\mathbf{C}} \lambda_c(\mathbf{C}) H_c(\mathbf{C}_f) - \sum_{\mathbf{S} \in \mathbf{A} \cup \mathbf{W}_f} H_c(\mathbf{S})$$

with

$$H_e(Y_f) = e(Y) - \sum_{S \in W} c_f(S) = e(Y) - \sum_{\mathbf{P} \in \mathbf{A}} \left(1 - \frac{1}{d_P}\right) - \sum_{\mathbf{Q} \in \mathbf{W}_f} \sum_{i=1}^{3} \frac{1}{2}\left(1 - \frac{1}{d_i}\right) - 2|\mathbf{W}_\infty^{rat}| \tag{4.10.2}$$

$$3H_\tau(Y_f) = 3\tau(Y) - \sum_{P\in \mathbf{A}} \left(3l_P + Tr(P) - \frac{e_P}{d_P} - \frac{e_{P}'}{d_P}\right)$$
$$- \sum_{Q\in \mathbf{W}_f} \sum_{i=1}^{3} \left(3l_i + Tr(P_i) - \frac{e_i}{d_i} - \frac{e_i'}{d_i} + \frac{3}{2} - \frac{3}{2d_i}\right)$$
$$- \sum_{\kappa\in \mathbf{W}_\infty} \left(Tr(\kappa) + \sum_j \left(3l_j - \frac{e_j}{d_j}\right)\right) \tag{4.10.3}$$

The advantage of this kind of orbital decomposition is that we delegated the complicated constant part not including any weights v, v_i to $H_c(Y_f)$. It enables us to prove the following theorem in a constructive manner:

Effective Finiteness Theorem. 4.10.3 *Let $\widehat{Y}$ be a compact normal surface with (at most) quasihomogeneous singularities and $\widehat{B} = \widehat{B}^1 + \widehat{B}^0$ a reduced arrangement on $\widehat{Y}$ (supp $\widehat{B} = \widehat{B}^1 \cup \widehat{B}^0$ is called the weight locus). Then, there exists only finitely many possibilities of weighting $\widehat{B}$ (the irreducible components of $\widehat{B}^1$ and points of $\widehat{B}^0$) such that $\widehat{\mathbf{Y}} = (\widehat{\mathbf{Y}}, \widehat{\mathbf{B}})$, $\widehat{\mathbf{B}}$ the corresponding weighted arrangement, is an orbital ball quotient surface $\widehat{\mathbb{B}/\Gamma}$. Knowing all geometric invariants of $\widehat{Y}$, $\widehat{B}$, all such possibilities of weighting $\widehat{B}$ can be calculated explicitly by means of an algorithm.*

It is clear that the equalities of (Prop ∞), (Prop 0), (Prop 1) and (Prop 2) form a system of diophantine equations in all the weights as variables with geometric data coming from $\widehat{Y}$ and $\widehat{B}$ as coefficients. At the first glance it seems to be hopeless to solve such a system. But fortunately the main equation (Prop. 2) can be transformed after choosing point weights into a special "quadratic equation", which can be solved completely under secondary conditions (Prop ∞), (Prop 0) and (Prop 1).

Proof of Theorem 4.10.3 We assume that $\widehat{\mathbf{B}}^1$ will be an honest curve arrangement, that means all components will be endowed with a weight $v \geq 2$. Then through each point $S \in \widehat{B}^0$ can go at most four branches of curves of B^1 because quotient points are (at most) triple points and cusp points are (at most) quadruple points.

Step 1. Choose cusp and weighted quotient points: We choose in B^0 such points which have to be weighted quotient points or cusp points. The remaining points will be abelian points. Obviously we have only a finite number of possibilities of such choice.

Step 2. Choose weights around the cusp points κ: Using (Prop ∞), which is the characteristic property for cusp points, we proved in Corollary 3.5.9, (ii) classifying the graphs of all cusp points, that the weights v_i around

κ are not greater than 6. By our convention the central weight is uniquely determined as $v_\kappa = \infty$. So there are only finitely many possibilities of weighting κ to become a cusp point $\boldsymbol{\kappa}$.

Step 3. Choose central weights at the non-abelian quotient points Q. Here we use condition (Prop 0). It can be written as

$$\left(\frac{1}{v_1 d_1} + \frac{1}{v_2 d_2} + \frac{1}{v_3 d_3} - 1\right) = \frac{2}{v}\left((E_Q^2) + \frac{e_1}{d_1} + \frac{e_2}{d_2} + \frac{e_3}{d_3}\right) < 0 \tag{4.10.4}$$

or

$$(-\tau(E_Q))\frac{2}{v} = \frac{1}{v_1 d_1} + \frac{1}{v_2 d_2} + \frac{1}{v_3 d_3} - 1. \tag{4.10.5}$$

By the choice of Q to be a triple point, we know that $-\tau(\mathbf{E}_Q) > 0$, hence $-\tau(E_Q) > 0$. For the isolated quotient points not lying on any component of $\widehat{B}^1$ we have $v_1 = v_2 = v_3 = 1$. Then v is uniquely determined by (4.10.5). In the general case we assume that $d_1 \leq d_2 \leq d_3$. If $\mathbf{Q}$ is not a twin point and $d_2 \geq 3$, then we have only the following possibilities by the classification of non-abelian quotient points 3.5.9,(i):

$$(v_1 d_1, v_2 d_2, v_3 d_3) = \begin{cases} (2,3,3) \\ (2,3,4) \\ (2,3,5) \end{cases}$$

For these cases there are obviously only finitely many possibilities to choose v_1, v_2, v_3. For each triple v is uniquely determined again by (4.10.5). For the dieder types $(2, 2, v_3 d_3)$ and twin points of type $(1, v_3 d_3, v_3 d_3)$ we have not a finite number of possibilities to choose $v_3 \geq 2$ and v. Since v_3 is the weight of a component of $\widehat{B}^1$ we can delegate the choice of v_3 determining also v to the next step. For this step it is important to know the precise relation between v_3 and v assuming that we choosed $v_1, v_2 \in \{1, 2\}$ in the dieder case. We obtain from (4.10.5)

$$-\frac{2}{v}\tau(E_Q) = \frac{1}{wd_3} \text{ or } -\frac{2}{v}\tau(E_Q) = \frac{2}{wd_3}, \; (w = v_3) \tag{4.10.6}$$

for dieder points or twin points, respectively. We set $\mathbf{P} = \mathbf{P}(w)$ and consider the functions of w

$$H_e(\mathbf{P}(w)), \; H_\tau(\mathbf{P}(w))$$

described in Table 4.10.1 for points which we choosed to be of dieder type, and for twin points.

Lemma 4.10.4 *Let P be a dieder point with fixed v_1, v_2 or a twin point $P \in \widehat{B}^0 \subset \widehat{Y}$. Then the functions $H_e(\mathbf{P}(w)), H_\tau(\mathbf{P}(w))$ of $w \in \mathbb{N}_+$ are (absolutely) restricted. Upper and lower bounds can be computed explicitly.*

Proof For $H_e(\mathbf{P}(w))$ this is immediately clear from Table 4.10.1. In the signature case we have by the same table only to restrict

$$\frac{v}{2}\mathbf{e}(\mathbf{E}_P) = \frac{v}{2}\left(\frac{1}{wd_3}\right) \text{ or } \frac{v}{2}\left(\frac{2}{wd_3}\right),$$

hence $w|vd_3$. But these values are constant by (4.10.6).

□

Step 4. Choose curve weights: In order to apply (Prop 2) we write down the orbital decompositions in 4.10.2 separately with obvious notations

$$H_e(\mathbf{Y}) = H_e(Y) - \sum_{i=1}^{m}\left(1 - \frac{1}{v_i}\right)H_e(\mathbf{C}_{if})$$
$$- \sum_{\mathbf{P}\in\mathbf{A}}{}' \frac{1}{d_\mathbf{P}}\left(1 - \frac{1}{v_\mathbf{P}}\right)\left(1 - \frac{1}{v'_\mathbf{P}}\right) - \sum_{\mathbf{Q}\in\mathbf{W}_f} H_e(\mathbf{Q})$$

$$3H_\tau(\mathbf{Y}) = 3H_\tau(Y) - \sum_{i=1}^{m}\left(v_i - \frac{1}{v_i}\right)3H_\tau(\mathbf{C}_{if}) - \sum_{\mathbf{Q}\in\mathbf{W}_f} 3H_\tau(\mathbf{Q})$$

The first summands are constants, see (4.10.2) and (4.10.3). The other summands are (absolutely) restricted functions of $\mathfrak{v} = (v_1, \ldots, v_m)$ by Lemma 4.10.4. Applying (Prop 2) and (Prop 1) we get

$$\frac{1}{2}\sum_i\left(1 - \frac{1}{v_i}\right)^2 v_i H_e(\mathbf{C}_{if}) - \sum_{\mathbf{A}}{}'\frac{1}{d_\mathbf{P}}\left(1 - \frac{1}{v_\mathbf{P}}\right)\left(1 - \frac{1}{v'_\mathbf{P}}\right) = f(\mathfrak{v}), \tag{4.10.7}$$

$f(\mathfrak{v})$ an absolutely restricted function for $\mathfrak{v} \in \mathbb{N}_+^m$. Using the Definition 4.7.3 of $H_e(\mathbf{C}_i) = \mathbf{e}_f(\mathbf{C}_i)$ and the inequality of (Prop 1) we obtain the estimation

$$0 > H_e(\mathbf{C}_{if}) = e\left(C'_i\right) - \sum_j{}'\left(1 - \frac{1}{v_{ij}d_{ij}}\right) \tag{4.10.8}$$

with obvious notations allowing some v_{ij} to be ∞. The EULER number of the smooth model C'_i of C_i is equal to 2 if C_i is rational or it is not positive. Now it is an elementary exercise to check that there does not exist an orbital curve $\mathbf{C}_i$ such that

$$0 > H_e(\mathbf{C}_{if}) > -\frac{1}{20}$$

So, for any admissible weighting we look for, one gets from (4.10.7) and $\mathbf{e}_f(\mathbf{C}_i) \leq -\frac{1}{20}$ the effective condition

$$\frac{1}{40}\sum_i \left(1 - \frac{1}{v_i}\right)^2 v_i \leq M = M\left(\widehat{Y}, \widehat{B}\right) \tag{4.10.9}$$

for the weights $v_i \in \mathbb{N}_+$, M an effectively calculable constant depending on the geometric data of $\widehat{Y}$ and the weight locus $\widehat{B}$. Now it suffices to solve (4.10.9) componentwise: $\left(1 - \frac{1}{v_i}\right)^2 v_i \leq 40M$. The Theorem 4.10.3 is proved.

□

4.10.1 **Table**

for point heights $H_c(\mathbf{S}) = \mathbf{c}_f(\mathbf{S}) - \mathbf{c}_f(S)$, $c = e,\ \tau$
(for notations, $\mathbf{c}_f(\mathbf{S})$, graphs and pictures see Table 4.8.1)

$$\mathbf{e}_f(P) = 1 - \tfrac{1}{d} \qquad H_e(\mathbf{P}) = \left(1 - \tfrac{1}{v}\right)\left(1 - \tfrac{1}{v'}\right)/d$$

$$3\boldsymbol{\tau}_f(P) = 3l_P + \mathrm{Tr}(P) - \tfrac{e}{d} - \tfrac{e'}{d}, \qquad H_\tau(\mathbf{P}) = 0$$

$$\mathbf{e}_f(Q) = e_f(Q) - \tfrac{1}{2}\mathbf{e}(E_Q) = 1 + \tfrac{1}{2}\sum_{i=1}^{3}\left(1 - \tfrac{1}{d_i}\right),$$

$$H_e(\mathbf{Q}) = \tfrac{1}{2v}\mathbf{e}(\mathbf{E}_Q) + \left(\tfrac{1}{2} - \tfrac{1}{v}\right)\mathbf{e}(E_Q)$$

$$\begin{aligned}3\boldsymbol{\tau}_f(Q) &= -\tfrac{3}{2}\mathbf{e}(E_Q) + \sum_{i=1}^{3}\left[3l_i + \mathrm{Tr}(P_i) - \tfrac{e_i}{d_i} - \tfrac{e'_i}{d_i}\right]\\ &= 3\left[\left(E_Q^2\right) + \sum_i \tfrac{e_i}{d_i}\right] + \sum_i\left[3l_i + \mathrm{Tr}(P_i) - \tfrac{e_i}{d_i} - \tfrac{e'_i}{d_i}\right]\\ &= 3\left(E_Q^2\right) + \sum_i\left[3l_i + \mathrm{Tr}(P_i) + \tfrac{e_i}{d_i}\right],\end{aligned}$$

$$3H_\tau(\mathbf{Q}) = \tfrac{3}{2}\mathbf{e}(E_Q) - \left(\tfrac{v}{2} + \tfrac{1}{v}\right)\mathbf{e}(\mathbf{E}_Q) - \sum_i\left(v_i - \tfrac{1}{v_i}\right)/vd_i$$

$$\mathbf{e}_f(\kappa) = e(E_\kappa) = 0 \ (\kappa \text{ elliptic}) \text{ or } 2 \ (\kappa \text{ rational}), \qquad H_e(\boldsymbol{\kappa}) = 0,$$

$$3\boldsymbol{\tau}_f(\kappa) = \mathrm{Tr}(\kappa) + \sum_{j=1}^{4}\left(3l_j - \tfrac{e_j}{d_j}\right), \qquad H_\tau(\boldsymbol{\kappa}) = 0,$$

$$\left(\mathrm{Tr}(\kappa) := \left(E_\kappa^2\right) + \sum_j \mathrm{Tr}(P_j)\right)$$

4.11 Leading Examples

The proof of the Finiteness Theorem 4.10.3 is constructive. By means of a computer program one can determine all ball quotient weights for $(\widehat{Y}, B = B^1 + B^0)$ if one knows the basic geometric data of $\widehat{Y}$, B^1 and B^0. This was the way of finding the ball quotient weights for linear arrangements on the projective plane $\mathbb{P}^2$ in the monograph [BHH] of BARTHEL, HIRZEBRUCH and HÖFER in a lot of interesting cases. These examples are closely connected with partial differential equations of PICARD-FUCHS type. For this connection we refer to the monograph [YO1] of YOSHIDA, but also to [BHH]. The connection with special algebraic curve families from a higher algebraic geometric point of view has been outlined in the monograph [HO4] of the author. There one can find a finer arithmetic study of the family of PICARD curves $Y^3 = X(X-1)(X-t_1)(X-t_2)$ in close connection with PICARD modular forms of the EISENSTEIN lattice of the ball. For this family the complete quadrihedral 3.7.1 on $\mathbb{P}^2$ plays an important role. Following HÖFER'S thesis [BHH], see also [BHH] and [HO3] we solve the following

Problem 4.11.1 Let L_{ij}, $1 \leq i,j \leq 4$, $i \neq j$, $L_{ij} = L_{ji}$, denote the six lines of the quadrihedral. Choose some of the four triple points P_i, for $i = 1, \ldots, k \in \{0,1,2,3,4\}$, say, to be cusp points. Then blow up the remaining points P_j, $j = k+1, \ldots, 4$. The blown up surface is denoted by $\widehat{Y}$. With our notations the canonical resolution $Y' = Y$ is obtained by blowing up also $P_1, \ldots, P_k \in \widehat{Y}$. As in Figure 3.7.6, we denote the exceptional lines on Y' by L_{0j}, $j = 1,2,3,4$. Furthermore, we set

$$B^1 = \sum_{0 \leq i < j \leq 4} L_{ij},\ B^0 = \text{ intersection points of the lines },\tag{4.11.1}$$

and let $\widehat{B}^1$ be its image on $\widehat{Y}$. The finite point part of $\widehat{B}^0_f$ consists of (nonsingular) abelian surface points. The problem is to find all ball quotient weights $v_{ij} = v(L_{ij})$.

In [HO3] we found a rather magic method of weighting in the compact case $k = 0$. This elementary geometric translation of the proportionality conditions (Prop 0), (Prop 1), (Prop 2) works also for $k \geq 1$. It leads to magic tetrahedrons and related squares. After this translation it will be easy to find all the ball quotient weights of $(\widehat{Y}, \widehat{B})$ we look for.

Let $\widehat{\mathbf{Y}}$ be an orbital weighting of $(\widehat{Y}, \widehat{B})$. We set

$$\nu_{ij} = \nu_{ji} = \frac{1}{v_{ij}} \in \mathbb{N}_+^{-1} \cup \{0\},\tag{4.11.2}$$

where ν_{ij} is the weight of L_{ij} and $\mathbb{N}_+$ denotes the set of positive integers. Only for some ν_{oj}'s we allow the weight to be ∞. Now remember the tetrahedron 3.7.6 with four cusp weights ∞. In analogy to Figure 3.7.6 we draw inside of a cube a weighted tetrahedron as described in Figure (4.11.1) with weights ν_{ij}. The weights $\mu_i \in \mathbb{Q}, i = 0, 1, 2, 3, 4$ will be explained later. We prove that
the conditions (Prop $*$) = (Prop ∞) *and* (Prop 1) *can be translated to* (Prop $\triangle$)

(0) $2\nu_{0j} + \nu_{1j} + \nu_{2j} + \nu_{3j} + \nu_{4j} = 1, \ j = 1, 2, 3, 4; \ \nu_{jj} := 0$
(1) $2\nu_{ij} + \nu_{0i} + \nu_{0j} + \nu_{kl} = 1, \ \{i, j, k, l\} = \{1, 2, 3, 4\}.$

Tetramagic Cube

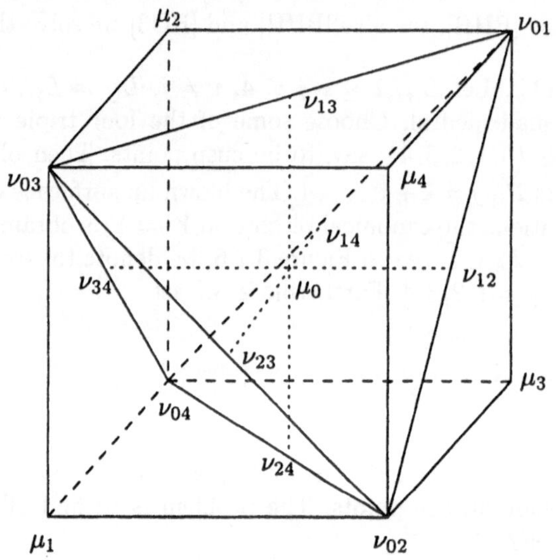

Figure 4.11.1

The relations (0), (1) have a nice magic interpretation on the tetrahedron. Each ν_{ij}-weighted point has three neighbours of the same kind connected directly with the starting point by a solid or dotted line segment. Then one has to take twice the weight of the starting point and to add the weights of its three neighbours.

Proof Since there are no triple points on $\mathbf{Y}$ we have only to take (Prop ∞) and (Prop 1) into account. For example, let $\mathbf{L}_{01}$ be the canonical resolution of the cusp point $\mathbf{P}_{01}$, where P_{0i}, $i = 1,2,3,4$ denote the triple points of the quadrihedral (κ_i in Figure 3.7.1). On $\mathbf{L}_{01}$ are three abelian points Q_{12}, Q_{13}, Q_{14}, the intersection points with L_{12}, L_{13}, L_{14}, respectively (see Figure 3.7.6). Then (Prop ∞) for the cusp point $\mathbf{P}_{01}$ can be written as

$$0 = \mathbf{e}(\mathbf{L}_{01}) = 2 - \left(1 - \frac{1}{v_{12}}\right) - \left(1 - \frac{1}{v_{13}}\right) - \left(1 - \frac{1}{v_{14}}\right)$$

With the convention $v_{01} = \infty$, hence $\nu_{01} = 0$, the identity appears as first equation in (Prop Δ).

Now assume that P_{04} has not been choosen as cusp singularity. Then $\mathbf{L}_{04}$ goes through the finite part $\mathbf{Y}_f$ of $\widehat{\mathbf{Y}}$. In this case we apply (Prop 1) and calculate

$$\mathbf{e}_f(\mathbf{L}_{04}) = 2 - \left(1 - \frac{1}{v_{14}}\right) - \left(1 - \frac{1}{v_{24}}\right) - \left(1 - \frac{1}{v_{34}}\right),$$

$$3\tau_f(\mathbf{L}_{04}) = \frac{1}{v_{04}}(-1).$$

We used again the notations of Figure 3.7.6 and the definitions in 4.7.3. The (Prop 1)-condition $\mathbf{e}_f = 6\tau_f$ yields the last identity of the firtst block (0) of (Prop Δ). All relations of (0) are proved.

If $1 \le i < j \le 4$, then $\mathbf{L}_{ij}$ goes through the finite part $\mathbf{Y}_f$ of $\widehat{\mathbf{Y}}$. The $\mathbf{P}_{0i}$-contribution of $\mathbf{e}_f(\mathbf{L}_{ij})$ is in any case $\left(1 - \frac{1}{v_{0i}}\right)$. Namely, if $\mathbf{P}_{0i} \in \widehat{\mathbf{Y}}_f$, then it is clear from the definition 4.7.3, (e). Otherwise the contribution is

$$1 = \left(1 - \frac{1}{\infty}\right) = \left(1 - \frac{1}{v_{0i}}\right).$$

Now the calculation is the same as that for L_{04}. Thus, the relations of the second block (1) are also verified. The converse conclusion is obvious now. □

The following elementary lemma shows that any of the equations in (Prop Δ) has only finitely many solutions with values in

$$\mathbb{N}_+^{-1} \cup \{0\} = \left\{0, \frac{1}{1}, \frac{1}{2}, \frac{1}{3}, \frac{1}{4} \cdots\right\}.$$

The proof shows how to find them algorithmically.

Lemma 4.11.2 *Let $q, q_1, \ldots, q_l$ be positive rational numbers. Then the diophantine equation*

$$\sum_{i=1}^{l} \frac{q_i}{x_i} = q$$

has at most finitely many solutions $\mathfrak{x} = (x_1, \ldots, x_l) \in (\mathbb{N}_+ \cup \{\infty\})^l$.

Proof This is obviously true for $l = 1$. Let $(m_1, \ldots, m_l) \in (\mathbb{N}_+ \cup \{\infty\})^l$ be a solution. For at least one summand we must have $m_i \leq lq_i/q$. By inductive hypothesis the set $\mathcal{L}_j$ of solutions $\mathfrak{x}$ with $x_j \leq lq_j/m, x_j \in \mathbb{N}_+$, is finite. But $\mathcal{L} = \mathcal{L}_1 \cup \mathcal{L}_2 \cup \ldots \mathcal{L}_l$ is the whole set of solutions. This set is finite. The lemma is proved.

□

In Table 4.A at the end of this section we list all solutions of the diophantine equation

$$\frac{2}{x_0} + \frac{1}{x_1} + \frac{1}{x_2} + \frac{1}{x_3} = 1, \; x_i \in \mathbb{N}_+ \cup \infty, \; \infty \neq x_1 \leq x_2 \leq x_3.$$

It contains the solutions of any equation of our system (Prop Δ) of diophantine equations. The following *tetramagic squares* yield a convenient method how to work with the table in order to get the complete solution of (Prop Δ) with non-negative ν_{ij}.

In picture (Squ) we arrange our weights ν_{ij} in a square putting the ν_{0j}'s into the middle part:

$$\begin{array}{cccc} \nu_{14} & \nu_{13} & \nu_{24} & \nu_{23} \\ \nu_{12} & \nu_{01} & \nu_{02} & \nu_{34} \\ \nu_{34} & \nu_{03} & \nu_{04} & \nu_{12} \\ \nu_{14} & \nu_{24} & \nu_{13} & \nu_{23} \end{array} \qquad \text{(Squ)}$$

The sums in the second block (1) of (Prop Δ) appear as weighted sums of the numbers in the two middle rows, middle columns or diagonals, respectively, following the direction of the arrows. The first summand has to be counted twice. The special square (Squ ∞)

$$\begin{array}{|c|}\hline \begin{array}{cccc} \frac{1}{3} & \frac{1}{3} & \frac{1}{3} & \frac{1}{3} \\ \frac{1}{3} & \multicolumn{1}{|c}{0} & \multicolumn{1}{c|}{0} & \frac{1}{3} \\ \frac{1}{3} & \multicolumn{1}{|c}{0} & \multicolumn{1}{c|}{0} & \frac{1}{3} \\ \frac{1}{3} & \frac{1}{3} & \frac{1}{3} & \frac{1}{3} \end{array} \\ \hline \end{array} \qquad \text{inverse:} \quad \begin{array}{|c|}\hline \begin{array}{cccc} 3 & 3 & 3 & 3 \\ 3 & \multicolumn{1}{|c}{\infty} & \multicolumn{1}{c|}{\infty} & 3 \\ 3 & \multicolumn{1}{|c}{\infty} & \multicolumn{1}{c|}{\infty} & 3 \\ 3 & 3 & 3 & 3 \end{array} \\ \hline \end{array} \qquad (\text{Squ } \infty)$$

yields a solution of (Prop Δ). It is the only solution of our problem allowing at least two cusp points. From (Prop Δ) it follows that for all weights it holds that

$$\frac{1}{\nu_{ij}} = v_{ij} > 2,$$

because ν_{kl}, $0 \leq k < l \leq 4$, is positive. The classification of cusp points in Corollary 3.5.9,(ii) allows only the weights $v_{12} = v_{13} = v_{14} = 3$ for the lines crossing L_{01}. Looking at the second row of (Squ) we have to solve

$$\frac{2}{3} + \frac{1}{\infty} + \frac{1}{\infty} + \frac{1}{v_{34}} = 1.$$

This is only possible for $v_{34} = 3$. Then the middle part of the third row of (Squ) is 0, 0. Now it is easy to fill the rest of the tetramagic square using the rules described above. The square (Squ ∞) discovers the ball quotient surface of the congruence EISENSTEIN lattice described in 3.7.4 and 4.6.5.

Together with Table 4.A at the end of this chapter one can easily find all squares with exactly one cusp contribution ∞. There are only two up to a change of numerations described in (4.11.3)

$$\begin{array}{|c|}\hline \begin{array}{cccc} 3 & 3 & 12 & 12 \\ 3 & \multicolumn{1}{|c}{\infty} & \multicolumn{1}{c|}{4} & 12 \\ 12 & \multicolumn{1}{|c}{4} & \multicolumn{1}{c|}{4} & 3 \\ 3 & 12 & 3 & 12 \end{array} \\ \hline \end{array} \qquad \begin{array}{|c|}\hline \begin{array}{cccc} 3 & 3 & 6 & 6 \\ 3 & \multicolumn{1}{|c}{\infty} & \multicolumn{1}{c|}{6} & 6 \\ 6 & \multicolumn{1}{|c}{6} & \multicolumn{1}{c|}{6} & 3 \\ 3 & 6 & 3 & 6 \end{array} \\ \hline \end{array} \qquad (4.11.3)$$

Squares without cusp contribution ∞ corresponding to compact ball quotient surfaces are listed in Table 4.11.3 . They can be found by the same puzzle as above.

We want to show by a special argument that condition (Prop 2) is automatically satisfied for our example. First we go back to the original definitions of orbital heights of locally abelian orbital surfaces of compact or open type given in 3.3.1, 3.4.1 and 3.6.8.

4.11.3 of all inverse tetramagic squares with (only) positive contributions

9	9	9	9
9	3	3	9
9	3	3	9
9	9	9	9

6	12	6	12
12	3	3	6
6	3	4	12
6	6	12	12

5	15	5	15
15	3	3	5
5	3	5	15
5	5	15	15

6	6	6	6
6	4	4	6
6	4	4	6
6	6	6	6

4	8	4	8
8	4	4	4
4	4	8	8
4	4	8	8

5	5	5	5
5	5	5	5
5	5	5	5
5	5	5	5

4	4	4	4
6	6	6	3
3	12	12	6
4	4	4	4

4	4	3	3
4	8	24	3
3	24	24	4
4	3	4	3

Lemma 4.11.4 *If $\mathbf{Y}_f$ is an open orbital surface with regular arrangement and smooth canonical compactification $\mathbf{Y}' = \mathbf{Y} \longrightarrow \widehat{\mathbf{Y}} \supset \mathbf{Y}_f, \mathbf{Y} = (Y, \mathbf{B})$, $\mathbf{B} = \mathbf{B}^1 + \mathbf{B}^0$ and $B^1 = \sum_{i \in I} C_i$, then*

$$H_e(\mathbf{Y}_f) = \mathbf{e}_f(\mathbf{Y}) = e(Y) - \sum_{i \in I} \left(1 - \frac{1}{v_i}\right) e_i + \sum_{P \in A} \left(1 - \frac{1}{v_2}\right)\left(1 - \frac{1}{v'_{\mathbf{P}}}\right) \tag{4.11.4}$$

$$3H_\tau(\mathbf{Y}_f) = 3\tau_f(\mathbf{Y}) = 3\tau(Y) - \sum_{i \in I} \left(1 - \frac{1}{v'_i}\right) s_i, \tag{4.11.5}$$

where $\mathbf{P} : \underset{v_{\mathbf{P}}}{\circ} \xrightarrow[\langle 1,0\rangle]{} \underset{v'_{\mathbf{P}}}{\circ}$, $v_i \in \mathbb{N}_+ \cup \{\infty\}$ *is the weight of* C_i, $e_i = e(C_i)$ *the* EULER *number of* C_i, $s_i = (C_i^2)$ *its selfintersection number and* $\mathbf{A} = \mathbf{B}^0$ *is the set of abelian points on* $\mathbf{Y}$.

Proof The orbital heights $H_e(\mathbf{Y})$ have by original definitions the same form as the right-hand sides of (4.11.4) and (4.11.5) with the only difference that $v_j = \infty$ for $j \in I_\infty$ corresponding to the canonical resolution of a cusp point. Subtracting $\frac{1}{v_j}\mathbf{e}(\mathbf{C}_j) = \frac{1}{v_j}\left(e_j - \sum_{\mathbf{P} \in \mathbf{C}_j}\left(1 - \frac{1}{v'_j}\right)\right)$ or $3\frac{1}{v_j}\tau(\mathbf{C}_j) =$

$\frac{1}{v_j^2}s_j$, respectively, we obtain the formulas of Lemma 4.11.4 with the new weights $v_j = \infty$ for $j \in I_\infty$. Notice that definition 4.7.3 yields for $i \in I_f = I \setminus I_\infty$

$$\mathbf{e}_f(\mathbf{C}_i) = e_i - \sum_{\mathbf{P}\in\mathbf{A}(\mathbf{C}_i)} \left(1 - \frac{1}{v'_{\mathbf{P}}}\right) \tag{4.11.6}$$

$$3\tau_f(\mathbf{C}_i) = s_i v_j \tag{4.11.7}$$

hence

$$3\tau_f(\mathbf{Y}) = 3\tau(Y) - \sum_{j\in I\setminus I_f} s_j - \sum_{i\in I_f} \left(1 - \frac{1}{v_i^2}\right) s_i$$

Observe also that $s_j = 1 \cdot s_j = \left(1 - \frac{1}{v_j}\right) s_j$ for $v_j = \infty$.

□

Corollary 4.11.5 *If the conditions of Proposition 4.11.4 and (Prop 1) are satisfied, then the condition (Prop 2) can be linearized with respect to the variables $\frac{1}{v_i}$ of inverse weights. More precisely, (Prop 2) can be written as*

(Prop 2′)

$$\sum_i \left(\frac{1}{2v_i} - 1\right) \left[-e_i + \left(C_i \cdot \sum_{j\neq i} C_j\right)\right] = \sum_i s_i - |\mathbf{A}| + e(Y) - 3\tau(Y),$$

where **A** *is the set of abelian points on* **Y** *and*

$$\left(C_i \cdot \sum_{j\neq i} C_j\right) = |\mathbf{A}_i| = |\mathbf{A} \cap \mathbf{C}_i|.$$

Proof The (Prop 1)-relation $\mathbf{e}_i = \mathbf{e}_f(\mathbf{C}_i) = 6\tau(\mathbf{C}_i) = 6\tau_i$ has by (4.11.6), (4.11.7) the form

(Prop 1′)

$$-\frac{2}{v_i} s_i + \left(C_i \cdot \sum_{j\neq i} \frac{1}{v_j} C_j\right) = |\mathbf{A}_i| - e_i, i \in I.$$

In (4.11.4) we change over in the e_i-part to $\mathbf{e}_i$ and s_i writing

$$\begin{aligned}\mathbf{e}_f(\mathbf{Y}) &= e(Y) - \sum_i \left(1 - \frac{1}{v_i}\right)\mathbf{e}_i - \sum_{\mathbf{P}} \left(1 - \frac{1}{v_{\mathbf{P}}}\right)\left(1 - \frac{1}{v'_{\mathbf{P}}}\right) \\ &= e(Y) - 2\sum_i \left(1 - \frac{1}{v_i}\right)\frac{s_i}{v_i} - \sum_{\mathbf{P}} \left(1 - \frac{1}{v_{\mathbf{P}}}\right)\left(1 - \frac{1}{v'_{\mathbf{P}}}\right)\end{aligned}$$

With (4.11.5) and (Prop 2) we get

$$\begin{aligned}\mathbf{e}_f(\mathbf{Y}) - 3\tau_f(\mathbf{Y}) &= e(Y) - 3\tau(Y) + \sum_i \left(1 - \frac{1}{v_i}\right)^2 s_i \\ &\quad - \sum_P \left(1 - \frac{1}{v_P}\right)\left(1 - \frac{1}{v'_P}\right) \\ &= e(Y) - 3\tau(Y) + \sum_i s_i - \sum_i \frac{2}{v_i} s_i \qquad (4.11.8) \\ &\quad + \sum_i \frac{1}{v_i^2} s_i - \sum_P \left(1 - \frac{1}{v_P}\right)\left(1 - \frac{1}{v'_P}\right) = 0.\end{aligned}$$

The second sum can be expressed by (Prop 1′):

$$-\sum_i \frac{2}{v_i} s_i = \sum_i |\mathbf{A}_i| - \sum_i e_i - \left(\left(\sum_i C_i\right) - \left(\sum_{j \neq i} \frac{1}{v_j} C_j\right)\right) \qquad (4.11.9)$$

The last sum is

$$\begin{aligned}\sum_{\mathbf{P} \in \mathbf{A}} \left(1 - \frac{1}{v_{\mathbf{P}}}\right)\left(1 - \frac{1}{v'_{\mathbf{P}}}\right) &= |\mathbf{A}| - \left(\left(\sum_i C_i\right) - \left(\sum_{j \neq i} \frac{1}{v_j} C_j\right)\right) \\ &\quad + \sum_{\mathbf{P}} \frac{1}{v_{\mathbf{P}} v'_{\mathbf{P}}}\end{aligned}$$

and

$$\begin{aligned}\sum_{\mathbf{P}} \frac{1}{v_{\mathbf{P}} v'_{\mathbf{P}}} &= \frac{1}{2} \sum_i \left(\frac{1}{v_i} C_i \cdot \sum_{j \neq i} \frac{1}{v_j} C_j\right) \\ &= \sum_i \frac{1}{v_i^2} s_i + \frac{1}{2} \sum_i \frac{1}{v_i} (|\mathbf{A}_i| - e_i)\end{aligned}$$

by (Prop 1′), hence

$$\sum_{\mathbf{P}} \frac{1}{v_{\mathbf{P}} \cdot v'_{\mathbf{P}}} - \sum \frac{1}{v_i^2} s_i = \sum_i (|\mathbf{A}_i| - e_i)/2v_i \tag{4.11.10}$$

Now, we get by substitution

$$\begin{aligned}
-\sum \frac{2}{v_i} s_i &= \sum_P \left(1 - \frac{1}{v_P}\right)\left(1 - \frac{1}{v'_P}\right) + \sum_i \frac{1}{v_i^2} s_i \\
&= \sum_i (|\mathbf{A}_i| - e_i) - |\mathbf{A}| - \sum_{\underline{P}} \frac{1}{v_P v'_P} + \sum \frac{1}{v_i^2} s_i \\
&= \sum_i (|\mathbf{A}_i| - e_i) - |\mathbf{A}| - \sum_i \frac{1}{2v_i} (|\mathbf{A}_i| - e_i)
\end{aligned}$$

Together with (4.11.8) we obtain the relation (Prop 2′).

□

Now it is easy to check the implication

$$\text{(Prop 0), (Prop 1)} \Longrightarrow \text{(Prop 2)} \tag{4.11.11}$$

for our example. The starting conditions are the equations (Prop Δ). Under these conditions (Prop 2) is equivalent with the "linear" diophantine equation (Prop 2′). Explicitly we have in our case

$$\begin{gathered}
e_i = 2,\ s_i = -1,\ |\mathbf{A}_i| = 3,\ i = 1, \ldots, 10, \\
|\mathbf{A}| = 3 \cdot 10/2 = 15,\ e(Y) = e(\mathbb{P}^2) + 4 = 7, \\
\tau(Y) = \tau(\mathbb{P}^2) - 4 = -3.
\end{gathered}$$

Working now with double indices again corresponding to the lines L_{ij}, $0 \le i < j \le 4$, we get

(Prop$_2\Delta'$)

$$\sum_{0 \le i < j \le 4} \frac{1}{v_{ij}} = 2 \cdot [10 + 2 \cdot 10(-1) - 15 + 7 - 3(-3)] = 2$$

If we add all equations of (Prop Δ) we get the same relation because each $\nu_{ij} = \dfrac{1}{v_{ij}}$ appears five times in this sum. This proves the special

But also the inverse implication

$$(\mathrm{Prop}_2\ \Delta) \Longrightarrow (\mathrm{Prop}\ \Delta), (\mathrm{Prop}_2\ \Delta') \tag{4.11.13}$$

can be verified immediately; it is left to the reader.

Remark 4.11.6 (see ([HO3], 5.12)) For **Y** as in 4.11.1 the condition (Prop 2) can be written as the vanishing of the sum of seven squares:

$$\begin{aligned}\sum_{j=1}^{4}\left[\frac{1}{v_{0j}} - \frac{1}{2}\left(1 - \sum_{1\le i\ne j}\frac{1}{v_{ij}}\right)\right]^2 \\ + \frac{1}{4}\left(\frac{1}{v_{12}} + \frac{1}{v_{34}} - \frac{1}{v_{13}} - \frac{1}{v_{24}}\right)^2 \\ + \frac{1}{4}\left(\frac{1}{v_{13}} + \frac{1}{v_{24}} - \frac{1}{v_{14}} - \frac{1}{v_{23}}\right)^2 \\ + \frac{1}{4}\left(\frac{1}{v_{14}} + \frac{1}{v_{24}} - \frac{1}{v_{12}} - \frac{1}{v_{34}}\right)^2 = 0\end{aligned} \tag{4.11.14}$$

For this purpose set $\beta = e - 3\tau$, $H_\beta(\mathbf{Y}_f) = \boldsymbol{\beta}_f(\mathbf{Y}) = \mathbf{e}_f(\mathbf{Y}) - 3\boldsymbol{\tau}_f(\mathbf{Y}) = H_e(\mathbf{Y}_f) - 3H_\tau(\mathbf{Y}_f)$. Then (Prop 2) is by definition the condition $H_\beta(\mathbf{Y}_f) = \boldsymbol{\beta}_f(\mathbf{Y}) = 0$. Now express $H_\beta(\mathbf{Y}_f)$ by means of (4.11.4) and (4.11.5) as rational function of v_{ij}.

Remark 4.11.7 One can prove also the implication 4.11.13 for **Y** of type 4.11.1.

Hint: (see [HO3], § 8). Namely, the relation (4.11.14) splits into seven linear equations in $\nu_{ij} = \dfrac{1}{v_{ij}}$. Show that this system of linear equations is equivalent to the system (Prop Δ).

With similar methods one can also solve the following three problems

Problem 4.11.8 Show that the equalities of (Prop 0) and (Prop 1) together are equivalent to the linear system (Prop Δ) allowing negative inverse integers v_{0j} corresponding to weighted quotient points and $v_{ij} = \infty$.

Problem 4.11.9 Solve the system (Prop Δ) completely with $\nu_{0j} \in \mathbb{Z}^{-1}$, $\nu_{ij} \in \mathbb{N}^{-1} \cup \{0\}$ for $i, j = 1, 2, 3, 4$, $i < j$. Check that beside the squares (4.11.15) there are only finitely many solutions.

$$\begin{array}{|cccc|}\hline 1 & 1 & d & d \\ 1 & \multicolumn{1}{|c}{-1} & \multicolumn{1}{c|}{-d} & d \\ d & \multicolumn{1}{|c}{-d} & \multicolumn{1}{c|}{-d} & 1 \\ 1 & d & 1 & d \\ \hline \end{array} \tag{4.11.15}$$

Hint: Use the method of tetramagic squares. Follow the line of the proof of the general Finiteness Theorem 4.10.3 using the classification of cusp and quotient points. Compare with the list of inverse tetramagic squares in Table **4**. B at the end of this chapter.

Problem 4.11.10 Select from all orbital surfaces $\mathbf{Y}$ corresponding to tetramagic squares those which are finite quotients $\mathbf{Y} = \mathbb{P}^2/G,\ G \subset Aut\ \mathbb{P}^2 = \mathbb{P}\mathbb{G}l_3(\mathbb{C})$ a finite group.

All other cases are ***honest orbital*** ball quotient surfaces $\mathbf{Y}_f = \mathbb{B}/\Gamma$. The corresponding solutions of Prop Δ are called *hyperbolic* . For a proof we refer to YOSHIDA'S monograph [YO2]. The author uniformizes the systems of partial differential equations of Fuchsian type on $Y\backslash B$. The complete list is given in Table **4**.B.

The quadrihedral example goes back to the pioneering work of PICARD at the end of the last century. PICARD could prove that the fundamental group of $Y\backslash B$ has a natural representation on the ball $\mathbb{B}$. He could not prove that the representing group $\Gamma \subset Aut_{hol}\mathbb{B}$ is a lattice. A positive solution of this problem was given by MOSTOW, DELIGNE [D-M]. Moreover, MOSTOW and DELIGNE proved that $\mathbb{B}/\Gamma = \mathbf{Y}_f$.

The basic reference for linear arrangements on $\mathbb{P}^2$ is the monograph [BHH]. There one can find a general proof for the equivalence of (Prop 2) and (Prop 0) + (Prop 1) for the corresponding (open) linearly arranged surfaces. The proof combines quadratic forms with EINSTEIN-KÄHLER arguments. Surprisingly, the proof of the Finiteness Theorem counts ball quotient arrangements with prescribed support only by solving the diophantine system (Prop 0), (Prop 1), (Prop 2) of equations. In the special case of the complete quadrihedral on $\mathbb{P}^2$ we transferred the diophantine equation system to relations between weights of points on a tetrahedron. The method of magic tetrahedron is convenient to find the complete solution. This method seems to be important for arbitrary cases. Namely, we correspond to any open orbital surface a molecular graph. Forgetting weights the molecular graph appears as a geometric object like the tetrahedron in our classical special case. Then we look for weights such that the proportionality conditions (Prop $*$), (Prop 1) and (Prop 2) are satisfied. The corresponding weighted graphs can be considered as "magic" ones. It is an open problem to find all such graphs coming from line configurations in $\mathbb{P}^2$. Do there exist only finitely many? Most of the known graphs (see [BHH], [YO2]) are closely connected with regular bodies or polyhedra with a remarkable symmetry group.

The classical approach to ball lattices initiated by PICARD [PI1], APPELL [AP1] and a little bit earlier by POCHHAMMER [POC] started from functions in several variables (APPELL series), in connection with their integral representations and systems of partial differential equations generalizing the GAUSS-SCHWARZ theory for functions of one variable. For a historical introduction for two variables we refer to [HO4], for the general theory to [YO1]. PICARD and APPELL [AP2] considered integrals of the following type

$$\int u^{\mu_1-1}(u-1)^{\mu_2-1}(u-t_1)^{\mu_3-1}(u-t_2)^{\mu_4-1}du, \ \mu_i \in \mathbb{Q} \text{ fixed },$$

along paths in $\mathbb{C} \setminus \{0,1,t_1,t_2\}$. These are multivalued analytic functions in two variables t_1, t_2. One has to exclude $t_1, t_2 = 0, 1$ and $t_1 = t_2$. These are five lines of the affine place $\mathbb{C}^2$. The complement of these lines in $\mathbb{C}^2$ is equiavlent with the complement of a complete quadrihedral in the complex plane $\mathbb{P}^2 = \mathbb{P}^2(\mathbb{C})$. The fundamental group $\pi_1(\mathbb{P}^2 \setminus \Delta)$ acts via paths on the integral functions. The functions generate a finite-dimensional vector space. So, one receives a linear representation of $\pi_1(\mathbb{P}^2 \setminus \Delta)$. The image group is called the *monodromy group* of the functions, or better, of the corresponding system of partial differential equations. This system solved by the above integrals can be established in a modern manner by means of an algebraic curve family over $\mathbb{P}^2 \setminus \Delta$ and its GAUSS-MANIN connection. At a point $t = (t_1, t_2) \in \mathbb{P}^2 \setminus \Delta$ one considers the following algebraic curve:

$$C_t : Y^n = u^{b_1}(u-1)^{b_2}(u-t_1)^{b_3}(u-t_2)^{b_4},$$

where n is the (smallest) common denominator of $\mu_1, \mu_2, \mu_3, \mu_4$ and

$$b_i = n \cdot (\mu_i - 1), \ i = 1, 2, 3, 4.$$

The structure of the GAUSS-MANIN connection of the family $\{Y_t\}$ has been intensively studied in the monograph [HO4] of the author in the framework of algebraic geometry. The structures produce the corresponding EULER-PICARD systems of partial differential equations. They are called of Fuchsian type by YOSHIDA [YO2].

Theorem 4.11.11 (PICARD, MOSTOW, DELIGNE, see [D-M])) *The monodromy group of our integral functions is a ball lattice or a finite subgroup of* Aut $\mathbb{P}^2$ *if and only if the relations*

$$1 - \mu_i - \mu_j \in \pm\mathbb{N}_+^{-1} \cup \{0\} \ \textit{for}\ 0 \leq i < j \leq 4, \qquad \text{(INT)}$$

with

$$\mu_0 + \mu_1 + \mu_2 + \mu_3 + \mu_4 = 2 \qquad (0)$$

defining μ_0 *are satisfied*

In these cases there exist a uniformization of the corresponding system of differential equations over $\mathbb{P}^2 \setminus \Delta$ on the ball $\mathbb{B}$ or on $\mathbb{P}^2$, where the solutions loose the multivalence. There is a classical paper [LEV] of LE VAVASSEUR solving (4.11.11) completely. This means that all pentuples $(\mu_0, \mu_1, \mu_2, \mu_3, \mu_4)$ satisfying (4.11.11) have been determined. This long article can be now substituted by the following

Proposition 4.11.12 ([HO3], § 8) *The system (4.11.11) is equivalent with (Prop* Δ*). More precisely, the following maps described in* (μ), (ν):

$$\left\{(\nu_{ij})_{0\leq i<j\leq 4},\ \textit{satisfying (Prop } \Delta),\ \nu_{ij} \in \pm\mathbb{N}^{-1} \cup \{0\}\right\}$$

$$\mu \downarrow\uparrow \nu$$

$$\{(\mu_0, \mu_1, \mu_2, \mu_3, \mu_4);\ \textit{satisfying}\ (4.11.11)\},$$

$$\mu_0 = \frac{1}{3}\left(2 - \sum_{j=1}^{4} \nu_{0j}\right)$$

$$\mu_i = \frac{1}{3}\left(1 - 3\nu_{0i} + \sum_{j=1}^{4} \nu_{0j}\right) = 1 - \nu_{0j} - \mu_0,\ i = 1, \ldots, 4, \qquad (\mu)$$

$$\nu_{ij} = 1 - \mu_k - \mu_l,\ \{i, j, k, l\} = \{1, 2, 3, 4\}, \qquad (\nu)$$

$$\nu_{0j} = 1 - \mu_0 - \mu_j,\ j = 1, 2, 3, 4,$$

establish a bijective correspondence between the solutions of (Prop Δ*) and the solutions of ([INT]). They correspond biuniquely to the set of magic tetrahedrons and to the set of tetramagic squares.*

Proof The last statement has been proved in the previous section. The link between (Prop Δ) and [INT] is the system of linear equations

$$\mu_k + \nu_{ij} + \mu_l = 1, \ \{i,j,k,l\} = \{1,2,3,4\},$$
$$\mu_j + \nu_{0j} + \mu_0 = 1, \ j = 1,2,3,4,$$

$$\mu_0 + \mu_1 + \mu_2 + \mu_3 + \mu_4 = 2. \tag{TMC}$$

The cube weighted with the rational numbers ν_{ij}, μ_k as described in Figure (4.11.1) is called a *tetramagic cube* if the equations (TMC) are satisfied. The left-hand sides of the first ten equations can be interpreted geometrically as sums along six surface diagonals complementary to the edges of the tetrahedron or as sums along the four diagonals of the cube, respectively.

(Prop Δ) $\Longrightarrow$ (TMC): We work with definition (μ) of $\mu_0, \mu_1, \mu_2, \mu_3, \mu_4$. Using relations of (Prop Δ) we find the following system of identities:

$$\begin{aligned} 1 - 2\nu_{01} + \nu_{02} + \nu_{03} + \nu_{04} &= 3\mu_1 \\ 1 + \nu_{01} - 2\nu_{02} + \nu_{03} + \nu_{04} &= 3\mu_2 \\ 2\nu_{12} + \nu_{34} + \nu_{01} + \nu_{02} &= 1 \\ -2\nu_{12} - 4\nu_{34} - 2\nu_{03} - 2\nu_{04} &= -2 \end{aligned}$$

Summing up we get

$$-3\nu_{34} = 3\mu_1 + 3\mu_2 - 3.$$

This is the first relation of (TMC). The next relations of the same type hold by symmetry. The others follow directly from definition (μ).

(TMC) $\Longrightarrow$ (Prop Δ): Assume that $(\nu_{ij})_{0\leq i<j\leq 4}$ is a solution of (TMC). This solution given in (ν) is unique. The left-hand sides of (Prop Δ) can be written on this way as linear combination of 1 and the μ_i's. These combinations are equal to 1 by the last relation of (TMC).

(Prop Δ), (TMC) $\Longleftrightarrow$ (4.11.11): We have to remember that (Prop Δ) is a diophantine system of equations for $v_{ij} = \dfrac{1}{\nu_{ij}}$. Let (ν_{ij}) be a solution of (Prop Δ). We know that $(\mu_i)_{i=0,\ldots,4}$ defined in (μ) satisfies the relations of (TMC). Now it is clear that the conditions of (4.11.11) are satisfied. If conversely $(\mu_i)_{0\leq i\leq 4}$ satisfies these conditions, then we define ν_{ij} by (ν). From [INT] follows immediately that $\nu_{ij} \in \pm\mathbb{N}^{-1} \cup \{0\}$. On the other hand (TMC) holds. Since (TMC) is equivalent with (Prop Δ) the proof is finished.

□

Remark 4.11.13 A tetramagic cube is uniquely determined by the weights ν_{0j}, j=1,2,3,4 of the vertices of the tetrahedron because they determine uniquely the weigths μ_j, $0 \leq j \leq 4$, by (μ) and the weights ν_{ij} by (TMC).

4.12 Towards the Count of Ball Metrics on Non-Compact Surfaces

Let $\widehat{\mathbf{Y}_f} = \mathbf{Y}_f$ be an open orbital surface. If $\mathbf{Y}_f$ is an orbital ball quotient surface $\mathbb{B}/\Gamma$, then we remove from Y the support B of the arrangement and obtain a non-compact surface $Y_f^0 = Y \backslash B \subset Y_f$. We denote by $\mathbb{B}^0$ the preimage of Y_f^0 along the projection $\mathbb{B} \longrightarrow \mathbb{B}/\Gamma = Y_f$. $\mathbb{B}^0$ is an open analytic part of $\mathbb{B}$, more precisely,

$$\mathbb{B}^0 = \mathbb{B} \setminus (\{\Gamma - \text{reflection discs }\} \cup \{\Gamma - \text{elliptic points }\}),$$

and $\mathbb{B}^0 \underset{\Gamma}{\to} Y_f^0$ is an unramified covering of two-dimensional analytic manifolds of infinite degree, because we have removed the branch locus and all singularities of Y_f. The EINSTEIN-KÄHLER metric of $\mathbb{B}$ with negative constant RICCI-curvature restricts to $\mathbb{B}^0$. Since it is Γ-invariant we receive an EINSTEIN-KÄHLER metric of the same quality on the quotient surface $Y_f^0 = \mathbb{B}^0/\Gamma$.

Definition 4.12.1 The EINSTEIN-KÄHLER metric on Y_f^0 defined above is called a *ball metric* on Y_f^0 or a *ball metric* on $Y, \widehat{Y}$ or Y_f *degenerated along* $B, \widehat{B}$ or B_f, respectively.

Finiteness Theorem 4.12.2 *On a compact complex normal algebraic surface $\widehat{Y}$ there exist at most finitely many ball metrics with prescribed locus $\widehat{B}$ of degeneracy. In any case for given $(\widehat{Y}, \widehat{B})$ a bound of the number of ball metrics on $\widehat{Y} \backslash \widehat{B}$ can be calculated effectively, if the basic geometric data of $\widehat{Y}, \widehat{B}$ are known.*

Proof Each solution of the system of diophantine equations (Prop $*$), (Prop 1), (Prop 2) yields an open orbital surface $\mathbf{Y}_f$. Only these open orbital surfaces have a chance to be ball quotients by the Proportionality Theorem 4.9.2. By the effective Finiteness Theorem 4.10.3 there exist only finitely many hyperbolic solutions of our system of diophantine equations. They can be calculated effectively. It remains to prove that each solution allows at most one ball metric on $Y_f^0 = \widehat{Y} \backslash \widehat{B}$. We fix a solution and let $\mathbf{Y}_f = \mathbb{B}/\Gamma = \mathbb{B}/\Gamma_1$ for two ball lattices $\overline{\Gamma}$, $\overline{\Gamma}_1$. We show that there is a commutative diagram:

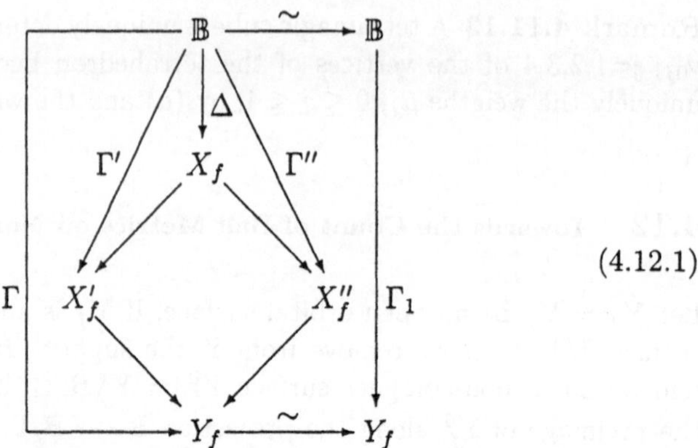

(4.12.1)

with quotient maps (indicated by the corresponding groups), where $\Delta, \Gamma', \Gamma''$ are neat normal subgroups of Γ. For this purpose let $\mathbb{V}$ be the universal cover of Y_f^0, $Y_f^0 = \mathbb{V}/\tilde{\Gamma}$. Omitting Γ-fixed points on $\mathbb{B}$ we get an unramified covering $\mathbb{B}^0 \underset{\Gamma}{\to} Y_f^0$. The universal covering of $\mathbb{B}^0$ is $\mathbb{V}$, say $\mathbb{B}^0 = \mathbb{V}/N$, $\Gamma = \tilde{\Gamma}/N$. Starting from Γ_1 we get in the same manner a subgroup $\widetilde{\Gamma}_1$ of Aut $\mathbb{V}$ such that Γ_1 is a quotient of $\widetilde{\Gamma}_1$ by a normal subgroup. Without loss of generality we can assume that $\widetilde{\Gamma}_1 = \widetilde{\Gamma}$ because both groups coincide with the fundamental group of Y_f^0 and we use the same universal cover $\mathbb{V}$. We choose a neat normal subgroup Γ' of Γ of finite index and set $X'_f = \mathbb{B}/\Gamma'$, $X_f'^0 = \mathbb{B}^0/\Gamma'$. The universal covering of $X_f'^0$ is $\mathbb{V}$, hence $\Gamma' = \widetilde{\Gamma}'/N$ for a normal subgroup $\widetilde{\Gamma}'$ of $\widetilde{\Gamma}$ of finite index. The quasiprojective surface X''_f is constructed in the same manner by means of Γ_1. The fundamental group $\tilde{\Gamma''}$ is also a normal subgroup of $\widetilde{\Gamma} = \widetilde{\Gamma}_1$ of finite index. The quasiprojective surface X_f^0 is defined as quotient surface $\mathbb{V}/\tilde{\Delta}$, where $\tilde{\Delta} = \tilde{\Gamma}' \cap \tilde{\Gamma}''$. The surface X''_f is defined by STEIN's extension theorem 1.12.2. It yields a commutative diagram (4.12.2):

Up to isomorphy X''_f is also defined by a neat normal subgroup of Γ_1 because $X_f''^0$ is, and STEIN's extension is unique. By assumption, the branch arrangements of g and h are equal. Comparing ramification indices we see that together with g' also h' must be unramified. The common universal cover of X'_f, X_f and X''_f is the ball $\mathbb{B}$. So we get the commutative diagram (4.12.1) up to the Γ_1-part. Remember that h is a GALOIS covering by construction via Γ_1 with GALOIS group H. This group can be lifted along the universal covering $\mathbb{B} \longrightarrow X''_f$. On this way we get Γ_1 back up to isomorphy. The Finiteness Theorem 4.12.2 is proved.

□

We did not touch the problem of the existence of a hyperbolic metric on

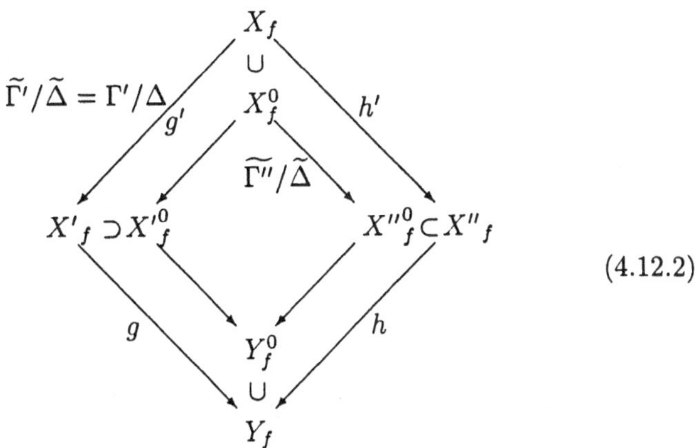

(4.12.2)

Y_f for a solution of the system of diophantine equations (Prop $*$), (Prop 1), (Prop 2). In order to exclude the possibility $\widehat{\mathbf{Y}} = \mathbb{P}^2/G$, $G \subset \mathrm{Aut}\left(\mathbb{P}^2\right)$ a finite group, one has to claim additionally that $\mathbf{Y}_f$ is logarithmically of general type. The Japanese mathematicians KOBAYASHI, NARUKI and SAKAI announced the positive solution of the existence problem in [KNS]. They worked with another language. The equivalence of their conditions with our proportionality conditions should be proved. Then one has both: The finiteness theorem and the existence theorem. They yield together in any case a precise solution of the problem of finding all EINSTEIN-KÄHLER metrics with constant negative RICCI curvature (hyperbolic metrics) on a surface $\widehat{Y}$ with prescribed locus B of degeneracy.

Problem 4.12.3 ([HO3],S. 8) The double cover X of $\mathbb{P}^2$ branched along the complete quadrihedral is a model of a K3–surface with 4 non-abelian quotient singularities (of quaternion group type) and 3 cyclic singularities of type $\langle 2, 1\rangle$.

One blows up first the four triple points of the quadrihedral. Then one takes the double cover X' of the arising surface Y' along the ten lines of the preimage of the quadrihedral.

Problem 4.12.4 (see [HO3],S. 7) Let $X' \longrightarrow \widehat{X} = \mathbb{P}^2$ be the canonical resolution of the four triple points of the quadrihedral Δ, see Figure 3.7.6, and $X_1 \to \widehat{X}$ a resolution of a part of them such that we have a decomposition $X' \to X_1 \to \widehat{X}$ with birational morphisms. Then one can calculate the number of different ball metrics on X_1 degenerated at most along the preimage B_1 of the quadrihedral on X_1.

For this purpose write the conditions (Prop 0), (Prop 1) as a system of diophantine equations of type (Prop Δ) buth with 2 instead of 1 on the right-hand sides of these equations. Solve this system by means of tetra-magic squares with sum 2.

Problem 4.12.5 Relating the solutions of the previous remark with the solutions of the classical problem for $(\mathbb{P}^2, \Delta)$ one can show that the former solutions correspond to subgroups of index 2 of the ball lattices corresponding to the classical case.

Problem 4.12.6 (see [YO1]) By means of formula (4.11.4) and Theorem 4.8.1 it is easy to calculate the c_2-volume of a Γ-fundamental domain for all ball lattices Γ corresponding to ball metrics on $\mathbb{P}^2 \setminus \Delta$.

4.A Table of all solutions of $\frac{2}{x_0} + \frac{1}{x_1} + \frac{1}{x_2} + \frac{1}{x_3} = 1$, $x_i \in \mathbb{N}_+ \cup \{\infty\}$, $x_i \geq 3$, up to S_3-symmetrie

x_0	x_1	x_2	x_3	x_0	x_1	x_2	x_3
3	∞	∞	3	3	9	9	9
3	∞	4	12	4	3	7	42
3	∞	6	6	4	3	8	24
4	∞	3	6	4	3	9	18
4	∞	4	4	4	3	10	15
6	∞	3	3	4	3	12	12
3	4	13	156	4	4	5	20
3	4	14	84	4	4	6	12
3	4	15	60	4	4	8	8
3	4	16	48	4	5	5	10
3	4	18	36	4	6	6	6
3	4	20	30	5	3	4	60
3	4	21	28	5	3	5	15
3	4	24	24	5	3	6	10
3	5	8	120	5	4	4	10
3	5	9	45	5	5	5	5
3	5	10	30	6	3	4	12
3	5	12	20	6	3	6	6
3	5	15	15	6	4	4	6
3	6	7	42	8	3	3	12
3	6	8	24	8	3	4	6
3	6	9	18	9	3	3	9
3	6	10	15	12	3	3	6
3	6	12	12	12	3	4	4
3	7	7	21	15	3	3	5
3	8	8	12	24	3	3	4

Table 4.B Inverse tetramagic squares corresponding to hyperbolic solutions of (TMC)

3	3	3	3
3	∞	∞	3
3	∞	∞	3
3	3	3	3

4	4	4	4
2	∞	∞	∞
∞	4	4	2
4	4	4	4

2	4	2	4
4	∞	∞	2
2	∞	-4	4
2	2	4	4

5	5	5	5
5	5	5	5
5	5	5	5
5	5	5	5

3	3	6	6
3	∞	6	6
6	6	6	3
3	6	3	6

3	6	3	6
2	∞	∞	∞
∞	3	6	2
3	3	6	6

2	2	6	6
3	-6	6	3
3	∞	∞	3
2	6	2	6

2	2	3	3
3	-6	∞	2
2	-6	-6	3
2	3	2	3

4	8	4	8
8	4	4	4
4	4	8	8
4	4	8	8

2	2	8	8
2	-4	8	8
8	8	8	2
2	8	2	8

2	2	4	4
4	-8	8	2
2	-8	-8	4
2	4	2	4

9	9	9	9
9	3	3	9
9	3	3	9
9	9	9	9

5	2	5	2
2	-10	-10	5
5	-10	5	2
5	5	2	2

6	6	6	6
6	4	4	6
6	4	4	6
6	6	6	6

4	4	4	4
6	6	6	3
3	12	12	6
4	4	4	4

6	12	6	12
12	3	3	6
6	3	4	12
6	6	12	12

3	3	12	12
3	∞	4	12
12	4	4	3
3	12	3	12

3	3	4	4
2	-12	∞	12
12	6	6	2
3	4	3	4

3	2	6	4
2	-6	12	12
12	12	4	2
3	6	2	4

3	3	3	3
2	-12	-12	6
6	12	12	2
3	3	3	3

6	2	6	2
2	-12	-12	6
6	-12	4	2
6	6	2	2

2	2	6	6
2	-4	12	6
6	12	12	2
2	6	2	6

2	2	3	3
2	-4	-12	3
3	-12	-12	2
2	3	2	3

5	15	5	15
15	3	3	5
5	3	5	15
5	5	15	15

9	2	9	2
2	-18	-18	9
9	-18	3	2
9	9	2	2

2	2	5	5
2	-4	20	5
5	20	20	2
2	5	2	5

4	4	3	3
4	8	24	3
3	24	24	4
4	3	4	3

5 Picard Modular Surfaces

5.1 Classification Diagram

This chapter is dedicated to a survey of the classification of PICARD *modular surfaces* $\widehat{\mathbb{B}/\Gamma}, \Gamma = \mathbb{U}((2,1), \mathcal{O}_K), K = \mathbb{Q}(\sqrt{-d})$ an imaginary quadratic number field. For the leading examples of EISENSTEIN and GAUSS numbers we classify completely the corresponding surfaces in sections 1 and 2. These examples clarify the way of the KODAIRA classification of PICARD modular surfaces corresponding to K with higher discriminants. There are some new results for the PICARD modular surfaces of EISENSTEIN numbers which are of number theoretic interest in close connection with 2–dimensional versions of HILBERT'S 12-th (class field) and 7-th (transendence) problem. We refer to the monograph [HO8].

A central role play the heights of the open orbital PICARD modular surfaces $\mathbf{Y}_f$. Its number theoretic formula will be established in the appendix. The most interesting aspect is that $H(\mathbf{Y}_f)$, $H = H_e$, splits p-adically in the geometric-arithmetical situation of PICARD modular surfaces in a natural and calculable manner: $H = \mathbf{c}_\infty \cdot \prod_p \mathbf{c}_p$. So we dispose on useful examples for heights of open orbital surfaces. Section 4 is devoted to classification results of PICARD modular surfaces corresponding to higher discriminants together with their meaning in surface geography. Before we start with the leading examples we present the general scheme for the KODAIRA classification of ball quotient surfaces $\widehat{\mathbb{B}/\Gamma}$ in (D-5):

(D-5)

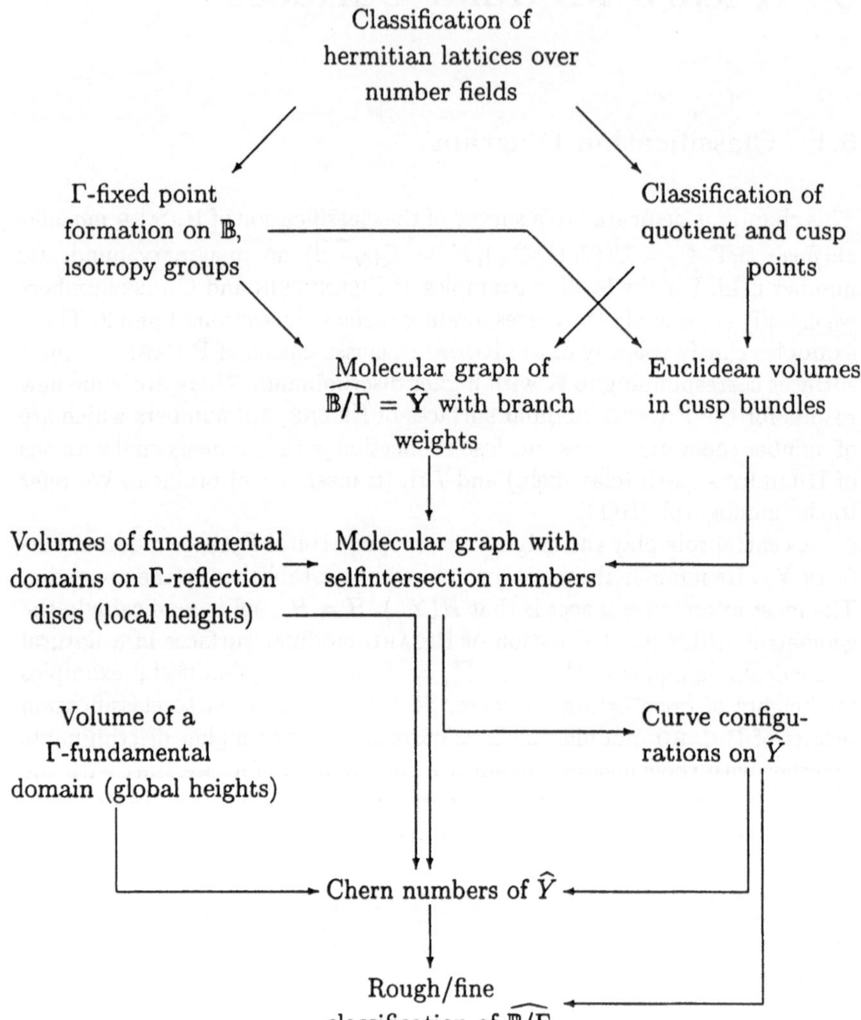

Classification of hermitian lattices over number fields
Γ-fixed point formation on $\mathbb{B}$, isotropy groups
Classification of quotient and cusp points
Molecular graph of $\widehat{\mathbb{B}/\Gamma} = \widehat{Y}$ with branch weights
Euclidean volumes in cusp bundles
Volumes of fundamental domains on Γ-reflection discs (local heights)
Molecular graph with selfintersection numbers
Volume of a Γ-fundamental domain (global heights)
Curve configurations on $\widehat{Y}$
Chern numbers of $\widehat{Y}$
Rough/fine classification of $\widehat{\mathbb{B}/\Gamma}$

5.2 Picard Modular Surface of the Field of Eisenstein Numbers

Let Γ' be the congruence subgroup $\Gamma = \mathbb{U}((2,1),\mathcal{O}_K)$, $K = \mathbb{Q}(\rho)$, $\rho = e^{2\pi i/3}$, of the prime ideal $\mathcal{O}(1-\rho)$, see Example 4.6.5. There are exactly six Γ'-reflection discs on $\mathbb{B}$ up to Γ'-equivalence drawn in Figure 4.6.1. The molecular graph of the open orbital quotient surface $\mathbf{Y}_f = \mathbb{B}/\Gamma'$ is the magic tetrahedron inside of the cube 4.11.1 with weight 0 at the vertices. From Remark 4.11.11 we know that this magic tetrahedron is uniquely determined by its weights at the vertices. Notice that we can read off this molecular graph only from the fixed point formation 4.6.1 and the knowledge of the stationary groups at elliptic points, cusps and along reflection discs. The structure of these groups has been found with the methods described in the authors monograph [HO4]. There we worked with the subgroup $\mathbb{S}\Gamma' = \Gamma' \cap \mathbb{S}l_3(\mathbb{C})$ of index 3 in Γ'. The aim of this section is to prove that $\widehat{\mathbb{B}/\Gamma'}$ is the projective plane $\mathbb{P}^2$. So we must avoid to work already on $\mathbb{P}^2$. The volume of a $\Gamma'_{\mathbb{D}}$-fundamental domain in a Γ'-reflection disc $\mathbb{D}$ has been calculated in [HO4], I.1.5, by the isometry circle method described in Remark 4.7.7. It holds that $c_1(\mathbb{D}/\Gamma_{\mathbb{D}}) = -2/3$.

Lemma 5.2.1 *The compactified quotient curve $L = \overline{\mathbb{D}/\Gamma'_{\mathbb{D}}}$ of any Γ'-reflection disc $\mathbb{D}$ is a smooth rational curve. Its selfintersection number on the canonical compactification $Y' = Y$ resolving cusp points is $(L^2) = -1$.*

Proof We use (4.7.7), (4.7.3) and the graph 4.7.2 in order to calculate the EULER number $e(L)$:

$$\frac{-2}{3} = e(L) - \left(1 - \frac{1}{3}\right) - 2.$$

It follows that $e(L) = 2$, $L \simeq \mathbb{P}^1$. From section 4.9.4 (Prop 1), see also Proposition 4.7.4,(ii), we get $3\tau_f(\mathrm{L}) = \dfrac{-1}{3}$, hence

$$\frac{-1}{3} = \frac{1}{3}(L^2) + 0 + 0$$

by 4.7.3,(τ) and the tetrahedron 4.6.9. The lemma is proved.

□

The selfintersection number of the cusp curves L_{0j}, $j = 1, \dots, 4$ can be calculated by means of formula (4.2.15) knowing the structure of the isotropy groups of the cusps. It holds that $(L_{0j}^2) = -1$. For details we refer to [HO4],

I, 1.2.2. We denote the compactified quotient curves of the Γ'-reflection discs by $L_{ij} = L_{ji}$, $0 \le i, j \le 4$, $i \ne j$. So we receive exactly a configuration of 10 smooth rational curves with selfintersection number -1 as described in Figure 3.7.6.

Lemma 5.2.2 *Let $Y' = Y \longrightarrow \hat{Y} = \widehat{\mathbb{B}/\Gamma'}$ be the canonical resolution of $\widehat{\mathbb{B}/\Gamma'}$. Then Y has the following basic invariants:*

$$e(Y) = 7,\ \tau(Y) = -3.$$

Proof We use the formula for volumes of fundamental domains of PICARD modular groups proved in the Appendix, 5A.4.7. For the full EISENSTEIN lattice $\Gamma = (\mathbb{U}(2,1), \mathcal{O})$ one finds that

$$c_2(\mathbb{B}/\Gamma) = \frac{1}{72}$$

knowing that the volume must be a rational number with low denominator. The factor group Γ/Γ' is isomorphic to the symmetric group S_4 of order 24, see [HO4], I,1.3.1. Therefore,

$$c_2(\mathbb{B}/\Gamma') = 24 \cdot \frac{1}{72} = \frac{1}{3}$$

is the volume we look for. This volume is equal to the orbital height $H_e(\mathbf{Y}_f) = 3H_\tau(\mathbf{Y}_f)$ by Theorem 4.8.1. Now we apply Lemma 4.11.4 to get with $v_{0j} = \infty$, $j = 1, 2, 3, 4$,

$$\frac{1}{3} = e(Y) - \sum_1^6 \left(1 - \frac{1}{3}\right) 2 - \sum_1^4 2 + \sum_1^3 \left(1 - \frac{1}{3}\right)\left(1 - \frac{1}{3}\right) + \sum_1^{12} \left(1 - \frac{1}{3}\right)$$

$$\frac{1}{3} = 3\tau(Y) - \sum_1^6 \left(1 - \frac{1}{9}\right)(-1) - \sum_1^4 (-1)$$

and finally $e(Y) = 7$, $\tau(Y) = -3$. □

Proposition 5.2.3 *(Fine classification of the orbital ball quotient surface of the congruence* EISENSTEIN *lattice) The* BAILY-BOREL *compactification $\hat{Y} = \widehat{\mathbb{B}/\Gamma'}$ is the projective plane. The ball quotient arrangement $\hat{\mathbf{B}}$ is the complete quadrihedral with 3–weighted lines and ∞-weighted triple points.*

Proof The canonical arrangement **B** on **Y** consists of ten (-1)-lines L_{ij}, $0 \leq i < j \leq 4$. We blow down the cusp lines L_{oj}, $j = 1, \ldots, 4$ and get the smooth surface $\widehat{Y}$. From Lemma 5.2.2 we conclude that

$$c_2\left(\widehat{Y}\right) = e\left(\widehat{Y}\right) = 3,\ \tau\left(\widehat{Y}\right) = 1,\ \chi\left(\widehat{Y}\right) = \frac{1}{4}(e+\tau) = 1,$$
$$c_1^2 = 12 - c_2 = 9$$

Moreover, on $\widehat{Y}$ exist rational curves L_{ij}. We finish the proof applying the following well-known

Proposition 5.2.4 (MIYAOKA-YAU) *Let X be a smooth compact complex algebraic surface. X is the projective plane $\mathbb{P}^2$ if and only if $c_1^2(X) = 9$, $c_2(X) = 3$ and X supports a rational or elliptic curve.*

Proof There are only two types of algebraic surfaces with positive CHERN numbers c_1^2, c_2, namely rational surfaces and surfaces of general type. This follows from the classification theory of algebraic surfaces, see e.g. [SC1]. Assume that $X \neq \mathbb{P}^2$. Then X cannot be rational because $\mathbb{P}^2$ is the only rational surface with $c_1^2 = 9$, $c_2 = 3$. Therefore, X is of general type. By a theorem of MIYAOKA [MI2] the canonical bundle of X is ample, if $c_1^2(X)/c_2(X) = 3$. YAU proved in [YAU] that a compact complex surface with ample canonical bundle, $c_1^2(X)/c_2(X) = 3$ and $c_2(X) > 0$ is a ball quotient surface $\mathbb{B}/\Gamma$, Γ a neat cocompact ball lattice. Now denote our rational or elliptic curve on X by C and let $\widetilde{C} \to C$ be its normalization. The universal covering of $\widetilde{C}$ is $\mathbb{P}^1$ or the affine line $\mathbb{A}^1 = \mathbb{A}^1(\mathbb{C})$, respectively. The holomorphic map $\widetilde{C} \to X$ can be lifted to the universal coverings:

$$\begin{array}{ccc} \mathbb{A}^1 \text{ or } \mathbb{P}^1 & \xrightarrow{\ \widetilde{f}\ } & \mathbb{B} \\ \downarrow & & \downarrow \\ \widetilde{C} & \xrightarrow{\ f\ } & X \end{array}$$

The restriction of $\widetilde{f}$ to $\mathbb{A}^1$ or $\mathbb{P}^1$ is bounded. Therefore, $\widetilde{f}$, hence also f, degenerates to a constant map by a classical theorem of LIOUVILLE. But then $C = f(\widetilde{C})$ is a point. This is a contradiction. One direction of the proposition is proved, the other is well-known.

□

Remark 5.2.5 A version of the proposition without elliptic curve part can also be deduced from CASTELNUOVO'S rationality criterion for algebraic surfaces, see Prop. 5.4.17, below.

Remark 5.2.6 In [HO4] we presented another proof of Proposition 5.2.3. There we classified first $X = \overline{\mathbb{B}/\mathbb{S}\Gamma'}$ as a blown up ruled surface over an elliptic curve. Then we proved that $\widehat{X}/Z_3$, $Z_3 = \mathbb{Z}/3\mathbb{Z} \simeq \Gamma'/\mathbb{S}\Gamma'$, is the projective plane. The most immediate way of fine classification of $\widehat{\mathbb{B}/\Gamma'}$ is the proof presented above. But in [HO4] it was very important to know the structure of $\mathbb{B}/\mathbb{S}\Gamma'$ because $\mathbb{S}\Gamma'$ is almost neat, hence the ring of automorphic forms appears as pluricanonical ring on this surface. In [HO4] one can also find the fine classification of $\widehat{\mathbb{B}/\Gamma}, \Gamma$ the full EISENSTEIN lattice. The birational relation between $\mathbb{P}^2$ and $\widehat{\mathbb{B}/\Gamma} = \widehat{\mathbb{B}/\Gamma'}/S_4$ is precisely described there.

Remark 5.2.7 ([HO2], 4.1) The ruled rational surfaces are the HIRZEBRUCH surfaces $\mathbb{F}_n = \mathbb{P}(\mathcal{O}_{\mathbb{P}^1} \oplus \mathcal{O}_{\mathbb{P}^1}(-n))$, $n = 0, 1, 2, \ldots$. The surfaces $\mathbb{F}_0, \mathbb{P}^1$,
$\mathbb{F}_2, \mathbb{F}_3, \ldots$ are the minimal models of the class of smooth rational surfaces. The basic invariants of the ruled surfaces are

$$c_1^2(\mathbb{F}_n) = 8, c_2(\mathbb{F}_n) = 4, n = 0, 1, 2, \ldots$$

The ruled surface $\mathbb{F}_n$ supports a configuration C_n (see 5.2.1):

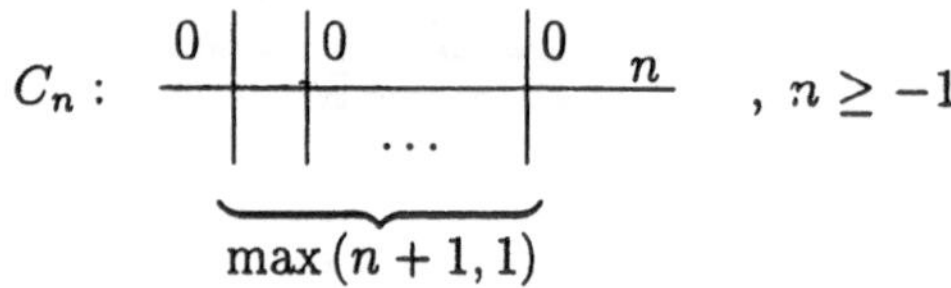

Figure 5.2.1

of projective lines with selfintersection numbers 0 (fibres) or n, respectively. It is not difficult to prove the following fine classification result:
A smooth (compact algebraic) surface is a k-th blowing up of $\mathbb{F}_n$ ($\mathbb{F}_{-1} = \mathbb{F}_1$) *if and only if there is a pair of integers* (n, k), $k \geq 0$, *such that*

(a) $c_1^2(X) = 8 - k$, $\chi(X) = 1$

(b) on X *there exists a k-th blowing up of a configuration* $C_n \cup \{$ *some points* $\}$.

Hint. If $n \geq 0$, blow up the intersection points of the fibres with the central line and blow then down the fibres. Then blow down the central line and apply Proposition 5.2.4.

5.3 Picard Modular Surface of the Field of Gauss-Numbers

Besides the general volume formulas in Theorem 4.8.3 for the basic invariants of ball quotient surfaces, the starting point for the fine classification of $\mathbf{Y} = \widehat{\mathbb{B}/\Gamma}$, $\Gamma = \mathbb{U}((2,1), \mathcal{O}_K)$, $K = \mathbb{Q}(i)$, $i = \sqrt{-1}$, is SHVARTSMAN'S fixed point formation described in Figure 4.6.2. SHIMURA [SH2] proved that $\widehat{Y}$ is a rational surface, and SHVARTSMAN [SV2] calculated the EULER number $e(\widehat{Y}) = 3$ by means of arithmetic-topological methods. The fine classification is due to the author [HO2] with the methods of this monograph using additionally only SHVARTSMAN's fixed point formation. From the fixed points we established the molecular graph of $\widehat{Y}$ described in 4.6.8. It can be translated into the configuration of curves on Y', the canonical resolution of $\widehat{Y}$, described in Figure 5.3.1:

Thus, the arrangement $\mathbf{B}'$ of $\mathbf{Y}'$ is supported by six smooth curves, which are rational by (4.7.9). On these curves lie the five cyclic singularities of the surface Y' indicated by a point with type descriptions. The selfintersection numbers of D_0, D_1, D_2 have been calculated in (4.7.10). For the resolution curve of the cusp point one has to use the volume formula (4.2.15) after the determination of the structure of the isotropy group $\Gamma_{\tilde{\kappa}}$ of a cusp $\tilde{\kappa} \in \partial_\Gamma \mathbb{B}$, say $\tilde{\kappa} = (1,0)$. The details are left to the reader as exercise. We refer to the analogous determination of the cusp group structure of the PICARD modular group of EISENSTEIN numbers carried out in [HO4] I.2. The selfintersection numbers of the resolution curves of the two non-abelian quotient points should be calculated by means of 4.9.4, (Prop 0) knowing the graphs of these points described in Table 4.6.19.

The invariants of Y' have been already calculated in (4.8.23)

$$\begin{gathered} e(Y') = 6,\ \tau(Y') = -2,\ \chi(Y') = \frac{1}{4}(e + \tau) = 1, \\ c_1^2(Y') = 12\chi - e = 6. \end{gathered} \tag{5.3.1}$$

On the minimal singularity resolution $\widetilde{Y}$ of Y' one finds the configuration (Figure 5.3.2) of smooth rational curves.

From (5.3.1) it follows that $e(\widetilde{Y}) = e(Y') + 6 = 12$. The surface $\widehat{Y}$ is received by contracting

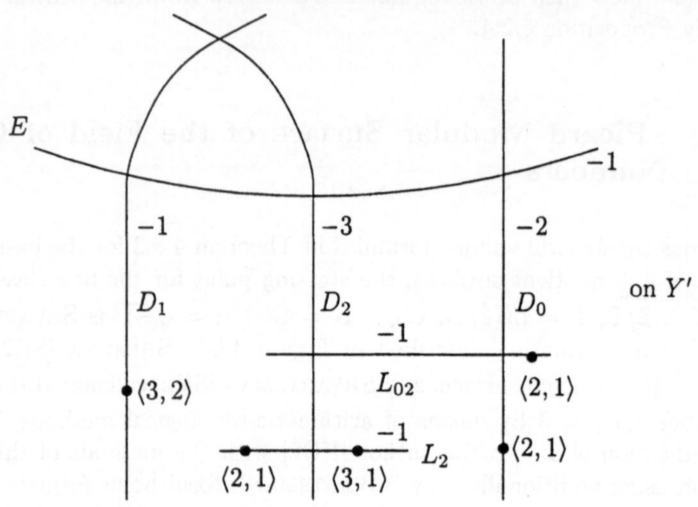

Figure 5.3.1

$$E + \left(L_1 + L_1'\right) + (L_2 + L_{22} + L_{23}) + \left(L_{02} + L_{02}'\right) + L_0.$$

So we rediscovered SHVARTSMAN'S result $e(\widehat{Y}) = 3$. The minimal singularity resolution $\widehat{Y}_{min}$ of $\widehat{Y}$ is received by blowing down

$$E + (L_2 + L_{22} + L_{23}) + \left(L_{02} + L_{02}'\right)$$

on $\tilde{Y}$, $e\left(\widehat{Y}_{\min}\right) = e\left(\tilde{Y}\right) - 6 = 6$. We are not able to classify immediately one of the above surfaces. We intervent the surface $\overline{Y}$ blowing down on $\widetilde{Y}$ only

$$L_2 + L_{22} + L_{23}.$$

Its EULER number is $e(\overline{Y}) = 9$. Altogether we have the following commutative diagram of birational morphisms 5.3.2. On $\overline{Y}$ we blow down

$$\left(D_1 + L_1 + L_1'\right) + (L_{02} + D_0).$$

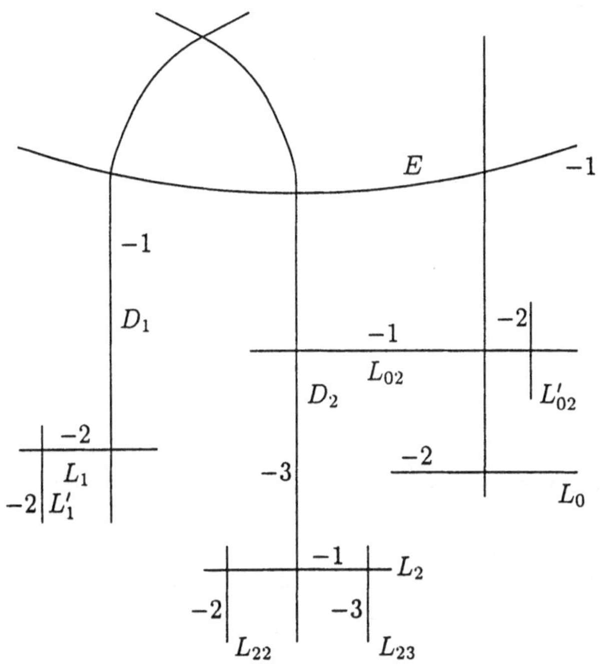

Figure 5.3.2

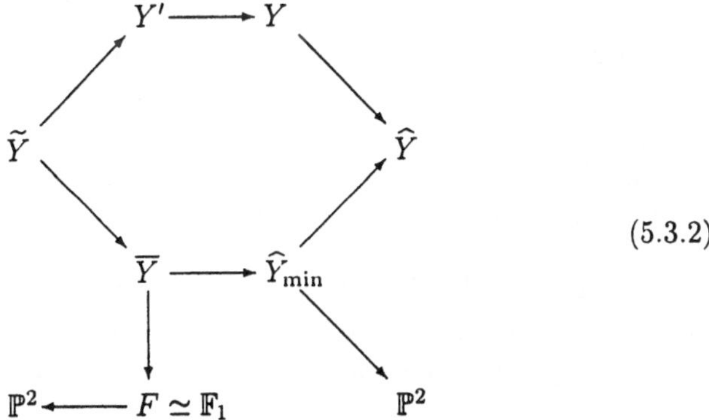

(5.3.2)

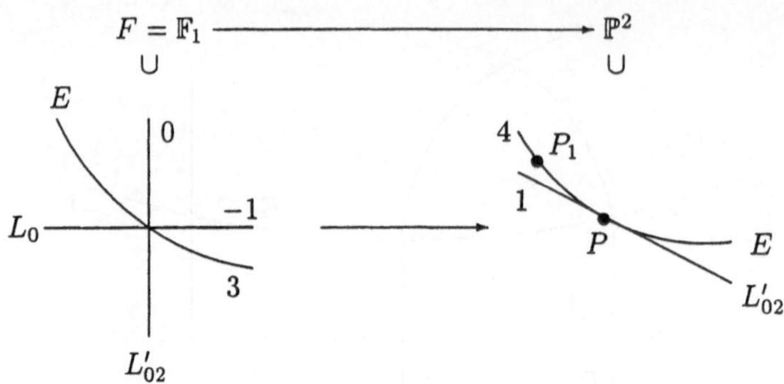

Figure 5.3.3

The resulting smooth surface F has EULER number $e(F) = 4$. Moreover, on F lies the configuration (5.3.3) including one of type C_{-1} with the notations of Figure (5.2.1)

We know that $e(F) + c_1^2(F) = 12\chi(F) = 12$, hence $c_1^2(F) = 8$. By Remark 5.2.7 or after blowing down L_0 on F and the use of Proposition 5.2.4 it follows that F is the ruled surface $\mathbb{F}_1$, $\mathbb{F}_1 \underset{L_0}{\longrightarrow} \mathbb{P}^2$. On $\mathbb{P}^2, E$ appears as smooth rational curve with $(E^2) = 4$ and L'_{02} touches E of order 2 in the common point, $(L'^2_{02}) = 1$. Therefore L'_{02} appears as tangent line of the quadric E on $\mathbb{P}^2$.

One knows the curve theory on $\mathbb{P}^2$ very well. We are now able to enrich the configuration (5.3.2) on $\widehat{Y}$ in order to classify $\widehat{Y}_{\min}$ immediately. Let $P_1 \in E \subset \mathbb{P}^2$ be the contraction point of $D_1 + L_1 + L'_1$ and S the line through P, P_1.

Going back to $\overline{Y}$ and forgetting D_2 we obtain the configuration (5.3.4). Blowing down $E + L_{02} + L'_{02}$ we find on $\widehat{Y}_{\min}$ the configuration (5.3.5):

On $\widehat{Y}_{\min}$ we blow down $S + L'_1 + L_1$. The EULER number of the arising surface is $e(\widehat{Y}_{\min}) - 3 = 6 - 3 = 3$. So we obtain $\mathbb{P}^2$ again together with three projective coordinate axis $L_0 + D_0 + D_1$.

Proposition 5.3.1 *The minimal singularity resolution $\widehat{Y}_{\min}$ of the* PICARD *modular surface $\widehat{Y} = \widehat{\mathbb{B}/\Gamma}$ of the full* GAUSS *lattice $\Gamma = \mathbb{U}((2,1), \mathbb{Z} + \mathbb{Z}i)$ is received by blowing up a point on the projective plane three times such that the exceptional curve of $\widehat{Y}_{\min} \longrightarrow \mathbb{P}^2$ has intersection graph* $\overset{-2}{\bullet}\!\!-\!\!\!-\!\!\overset{-2}{\bullet}\!\!-\!\!\!-\!\!\overset{-1}{\bullet}$.

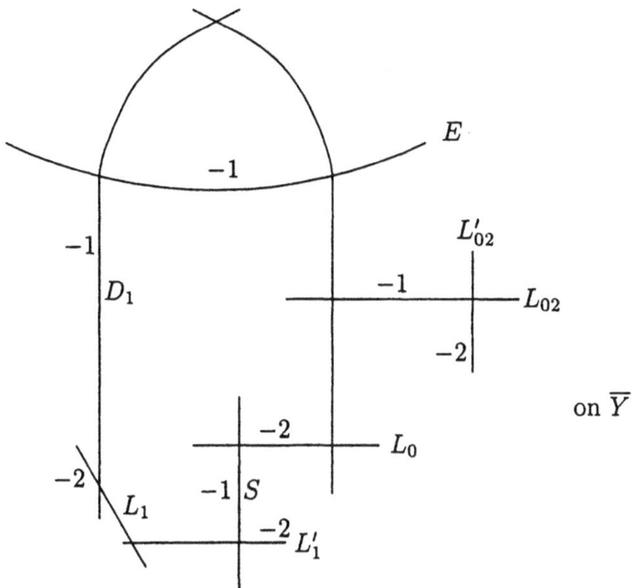

Figure 5.3.4

The PICARD *modular surface $\widehat{Y}$ itself is received by contracting the (-2)-curves L_0 and L_1, L'_1 drawn in Figure 5.3.5 to cyclic singularities of type $\langle 2,1\rangle$ or $\langle 3,2\rangle$, respectively. The arrangement $\widehat{\mathbf{B}}$ of $\widehat{\mathbf{Y}}$ is supported by two smooth rational curves D_0, D_1 and a rational curve D_2 with a cuspidal singularity.*

The configuration of $\widehat{B}$ is drawn in Figure 5.3.6

The notations of points and curves correspond to the notations of Γ-reflection discs, Γ-elliptic points and the Γ-cusp in the fixed point formation 4.6.2. The latter objects are projected onto the former ones along $\mathbb{B} \longrightarrow \mathbb{B}/\Gamma = Y_f = \widehat{Y} \setminus \{\kappa\}$.

Proof We have only to fill in the curve D_2. For the convenience of the reader we summarize in (5.3.3) all contradictions we used along the lower part of diagram (5.3.2).

Now we calculate the selfintersection number (D_2^2) on $\mathbb{P}^2 = \mathbb{F}_1 / L_0$ starting on $\widehat{Y}$, see Figure 5.3.2

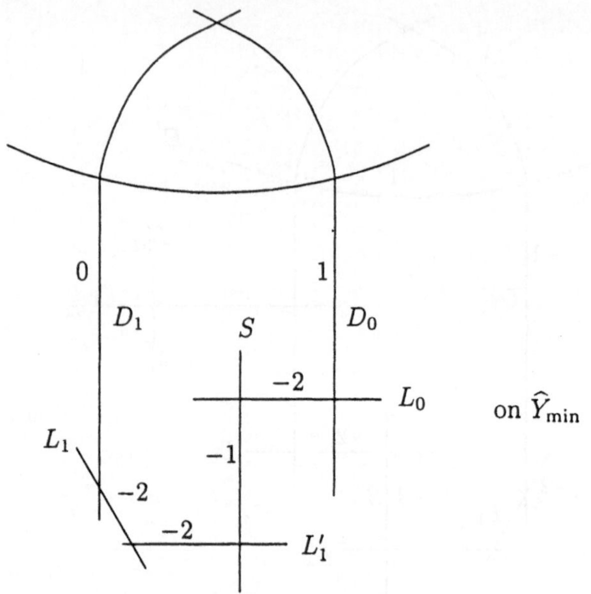

Figure 5.3.5

$$(D_2^2)_{\overline{Y}} = (D_2^2)_{\widetilde{Y}} + 1 + 1 + 4 = 3,$$
$$(D_2^2)_{\mathbf{F}_1} = (D_2^2)_{\overline{Y}} + 1 + 1 + 1 + 1 + 1 = 8,$$
$$(D_2^2)_{\mathbb{P}^2} = (D_2^2)_{\mathbf{F}_1} + 1 = 9.$$

Therefore D_2 is an irreducible curve of degree 3 on $\mathbb{P}^2$. This is a rational curve. We left to the reader to calculate the genus $g(D_2) = 0$ by means of simple standard formulas (see [HAR], proof of 5.3.7 and example 3.9.1). Now we fill in D_2 in the configuration $D_0 + D_1 + L_0$ on the lower $\mathbb{P}^2$ of diagram (5.3.3). The intersection points of D_2 with D_0 can be found already on $\widehat{Y}_{\min}$. Starting from configuration 5.3.2 on $\widetilde{Y}$ we observe D_2 and D_0 along $\widetilde{Y} \to \overline{Y} \to \widehat{Y}_{\min}$. It turns out that D_2 crosses D_0 in one point and touches D_0 in another point such that

$$(D_2 \cdot D_0)_{\widehat{Y}_{\min}} = (D_2 \cdot D_0)_{\mathbb{P}^2} = 1 + 2 = 3,$$

hence $(D_2 \cdot D_1)_{\mathbb{P}^2} = (D_2 \cdot L_0)_{\mathbb{P}^2} = 3$ because $D_0 \equiv D_1 \equiv L_0$ are projective lines on $\mathbb{P}^2$. One finds the following configuration 5.3.7

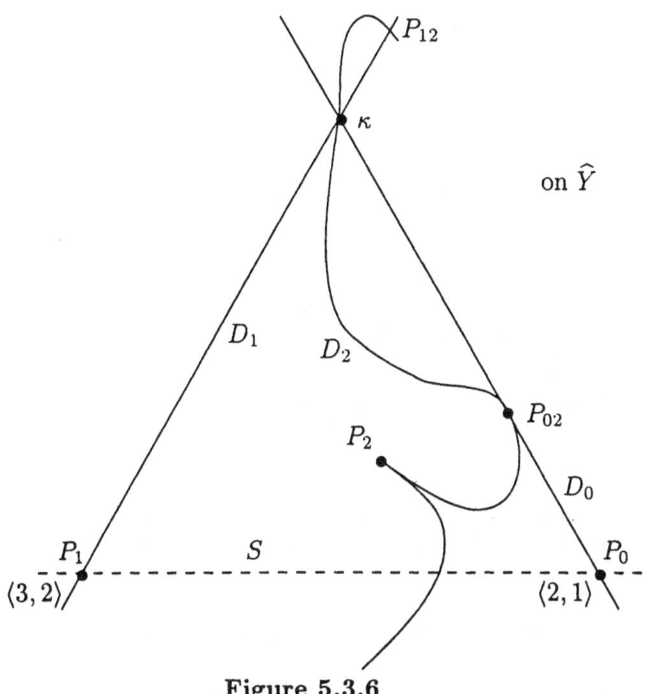

Figure 5.3.6

This configuration can be transformed along $\mathbb{P}^{2\cdot} \leftarrow \widehat{Y}_{\min} \rightarrow \widehat{Y}$ to the configuration 5.3.6 on Y. Forgetting S we found the support $\widehat{B}$ of the arrangement $\widehat{\mathbf{B}}$ of the PICARD modular surface $\mathbb{B}/\Gamma, \Gamma$ the full GAUSS lattice, with one (regular) cusp point of type (2,4,4), two singularities P_0, P_1 of type $\langle 2,1\rangle$ or $\langle 3,2\rangle$, respectively, and weights 4,4,2 for D_0, D_1 or D_2, respectively. □

The arrangement is completely determined if we know the equation of D_2 on $\mathbb{P}^2$. We choose D_0, D_1, L_0 as projective coordinate axis, see 5.3.7. Since any two quadruples of points on $\mathbb{P}^2$ in general position are projectively equivalent, we have the freedom to choose the coordinates of the cuspidal point P_2 of D_2. Altogether we choose the following coordinates:

$$\kappa = (1:0:0),\ Q_0 = (0:1:0),\ Q_1 = (0:0:1),\ P_2 = (1:1:4)$$

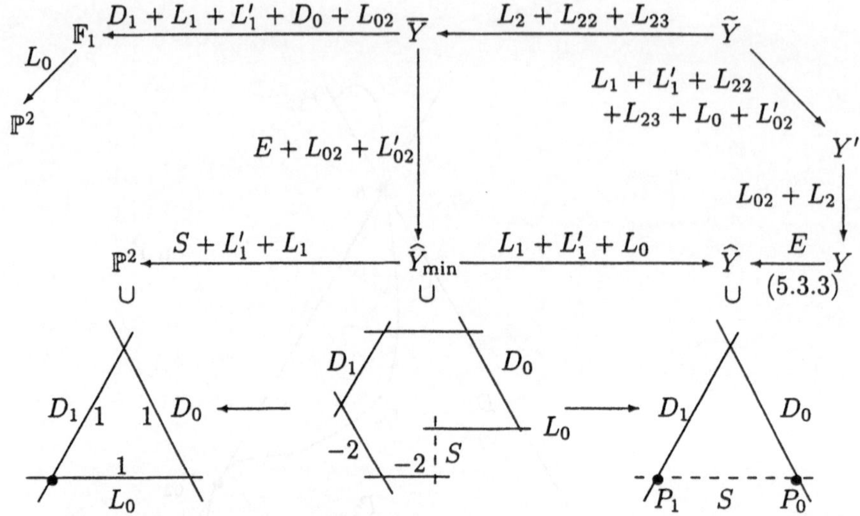

Corollary 5.3.2 *[FE6]. With the above choice of coordinates the equation for the curve D_2 on $\mathbb{P}^2$ is uniquely determined by the configuration 5.3.7. Its equation is*

$$Y^3 + 6Y^2X + 9YX^2 - 2X^2Z - 6XYZ + Z^2X = 0.$$

Hints for proof: Check first that the curve cusp P_2 of D_2 lies really outside of the three coordinate lines in 5.3.7. Then start with the general equation $F(X, Y, Z) = 0$ for a cubic in $\mathbb{P}^2$ with variable coefficients. The coefficients can be determined uniquely (up to a common constant factor) using the following conditions:

(i) the intersection behaviour of D_2 with the lines $D_0 : Z = 0$, $D_1 : Y = 0$, $L_0 : X = 0$, described in 5.3.7,

(ii) F and the derivatives F_x, F_y, F_z have to vanish at the singular point $P_2 = (1 : 1 : 4)$,

(iii) The HESSE matrix

$$\begin{pmatrix} F_{xx} & F_{xy} & F_{xz} \\ F_{yx} & F_{yy} & F_{yz} \\ F_{zx} & F_{zy} & F_{zz} \end{pmatrix}$$

has rank 1 in the cusp P_2 of D_2 (compare with the standard cusp of the curve $Y^3 - X^2Z = 0$ at $(0 : 0 : 1)$).

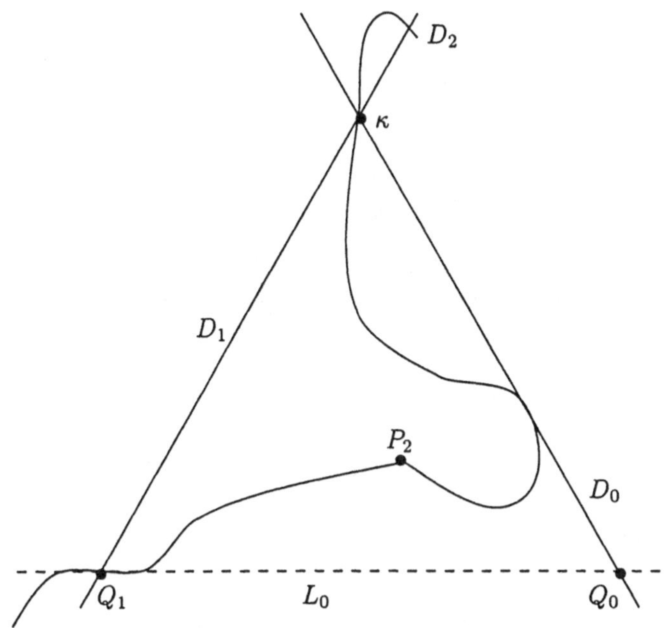

Figure 5.3.7

Remark 5.3.3 Conversely, it is clear that one can check directly that the arrangement $\widehat{\mathbf{B}}$ on $\widehat{Y}$ supported by the three curves in 5.3.6 which are constructed on $\mathbb{P}^2$ (see 5.3.7) in the explicit manner described above satisfies the proportionality conditions for ball quotients (Prop 0), (Prop 1) and (Prop 2) of Theorem 4.9.2.

Hint: Construct the molecular graph of $\widehat{\mathbf{B}}$ going back to $\mathbf{Y}'$. Compare it with the molecular graph 4.6.8. The central weights (point weights) of the non-abelian quotient points are determined by (Prop 0). The local heights $H_e(\mathbf{D}_{if})$, $H_\tau(\mathbf{D}_{if})$, $i = 0, 1, 2$, can be calculated by means of definition 4.7.3. For the calculation of the global heights $H_e(\hat{\mathbf{Y}}_f)$, $H_\tau(\hat{\mathbf{Y}}_f)$ one can use the formula (4.8.12) together with Table 4.8.1. The local contributions have been already calculated in 4.8.5.

Since $H_e(\hat{\mathbf{Y}}_f)$ coincides with the c_2-volume of a Γ-fundamental domain, $\widehat{\mathbf{Y}_f} = \mathbb{B}/\Gamma$, this is a purely geometric way to calculate this volume.

5.4 Kodaira Classification of Picard Modular Surfaces

In this section we will give an outline on the rough classifications of PICARD modular surfaces. We introduced (open) orbital surfaces $\mathbf{Y}_f$ and their rational heights $H(\mathbf{Y}_f)$. Any lattice Γ of the ball defines an (open) orbital surface $\mathbf{Y}_f = \mathbb{B}/\Gamma$. The scheme (4.8.3) shows roughly how the degree formula for the rational heights together with HIRZEBRUCH-MUMFORD'S proportionality theorem and RIEMANN-ROCH theory is joint with volumes of fundamental domains of ball lattices: The numbers $H_e(\mathbf{Y}_f), H_\tau(\mathbf{Y}_f)$ are the non-euclidean volumes $c_2(\mathbb{B}/\Gamma)$ or $\tau(\mathbb{B}/\Gamma) = \frac{1}{3}c_2(\mathbb{B}/\Gamma)$, respectively, of a Γ-fundamental domain with respect to the EINSTEIN-KÄHLER ball metric and to the CHERN forms of c_2 or τ, respectively (Theorem 4.8.1). The orbital decomposition of orbital heights give rise to establish the classification scheme (D-5). It appears in Proposition 4.10.2:

$$H_c(\mathbf{Y}_f) = H_c(Y_f) - \sum_{\mathbf{C}} \lambda_c(\mathbf{C}) H_c(\mathbf{C}_f) - \sum_{\mathbf{S}} \mathbf{c}_{\mathbf{f}}(\mathbf{S}), \; c = e, \tau$$

$$H_c(Y_f) = c(Y) - \sum_{S} c_f(S), \; ((4.10.2), (4.10.3)).$$

The integral CHERN numbers $c(Y)$ appear as an integral part of the rational heights $H_c(\mathbf{Y}_f)$. The first sum runs over orbital curves $\mathbf{C}$ coming from Γ-reflection discs, the second sum over orbital points of abelian or non-abelian quotient type and cusp type. More explicitly, the formulas of orbital decomposition have been written in Theorem 4.8.3. Now, we assume that Γ is a PICARD modular group $\mathbb{P}U((2,1),\mathcal{O}_K) \simeq \mathbb{P}SU((2,1),\mathcal{O}_K)$ of an imaginary quadratic number field $K = \mathbb{Q}(\sqrt{-d})$, d a squarefree natural number excluding $d = 1, 3$ because these cases have already been investigated in sections 1 and 2. Then the formulas 4.8.3 (e),(τ) can be simplified.

Corollary 5.4.1 *(of Theorem 4.8.3) If $\Gamma = \Gamma_d$ is the* PICARD *modular group of $K = \mathbb{Q}(\sqrt{-d})$, $d \neq 1, 3$, as above and $Y' \to Y \to \widehat{Y}$ the canonical resolution of $\widehat{Y} = \widehat{\mathbb{B}/\Gamma}$, then the following volume formula is valid:*

$$\begin{aligned} e(Y') &= c_2(\mathbb{B}/\Gamma) + \frac{1}{2} \sum_{\mathbb{D} \bmod \Gamma} c_1(\mathbb{D}/\Gamma_{\mathbb{D}}) \\ &+ \sum_{\mathbf{P} \in \mathbf{A} \cup \mathbf{W}_f} \mathbf{e}_f(\mathbf{P}) + 2 \cdot h(d) \end{aligned} \qquad (5.4.1.1)$$

$$\begin{aligned}\tau(Y') &= \frac{1}{3}c_2(\mathbb{B}/\Gamma) + \frac{1}{4}\sum_{\mathbb{D} \bmod \Gamma} c_1(\mathbb{D}/\Gamma_{\mathbb{D}}) \\ &+ \sum_{\mathbf{P}\in\mathbf{A}\cup\mathbf{W}_f} \tau_f(\mathbf{P}) + \sum_{\kappa\in\mathbf{W}_\infty} \tau_f(\kappa),\end{aligned} \tag{5.4.1.2}$$

where $\mathbb{D}$ *runs over all* Γ*-reflection discs modulo* Γ *and* $\mathbf{A}$, $\mathbf{W}_f$, $\mathbf{W}_\infty$ *are the sets of abelian, non-abelian quotient points or cusp points, respectively, of* $\widehat{\mathbf{Y}}$. *The number of cusp points of* $\widehat{Y}$ *is denoted by* $h(d)$.

5.4.2 *All reflections of the* PICARD *modular group* Γ_d, $d \neq 1, 3$, have order 2.

Proof The only unit roots in $K = \mathbb{Q}(\sqrt{-d})$ are 1 and -1. If $\sigma \in \Gamma_d$ is a reflection, then it has exactly two different eigenvalues as element of $\mathbb{G}l_3(\mathcal{O}_K)$. The trace and the determinant of σ lie in $\mathcal{O}_K$.

□

5.4.3 *The boundary point* $\tilde{\kappa} = (0,1)$ *of* $\mathbb{B}$ *is a* Γ_d*-cusp for all d. The corresponding cusp singularity* $\kappa \in \widehat{Y}_d = \widehat{\mathbb{B}/\Gamma_d}$ *is rational.*

Hint for proof. Check that the disc $\mathbb{D}$ "joining" $O = (0,0)$ and $\tilde{\kappa}$ in the ball $\mathbb{B}$ is a Γ_d-reflection disc and look at the classification list of cusp points.

After this preparation we need for the proof of Corollary 5.4.1 only the first part of the following

Proposition 5.4.4 (FEUSTEL) *Each cusp singularity of any* PICARD *modular surface* $\widehat{Y}_d = \widehat{\mathbb{B}/\Gamma}_d$ *is rational. For fixed d the cusp points of* $\widehat{\mathbf{Y}}_d$ *are of the same type. On* PICARD *modular surfaces appear exactly four types of cusp points. The types depend on the residue of d modulo 8. They are listed in Table 5.4.1*

It is not difficult to find the cusp types of κ coming from the special cusp $\tilde{\kappa} = (0,1) \in \partial\mathbb{B}$. FEUSTEL succeeded to compare the types of cusp points of the fixed PICARD modular surface of the number field $K = \mathbb{Q}(\sqrt{-d})$ using the $\mathbb{U}((2,1), K)$-equivalence of the cusps. He proved on this way that the types have to be the same on this surface.

The Corollary 5.4.1 teaches us how to fulfil the program of classification of PICARD modular surfaces described in scheme (D-5). The next step is to classify the types of quotient points on the PICARD modular surfaces $\widehat{Y_d}$, d arbitrary.

Proposition 5.4.5 *([FE5], FEUSTEL [FE4]). There are only finitely many types of quotient points on PICARD modular surfaces. The occurring types for $d \neq 1,2,3,7$ are completely listed in Table 5.4.1 together with the corresponding isotropy groups at elliptic points.*

It is not difficult to see that the number of occurring types is restricted. Let $\gamma \in \Gamma_d$ be an elliptic element, that means that γ has an isolated fixed point on $\mathbb{B}$. Each eigenvalue of $\gamma \in \mathbb{G}l_3(\mathcal{O}_K)$ is a unit root ζ and $[K(\zeta):K] \leq 3$ because ζ is a root of the characteristic polynomial of γ. Therefore $\mathbb{Q}(\zeta)$ is at most of degree 6 over $\mathbb{Q}$. There are only finitely many such unit roots. Therefore, the orders of all elliptic elements γ are universally restricted. All abelian isotropy groups have at most two generators because of their diagonal representation in the tangent space $T_P(\mathbb{B}) \simeq \mathbb{C}^2$, P the fixed point of the isotropy group. So, up to conjugation, there are only finitely many abelian isotropy groups contained in PICARD modular groups. There are also only finitely many types of graphs

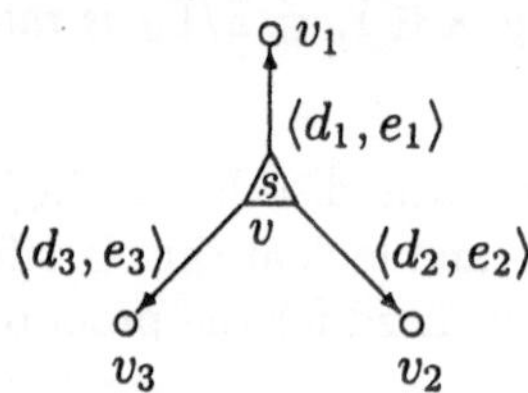

of non-abelian quotient points on PICARD modular surfaces. Namely, the weights v, v_1, v_2, v_3 and also the numbers d_1, d_2, d_3 are orders of cyclic subgroups of the corresponding isotropy groups at elliptic points; e.g. for $d = 1,2,3,7$ these orders are not greater than 6. The selfintersection number s is determined by the other data in the graph by (Prop 0):

$$0 < \frac{1}{v_1 d} + \frac{1}{v_2 d_2} + \frac{1}{v_3 d_3} - 1 = \mathbf{e}(\mathbf{E}) = -6\tau(\mathbf{E})$$
$$= -2\left(s + \frac{e_1}{d_1} + \frac{e_2}{d_2} + \frac{e_3}{d_3}\right)/v,$$

$\mathbf{E}$ the canonical resolution of the quotient point (see section 3.5). The restricted number of graphs imply a restricted number of types of isotropy groups Γ_P. Namely, the order of Γ_P can be expressed in terms of the graph data. This can be read off from (3.6.10) and (3.6.14):

$$\frac{1}{|\Gamma_P|} = \mathbf{e}(\mathrm{E})/2v = \left(\frac{1}{v_1 d_1} + \frac{1}{v_2 d_2} + \frac{1}{v_3 d_3} - 1\right)/2.$$

There are finitely many abstract groups of restricted finite order and these groups have only finitely many representations of degree 2. That means that there are up to conjugation only finitely many subgroups of $\mathbb{G}l_2(\mathbb{C})$ of these orders. Only some of them appear really as isotropy groups of PICARD modular groups. Now establish (for $d \neq 1, 2, 3, 7$) a list of all possible graphs and find the corresponding finite subgroups of $\mathbb{G}l_2(\mathbb{C})$. Such a provisorial list for all d has been established by the author in [HOI]. It needs much more effort to pick out the groups which really occur as isotropy subgroups of PICARD modular groups. This work has been done by FEUSTEL in [FE4] using arithmetic classifications of hermitian lattices over number fields.

For the connection with lattices we remark that there are bijective correspondences

$$\{\text{ points of } \mathbb{B}\} \longleftrightarrow \left\{\begin{matrix}\text{negative lines}\\ \mathbb{C}\mathfrak{a} \text{ in } \mathbb{C}^3\end{matrix}\right\} \longleftrightarrow \left\{\begin{matrix}\text{positive}\\ \text{subplanes of } \mathbb{C}^3\end{matrix}\right\}$$

$$\begin{matrix} P = \mathbb{P}\mathfrak{a} & \longleftarrow & \mathfrak{a} & \longmapsto & \mathfrak{a}^{\perp} \\ \Cap & & & & \\ \mathbb{B} \subseteq \mathbb{P}^2 & & & & \end{matrix}$$

with restrictions

$$\left\{\begin{matrix}\Gamma\text{-elliptic}\\ \text{points}\end{matrix}\right\} \longrightarrow \left\{\begin{matrix}\mathfrak{O}_L\mathfrak{a};\ \mathfrak{a} \in \mathfrak{O}_L^3\\ \overline{L} = L/K\\ \text{number field}\end{matrix}\right\} \longrightarrow \left\{\begin{matrix}\text{positive } \mathfrak{O}_L-\\ \text{sublattices of dim 2}\\ \overline{L} = L/K \text{ number field}\end{matrix}\right\}$$

$$\mathfrak{a} \longmapsto \mathfrak{O}_L^3 \cap \mathfrak{a}^{\perp}$$

where $^-$ means complex conjugation, $K = \mathbb{Q}(\sqrt{-d})$. If P is a Γ_d-elliptic point, say $\gamma P = P$, then $P = \mathbb{P}\mathfrak{a}$ for an eigenvector of γ, which can be obviously represented by a vector $\mathfrak{a} \in \mathbb{C}^3$ with coordinates in a number field L of small degree over K.

Tables 5.4.1

5.4.1.κ Cusp points κ of $\widehat{\mathbb{B}/\Gamma_d}$

	Graph of κ	Graph of minimal resolution of κ	$\mathbf{e}_f(\kappa)$	$\tau_f(\kappa)$	$h_\kappa(d)$
$d = 1$	2, −1, ∞	× (regular point)	2	$-\frac{1}{3}$	1
$d \equiv 1, 2 \bmod 4$, $d \neq 1$	2, 2, −2, ∞, 2, 2	• −2	2	$-\frac{2}{3}$	$h(d)$
$d \equiv 3 \bmod 8$	2, ⟨2, 1⟩, −2, ⟨2, 1⟩, ∞, ⟨2, 1⟩	−2, −2, −2, −2	2	$-\frac{1}{6}$	$h(d)$
$d \equiv 7 \bmod 8$	2, −1, 2, ∞, 2, ⟨2, 1⟩	×	2	$-\frac{1}{6}$	$h(d)$

5.4.1.Q Quotient points **Q** of $\mathbb{B}/\Gamma_d$

Type of **Q**	Graph of **Q**	Graph of minimal resolution of **Q**	$\mathbf{e}_f(\mathbf{Q})$	$\tau_f(\mathbf{Q})$	$h_{\mathbf{Q}}(d)$
$\langle 2,1\rangle$	o——o $\langle 2,1\rangle$	$\bullet$ -2	$\frac{1}{2}$	0	$-\frac{h^2}{4}\left(\left(\frac{D}{2}\right)^2-2\left(\frac{D}{2}\right)+7\right)$ $+\frac{h}{6}\lvert D\rvert+\frac{h}{6}\left(\left(\frac{D}{3}\right)+3\right)$
$\langle 4,2\rangle$	o——o $2\ \langle 2,1\rangle$	$\bullet$ -2	$\frac{1}{2}$	0	$\frac{3h}{2}\left(1-\left(\frac{D}{2}\right)\right)$
$\langle 6,2\rangle$	o——o $2\ \langle 3,1\rangle$	$\bullet$ -3	$\frac{2}{3}$	$-\frac{2}{9}$	$\frac{h}{2}\left(1-\left(\frac{D}{3}\right)\right)$
K_4	o——o $2\quad 2$	$\times$	$\frac{1}{4}$	0	$\frac{h^2}{4}\left(\left(\frac{D}{2}\right)^2-2\left(\frac{D}{2}\right)+7\right)$ $-\frac{h}{2}\left(\left(\frac{D}{2}\right)+\left(\frac{D}{3}\right)+4\right)$
D_3	2; −1; 1; $\langle 3,1\rangle$; $\langle 2,1\rangle$	$\times$	$\frac{4}{3}$	$-\frac{8}{9}$	$\frac{h}{2}\left(3-\left(\frac{D}{3}\right)\right)$
D_4	$\langle 2,1\rangle$; −1; 2; 2; 2	$\times$	$\frac{11}{8}$	$-\frac{3}{4}$	$\frac{h}{2}\left(3+\left(\frac{D}{2}\right)\right)$
D_6	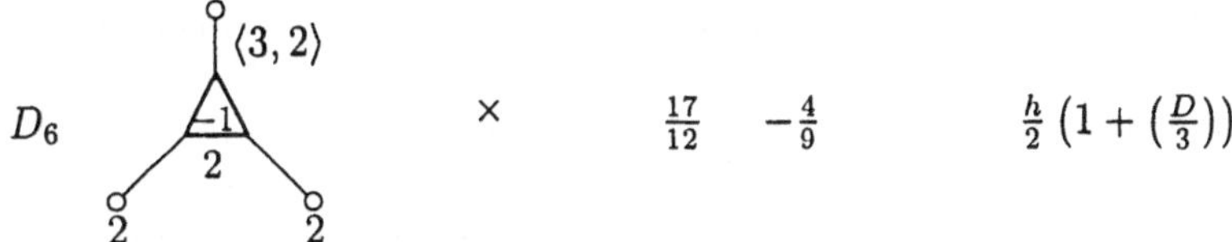	$\times$	$\frac{17}{12}$	$-\frac{4}{9}$	$\frac{h}{2}\left(1+\left(\frac{D}{3}\right)\right)$

In the formulas (5.4.1.1), (5.4.1.2) we know the contribution of any arranged point by the Table 4.8.1, where $\mathbf{c}_f(\mathbf{P}), \mathbf{c}_f(\kappa)$ are expressed in terms of graph data. Since there are only a few types of such points, we can establish a special table with concrete contributions of the different orbital points. This has been done in 5.4.1.

It is now necessary to find the precise number of each point type on any PICARD modular surface. This difficult problem has been solved by FEUSTEL, too. As a preparation for the cusp points the reader should observe that

A point $\tilde{\kappa} \in \partial\mathbb{B}$ is a Γ_d-cusp if and only if the affine coordinates of $\tilde{\kappa}$ lie in the field $K = \mathbb{Q}(\sqrt{-d})$.

For the easy proof see [HOI].

Theorem 5.4.6 *(FEUSTEL, [FE1]). The number $h(d)$ of cusp points of the PICARD modular surface $\widehat{Y}_d = \widehat{\mathbb{B}/\Gamma_d}$ is equal to the class number $h(K)$ of the field $K = \mathbb{Q}(\sqrt{-d})$.*

For the proof one corresponds to each cusp singularity κ of $\widehat{Y}_d$ with preimage $\tilde{\kappa} = \mathbb{P}\mathfrak{a}$ the class of the fractional ideal generated by the coordinates of $\mathfrak{a} \in K^3$. It is easy to see that the definition is correct. FEUSTEL proved in a constructive manner that this correspondences induce a bijection

$$\{\Gamma_d - \text{cusps}\}/\Gamma_d \Longleftrightarrow \left\{\text{ideal classes of } \mathbb{Q}\left(\sqrt{-d}\right)\right\}.$$

A more intrinsic lattice-theoretic proof with a generalization to higher dimensions has been found by Th. ZINK [ZIN].

In order to find the number $h_{\mathbf{Q}}(d)$ of quotient points on $\widehat{Y}_d$ of a fixed type $\mathbf{Q}$ (see Table 5.4.1) it is necessary to count first the components of the arrangement of $\hat{\mathbf{Y}}_d$. The problem of classifying the Γ_d-reflection discs up to Γ_d-equivalence can also be solved by means of lattice theory, more precisely, by classifying indefinite (non-degenerated) hermitean $\mathcal{O}_K$-lattices of dimension 2 in $\mathbb{C}^3$. The connection with such lattices can be described as follows. If $\mathbb{D} = \mathbb{D}(\mathfrak{a})$ is a reflection disc of Γ_d, then $L(\mathbb{D}) = \mathfrak{a}^\perp \cap \mathcal{O}^3, K = \mathbb{Q}(\sqrt{-d})$, is a 2–dimensional indefinite hermitean $\mathcal{O}_K$-lattice and $\mathfrak{a}$ a positive vector. It is easy to prove that for $\mathbb{D} = \mathbb{D}(\mathfrak{a})$ as above, $\mathfrak{a}$ can be choosen as vector in $\mathcal{O}_K^3$ with $\mathfrak{a}^2 = \langle\mathfrak{a},\mathfrak{a}\rangle = 1$ or 2. The classification of indefinite hermitean $\mathcal{O}_K$-lattices can be managed by SHIMURA's "Arithmetic of Unitary Groups" [SH1] together with JACOBOWITZ' local (p-adic) classifications [JAC], which allow to apply the Local-Global Principle. After the abstract lattice classification one has to check which of the lattices can be represented as hermitean sublattice of $\mathcal{O}^3$.

The volumes of fundamental domains in the reflection discs have been also calculated by FEUSTEL. Following the idea of the number theoretic proof of the volume formula for $c_2(\mathbb{B}/\Gamma_d)$ he found an explicitly calculable infinite product formula (see [FE2], p. 22). Altogether one gets a class number formula for the 1–dimensional contribution to the CHERN numbers.

Theorem 5.4.7 FEUSTEL, *[FE3] For $d \neq 1, 3$ it holds that*

$$\sum_{\mathbb{D} \bmod \Gamma_d} c_1(\mathbb{D}/\Gamma_{\mathbb{D}}) = \frac{1}{12} h \cdot \left[40 - 1 - 3\left(\frac{D}{2}\right)\right],$$

where h is the class number of $K = \mathbb{Q}(\sqrt{-d}), D = D_{K/\mathbb{Q}}$ the discriminant of K over $\mathbb{Q}$ and $\left(\frac{D}{2}\right) = 0, \pm 1$ the JACOBI *symbol.*

Theorem 5.4.8 FEUSTEL *[F-H] There are class number formulas for the number $h_{\mathbf{Q}}(d)$ of quotient points of fixed type $\mathbf{Q}$ on the* PICARD *modular surfaces $\widehat{Y}_d = \widehat{\mathbb{B}/\Gamma_d}$. The explicit formulas are listed in Table 5.4.1.Q.*

Ideas of Proof. Up to the type $\langle 2, 1\rangle$ all quotient points $\mathbf{Q}$ lie on a quotient curve of a reflection disc. Points of type K_4, D_4, D_6 appear as intersection points of reflection discs. In order to distinguish the types $\langle 4, 2\rangle$ and $\langle 6, 2\rangle$ or K_4, D_4 and D_6, respectively, FEUSTEL developed a classification theory of hermitean GALOIS modules. GALOIS extensions of K come in by elements of order 3, 4 or 6 of the isotropy groups, their eigenvalues and the coordinates of the eigenvectors. The most difficult problem is the counting of the isolated quotient points of type $\langle 2, 1\rangle$. These points, or their preimages on $\mathbb{B}$, are called *symmetry points*, because the stationary groups are generated by a symmetry of $\mathbb{B}$. If $P = \mathbb{P}\mathfrak{a}$ is a symmetry point, then $\mathfrak{a}^{\perp} \cap \mathcal{O}_K^3$ is a positive definit $\mathcal{O}_K$-lattice of dimension 2. The author used the hermitian analogon of the MINKOWSKI-SIEGEL Theorem for positive definit lattices due to HEL BRAUN and S. BÖGE (see [BOEG]) in order to estimate the number of symmetry points on $\widehat{Y}_d$. The rough estimation

$$h_{\langle 2,1\rangle}(d) \leq h(d) \cdot |D_{K/\mathbb{Q}}|^{3/2}, \ K = \mathbb{Q}\left(\sqrt{-d}\right)$$

was sufficient to prove that $c_1^2(\widehat{Y}_{d,\min})/c_2(\widehat{Y}_{d,\min})$ come arbitrary near to the extreme ratio 3 for $d \gg 0$, see Theorem 5.4.12 below.

Up to the D_3-type also all other non-trivial isotropy groups contain a symmetry. Knowing already the number the corresponding quotient points FEUSTEL succeeded to count the number of symmetry points precisely. He wrote down the complete proof including the MINKOWSKI-SIEGEL argument in § 2 of [F-H].

Now all contributions appearing in the formulas (5.4.1.1) and (5.4.1.2) are explained. They are listed in Table 5.4.2.A, which is complete for $d \neq 1,2,3,7$, $|D_{K/\mathbb{Q}}| < 200$. By means of this table and the formulas (5.4.1.1) and (5.4.1.2) one can calculate easily the CHERN numbers $e(Y_d')$ and $\tau(Y_d')$,

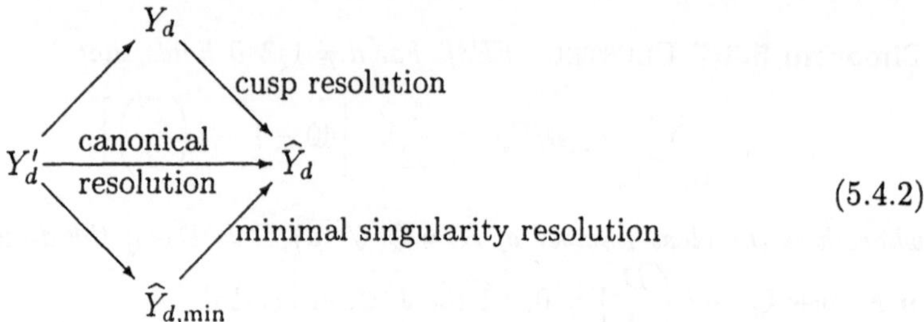

(5.4.2)

We prefer to calculate the invariants of the minimal singularity resolution $\widehat{Y}_{d,\min}$ of $\widehat{Y}_d$. The invariants of Y_d', Y_d, $\widehat{Y}_d$ are trivially related with those of $\widehat{Y}_{d,\min}$ via the intersection graphs of the minimal resolutions of the occurring singular points on $\widehat{Y}_d$ listed in Table 5.4.1. For instance, the relations between the basic invariants of Y_d' and $\widehat{Y}_{d,\min}$ are

$$e\left(\widehat{Y}_{d,\min}\right) = e(Y_d') + h_{\langle 2,1\rangle}(d) + h_{\langle 4,2\rangle}(d) + h_{\langle 6,2\rangle}(d) - \sum_{i=3}^{6} h_{D_i}(d) + \begin{cases} 0 & d \equiv 1,2 \bmod 4 \\ 3h(d) & d \equiv 3 \bmod 8 \\ -h(d) & d \equiv 7 \bmod 8 \end{cases} \tag{5.4.3}$$

$$\tau\left(\widehat{Y}_{d,\min}\right) = \tau\left(Y_d'\right) - h_{\langle 2,1\rangle}(d) - h_{\langle 4,2\rangle}(d) - h_{\langle 6,2\rangle}(d) + \sum_{i=3}^{6} h_{D_i}(d) + \begin{cases} 0 & d \equiv 1,2 \bmod 4 \\ -3h(d) & d \equiv 3 \bmod 8 \\ +h(d) & d \equiv 7 \bmod 8 \end{cases} \tag{5.4.4}$$

Theorem 5.4.9 *[F-H]. Let $\widehat{Y}_{d,\min}$ be the minimal singularity resolution of the* BAILY-BOREL *compactification $\widehat{Y}_d = \widehat{\mathbb{B}/\Gamma_d}$, Γ_d the (full)* PICARD *modular group of the imaginary quadratic number field $K = \mathbb{Q}(\sqrt{-d})$, $d \neq 1,2,3,7$ a squarefree natural number. For any* CHERN *symbol $c = ac_1^2 + bc_2 \in \mathbb{Q}c_1^2 + \mathbb{Q}c_2$ there exists a universal polynomial $\widehat{P}_c(T,H,T_2,T_3)$ with rational coefficients and a number r_c such that*

$$c\left(\widehat{Y}_d\right) = r_c \cdot \Delta_d^5 + \widehat{P}_c\left(\Delta_d, h(d), \left(\frac{D}{2}\right), \left(\frac{D}{3}\right)\right),$$

where $\Delta_d = |D|^{1/2}, D = D_{K/\mathbb{Q}}$ *the discriminant of* $K = \mathbb{Q}(\sqrt{-d}), h(d)$ *the class number of* K,

$$\left(\frac{D}{2}\right) = \begin{cases} 0, & d \equiv 1,2 \bmod 4 \\ -1, & d \equiv 3 \bmod 8 \\ 1, & d \equiv 7 \bmod 8 \end{cases}, \quad \left(\frac{D}{3}\right) = \begin{cases} 0, & d \equiv 0 \bmod 3 \\ 1, & d \equiv 2 \bmod 3 \\ -1, & d \equiv 1 \bmod 3 \end{cases},$$

and $r_c \cdot \Delta_d^5 = \left(a \cdot r_{c_1^2} + b \cdot r_{c_2}\right)\Delta_d^5 = \mathrm{vol}_c\,(\mathbb{B}/\Gamma_d)$ *the c-volume of a* Γ_d-*fundamental domain.*

$$r_c = (3a+b)r_{c_2} = (3a+b)\frac{L(3,(\frac{D}{\cdot}))}{32\pi^3} \tag{5.4.5}$$

$$L\left(s,\left(\frac{D}{\cdot}\right)\right) = \sum_{n=1}^{\infty}\left(\frac{D}{n}\right)/n^s = \prod_p\left(1-\left(\frac{D}{p}\right)p^{-s}\right)^{-1}$$

is the L-series (of the field K) with DIRICHLET *character* $(\frac{D}{\cdot}) : n \mapsto (\frac{D}{n})$, $(\frac{D}{n}) \in \{0,1,-1\}$ *the* JACOBI *symbol.*

Remark. The same statements are true for the CHERN numbers $c(Y_d')$, $c(Y_d)$, $c(\widehat{Y}_{d,\min})$ of Y_d', Y_d, $\widehat{Y}_{d,\min}$ as in (5.4.2) with universal polynomials $P_c', P_c, \widehat{P}_{c,\min} \in \mathbb{Q}[T,H,T_2,T_3]$, respectively, instead of $\widehat{P}_c$.

Proof We set $\widehat{P}_c = a\widehat{P}_{c_1^2} + b\widehat{P}_{c_2}$. Then it suffices by linearity to construct $\widehat{P}_{c_1^2}$ and $\widehat{P}_{c_2}$. By the same argument we can change over to $\widehat{P}_\tau$ and $\widehat{P}_e = \widehat{P}_{c_2}$. The difference of signatures (EULER numbers) of $Y' = Y_d'$, Y, $\widehat{Y}_{\min}$, $\widehat{Y}$ can be expressed explicitly and universally by polynomials in $h = h(d), (\frac{D}{2}), (\frac{D}{3}), |D|$ because these differences come from the singular points of $\widehat{Y}_d$ and their resolutions, and the number of any type of these singularities is a polynomial of this type (see Table 5.4.1). For example the differences $e(\widehat{Y}_{\min}) - e(Y')$, $\tau(\widehat{Y}_{\min}) - \tau(Y')$ can be read off from (5.4.3) and (5.4.4). So, it suffices to find P_τ' and P_e'. Table 5.4.1 teaches us how to write the 0–dimensional part of the formulas (5.4.3) and (5.4.4) for $e(Y')$ and $\tau(Y')$ as polynomials in h, $(\frac{D}{2})$ and $(\frac{D}{3})$. For the one-dimensional part we refer to Theorem 5.4.7. The c_2-volume of a Γ_d-fundamental domain is

$$c_2(\mathbb{B}/\Gamma_d) = \frac{\Delta}{32\pi^3}L\left(3,\left(\frac{D}{\cdot}\right)\right), \Delta = \Delta_d = |D|^{1/2}, D = D_{K/\mathbb{Q}},$$

by Theorem 5.4.8 of the appendix. From (Prop 2) we know that $c_2(\mathbb{B}/\Gamma_d) = 3\tau(\mathbb{B}/\Gamma_d)$ or, equivalently, $c_1^2(\mathbb{B}/\Gamma_d) = 3c_2(\mathbb{B}/\Gamma_d)$. Therefore r_c is correctly expressed in (5.4.5). The theorem is verified.

□

The simplest polynomials are the polynomials P'_c. they can be easily found now by the proof of the above theorem. We get on this way the following class number/L-value formulas for the basic invariants of $Y' = Y'_d$.

Theorem 5.4.10 *Let $Y' = Y'_d \to \widehat{Y}_d$ be the canonical resolution of the* BAILY-BOREL *compactification $\widehat{Y}_d$ of the open* PICARD *modular surface $\mathbb{B}/\Gamma_d, d \neq 1,2,3,7$. Then it holds that*

$$e(Y') = \frac{L(3,(\frac{D}{\cdot}))}{32\pi^3}|D|^{5/2} - \frac{1}{12}h\cdot|D| + h\cdot\left[\frac{119}{16} - \frac{5}{16}\left(\frac{D}{2}\right) - \frac{1}{3}\left(\frac{D}{3}\right)\right]$$

$$\tau(Y') = \frac{L(3,(\frac{D}{\cdot}))}{96\pi^3}|D|^{5/2} - \frac{1}{12}h\cdot|D| + h\cdot\left[\frac{1}{2}\left(\frac{D}{2}\right)^2 - \frac{7}{16}\left(\frac{D}{2}\right) + \frac{1}{3}\left(\frac{D}{3}\right) - \frac{169}{48}\right]$$

$$\chi(Y') = \chi\left(\widehat{Y}\right) = \frac{L(3,(\frac{D}{\cdot}))}{96\pi^3}|D|^{5/2} - \frac{1}{24}h\cdot|D| + h\cdot\left[\frac{49}{48} + \frac{1}{8}\left(\frac{D}{2}\right)^2 - \frac{3}{16}\left(\frac{D}{2}\right)\right]$$

$$c_1^2(Y') = \frac{3\cdot L(3,(\frac{D}{\cdot}))}{32\pi^3}|D|^{5/2} - \frac{5}{12}h\cdot|D| + h\cdot\left[\frac{75}{16} + \frac{3}{2}\left(\frac{D}{2}\right)^2 - \frac{31}{16}\left(\frac{D}{2}\right) + \frac{1}{3}\left(\frac{D}{3}\right)\right].$$

with the notations of Theorem 5.4.9.

Corollary 5.4.11 *([F-H], Th. 8.) With the assumptions and notations of Theorem 5.4.10 it holds that*

$$\tau\left(\widehat{Y}_{\min}\right) = \frac{L(3,(\frac{D}{\cdot}))}{96\pi^3}|D|^{5/2} - \frac{1}{4}h\cdot|D| + h^2\cdot\left[\frac{7}{4} + \frac{1}{4}\left(\frac{D}{2}\right)^2 - \frac{1}{2}\left(\frac{D}{2}\right)\right] - h\cdot\left[\frac{119}{48} + \frac{1}{2}\left(\frac{D}{2}\right)^2 - \frac{57}{16}\left(\frac{D}{2}\right) - \frac{2}{3}\left(\frac{D}{3}\right)\right]$$

□

The invariants $c_1^2(\widehat{Y}_{\min})$, $\chi(\widehat{Y}_{\min})$, $c_2(\widehat{Y}_{\min})$, $\tau(\widehat{Y}_{\min})$ are listed in Table 5.4.2.A for all discriminants D with $|D| < 200$, $|D| \neq 3,4,7$. The table is extended for the signatures until $|D| = 503$.

Theorem 5.4.12 *For discriminants $D \ll 0$ the corresponding* PICARD *modular surfaces are of general type. Their minimal models Y_{Min} have positive signature. Moreover, it holds that*

$$\lim_{d\to\infty} \frac{c_1^2\left(Y_d'\right)}{c_2\left(Y_d'\right)}, \frac{c_1^2\left(\widehat{Y}_{d,\min}\right)}{c_2\left(\widehat{Y}_{d,\min}\right)}, \frac{c_1^2(Y_{d,Min})}{c_2(Y_{d,Min})} = 3.$$

Remark 5.4.13 (YAU, MIYAOKA, HIRZEBRUCH [HI1]) The Theorem of BOGOMOLOV-MIYAOKA-YAU states that there exist no smooth surfaces X of general type with $c_1^2(X) > 3c_2(X)$. Those surfaces X of general type satisfying the extreme relation $c_1^2(X) = 3c_2(X)$ are ball quotient surfaces $X = \mathbb{B}/\Gamma, \Gamma$ a compact ball lattice acting freely on $\mathbb{B}$, and vice versa.

Proof of Theorem 5.4.12. Minimal surfaces of general type have positive invariants c_1^2, c_2 (see e.g. Table 10 in [BPV], IV.1). The only exceptions with the same property not being of general type are the HIRZEBRUCH surfaces $\mathbb{F}_n$ (ruled surfaces over $\mathbb{P}^1$) and the projective plane $\mathbb{P}^2$. Therefore, the condition

$$c_1^2 > 0, \chi > 1 \tag{5.4.6}$$

is a sufficient criterion for a surface to be of general type. If (5.4.6) is satisfied for a surface X, then it is also satisfied for its minimal model X_{Min} which is unique because X_{Min} cannot be a rational or ruled surface (see [BPV] III.4, Prop. 5.4.1). Since

$$\tau(X_{\mathrm{Min}}) \geq \tau(X), \ \frac{c_1^2(X)}{c_2(X)} \leq \frac{c_1^2(X_{\mathrm{Min}})}{c_2(X_{\mathrm{Min}})} \leq 3,$$

it suffices to prove first the relations (5.4.6) for the models Y_d', then $\tau(Y_d') > 0$ for $d >> 0$, and finally that $c_1^2(Y_d')/c_2(Y_d')$ thends to 3. All these statements follow immediately from the formulas of Theorem 5.4.15 and the following.

Lemma 5.4.14 *For the class numbers $h = h(K)$ of the imaginary quadratic numbere field $K = \mathbb{Q}(\sqrt{-d})$ with discriminant $D = D_{K/\mathbb{Q}}$ it holds that*

$$h \leq \frac{1}{4}|D| \ (d \neq 1, 3).$$

Proof The lemma is an immediate consequence of the well-known class number formula for imaginary quadratic number fields

$$h = \left(2 - \left(\frac{D}{2}\right)\right)^{-1} \cdot \sum_{\substack{0 < m < |D|/2 \\ (m, D) = 1}} \left(\frac{D}{m}\right) \quad (D \neq -3, -4)$$

(see e.g. [BOS], V, § 4, Theorem 3) and of another version in the case of $d \equiv 7 \bmod 8$ found by BÖLLING [BOEL]:

$$h = \sum_{\substack{0 < a < d \\ a \equiv 0 \bmod 4}} \left(\frac{a}{d}\right), \quad \left(\left(\frac{D}{2}\right) = 1\right).$$

□

We are now able to pick out all PICARD modular surfaces which are not of general type and to determine their KODAIRA dimension.

Theorem 5.4.15 *A* PICARD *modular surface* $\widehat{Y}_d = \widehat{\mathbb{B}/\Gamma_d}$ *is of general type if*

$$d \neq 1, 2, 3, 5, 6, 7, 11, 15, 19, 23, 31, 39, 47, 71 \tag{5.4.7}$$

Outline of Proof. Step 1. One works with the inequality of Lemma 5.4.14. This weak inequality suffices to prove that

$$c_1^2\left(Y_d'\right) > 0, \ \chi\left(Y_d'\right) > 1 \text{ for } |D| > 2000.$$

We only check the first of these relations. The other is an easy exercise for the reader.

First we estimate the value $L(3, (\frac{D}{\cdot}))$ of the L-series at 3.

$$L\left(3, \left(\frac{D}{\cdot}\right)\right) = 1 + \sum_{n=2}^{\infty} \left(\frac{D}{2}\right)/n^3 \geq 1 - \sum_{n=2} \frac{1}{n^3} = 2 - \zeta(3) > 0,79$$

because the value of RIEMANN'S Zeta-function at 3 is

$$\zeta(3) = 1,202056903\ldots$$

(see [DWI], formula 48.003). Together with Lemma 5.4.14 it follows immediately from the last formula of Theorem 5.4.10 that

$$c_1^2\left(Y_d'\right) > \frac{0,79 \cdot 3}{32\pi^3}|D|^{5/2} - \frac{5}{48}|D|^2 > \frac{0,076}{32}|D|^{5/2} - 0,105|D|^2$$
$$= |D|^2\left(\frac{19}{8}|D|^{1/2} - 105\right) \cdot 10^{-3} > 0$$

if $|D| > 2000$.

Step 2. Now one has to use a big table of class numbers of imaginary quadratic fields. There one picks out all discriminants D between -200 and -2000 such that $h^2(d) > |D|$. there exist only very few of them. For these numbers one checks immediately that $c_1^2(Y_d') > 0$ and $\chi(Y_d') > 1$ by means of the formulas of Theorem 5.4.10.

Step 3. After the first two steps we can assume now that

$$h(d) \leq |D|^{1/2}.$$

With similar calculations as in step 1 we get

$$c_1^2\left(Y_d'\right) > |D|^2\left(\frac{19}{8}|D| - 420\right) \cdot 10^{-3} > 0$$

if $|D| > 200$.

Step 4. Enter the Table 5.4.2.A now and look for discriminants with $c_1^2(\widehat{Y}_{d,\min}) > 0$ and $\chi(Y_d') = \chi(\widehat{Y}_{d,\min}) > 1$. Except for the numbers d listed in (5.4.7) you find only $d = 10, 14, 35$ with $c_1^2(Y_{d,\min}) > 0$. The arithmetic genus is bigger than 1.

Step 5 (FEUSTEL, [FE7]). Find for $d = 10, 14, 35$ a curve configuration on Y_d' which can be contracted such that the arising surface X_d is smooth and $c_1^2(X_d) > 0$. In this part of the proof the molecular graph of the open arranged surface $\widehat{\mathbf{Y}}_d$ plays an important role, see scheme (D-5). The knowledge of this graph is sufficient to find the contractions $Y_d' \to X_d$. The Theorem 5.4.15 is verified.

It happens that up to $d = 71$ all PICARD modular surfaces of the exceptional list (5.4.7) have arithmetic genus equal to 1. For $d = 71$ we only know that $\widehat{Y}_{71,\min}$ is a surface of general type or an elliptic surface over $\mathbb{P}^1$ without multiple fibres (FEUSTEL, [FE7], Prop. 4.1). It is diifcult to establish the molecular graph and to work with it. May be the molecular graph is not sufficient to decide which of both surface types is submitted. For the other cases we have the following

Theorem 5.4.16 *([FE5], Theorem 1.) The following conditions for* PICARD *modular surfaces* $\widehat{Y}_d = \widehat{\mathbb{B}/\Gamma_d}$ *are equivalent:*

(i) $\widehat{Y}_d$ *is a rational surface*

(ii) $\chi(\widehat{Y}_d) = 1$

(iii) $d \in \{1, 2, 3, 5, 6, 7, 11, 15, 19, 23, 31, 39, 47\}$.

Idea of Proof. The first step is to construct the molecular graph of Y_d' at least partly. Then one blows down some curves: $Y_d' \to X_d$, X_d a smooth model of Y_d'. On a suitable model X_d one can find a smooth rational curve C coming from the curve configuration of the molecular graph with non-negative selfintersection number. Then one can use a special version of CASTELNUOVO'S criterion for rationality of algebraic surfaces as has been observed by FEUSTEL.

Proposition 5.4.17 *If X is a smooth algebraic surface supporting a smooth rational curve C such that $\chi(X) = 1$, $(C^2) \geq 0$, then X is rational.*

Proof (see [FE7]). It suffices to prove that all plurigenera $P_m = P_m(X)$ are trivial for $m \geq 1$. Namely, from $\chi(X) = 1$ and $P_1 = 0$ it follows that the irregularity $q = q(X)$ is trivial by the relation

$$\chi(X) = 1 - q + P_1$$

defining $\chi(X)$. From $q = 0$ and $P_2 = 0$ it follows that X is rational by CASTELNUOVO'S theorem (see e.g. [BPV], Theorem IV.2.1).

In order to prove $P_m = 0$ for all $m \geq 1$ we remark first that

$$(K \cdot C) < 0 \tag{5.4.8}$$

for a canonical divisor K on X. this follows immediately from the adjunction formula

$$K \cdot C + (C^2) = 2g(C) - 2$$

and the assumptions $(C^2) \geq 0$, $g(C) = 0$. If $P_m > 0$, then there exists an effective divisor $D \in |mK|$, say

$$D = nC + R,\ (R \cdot C) \geq 0,\ n \geq 0.$$

This leads to a contradiction with (5.4.8), namely

$$m(K \cdot C) = (mK \cdot C) = (D \cdot C) = n(C^2) + (R \cdot C) \geq 0.$$

□

In addition to the PICARD modular surfaces of GAUSS and EISENSTEIN numbers ($d = 1, 3$) we get immediately two further fine classifications, for $d = 15$ and $d = 23$.

Corollary 5.4.18 *The* PICARD *modular surface $\widehat{Y}_{23,\min}$ is the projective plane $\mathbb{P}^2$. The* PICARD *modular surface $\widehat{Y}_{15,\min}$ is the* HIRZEBRUCH *surface $\mathbb{F}_3$.*

Proof Both surfaces aree rational by Theorem 5.4.16. For $d = 23$ we read off from Table 5.4.2.A that $c_1^2(\widehat{Y}_{23,\mathrm{min}}) = 9$. There is only one smooth rational surface with this property, namely $\mathbb{P}^2$.

For $d = 15$ the surface $F = \widehat{Y}_{15,\mathrm{min}}$ has invariants $c_1^2(F) = 8$, $c_2(F) = 4$ (see Table 5.4.2.A). By classification of minimal models of rational surfaces, F is one of the HIRZEBRUCH surfaces $\mathbb{F}_n, n \geq 0$. The PICARD modular group $\Gamma = \Gamma_{15}$ has isotropy groups of cyclic type $\langle 6, 2\rangle$ (see Table 5.4.2.A). The image of a corresponding elliptic point is a cyclic singularity on $\widehat{\mathbb{B}/\Gamma}$ of type $\langle 3, 1\rangle$. Therefore, the minimal resolution $F = \mathbb{F}_n$ of singularities of $\widehat{\mathbb{B}/\Gamma}$ supports a smooth rational curve S with $(S^2) = -3$. This is a characteristic property for $\mathbb{F}_3$ among all $\mathbb{F}_n$ (see e.g. [BPV], V, p. 141).

□

5.4.2 Table of Invariants

5.4.2.A Volumes, number of quotient points, CHERN invariants of $\left(\widehat{\mathbb{B}/\Gamma_d}\right)_{\min}$ for $|D| < 200$

$\lvert D_{K/\mathbb{Q}}\rvert$	$c_2(\mathbb{B}/\Gamma_d)$	$\langle 6,2\rangle$	D_3	D_6	$\langle 4,2\rangle$	D_4	K_4	$\langle 2,1\rangle$	χ	c_2	τ
11	3/8	0	1	1	3	1	3	0	1	10	-6
15	1	1	3	1	0	4	7	0	1	4	0
19	11/8	1	2	0	3	1	4	1	1	13	-9
20	15/8	0	2	2	3	3	9	1	1	10	-6
23	3	0	3	3	0	6	18	0	1	3	1
24	23/8	1	3	1	3	3	10	2	1	13	-9
31	6	3	6	0	0	6	21	3	1	12	-8
35	27/4	0	2	2	6	2	16	3	2	24	-16
39	11	2	6	2	0	8	38	4	1	13	-9
40	79/8	2	4	0	3	3	11	7	2	24	-16
43	83/8	1	2	0	3	1	4	5	3	24	-12
47	18	0	5	5	0	10	60	5	1	13	-9
51	67/4	1	3	1	6	2	17	8	4	39	-23
52	151/8	2	4	0	3	3	11	11	4	35	-19
55	25	4	8	0	0	8	40	14	3	35	-23
56	99/4	0	4	4	6	6	46	12	3	39	-27
59	201/8	0	3	3	9	3	39	9	5	51	-31
97	251/8	1	2	0	3	1	4	9	9	47	-9
68	159/4	0	6	4	6	6	46	20	6	58	-34
71	51/5	0	7	7	0	14	126	14	3	37	-25
79	62	5	10	0	0	10	65	30	9	77	-41
83	465/8	0	3	3	9	3	39	21	13	90	-38
84	263/4	2	6	2	6	6	48	30	12	92	-44
87	81	3	9	3	0	12	93	36	11	90	-46
88	559/8	2	4	0	3	3	11	23	18	92	-20
61	275/4	2	4	0	6	2	18	21	18	99	-27
95	105	0	8	8	0	16	168	36	11	93	-49
103	119	5	10	0	0	10	65	50	23	144	-52
104	933/8	0	6	6	9	9	111	45	19	141	-65
107	873/8	0	3	3	9	3	39	33	27	147	-39
111	151	4	12	4	0	16	172	56	21	155	-71
115	491/4	2	4	0	6	2	18	29	34	157	-21
116	1221/8	0	6	6	9	9	111	57	28	183	-71
119	186	0	10	10	0	20	270	55	22	161	-73
120	635/4	2	6	2	6	6	48	54	37	197	-49
123	595/4	1	3	1	6	2	17	32	42	183	-15
127	200	5	10	0	0	10	65	70	45	235	-55
131	1497/8	0	5	5	15	5	115	50	41	233	-69
132	803/4	2	6	2	6	6	48	62	49	243	-47
136	847/4	4	8	0	6	6	50	64	52	258	-50
139	1593/8	3	6	0	9	3	42	48	53	249	-37
143	291	0	10	10	0	20	270	95	47	286	-98
148	2047/8	2	4	0	3	3	11	43	75	288	12
151	313	7	14	0	0	14	133	105	67	355	-87
152	2373/8	0	6	6	9	9	111	93	67	345	-77

$\lvert D_{K/\mathbb{Q}}\rvert$	$c_2(\mathbb{B}/\Gamma_d)$	$\langle 6,2\rangle$	D_3	D_6	$\langle 4,2\rangle$	D_4	K_4	$\langle 2,1\rangle$	χ	c_2	τ
155	555/2	0	4	4	12	4	72	66	72	334	-46
159	371	5	15	5	0	20	275	120	67	386	-118
163	2315/8	1	2	0	3	1	4	25	91	313	51
164	729/2	0	8	8	12	12	204	112	75	409	-109
167	429	0	11	11	0	22	330	132	77	429	-121
168	1463/4	2	6	2	6	6	48	86	98	420	-28
179	3207/8	0	5	5	15	5	115	90	103	469	-57
183	517	4	12	4	0	16	172	152	119	569	-93
184	1795/4	4	8	0	6	6	50	96	123	511	-19
187	1643/4	2	4	0	6	2	18	53	124	457	39
191	606	0	13	13	0	26	468	169	111	593	-149
195	955/2	2	6	2	12	4	74	92	132	550	-22
199	627	9	18	0	0	18	225	180	143	690	-118

5.4.2.B Volumes and Signatures for $200 < |D| \le 503$

$\lvert D_{K/\mathbb{Q}}\rvert$	$c_2(\mathbb{B}/\Gamma_d)$	τ	$\lvert D_{K/\mathbb{Q}}\rvert$	$c_2(\mathbb{B}/\Gamma_d)$	τ	$\lvert D_{K/\mathbb{Q}}\rvert$	$c_2(\mathbb{B}/\Gamma_d)$	τ
203	1083/2	-6	299	1455	0	407	3957	95
211	4497/8	30	303	1823	6	408	13391/4	726
212	5421/8	-40	307	11361/8	244	411	12393/4	467
215	813	-170	308	3459/2	58	415	3889	408
219	1279/2	-28	311	2061	-225	419	27027/8	333
223	820	-44	312	6851/4	277	420	7295/2	468
227	5751/8	-11	319	2029	28	424	28941/8	614
228	3131/4	51	323	3447/2	268	427	12899/4	857
231	942	-156	327	2217	-19	431	4638	-29
232	6295/8	147	328	7531/4	315	435	7051/2	754
235	2915/4	121	331	13833/8	329	436	31029/8	683
239	1062	-186	335	2463	-178	439	4539	203
244	7317/8	-17	339	7665/4	181	440	17049/4	331
247	1053	34	340	8263/4	364	443	30399/8	746
248	2019/2	-62	344	18459/8	66	447	4843	352
251	7509/8	-45	347	16527/8	288	451	15105/4	629
255	1200	-142	355	4115/2	342	452	8991/2	692
259	1883/2	36	356	10113/4	5	455	5280	110
260	2277/2	-43	359	2937	-161	456	8993/2	609
263	1329	-142	367	2850	245	463	5062	950
264	2309/2	301	371	2487	200	467	34893/8	718
267	4123/4	207	372	10631/4	532	471	5573	367
271	1358	-112	376	5395/2	234	472	37485/8	898
276	2579/2	-30	379	19401/8	525	479	6072	-1
280	5095/4	160	383	3414	-35	483	9115/2	1050
283	9273/8	175	388	11455/4	582	487	5743	1135
287	1656	-141	391	3398	57	488	43779/8	761
291	2599/2	156	395	2895	288	491	40011/8	712
292	5635/4	193	399	3692	28	499	38577/8	1234
295	1655	57	403	11147/4	723	503	6759	300
296	12699/8	-54	404	27801/8	62			

5.5 Special Results and Examples

5.5.1 A blown up abelian surface (HIRZEBRUCH [HI4]). We consider the elliptic curve $E = \mathbb{C}/\mathcal{O}, \mathcal{O} = \mathbb{Z} + \mathbb{Z}\rho$ the ring of EISENSTEIN integers, $\rho = e^{2\pi i/3}$. E has complex multiplication. The isotropy group at O of the automorphism group of E is the cyclic group generated by $-g$, $g : z \longmapsto \rho z$. The automorphism g has exactly three fixed points $Q_0 = O$, Q_1, Q_2 on E. On the abelian surface $E \times E$ we consider the curves $O \times E$, $E \times O$, $\Delta = \{(Q,Q); Q \in E\}$ and $(-g \times 1)\Delta$. We get the following curve configuration 5.5.1 on $E \times E$:

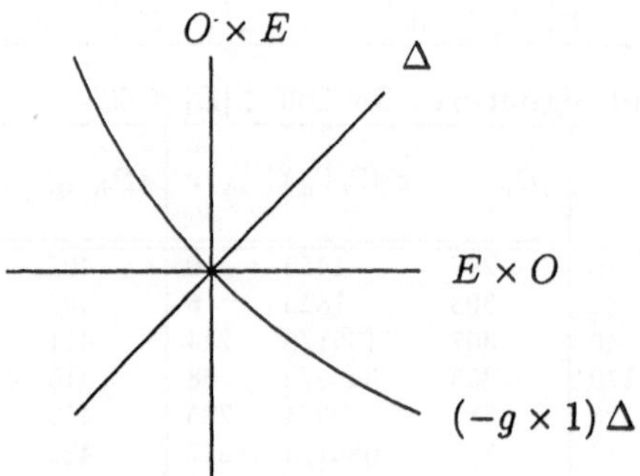

Figure 5.5.1

with only one intersection point $O = O_{E \times E} = O_E \times O_E$. We blow up this intersection point and denote the exceptional line by L. The proper transform of the curves of configuration (5.5.1) are denoted by the same symbol. They have selfintersection number -1 and no common point on the blown up surface $\widetilde{E \times E}$. We contract these elliptic curves and obtain an open surface $\widehat{E \times E}_f = \widetilde{E \times E}_f$ omitting the corresponding elliptic singularities:

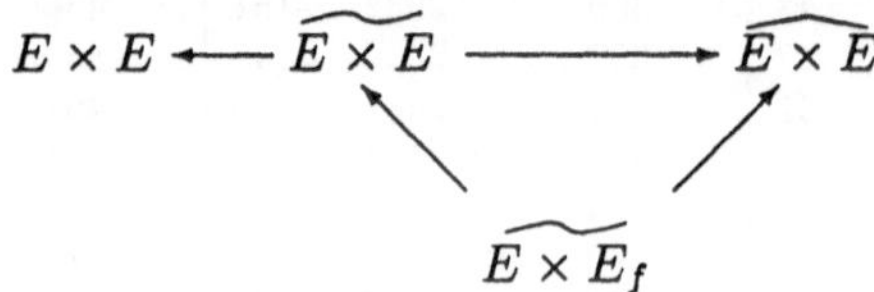

We regard $\widetilde{E \times E_f}$ as reduced open orbital surface and calculate its heights by means of (4.10.2) and (4.10.3)

$$H_e\left(\widetilde{E \times E_f}\right) = e\left(\widetilde{E \times E}\right) - 0 = e(E \times E) + 1 = 1$$

$$3H_\tau\left(\widetilde{E \times E_f}\right) = 3\tau\left(\widetilde{E \times E}\right) + 4 = 3\tau(E \times E) - 3 + 4 = 1$$

The proportionality conditions for ball quotient surfaces are satisfied. In our case they are reduced essentially to $H_e = 3H_\tau$. By a theorem of KOBAYASHI [KOB] one can show that $\widetilde{E \times E_f}$ is a ball quotient surface $\mathbb{B}/\Gamma$. In [HO5] the author proved that Γ can be chosen as subgroup of index 9 of the congruence sublattice Γ' of the full EISENSTEIN lattice $\mathbb{U}((2,1),\mathcal{O})$ considered in section 5.1. The connection is given by two cyclic GALOIS coverings φ, φ' of degree 3 described in:

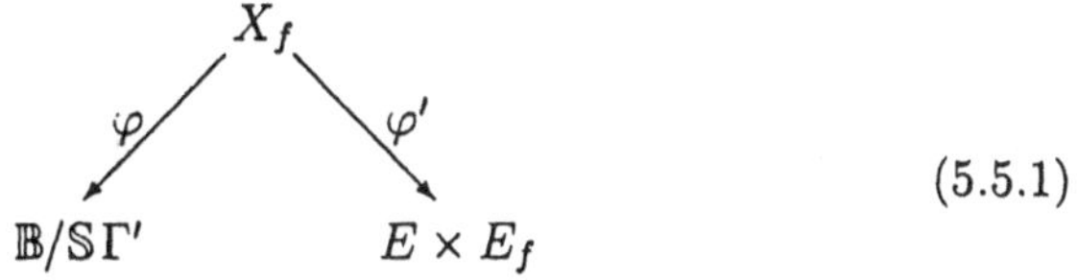

(5.5.1)

(see [HO5], (IV.3.)). The covering φ' is unramified and φ is unramified in codimension 1. Omitting the 3 singularities of $\mathbb{B}/\mathbb{S}\Gamma'$ and their preimages on X_f one gets an unramified covering. The common universal covering is the infinitely and discretely punctured ball $\mathbb{B} \setminus \mathbb{U}((2,1),\mathcal{O}) \cdot O$. Thus, the ball covers X_f and $\widetilde{E \times E_f}$ universally.

5.5.2 Kummer coverings of $\widetilde{E \times E}$ (HIRZEBRUCH [HI4]). The curve configuration 5.5.1 on $E \times E$ can be shifted by means of translations by n-division points on $E \times E$. Let U_n be the subgroup of $\mathrm{Aut}(E \times E)$ of all these translations. The order of U_n is n^4. The n-division points are the only intersection points of two of the $4n^2$ components of the divisor

$$D_n = U_n(E \times 0) + U_n(0 \times E) + U_n\Delta + U_n(-g \times 1)\Delta \tag{5.5.2}$$

We blow up all the n^4 n-division points and obtain a surface $\widetilde{E \times E}^{(n)}$. Omitting the proper preimages of the components of D_n we get an open surface $\widetilde{E \times E_f}^{(n)}$. We calculate its heights in the same manner as we have done above for $\widetilde{E \times E_f}$. The selfintersection numbers of the omitted components are equal to $-n^2$ because on each of the components lie exactly n^2 division points which have been blown up. So we can calculate

$$H_e\left(\widetilde{E \times E_f^{(n)}}\right) = e\left(\widetilde{E \times E}^{(n)}\right) - 0 = e(E \times E) + n^4 = n^4$$

$$3H_\tau\left(\widetilde{E \times E_f^{(n)}}\right) = 3\tau\left(\widetilde{E \times E}^{(n)}\right) - 4n^2 \cdot (-n^2) =$$

$$3\tau(E \times E) - 3n^4 + 4n^4 = n^4.$$

The proportionality condition $H_e = 3H_\tau$ is satisfied.

Proposition 5.5.3 *The open surfaces* $\widetilde{E \times E}_f^{(n)}$ *are ball quotient surfaces* $\mathbb{B}/\Lambda^{(n)}$. *The ball lattices* $\Lambda^{(n)}$ *are sublattices of the full* EISENSTEIN *lattice* $\mathbb{U}((2,1),\mathcal{O})$.

Proof HIRZEBRUCH constructed in [HI4] KUMMER extensions

$$\mathbb{C}(E \times E)\left(\sqrt[n]{f_1/f_0},\ \sqrt[n]{f_2/f_0},\ \sqrt[n]{f_3/f_0}\right)$$

of the function field $\mathbb{C}(E \times E)$ containing f_0, f_1, f_2, f_3. The normalization of $\widetilde{E \times E} = \widetilde{E \times E}^{(1)}$ in this KUMMER extension is denoted by $X^{(n)}$. On this way we receive KUMMER coverings $\pi^{(n)} : X^{(n)} \longrightarrow \widetilde{E \times E}$. The functions f_0, f_1, f_2, f_3 allow only branching along the divisor D_n which is the divisor at infinity of $\widetilde{E \times E}^{(n)}$. So we get on the finite parts unramified coverings of degree n^3

$$\pi_f^{(n)} : X_f^{(n)} \longrightarrow \widetilde{E \times E}_f^{(n)}.$$

On the other hand the group U_n of n-division points on $E \times E$ induces an isogeny $E \times E \longrightarrow E \times E$ of abelian surfaces. Blowing up the n-torsion points or 0, respectively, this isogeny extends to an etale covering of degree n^4

$$u^{(n)} : \widetilde{E \times E}^{(n)} \longrightarrow \widetilde{E \times E} = \widetilde{E \times E}^{(1)}$$

with restriction to the open parts

$$u_f^{(n)} : \widetilde{E \times E}_f^{(n)} \longrightarrow \widetilde{E \times E}_f$$

Composing $\pi^{(n)}$ and $u^{(n)}$ we receive coverings of degree n^7

$$\begin{aligned} p^{(n)} &: X^{(n)} \xrightarrow{\pi^{(n)}} \widetilde{E \times E}^{(n)} \xrightarrow{u^{(n)}} \widetilde{E \times E}, \\ p_f^{(n)} &: X_f^{(n)} \longrightarrow \widetilde{E \times E}_f^{(n)} \longrightarrow \widetilde{E \times E}_f, \end{aligned} \tag{5.5.3}$$

surjective and unramified on the open part. We know already that $\widetilde{E \times E}_f = \widetilde{E \times E}_f^{(1)} = \mathbb{B}/\Gamma^{(1)}$, $\Gamma^{(1)}$ a sublattice of $\mathbb{U}((2,1),\mathcal{O})$. Since $p_f^{(n)}$ is unramified, the surface $\widetilde{E \times E}_f$, $X_f^{(n)}$ and $\widetilde{E \times E}_f^{(n)}$ have the same universal covering $\mathbb{B}$. Therefore $X_f^{(n)}$ and $\widetilde{E \times E}_f^{(n)}$ are ball quotient surfaces

$$\widetilde{E \times E}_f^{(n)} = \mathbb{B}/\Lambda^{(n)}, \; X_f^{(n)} = \mathbb{B}/\Gamma^{(n)}, \; \Gamma^{(n)} \subseteq \Lambda^{(n)} \subseteq \Lambda^{(1)} = \Gamma^{(1)}.$$

□

Corollary 5.5.4 HIRZEBRUCH'S KUMMER *covers* $X_f^{(n)}$ *of* $\widetilde{E \times E_f}$ *are ball quotient surfaces with ball lattices* $\Gamma^{(n)} \subseteq \Gamma^{(1)} \subset \mathbb{U}((2,1),\mathcal{O})$.

5.5.4A $K3$-Surface. ([HO5], VI). Singular $K3$ surfaces have by definition maximal PICARD number 20. Among them there are two special $K3$ surfaces, which are called "the most algebraic $K3$ surfaces" (VINBERG [VIN], INOSE [I-S]). They are constructed as quotients of $E \times E$, E an elliptic curve with complex multiplication, by a cyclic automorphism group of order 3 or 4, respectively. We restrict ourselves to the case of the elliptic curve $E = \mathbb{C}/\mathcal{O}, \mathcal{O} = \mathbb{Z} + \mathbb{Z}\rho$ the ring of EISENSTEIN integers. The automorphism group G is generated by $\gamma = (g, g^2)$ with the notations of 5.5.1. We consider on $E \times E$ the curve configuration drawn in Figure (5.5.2), $Q_0 = O, Q_1, Q_2$ the fixed points of g on E.

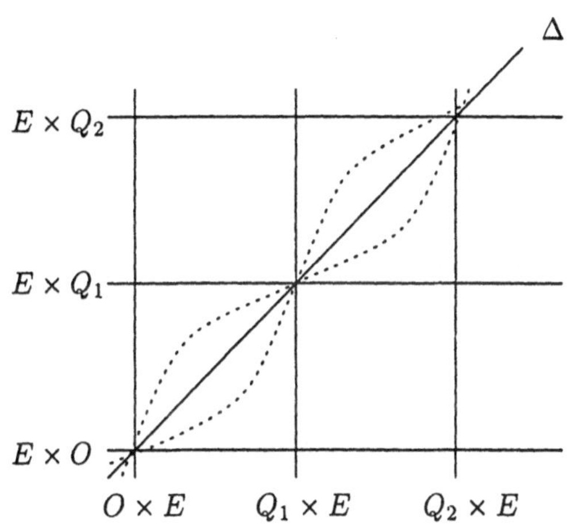

Figure 5.5.2

The dotted curves are $\gamma\Delta$ and $\gamma^2\Delta$. The diagonal curve Δ has exactly three G-cross points, namely $Q_i \times Q_i$, $i = 0, 1, 2$. We blow them up and denote the new surface by $\overline{E \times E}$. As open part we choose

$$\overline{E \times E}_f = \overline{E \times E} \setminus \left(\sum_{i=0}^{2} \gamma^i \Delta + \sum_{i=0}^{2} E \times Q_i \right).$$

We calculate by means of (4.10.2) and (4.10.3) the orbital heights

$$H_e(\overline{E \times E_f}) = e(\overline{E \times E}) = e(E \times E) + 3 = 3$$
$$3H_\tau(\overline{E \times E_f}) = 3\tau(\overline{E \times E}) - (-3 - 3 - 3 - 3)$$
$$= 3\tau(E \times E) - 9 + 12 = 3$$

It is not difficult to see that

$$\overline{E \times E_f}/G \times Z_3 \simeq \mathbb{P}^2 \setminus \Delta = \mathbb{B}/\Gamma'$$

where Z_3 is the cyclic subgroup of $\mathrm{Aut}(E \times E)$ generated by (g, g), and Γ' is the congruence EISENSTEIN lattice considered in section 5.1. There is a commutative diagram (5.5.4) consisting of cubic GALOIS coverings

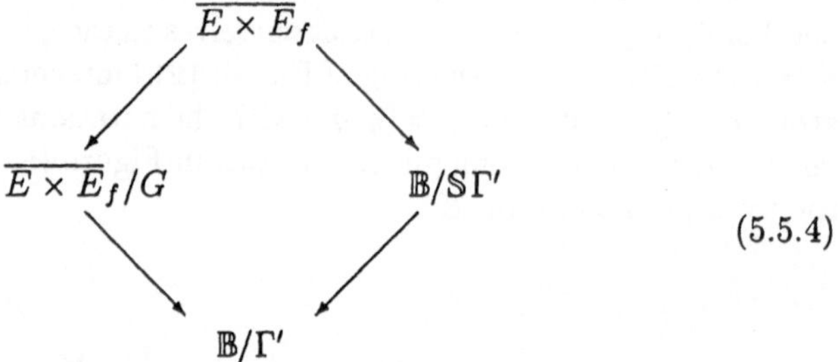

(5.5.4)

The upper coverings are unramified in codimension 1. So the punctured ball is the common universal covering of the upper three surfaces. $\overline{E \times E}/G$ is a model of one of the two "most algebraic $K3$ surfaces", mentioned above which is precisely defined as the minimal singularity resolution X of $E \times E/G$. So we get the following

Proposition 5.5.6 *The model $\overline{E \times E}$ of the $K3$ surface X is the canonical compactification of a ball quotient surface $\mathbb{B}/\Gamma_1'$, Γ_1' a normal sublattice of index 3 of the congruence* EISENSTEIN *lattice Γ'.*

5.5.7 Sequences of algebraic surfaces with extreme limit of CHERN quotients. We consider a smooth surface $X = X_0$ being the toroidal compactification of an open surface X_f. The BAILY-BOREL compactification of X_f is denoted by $\widehat{X}$. We assume that all cusp points are elliptic singularities, that means that their minimal resolutions are elliptic curves with negative selfintersection numbers. These curves lie on X because $X \to \widehat{X}$ is the minimal singularity resolution of $\widehat{X}$. Furthermore, we consider a sequence of surfaces $(X^{(n)})_{n \geq 0}$ with the following properties: X_n is a GALOIS cover of the base surface $X = X_0$ such that the corresponding GALOIS covering $\varphi_n : X^{(n)} \to X$ is unramifiedd over X_f. The normalization of X in the function field $\mathbb{C}(X^{(n)})$ is denoted by $\widehat{X}^{(n)}$. The boundary $\widehat{X}^{(n)} \setminus X_f^{(n)}$, $X_f^{(n)} = \varphi_n^{-1}(X_f)$, consists of elliptic singularities because cusp singularities are preserved by finite coverings (see Prop. 3.5.12).

Proposition 5.5.8 *With the above notations the following conditions are equivalent* $(e(X) \neq 0)$:

(i) $\lim\limits_{n\to\infty} \dfrac{\tau(X^{(n)})}{e(X^{(n)})} = \dfrac{H_\tau(X_f)}{H_e(X_f)} = \dfrac{\tau(X) - \frac{1}{3}\sum_{i=1}^{k}(E_i^2)}{e(X)}$

(ii) The ramification indices v_{ni} of φ_n at each boundary component E_i of $X \setminus X_f$ tend to ∞ for $n \to \infty$.

Proof The orbital heights $H_\tau(X_f)$, $H_e(X_f)$ of the open surface X_f (logarithmic CHERN numbers) are defined in (4.10.2) and (4.10.3). The invariants $\tau(X^{(n)})$, $e(X^{(n)})$ can be considered as heights $H_\tau(X_f^{(n)})$ or $H_e(X_f^{(n)})$, respectively. The GALOIS coverings $\varphi_n : X^{(n)} \to X$ can be "arranged". So we change over to the orbital morphisms $\varphi_n : X^{(n)} \to \mathbf{X}_n$, $\mathbf{X}_n = (X, \Sigma\mathbf{E}_i^{(n)})$, $\mathbf{E}_i^{(n)} = (v_{ni}E_i; \emptyset)$. By the orbital property (degree formula) of H_τ, H_e we get

$$\begin{aligned} \tau\left(X^{(n)}\right)/e\left(X^{(n)}\right) &= H_\tau\left(X_f^{(n)}\right)/H_e\left(X_f^{(n)}\right) \\ &= H_\tau(\mathbf{X}_n)/H_e(\mathbf{X}_n). \end{aligned} \tag{5.5.5}$$

By the orbital decompositions 3.3.3 and 3.4.2 we have

$$\begin{aligned} H_\tau(\mathbf{X}_n) &= \tau(X) - \frac{1}{3}\sum_{i=1}^{k}\left(1 - \frac{1}{v_{ni}^2}\right)(E_i^2) \\ H_e(\mathbf{X}_n) &= H_e(X) = e(X) \end{aligned} \tag{5.5.6}$$

because there are no orbital points and $H_e(E_i^{(n)}) = e(E_i) = 0$. Now the statement of the proposition follows immediately from (5.5.5) and (5.5.6). □

Corollary 5.5.9 *If additionally the proportionality condition (Prop 2) for X_f is satisfied, especially, if $X_f = \mathbb{B}/\Gamma$ is a neat ball quotient surface, then it holds that*

$$\lim \frac{c_1^2(X^{(n)})}{c_2(X^{(n)})} = 3$$

if and only if $\lim_{n\to\infty} v_{ni} = \infty$ for all E_i.

Proof The proportionality condition (Prop 2) means that $H_e(X_f) = 3H_\tau(X_f)$. On the other hand we know that $c_1^2 = 2e + 3\tau$. Together with (i) we obtain

$$\frac{c_1^2(X^{(n)})}{c_2(X^{(n)})} = \frac{(2e+3\tau)(X^{(n)})}{e(X^{(n)})} = 2 + \frac{3\tau(X^{(n)})}{e(X^{(n)})} \underset{n\to\infty}{\longrightarrow} 2 + \frac{3H_\tau(X_f)}{H_e(X_f)}$$
$$= 2 + 1 = 3.$$

□

Example 5.5.10 HIRZEBRUCH [HI4]). HIRZEBRUCH'S KUMMER covers $X^{(n)}$ of $E \times E$, see (5.5.3), together with the morphisms $p^{(n)} : X^{(n)} \to \widetilde{E \times E}$ satisfy the last condition of the corollary because the ramification index is equal to n over each component of D_n defined in (5.5.2). Therefore, the CHERN quotients $c_1^2(X^{(n)})/c_2(X^{(n)})$ tend to the extreme ratio 3.

Example 5.5.11 [HO8] Let $\Gamma = \mathbb{U}((2,1), \mathcal{O}_K), K = \mathbb{Q}(\sqrt{-d})$, the full PICARD modular group of the imaginary quadratic number field K and $\Gamma(n)$ the principal congruence subgroup defined by the exact sequence

$$1 \longrightarrow \Gamma(n) \longrightarrow \Gamma \longrightarrow \mathbb{U}((2,1), \mathcal{O}_K/n\cdot\mathcal{O}_K)$$

and $X(n)$ the toroidal compactification of $\mathbb{B}/\Gamma(n)$.

Then

$$\lim_{n\to\infty} \frac{c_1^2(X(n))}{c_2(X(n))} = 3.$$

Proof It suffices to prove that $3\tau(X(n))/e(X(n))$ tends to 1. We exclude $n = 2$. Then it is easy to see that $\Gamma(n), n > 2$, acts freely on $\mathbb{B}$. Consequently the branch arrangement of $X(n)_f \to X_f = X(1)_f$ on X_f does not depend on n. It coincides with that of $\mathbb{B} \to X_f$. We get commutative diagrams of orbital morphisms

$$\begin{array}{ccc} X(n) & \longrightarrow & \mathbf{X}_n = X(n)/\left(\Gamma/\Gamma(n)\right) \\ \uparrow & & \uparrow \\ X(n)_f & \longrightarrow & \mathbf{X}_f = \mathbb{B}/\Gamma = X(n)_f/\left(\Gamma/\Gamma(n)\right) \end{array}$$

We know that $3H_\tau(\mathbf{X}_f) = H_e(\mathbf{X}_f)$ because $\mathbf{X}_f = \mathbb{B}/\Gamma$ is an orbital ball quotient surface. By the definition of heights of open orbital surfaces in section 4.10 we get with obvious notations

$$H_e(\mathbf{X}_n) = 3H_\tau(\mathbf{X}_n) - 3\sum_\kappa \frac{1}{v_{\kappa,n}} \cdot H_\tau(\mathbf{E}_{\kappa,n})$$

or, via orbitality,

$$[\Gamma : \Gamma(n)]^{-1} e(X(n)) = 3[\Gamma : \Gamma(n)]^{-1} \tau(X(n)) - \sum_\kappa \frac{1}{v_{\kappa,n}^2} \cdot \left(E^2 + \sum_{i=1}^4 \frac{e_i}{d_i}\right),$$

hence

$$1 = \frac{3\tau(X(n))}{e(X(n))} - H_e(\mathbf{X}_f) \cdot \sum_\kappa \frac{1}{v_{\kappa,n}^2} \left(E^2 + \sum_{i=1}^4 \frac{e_i}{d_i}\right)$$

We have to check that $\lim\limits_{n\to\infty} v_{\kappa,n} = \infty$. For this purpose we look back to the construction of the canonical resolution curve E_κ of the Γ-cusp point κ on $\widehat{X} = \widehat{\mathbb{B}/\Gamma}$ living in the cusp bundle $\mathbb{F}_{\tilde\kappa}(\Gamma), \tilde\kappa \in \partial\mathbb{B}$, see diagram (4.2.8).

Let $E_\kappa(n)$ be a cusp curve on $X(n)$ lying over E_κ. The stationary group $\Gamma_{\tilde\kappa}/\Gamma_{\tilde\kappa}(n)$ acts on $E_\kappa(n)$. The elements which act trivially on $E_\kappa(n)$ belong to the quotient $\Delta_{\tilde\kappa}/\Delta_{\tilde\kappa}(n)$ of the center $\Delta_{\tilde\kappa}$ of $\Gamma_{\tilde\kappa}$ by $\Delta_{\tilde\kappa}(n) = \Delta_{\tilde\kappa} \cap \Gamma(n)$. Thus, the ramification index $v_{\kappa,n}$ is equal to $[\Delta_{\tilde\kappa} : \Delta_{\tilde\kappa}(n)]$. It is easy to see that this index tends to ∞ for $n \to \infty$.

□

Obviously, for $n >> 0$ the surfaces $X(n) = \overline{\mathbb{B}/\Gamma(n)}$ have positive signature. A more careful calculation using the Table 5.4.1 shows that this happens already for rather small n. For example, if $|D_{K/\mathbb{Q}}| \leq 56, D_{K/\mathbb{Q}} \neq -4$, then $X(n)$ has positive signature for $n \geq 6$ (see [HO8], Table 1). Moreover, these surfaces are minimal ([HO8], 4.16).

5A Volumes of Fundamental Domains of Picard Modular Groups

In this appendix we present the complete proof of the L-value formula for volumes of fundamental domains of PICARD modular groups acting on the complex two-dimensional ball $\mathbb{B}$ with respect to CHERN forms of the invariant EINSTEIN-KÄHLER (BERGMANN) metric on $\mathbb{B}$.

Let $\Gamma = \mathbb{S}U((2,1),\mathcal{O}) = \mathbf{G}(\mathbb{Z})$ be the PICARD modular group of the field $K = \mathbb{Q}(\sqrt{-d})$, d a squarefree natural number, $\mathcal{O} = \mathcal{O}_K$ the ring of integers in K. The group Γ acts in a natural manner on the complex 2–ball

$$\mathbb{B} = \left\{(z_1, z_2) \in \mathbb{C}^2 ; |z_1|^2 + |z_2|^2 < 1\right\} \subset \mathbb{P}^2 = \mathbb{P}^2\,(\mathbb{C}).$$

It suffices to calculate the volume $c_2(\mathbb{B}/\Gamma)$ of a Γ–fundamental domain with respect to the EULER-CHERN form c_2. In our geometric-arithmetic situation one can expect that $c_2\,(\mathbb{B}/\Gamma)$ has an infinite multiplicative decomposition

$$c_2\,(\mathbb{B}/\Gamma) = \gamma_\infty \cdot \prod_p \gamma_p, p \in \operatorname{Spec}\mathbb{Z}$$

and γ_p is naturally connected with the groups $\mathbf{G}(\mathbb{Z}/p^k\mathbb{Z})$ approximating $\mathbf{G}(\mathbb{Z}_p)$. On the other hand the inductive construction of orbital surface invariants should be reflected in some sense. We consider the “flag”

$$0 \in \mathbb{D} \subset \mathbb{B},\ \mathbb{D}:\ z_2 = 0 \text{ (disc)}.$$

It should serve as a geometric guide for the following plan of volume calculation. We will arithmetically define an invariant measure (volume form) $\omega^{(d)}$ on $G = \mathbf{G}(\mathbb{R})$ disposing on a group scheme $\mathbf{G} \longrightarrow \operatorname{Spec}\mathbb{Z}$ defined over $\mathbb{Z}$, $\mathbf{G} = \mathbf{G}^{(d)}$. Looking at fibres over points we are able to calculate p-adically the volume $\omega^{(d)}(G/\Gamma)$ of a Γ-fundamental domain in G. It turns out that $\omega^{(d)}(G/\Gamma)$ splits into a “relative volume” $\omega^{(d)}(\mathbb{B} \mid \mathbb{D})$ and $\alpha_{\mathbb{D}} = \zeta(2)$. On the other hand we can relate $\omega^{(d)}$ with $\omega = \omega^{(1)}$, and $\omega(G/\Gamma)$ with a volume of the isotropy group $G_0 = \mathbb{S}(\mathbb{U}(2) \times \mathbb{U}(1))$ and with $c_2\,(\mathbb{B}/\Gamma)$. In order to present a scheme of these connections we use for a moment the symbol $a \approx_f b$. It means that $a \approx f \cdot b$, defined by: $a = r \cdot f \cdot b$ with a trivial rational constant $r \neq 0$ calculated later during the proof of the final c_2-volume formula (5A.4.3). If f is rational, then we omit it. The connections are listed in the following scheme:

$$
\begin{array}{llll}
 & & \pi^3 & \\
 & & \wr & \\
\omega^{(1)}(G/\Gamma) \approx \kappa(G_0)\lambda(\mathbb{B}/\Gamma) & & \approx_{\pi^2} \kappa(G_0)c_2(\mathbb{B}/\Gamma) & \\
\quad \wr \; |D|^{5/2} & & & \\
\omega^{(d)}(G/\Gamma) \approx \alpha^{(d)}(\mathbb{B}|\mathbb{D})\alpha_{\mathbb{D}} & & \approx \prod_p \alpha_p^{(d)}(\mathbb{B}|\mathbb{D}) \prod_p \alpha_{p,\mathbb{D}} & \\
 & & \wr & \\
L(3,\chi)\zeta(2) = \prod_p \left(1 - \left(\frac{D}{p}\right) p^{-3}\right)^{-1} \prod_p (1-p^{-2})^{-1} & & & \\
\quad \wr & & & \\
\quad \pi^2 & & &
\end{array}
$$

where $D = D_{K/\mathbb{Q}}$ is the discriminant of $K/\mathbb{Q}$, ζ the RIEMANN Zeta Function, $L(s,\chi)$ is the L-series of the DIRICHLET character $\chi : n \mapsto \left(\frac{D}{n}\right)$ (JACOBI symbol) and κ is an invariant measure (volume form).

The whole proof joins HIRZEBRUCH'S proportionality principle with results on hermitian lattices of HEL BRAUN'S school and SHVARTSMAN'S calculation of the volume for the field of GAUSS numbers via TAMAGAWA number in [SV1].

5A.1 The Order of Finite Unitary Groups

Let R be a commutative ring with unit 1 and an involution $\iota : R \xrightarrow{\sim} R$ $(r \mapsto \overline{r})$. It can be extended componentwise to a $R^{\langle\iota\rangle}$ -linear isomorphism $V \xrightarrow{\sim} V$, $V = R^n, R^{\langle\iota\rangle} = \{r \in R; r = \overline{r}\}$.

Definition 5A.1.1 An $R^{\langle\iota\rangle}$ -linear map

$$\Phi : V \times V \longrightarrow R$$

is a *hermitian form* on V, if it is R-linear in the first argument and

$$\Phi(v,w) = \overline{\Phi(w,v)} \text{ for all } v,w \in V.$$

We will use the same symbol Φ for the hermitian matrix defined by Φ with respect to the canonical base of $V = R^n$. Φ-unitary matrices X are defined by the relation $X^t\Phi\overline{X} = \Phi$. The group of Φ-unitary matrices is denoted by $\mathbb{U}(\Phi,\mathbb{R})$, and $\mathbb{S}U(\Phi,R) = \mathbb{S}\mathrm{l}_n(R) \cap \mathbb{U}(\Phi,R)$ denotes the special unitary group.

Let $K = \mathbb{Q}(\sqrt{-d})$ be an imaginary quadratic number field, $\mathcal{O} = \mathcal{O}_K$ the ring of integers in K, p a natural prime and k a natural number. We consider the ARTIN ring $R = \mathcal{O}/p^k\mathcal{O}$ with involution induced by the conjugation. The unitary group $\mathbb{U}(\Phi, \mathcal{O}/p^k\mathcal{O})$, Φ a hermitian form on R^n, is finite; $A_{p^k}(\Phi)$ denotes the number of its elements. Any hermitian form Φ on $\mathcal{O}^n$ defines a hermitian form on $(\mathcal{O}/p^k\mathcal{O})^n$, which is denoted by the same symbol. We want to determine $A_{p^k}(\Phi)$ for $\Phi = \mathrm{diag}(1,1,-1)$ (diagonal matrix) for all p and almost all k.

Proposition 5A.1.2 *With the above notations it holds that*

$$A_{p^k}(\Phi) = \begin{cases} p^{9k}\left(1-\left(\frac{D}{p}\right)p^{-3}\right)(1-p^{-2})\left(1-\left(\frac{D}{p}\right)p^{-1}\right), & \text{for } k \geq 1,\ p \text{ not ramified in } K \\ 2p^{9k}(1-p^{-2}), \quad \text{for } \begin{cases} k \geq 1, & 2 \neq p \text{ ramified in } K \\ k \geq 3, & 2 = p \text{ ramified in } K \end{cases} \end{cases} \tag{5A.1.1}$$

where $D = D_{K/Q}$ denotes the discriminant.

Corollary 5A.1.3

$$\#\mathbb{S}U(\Phi, \mathcal{O}/p^k\mathcal{O}) = \begin{cases} p^{8k}\left(1-\left(\frac{D}{p}\right)p^{-3}\right)(1-p^{-2}), & \text{for } k \geq 1, p \text{ not ramified} \\ p^{8k}(1-p^{-2}), \quad \text{for } \begin{cases} k \geq 1, & 2 \neq p \text{ ramified} \\ k \geq 3, & 2 = p \text{ ramified} \end{cases} \end{cases} \tag{5A.1.2}$$

For the proof we need a generalization of $A_{p^k}(\Phi)$ and related local invariants not depending on k. Let (R^n, Φ) and (R^m, Ψ) be two hermitian modules, $R = \mathcal{O}/\mathcal{O}p^k$ as above. We set

$$A_{p^k}(\Phi, \Psi) = \#\{M \in Mat_{m,n}(R); M^t\Phi\overline{M} = \Psi\}.$$

We have $A_{p^k}(\Phi) = A_{p^k}(\Phi, \Phi)$. The *local densities* (of unitary representations) are defined by

$$\alpha_p(\Phi, \Psi) = p^{km(m-2n)}A_{p^k}(\Phi, \Psi), k \gg 0. \tag{5A.1.3}$$

This definition is correct by a result of HEL BRAUN [BRA], see also OTREMBA [OTR]. More precisely, it holds that

5A.1.4 *The identity (5A.1.3) is correct for*

$$k \geq 2v_p(det\Psi) + 3,\ 2 = p \text{ ramified in } K,$$
$$k \geq 2v_p(det\Psi) + 1, \text{ otherwise,}$$

where $v_p : \mathbb{Q}^* \to \mathbb{Z}, p \mapsto 1$, *is the normalized p-valuation.*

The local densities have been originally introduced for the calculation of class numbers of genera of positive definit hermitian forms (BÖGE [BOEG], OTREMBA [OTR]) in analogy with the SIEGEL-MINKOWSKI theory of quadratic forms. The local densities have nice multiplicative properties, which are useful for the calculation of $A_{p^k}(\Phi)$ in a recursive manner. For instance, one knows for diagonal matrices $D = diag(d_1, ..., d_n)$, especially for the unit matrix E_n, that

$$\alpha_p(D) = \alpha_p(D, d_i) \cdot \alpha_p(D_i), \text{ if } p \not| \, 2D_{K/Q} \cdot detD, \tag{5A.1.4}$$

where $D_i = diag(d_1, ..., \hat{d}_i, ..., d_n)$.

$$\alpha_p(E_n) = \alpha_p(E_n, 1) \cdot \alpha_p(E_{n-1}), \text{ if } p \neq 2 \text{ or } 2 = p \text{ not ramified,} \tag{5A.1.5}$$

We refer to [OTR] (3.1) or Lemma 3.4, respectively. Unfortunately, the multiplicative property is not generally liftable from α_p to A_{p^k}. But we dispose on the following kind of lifting:

Lemma 5A.1.5 OTREMBA *([OTR], Lemma 3.1) For D as above,* $d_i \in \mathbb{Z}$ *, it holds that*

$$A_{p^k}(D) = \tilde{A}_{p^k}(D, d_i) \cdot A_{p^k}(D_i) \tag{5A.1.6}$$

where $\tilde{A}_{p^k}(D, d_i)$ *denotes the number of "vectors"* $\mathfrak{a} \in R^n = \left(\mathcal{O}/p^k\mathcal{O}\right)^n$ *,* $D(\mathfrak{a}, \mathfrak{a}) = d_i$ *which can be extended to a D–unitary matrix* $U \in Mat_n(R)$.

Lemma 5A.1.6 *If p is not ramified in* $K, \Phi = diag(1, 1, -1)$, *then*

$$\alpha_p(\Phi) = \alpha_p(\Phi, -1) \cdot \alpha_p(E_2)$$

Proof For $p = 2$, $k = 1$ the definition (5A.1.3) together with 5A.1.4 yields

$$\alpha_2(\Phi) = 2^{-9} A_2(\Phi) = 2^{-9} A_2(E_3) = \alpha_2(E_3).$$

Now apply (5A.1.5) together with (5A.1.3), 5A.1.4 □

Remark. *According to (5A.1.3), 5A.1.4 and (5A.1.6) the statement is equivalent to*

$$\tilde{A}_{p^k}(\Phi,-1) = A_{p^k}(\Phi,1) \text{ for } k \geq 1. \tag{5A.1.7}$$

Lemma 5A.1.7 *If* $2 \neq p$ *is ramified in* K*, then*

$$\alpha_p(\Phi) = \alpha_p(\Phi,-1)\alpha_p(E_2).$$

We need classification results of JACOBOWITZ [JAC] for local hermitian lattices. Let E, F be local fields (complete w.r.t. the discrete valuation), F/E an extension of degree 2, $^{-} : F \longrightarrow F,\ a \mapsto \overline{a}$, the generator of the GALOIS group. The rings of integers are denoted by $\mathcal{O}_E$ or $\mathcal{O}_F$, respectively. A lattice L (in V) is an $\mathcal{O}_F$-module of finite rank in a F-vector space V of the same dimension. L is a *hermitian lattice*, if L is a lattice over $\mathcal{O}_F$ endowed with a map $L \times L \longrightarrow F, (\mathfrak{a},\mathfrak{b}) \mapsto \mathfrak{a}\cdot\mathfrak{b}$, which is $\mathcal{O}_F$-linear in the first argument, such that $\mathfrak{b}\cdot\mathfrak{a} = \overline{\mathfrak{a}\cdot\mathfrak{b}}$ for all $\mathfrak{a},\mathfrak{b} \in L$. The lattice L has an $\mathcal{O}_F$-basis because each ideal of $\mathcal{O}_F$ is principal. If $L = \mathcal{O}_F\mathfrak{a}_1 \oplus \ldots \oplus \mathcal{O}_F\mathfrak{a}_n$, then we call $dL = det(\mathfrak{a}_i\cdot\mathfrak{a}_j)$ the *discriminant* of L. It is uniquely determined up to multiplication with norm units belonging to $N_{F/E}(\mathcal{O}_F^*)$ by definition. We write

$$L = M_1 \textcircled{\perp} \cdots \textcircled{\perp} M_k$$

if $L = M_1 + \ldots + M_k$ for sublattices M_i of L such that $M_i \perp M_j$ for all $i \neq j$. That means that $\mathfrak{a}_i \perp \mathfrak{a}_j$ (or $\mathfrak{a}_i\cdot\mathfrak{a}_j = 0$) for all elements $\mathfrak{a}_i \in M_i, \mathfrak{a}_j \in M_j$. If $L = M \textcircled{\perp} N$, then $dL \simeq dM \cdot dN$ (multiplicative equivalence with respect to $N_{F/E}(\mathcal{O}_F^*)$). For isometric lattices L, M we write $M \simeq N$, too.

$$L \simeq \langle d_1 \rangle \textcircled{\perp} \cdots \textcircled{\perp} \langle d_k \rangle$$

means that $L \simeq \mathcal{O}_F\mathfrak{a}_1 \textcircled{\perp} \cdots \textcircled{\perp} \mathcal{O}_F\mathfrak{a}_k$ and $\mathfrak{a}_i^2 = d_i \in \mathcal{O}_E$.

The *trace* sL of the hermitian lattice L is the F-ideal generated by all products $\mathfrak{a}\cdot\mathfrak{b},\ \mathfrak{a},\mathfrak{b} \in L$. The *norm* nL is the E-ideal generated by all squares $\mathfrak{a}^2 = \mathfrak{a}\cdot\mathfrak{a}, \mathfrak{a} \in L$. An element $\mathfrak{a} \in L$ is called *maximal* if it cannot be written as $\mathfrak{a} = b\cdot\mathfrak{b}, \mathfrak{b} \in L, b \in \mathcal{O}_F\pi$, where π is a prime element of $\mathcal{O}_F$. A hermitian lattice L is called *normal* , if $nL = sL$ (in F). It is called π^i-*modular* , if $\mathfrak{a}\cdot L = (\pi^i)$ for all maximal elements $\mathfrak{a} \in L$. The plane lattice $H(i)$ is defined by $H(i) \simeq \mathcal{O}_F\mathfrak{a} \oplus \mathcal{O}_F\mathfrak{b},\ \mathfrak{a}\cdot\mathfrak{b} = \pi^i, \mathfrak{a}^2 = \mathfrak{b}^2 = 0$. Now we are able to formulate the following classification results due to JACOBOWITZ.

Proposition 5A.1.8 *([JAC], Prop. 8.1.a) If F/E is ramified with prime elements π,p of F or E, respectively, $p \not| 2$, L a π^i-modular lattice of rank n, i even, then*

$$L \simeq \left\langle p^{i/2} \right\rangle \oplus \cdots \oplus \left\langle p^{i/2} \right\rangle \oplus \left\langle p^{-(n-1)i/2} \right\rangle dL.$$

□

Proposition 5A.1.9 *([JAC], Prop. 10.2.a) With the notations of 5A.1.8 let L be a normal π^i—modular plane lattice, $i = 0$ or 1, $nL \supsetneq nH(i)$, $p|2$. Then it holds that*

$$L \simeq \langle 1 \rangle \oplus \langle dL \rangle.$$

□

We apply the propositions to hermitian lattices over $\mathcal{O}_p = \mathcal{O}_K \otimes \mathbb{Z}_p, K_p = K \otimes_{\mathbb{Q}} \mathbb{Q}_p$, p ramified in $K = \mathbb{Q}(\sqrt{-d})$. In the hermitian lattice (L, Φ), $L = \mathcal{O}_p^3. \Phi = \operatorname{diag}(1, 1, -1)$, we consider for any $\mathfrak{a} \in L$ the sublattices $L_\mathfrak{a} = \{\mathfrak{b} \in L; \mathfrak{a} \cdot \mathfrak{b} = 0\}$.

Lemma 5A.1.10 *If $\mathfrak{a}^2$ is a unit, then $L_\mathfrak{a}$ is a π^0-modular hermitian $\mathcal{O}_p$—lattice of rank 2 and*

$$L = \mathcal{O}_p\mathfrak{a} \oplus L_\mathfrak{a}, \ dL_\mathfrak{a} \simeq -1/\mathfrak{a}^2.$$

Proof We show that $L \subset \mathcal{O}_p\mathfrak{a} + L$. Obviously, any element $\mathfrak{b} \in L$ can be decomposed into

$$\mathfrak{b} = (\langle \mathfrak{b}, \mathfrak{a} \rangle / \mathfrak{a}^2)\mathfrak{a} + (\mathfrak{b} - (\langle \mathfrak{b}, \mathfrak{a} \rangle / \mathfrak{a}^2)\mathfrak{a}) \in \mathcal{O}_p\mathfrak{a} + L_\mathfrak{a}.$$

Therefore $L = \mathcal{O}_p\mathfrak{a} \oplus L_\mathfrak{a}$ and

$$-1 = \det \Phi = dL \simeq d(\mathcal{O}_p\mathfrak{a}) \cdot dL_\mathfrak{a} = \mathfrak{a}^2 \cdot dL_\mathfrak{a}.$$

If $\mathfrak{b}$ is a maximal vector in $L_\mathfrak{a}$, then $\mathfrak{b}$ is maximal in L, hence the components b_1, b_2, b_3 of $\mathfrak{b}$ are relative prime. We can choose c_1, c_2, c_3 in $\mathcal{O}_p$ such that $b_1\bar{c}_1 + b_2\bar{c}_2 - b_3\bar{c}_3 = 1$, hence $\mathfrak{c}$ in L such that $\mathfrak{b} \cdot \mathfrak{c} = 1$. Then L is a π^0-modular lattice because

$$\mathfrak{c}' \cdot L_\mathfrak{a} = \mathfrak{c} \cdot L_\mathfrak{a} = \mathcal{O}_p = \mathcal{O}_p \cdot \pi,$$

$\mathfrak{c}'$ the projection of $\mathfrak{c}$ in $L_\mathfrak{a}$. □

Proof of 5A.1.7 We show that

$$\tilde{A}_{p^k}(\Phi) = A_{p^k}(\Phi). \tag{5A.1.8}$$

The statement follows then by means of (5A.1.6), (5A.1.3) and 5A.1.4. The reductions $\mathcal{O}_p \longrightarrow \mathcal{O}_p/\mathcal{O}_p p^k \simeq \mathcal{O}/\mathcal{O}p^k, L = \mathcal{O}_p^3 \longrightarrow (\mathcal{O}/\mathcal{O}p^k)^3$ are surjective. For $\bar{\mathfrak{a}} \in (\mathcal{O}/\mathcal{O}p^k)^3$ with $\Phi(\bar{\mathfrak{a}}, \bar{\mathfrak{a}}) = -1$ in $\mathcal{O}/\mathcal{O}p^k$ we choose a preimage $\mathfrak{a} \in L$. By Lemma 5A.1.10 and Proposition 5A.1.8 we have a decomposition

$$L_\mathfrak{a} \simeq \langle 1 \rangle \oplus \langle -1/\mathfrak{a}^2 \rangle.$$

We choose two vectors $\mathfrak{b}, \mathfrak{c} \in L_\mathfrak{a}$ such that $\mathfrak{b}^2 = 1, \mathfrak{c}^2 = -1/\mathfrak{a}^2$ and denote their images in $(\mathcal{O}/\mathcal{O}p^k)^3$ by $\hat{\mathfrak{b}}$ or $\hat{\mathfrak{c}}$, respectively. In canonical coordinates we obtain a Φ-unitary matrix with columns $\hat{\mathfrak{b}}, \hat{\mathfrak{c}}, \hat{\mathfrak{a}}$. Hence $\tilde{A}_{p^k}(\Phi, -1) = A_{p^k}(\Phi, -1)$ by definitions. Lemma 5A.1.7 is proved.

□

The connection between $\tilde{A}_{2^k}(\Phi, -1)andA_{2^k}(\Phi, -1)$ is more complicated if 2 is ramified in $K = \mathbb{Q}(\sqrt{-d})$. We can prove that

$$\tilde{A}_{2^3}(\Phi, -1) = \begin{cases} A_{2^3}(\Phi, -1), & d \equiv 1 \bmod 4 \\ A_{2^3}(\Phi, -1) - 2^{12}, & d \equiv 2 \bmod 8 \\ A_{2^3}(\Phi, -1) - 3 \cdot 2^{12}, & d \equiv 6 \bmod 8 \end{cases} \tag{5A.1.9}$$

For preparation we call L an *even* $\mathcal{O}_2$-lattice, if 2 divides nL. Otherwise L is called *odd*. We choose $\hat{\mathfrak{a}} \in (\mathcal{O}/\mathcal{O} \cdot 2^3)^3$, $\hat{\mathfrak{a}}^2 = -1$ and $\mathfrak{a} \in \mathcal{O}_2^3$ a preimage of $\hat{\mathfrak{a}}$.

5A.1.11 *There exists a Φ-unitary matrix with coefficients in $\mathcal{O}/8\mathcal{O}$ extending $\hat{\mathfrak{a}}$ if and only if $L_\mathfrak{a}$ is an odd lattice.*

Proof If $L_\mathfrak{a}$ is an odd lattice, then $nL = \mathcal{O} = sL$. Hence $L_\mathfrak{a}$ is normal and π^0-modular by Lemma 5A.1.10. From Proposition 5A.1.9 it follows that $L_\mathfrak{a} \simeq \langle 1 \rangle \perp \langle dL_\mathfrak{a} \rangle$. As in the proof of (5A.1.8) we can extend $\hat{\mathfrak{a}}$ to a unitary matrix.

The reduction $L_\mathfrak{a} \longrightarrow \hat{L}_{\hat{\mathfrak{a}}}$ is surjective because $L \longrightarrow \hat{L} = (\mathcal{O}/\ 8\ \mathcal{O})^3$ is, and L is mapped to $\hat{L}_{\hat{\mathfrak{a}}}$. If $\hat{\mathfrak{a}}$ is extendable to a Φ—unitary matrix with columns $\hat{\mathfrak{b}}, \hat{\mathfrak{c}}, \hat{\mathfrak{a}}$, and $\mathfrak{b}$ is a preimage of $\hat{\mathfrak{b}}$ in $L_\mathfrak{a}$, then $\mathfrak{b}^2$ is a unit, and $L_\mathfrak{a}$ is odd.

□

The identities (5A.1.9) can be derived from the following relation:

$$\tilde{A}_{2^3}(\Phi,-1) - \tilde{A}_{2^3}(\Phi,-1) \tag{5A.1.10}$$
$$= \#\left\{ \widehat{\mathfrak{a}} \in (\mathcal{O}/\, 8\, \mathcal{O})^3;\ \widehat{\mathfrak{a}}^2 = -1, \widehat{\mathfrak{a}} \equiv \begin{pmatrix}1\\1\\1\end{pmatrix} \bmod \pi \right\}.$$

By definition of $\tilde{A}_{2^3}$ and by 5A.1.11 the difference in (5A.1.10) is equal to the number of vectors $\widehat{\mathfrak{a}}$, $\widehat{\mathfrak{a}}^2 = -1$, such that $L_{\mathfrak{a}}$ is an odd lattice. This can be decided by reduction modulo 2. Now (5A.1.10) and the calculation of the difference is a simple exercise which is left to the reader. On this way one finds the identities (5A.1.9).

For the sake of completeness we calculate now all local densities $\alpha_p(\Phi)$.

Proposition 5A.1.12 *For $\Phi = diag(1,1,-1)$ it holds that*

$$\alpha_p(\Phi) = \begin{cases} \left(1-\left(\frac{D}{p}\right)p^{-3}\right)\left(1-p^{-2}\right)\left(1-\left(\frac{D}{p}\right)p^{-1}\right), & \text{p not ramified} \\ 2\left(1-p^{-2}\right), & \text{p ramified in K.} \end{cases}$$

Proof Assume first that p is not ramified. From 5A.1.6 and (5A.1.5) we get the decomposition

$$\alpha_p(\Phi) = \alpha_p(\Phi,-1)\cdot\alpha_p(E_2,1)\cdot\alpha_p(1) \tag{5A.1.11}$$

The three values on the right hand side have been calculated in [OTR], (3.5)-(3.7), 4.2. These factors are exactly those appearing in the first identity of 5A.1.12.

Now let $2 \neq p$ be ramified in K. We get the decomposition (5A.1.11) again via Lemma 5A.1.7 and (5A.1.5). The calculation of the factors can be found also in [OTR] on the places mentioned above. In the order of the right-hand side of (5A.1.11) the factors are $1+\left(\frac{-1}{p}\right)p^{-1}, 1-\left(\frac{-1}{p}\right)p^{-1}$ or 2, respectively.

For $p=2$ ramified in $K=\mathbb{Q}(\sqrt{-d})$ we use the decomposition (5A.1.6), the difference formulas (5A.1.9) and the definition of α_2, via $A_{2^3}(\Phi)$ (see (5A.1.3) and 5A.1.4). The calculation of $A_{2^3}(E_2)$ and $A_{2^3}(\Phi,-1)$ is left to the reader. The latter number can be calculated, besides of a more elementary method, by means of GAUSS sums (see [OTR] (3.33), [HOV], Proof of 1.9). □

Proof of Proposition 5A.1.2 One has only to multiply $\alpha_p(\Phi)$ in 5A.1.12 with p^{9k} to obtain (5A.1.1) by definition (5A.1.3) and 5A.1.4.

□

Proof of Corollary 5A.1.3 Use the exact sequence

$$1 \longrightarrow S\mathbb{U}(\Phi, \mathcal{O}/p^k\mathcal{O}) \longrightarrow \mathbb{U}(\Phi, \mathcal{O}/p^k\mathcal{O}) \longrightarrow U(1, \mathcal{O}/p^k\mathcal{O}) \longrightarrow 1,$$

$$\#\mathbb{U}(1, \mathcal{O}/p^k\mathcal{O}) = A_{p^k}(1) = p^k\alpha_p(1) = \begin{cases} 1 - \left(\frac{D}{p}\right)p^{-1}, & p \text{ not ramified} \\ 2, & p \text{ ramified} \end{cases}$$

(see [OTR]), $\#\mathbb{U}(\Phi, \mathcal{O}/p^k\mathcal{O}) = A_{p^k}(\Phi)$ and Proposition 5A.1.2.

□

5A.2 Index of Congruence Subgroups

We look for a precise definition of a group scheme $\mathbf{G}$ over $\mathbb{Z}$ such that for $\Phi = diag(1, \cdots, -1), K = \mathbb{Q}(\sqrt{-d}), \mathcal{O} = \mathcal{O}_K$:

$$\begin{aligned} &G = \mathbf{G}(\mathbb{R}) = S\mathbb{U}(\Phi, \mathbb{C}), \mathbf{G}(\mathbb{Q}) = S\mathbb{U}(\Phi, K) \\ &\mathbf{G}(\mathbb{Z}) = S\mathbb{U}(\Phi, \mathcal{O}) = \mathbf{G}(\mathbb{Q}_p) = S\mathbb{U}(\Phi, K_p) \\ &G_p = \mathbf{G}(\mathbb{Z}_p) = S\mathbb{U}(\Phi, \mathcal{O}_p), \mathbf{G}(\mathbb{Z}_p/p^k\mathbb{Z}_p) = S\mathbb{U}(\Phi, \mathcal{O}_p/p^k\mathcal{O}_p). \end{aligned}$$

For this purpose we fix in $\mathcal{O}$ the $\mathbb{Z}$ -basis 1, ε,

$$\varepsilon = \begin{cases} \sqrt{-d}, & \text{if } d \equiv 1 \bmod 4 \\ (1+\sqrt{-d})/2, & \text{if } d \equiv 3 \bmod 4 \end{cases}$$

It defines a ring homomorphism

$$\begin{aligned} \rho: \ &\mathrm{Mat}_n(\mathbb{C}) \longrightarrow \mathrm{Mat}_{2n}(\mathbb{C}) \\ &A + \varepsilon B \longrightarrow \begin{pmatrix} A & O \\ O & A \end{pmatrix} + \begin{pmatrix} B & O \\ O & B \end{pmatrix}\xi, \end{aligned} \tag{5A.2.1}$$

where $A,\ B \in \mathrm{Mat}_n(\mathbb{R}),\ \xi = \begin{pmatrix} O & -\varepsilon\bar{\varepsilon}E_n \\ E_n & (\mathrm{Tr}\ \varepsilon)E_n \end{pmatrix},\ \mathrm{Tr}\ \varepsilon = \varepsilon + \bar{\varepsilon},\ E_n =$ $\mathrm{diag}(1, \cdots, 1)$. The image of ρ is

$$\mathbf{M}(\mathbb{R}) = \left\{ \begin{pmatrix} X & O \\ O & X \end{pmatrix} + \begin{pmatrix} Y & O \\ O & Y \end{pmatrix}\xi;\ X, Y \in \mathrm{Mat}_n(\mathbb{R}) \right\}$$

$\mathbf{M}/\mathbb{Z}$ is understood as subscheme of Mat_{2n} defined by linear polynomials in $\mathbb{Z}[X,Y,U,V] = \mathbb{Z}[X_{ij}, Y_{ij}, U_{ij}, V_{ij}]$ in the following manner:

$$\mathbf{M}(\mathbb{R}) = \left\{ \begin{pmatrix} X & U \\ Y & V \end{pmatrix};\ V - X - (\mathrm{Tr}\ \varepsilon \cdot Y) = O,\ U + \varepsilon\bar{\varepsilon}Y = X \right\}.$$

The multiplication in Mat_{2n} restricts to a $\mathbb{Z}-$ morphism $\mathbf{M} \times \mathbf{M} \longrightarrow \mathbf{M}$. The subscheme $\mathbf{G}$ is defined by

$$^t\,\overline{(A+\varepsilon B)}\Phi(A+\varepsilon B)-\Phi=0,\ \det(A+\varepsilon B)-1=0.$$

It can be easily translated to a system of polynomial equations with polynomials in $\mathbb{Z}[X,Y,U,V]$. The multiplication $\mathbf{G} \times \mathbf{G} \longrightarrow \mathbf{G}$ is defined by restriction. The inversion $\mathbf{G} \longrightarrow \mathbf{G}$ is also a $\mathbb{Z}$-morphism because $^t\,\overline{(A+\varepsilon B)}\Phi(A+\varepsilon B)=\Phi$ implies

$$\begin{aligned}(A+\varepsilon B)^{-1} &= \Phi^t\overline{(A+\varepsilon B)}\Phi = \Phi\left({}^tA+\bar{\varepsilon}^tB\right)\Phi\\ &= \Phi\left[\left({}^tA+(\varepsilon+\bar{\varepsilon})^tB\right)-\varepsilon^tB\right]\Phi\end{aligned}$$

So the inversion can be defined by means of $\mathbb{Z}$-polynomials.

Definition 5A.2.1 We denote by $red_k : G_p \longrightarrow \mathbf{G}(\mathbb{Z}_p/p^k\mathbb{Z}_p)$ the reduction map modulo p^k. The p^k-th congruence subgroup $G_p(k)$ of G_p is defined as kernel $Ker(red_k)$.

We want to calculate the local indices $[G_p : G_p(k)]$ and later the global indices $[\mathbf{G}(\mathbb{Z}) : \mathbf{G}(\mathbb{Z})(k)], k \in \mathbb{N}_+$, as far as possible. The next two propositions place the method at disposal.

Proposition 5A.2.2 *(*Neron *[NER], Prop. 20) Let R be a henselian discrete valuation ring, $m = Rt$ the maximal ideal, $K = Quot R$ the quotient field, $char(K) = 0, F = (F_1, ..., F_r)$ a system of polynomials $F_i \in R[X_1, \cdots, X_n] = R[X], FR[X]$ the ideal in $R[X]$ generated by $F_1, ..., F_n$. There are integers $k_o \geq 1, e \geq 0$ depending on $FR[X]$ such that for all $k \geq k_o$ and all $x \in R^n$ satisfying*

$$F(x) \equiv 0 \bmod \left(t^k\right)$$

there exists a solution $y \in R^n$ of $F = 0$ with $y \equiv x \mod (t^{k-e})$.

An immediate consequence is the following

Proposition 5A.2.3 *Let $\mathbf{G}$ be an irreducible linear group scheme (subscheme of* $\mathbf{Mat}_n$*) over $\mathbb{Z}$, p a prime. There are integers $k_0 \geq 1, e \geq 0$ such that for all $k \geq k_0$ the diagram (5A.2.2) of group homomorphisms*

$$\begin{array}{ccccccccc} 1 & \longrightarrow & G_p(k-e) & \longrightarrow & G_p & \longrightarrow & \mathrm{red}_{k-e}\,(G_p) & \longrightarrow & \mathbf{G}\,(\mathbb{Z}_p/p^{k-e}\mathbb{Z}_p) \\ & & \uparrow & & \| & & \nwarrow & & \uparrow \mathrm{red}^k_{k-e} \\ 1 & \longrightarrow & G_p(k) & \longrightarrow & G_p & \longrightarrow & & \longrightarrow & \mathbf{G}\,(\mathbb{Z}_p/p^{k}\mathbb{Z}_p) \end{array} \tag{5A.2.2}$$

has a commutative extension along the dotted arrow. Especially it holds that

$$[G_p : G_p(k-e)] = \left[\mathbf{G}(\mathbb{Z}_p/p^k\mathbb{Z}) : \mathrm{Ker}(\mathrm{red}^k_{k-e})\right] \tag{5A.2.3}$$

for the relative reduction red^k_{k-e}.

□

With finer methods we want to determine precisely e and k_o as small as possible for our special unitary groups $\mathbf{G}$. Notice that e and k_o depend on p. Then we will calculate the indices in (5A.2.3). The next result is well-known.

Proposition-Definition. 5A.2.4 *([BOU], § 4.5, Cor. 1 of Prop. 4). Let R be an integral domain with only principal ideals, $X \in Mat_{n,m}(R)$ of rank $r > 0$. There exists an equivalent matrix*

$$X_0 = \left(\begin{array}{ccc|c} \alpha_1 & & 0 & \\ & \ddots & & 0 \\ 0 & & \alpha_r & \\ \hline & 0 & & 0 \end{array}\right) \in \mathrm{Mat}_{n,m}\,(R) \tag{5A.2.4}$$

with respect to $Gl_n(R)$- or $Gl_m(R)$-multiplication from the right- or left-hand side, respectively, such that

$$\alpha_1 \mid \alpha_2 \cdots \mid \alpha_r \neq 0.$$

The ideals $R\alpha_1, ..., R\alpha_r$ are uniquely determined by X; α_r is called the *highest invariant factor* of X.

Let $F = (F_1, ..., F_m)$ be a system of polynomials in $R[X] = R[X_1, ..., X_n], R$ as in 5A.2.6, $K = \text{Quot}R$. It defines a "polynomial map"

$$F : K^n \longrightarrow K^m.$$

By means of TAYLOR series at $\mathfrak{a} \in R^n$ we can define the "differential" $d_\mathfrak{a}F : K^n \longrightarrow K^m$ as a linear map which can be restricted R-linearly to $R^n \longrightarrow R^m$. With respect to the canonical basis we obtain a matrix representation $M(d_\mathfrak{a}F) \in \text{Mat}_{n,m}(R)$ of $d_\mathfrak{a}F$. Via base change in R^n, R^m we can transform $M(d_\mathfrak{a}F)$ to an equivalent form (5A.2.4).

Definition 5A.2.5 With the above notations we call α_r in (5A.2.4) the highest invariant factor of the polynomial map $F : R^n \longrightarrow R^m$ at $\mathfrak{a} \in R^n$. If R is a discrete valuation ring with maximal ideal $Rt, R\alpha_r = Rt^e$, then we call $e = e_\mathfrak{a}(F) = e_{t,\mathfrak{a}}(F)$ the *highest exponent of* F *at* $\mathfrak{a}$.

The following result generalizes HENSEL'S lemma.

Proposition 5A.2.6 SHVARTSMAN, *([SV1]) Let* $F = (F_1, ..., F_m)$ *be a system of polynomials in* $\mathbb{Z}_p[X_1, \cdots, X_n], \mathfrak{a} \in \mathbb{Z}_p^n$ *and* $e = e_{p,\mathfrak{a}}(F)$ *the highest exponent of the polynomial map* $F : \mathbb{Z}_p^n \longrightarrow \mathbb{Z}_p^m$ *at* $\mathfrak{a}$. *If the residue of* $\mathfrak{a}$ *is a solution of* $F \equiv 0$ *modulo* p^{k+2e} *and* $k \geq 1$, *then there exists a solution* $\tilde{\mathfrak{a}} \in \mathbb{Z}_p^n$ *of* $F = 0$ *with* $\tilde{\mathfrak{a}} \equiv \mathfrak{a}$ *mod* p^{k+e}.

Proof For $m = 1$ a proof can be found in SERRE [SER], II, § 2, Theorem 1. The generalization to $m > 1$ can be managed by induction.

□

We will apply Proposition 5A.2.6 to the special unitary group $\mathbf{G}, \mathbf{G}(\mathbb{Z}) = \mathbb{S}U(\Phi, \mathcal{O}), \Phi = \text{diag}(1, 1, -1)$. The defining system of equations can be written formally as

$$F = \left({}^t\,\overline{X}\Phi X - X, \det X - 1\right) = 0,$$

where X denotes the matrix $(X_{ij} + \varepsilon Y_{ij})$ and F is constructed by means of the $\mathbb{Z}$ -basis $1, \varepsilon$ of $\mathcal{O}$. We refer to the construction of ρ in (5A.2.1). We consider F as function around E writing $X = E + H$, $H = (h_{ij} + \varepsilon h'_{ij})$.

Applying F to $E + H$ we obtain the matrix $M(d_E F)$ of the differential $d_E(F)$ looking at the degree 1 part in the TAYLOR series of X at E. It holds that

$$\begin{aligned} {}^t(\overline{E+H}) \cdot \Phi \cdot (E+H) - \Phi &= {}^t\overline{H}\Phi + \Phi H + O(2) \\ \det(E+H) - 1 &= \text{Tr}(H) + O(2), \end{aligned} \qquad (5A.2.5)$$

where $O(2)$ stands for the sum of monomials in h_{ij}, h'_{kl} of degree greater than 1. Explicitly we can write

$$F(E+H)=$$

$$\left(\begin{pmatrix} F_{11} & F_{12} & F_{13} \\ F_{21} & F_{22} & F_{23} \\ F_{31} & F_{32} & F_{33} \end{pmatrix} + \varepsilon \begin{pmatrix} F'_{11} & F'_{12} & F'_{13} \\ F'_{21} & F'_{22} & F'_{23} \\ F'_{31} & F'_{32} & F'_{33} \end{pmatrix}, D(h,h') + \varepsilon D'(h,h')\right),$$

$$F_{mn} = F_{mn}(h,h'),\ F'_{mn} = F'_{mn}(h,h'),\ (h,h') = \left(h_{ij}, h'_{kl}\right).$$

We compare with (5A.2.5) and omit the terms in h_{ij}, h'_{kl} of degree greater than 1. There appear 20 functions in 18 variables h_{ij}, h'_{kl}. Therefore, $M(d_E(F))$ is a matrix with 20 rows and 18 columns. It is easy to find the equivalent "diagonal form":

$$M(d_E(F)) \sim \begin{cases} \begin{pmatrix} 1 & & & & & & & & & \\ & 2 & & & & & & & & \\ & & 2 & & & & & & & \\ & & & 1 & & & & & & \\ & & & & \ddots & & & & & \\ & & & & & 1 & & & & \\ & & & & & & 0 & & & \\ & & & & & & & \ddots & & \\ & & & & & & & & 0 \end{pmatrix} & \text{if } d \equiv 1,2 \pmod 4 \\ \begin{pmatrix} 1 & & & & & \\ & \ddots & & & & \\ & & 1 & & & \\ & & & 0 & & \\ & & & & \ddots & \\ & & & & & 0 \end{pmatrix} & \text{if } d \equiv 3 \pmod 4 \end{cases}$$

Therefore

$$e_{p,E}\left({}^t\,\overline{\square}\Phi\square - \Phi, \det\Phi - 1\right) = \begin{cases} 1, & \text{if 2=p ramified in K} \\ 0 & \text{otherwise} \end{cases}. \tag{5A.2.6}$$

Proposition 5A.2.7 *Let* $K = \mathbb{Q}(\sqrt{-d}), \mathcal{O} = \mathcal{O}_K, \Phi = \operatorname{diag}(1,1,-1), \mathbf{G}$ *be the corresponding special unitary group scheme over* $\mathbb{Z}$ *. If* p *is odd or* $p=2$ *not ramified in* K*, then the sequences*

$$1 \longrightarrow G_p(k) \longrightarrow G_p \longrightarrow \mathbf{G}(\mathbb{Z}_p/p^k\mathbb{Z}_p) \longrightarrow 1$$

are exact for all $k \geq 1$. If $p = 2$ is ramified in K, then there are commutative diagrams (5A.2.7) with exact rows for all $k \geq 2$:

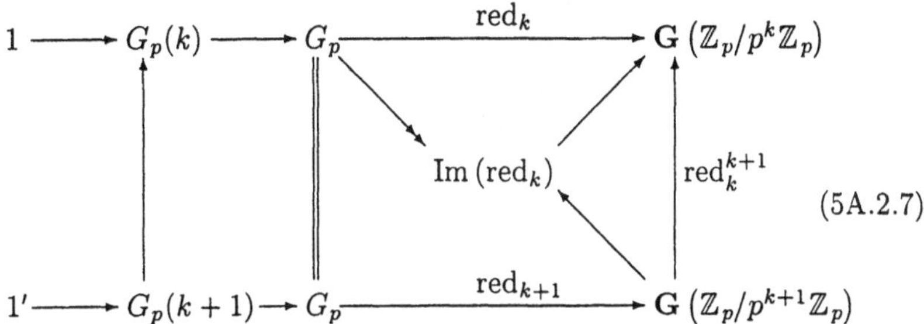

(5A.2.7)

Proof We know that the highest exponent of F at E is equal to $e = e_{p,E}(F) = 0$ or 1 according to (5A.2.6). Let $\overline{A}$ be an element of $\mathbf{G}(\mathbb{Z}/p^{k+2e}\mathbb{Z})$, A a preimage in $\mathbf{M}(\mathbb{Z}/p^{k+2e}\mathbb{Z})$. Then A is a solution of $F \equiv 0 \bmod p^{k+2e}$. There exists a solution $\tilde{A} \in \mathbf{G}(\mathbb{Z}_p) = G_p$ (of $F = 0$) with $\tilde{A} \equiv A \bmod p^{k+e}$ by Proposition 5A.2.6. Since $\text{red}_{k+e}\tilde{A} = \text{red}\,^{k+2e}_{k+e}\overline{A}$ we have the inclusion

$$\text{red}\,^{k+2e}_{k+e}\big(\mathbf{G}(\mathbb{Z}_p/p^{k+2e}\mathbb{Z}_p)\big) \subseteq \text{Im}\,(\text{red}_{k+e}).$$

If $e = 0$, then red^{k+2e}_{k+e} is the identity. It follows that $\mathbf{G}(\mathbb{Z}_p/p^k\mathbb{Z}_p) = \text{Im}\,(\text{red}_k)$. This proves the first part of 5A.2.7. If $e = 1$, then we obtain the commutative diagrams (D.2.20) for red^{k+2}_{k+1}, $k \geq 1$. This finishes the proof of Proposition 5A.2.7. □

Lemma 5A.2.8 *If $p = 2$ is ramified in K, then*

$$\ker\left(red_k^{k+1}\right) = 2^{10} \textit{ for all } k \geq 1.$$

Proof We work with

$$\mathbf{G}(\mathbb{Z}_p/p^{k+1}\mathbb{Z}_p) \cong \mathbb{S}U(\Phi, \mathcal{O}/p^{k+1}\mathcal{O})$$

and write $A \in \ker(red_k^{k+1})$ as $A = E + p^kH$, E the unit matrix. It follows that ${}^t\,\overline{H}\Phi + \Phi H \equiv 0 \bmod p$, $\text{Tr}(H) \equiv 0 \bmod p$. We have to count matrices $H \in Mat_3(\mathcal{O}/p\mathcal{O})$ with this property. This is an easy exercise. One finds exactly 2^{10} solutions.

□

Proposition 5A.2.9 *For the* PICARD *modular group* $\mathbf{G}(\mathbb{Z}) = \mathbb{S}U(\Phi, \mathcal{O})$ *of the field* $K = \mathbb{Q}(\sqrt{-d})$ *with discriminant* $D = D_{K/\mathbb{Q}}$ *it holds that*

$$[G_p : G_p(k)] = p^{8k}\left(1 - \left(\frac{D}{p}\right)p^{-3}\right)(1 - p^{-2})\delta_p, \ \delta_p := \begin{cases} \frac{1}{4}, & p = 2|D \\ 1, & \textit{otherwise} \end{cases} \tag{5A.2.8}$$

for

$$k \geq \begin{cases} 2, & \textit{if } p = 2|D \\ 1, & \textit{otherwise} \end{cases}.$$

Proof With the convention $\left(\frac{D}{p}\right) = 0$ for p ramified in K we apply (5A.1.2). For $k \geq 1$, p odd or $2 = p \not| D$ the formula (5A.2.14) then follows immediately from the first part of Prop. 5A.2.7.

Now let $2 = p|D$ and $k \geq 2$. Then we use diagram (5A.2.7), Lemma 5A.2.8 and (5A.1.2) to get

$$\begin{aligned} [G_p : G_p(k)] &= \#\mathrm{Im}\ \mathrm{red}_k \\ &= \left[\mathbf{G}(\mathbb{Z}_p/p^{k+1}\mathbb{Z}_p) : \mathrm{Ker}\ \mathrm{red}_k^{k+1}\right] \\ &= \#\mathbb{S}U(\Phi, \mathcal{O}/p^{k+1}\mathcal{O})/\#\mathrm{Ker}\ \mathrm{red}_k^{k+1} \\ &= p^{8(k+1)}(1-p^{-2})\left(1 - \left(\frac{D}{p}\right)p^{-3}\right)/2^{10} \\ &= \frac{1}{4}p^{8k}(1-p^{-2})\left(1 - \left(\frac{D}{p}\right)p^{-3}\right). \end{aligned} \tag{5A.2.9}$$

□

Definition 5A.2.10 Let $\mathbf{G}$ be an algebraic group defined over $\mathbb{Q}$. We say that $\mathbf{G}$ has (absolutely) *strong approximation* if the following two equivalent conditions are satisfied:

(i)

$$\mathbf{G}(\mathbb{A}) = \overline{G_\infty \cdot \mathbf{G}(\mathbb{Q})},$$

where $\mathbf{G}(\mathbb{A})$ is the adele group, $\mathbb{A} = \mathbb{A}_\mathbb{Q}$ the adele ring of $\mathbb{Q}$, $G_\infty = \mathbf{G}(\mathbb{R})$ and $^-$ denotes the topological closure in $\mathbf{G}(\mathbb{A})$.

(ii) For given primes p_i, natural numbers m_i and elements $A_i \in \mathbf{G}(\mathbb{Q}_{p_i})$, $i = 1, \dots, k$, there exists a global element $A \in \mathbf{G}(\mathbb{Q})$ such that $A \equiv A_i \bmod p_i^{m_i}, i = 1, \dots, k$ and $A \in \mathbf{G}(\mathbb{Z}_p)$ for all primes $p \neq p_1, \dots, p_k$.

An immediate consequence is the following

Lemma 5A.2.11 *Let* $\mathbf{G}$ *be an algebraic group defined over* $\mathbb{Z}$ *with strong approximation,* m *a natural number with prime decomposition* $m = p_1^{m_1} \cdot \ldots \cdot p_k^{m_k}$. *Then the sequence*

$$1 \to \mathbf{G}(\mathbb{Z})(m) \to \mathbf{G}(\mathbb{Z}) \underset{q}{\to} \prod_{i=1}^{k} G_{p_i}/G_{p_i}(m_i) \to 1$$

is exact.

Proof Obviously, the congruence subgroup $\mathbf{G}(\mathbb{Z})(m)$ is the kernel of q. We have to show that q is surjective. Let $A_i \in \mathbf{G}(\mathbb{Z}_{p_i}) = G_{p_i}$, $i = 1, \dots, k$. There exists $A \in \mathbf{G}(\mathbb{Q})$ such that $A \equiv A_i \bmod p_i^{m_i}, A \in \mathbf{G}(\mathbb{Z}_p), p \neq p_i$, for $i = 1, \dots, k$ by (ii). Then $A \in \mathbf{G}(\mathbb{Z}_p)$ for all primes p, hence $A \in \mathbf{G}(\mathbb{Z})$. Moreover,

$$E = A^{-1}A \equiv A^{-1}A_i \bmod p^{m_i}, i = 1, \dots, k,$$

hence $A^{-1}A_i \in G_{p_i}(m_i)$ and

$$q(A) = (A_1, \dots, A_k) \bmod^{\times} \prod_{i=1}^{k} G_{p_i}(m_i).$$

□

Theorem 5A.2.12 PLATONOV *([PLA], Theorem A) Let* $\mathbf{G}$ *be a simple simply-connected algebraic group defined over* $\mathbb{Q}$ *such that* $G_\infty = \mathbf{G}(\mathbb{R})$ *is not compact. Then* $\mathbf{G}$ *has strong approximation.*

□

Corollary 5A.2.13 *The special unitary group* $\mathbf{G}$ *with* $\mathbf{G}(\mathbb{Z}) = \mathbb{S}U(\Phi, \mathcal{O})$, $\mathcal{O} = \mathcal{O}_K, K = \mathbb{Q}(\sqrt{-d})$ *we work with, has strong approximation.*

Proof Indeed, $\mathbf{G}$ is simple (see HELGASON [HEL], IX, § 4, Lemma 4.4), simply-connected according to KNESER [KNE], Section 4, and defined over $\mathbb{Q}$. Moreover, $G_\infty = \mathbf{G}(\mathbb{R}) = \mathbb{S}U(\Phi, \mathbb{C}) = \mathbb{S}U((2,1), \mathbb{C})$ is not compact.

□

Proposition 5A.2.14 *For the* PICARD *modular group* $\mathbf{G}(\mathbb{Z})$ *of the field* $K = \mathbb{Q}(\sqrt{-d})$ *with discriminant* $D = D_{K/\mathbb{Q}}$ *and for natural numbers* $m > 1$ *with prime decomposition*

$$m = p_1^{m_1} \cdot \ldots \cdot p_k^{m_k}, p_1 < \ldots < p_k,$$

$m_1 \neq 1$ *if* 2 *is ramified in* K, *the index of the* m-*th congruence subgroup in* $\mathbf{G}(\mathbb{Z}) = \mathbb{S}U(\Phi, \mathcal{O}_K)$ *is*

$$[\mathbf{G}(\mathbb{Z}) : \mathbf{G}(\mathbb{Z})(m)] = \delta \cdot m^8 \cdot \prod_{i=1}^{k} (1 - p_i^{-2}) \left(1 - \left(\frac{D}{p_i}\right) p_i^{-3}\right),$$

$$\text{where } \delta = \begin{cases} 1/4, & \text{if } p_1 = 2 \text{ is ramified in } K \\ 1, & \text{otherwise} \end{cases}$$

Proof By Corollary 5A.2.13, the algebraic group $\mathbf{G}$ has strong approximation. It is also defined over $\mathbb{Z}$. Therefore, we can apply Lemma 5A.2.10. The exact sequence there and the local index formulas in Proposition 5A.2.9 yield the result.

□

5A.3 Local Volumina

We first define a $\mathbb{Q}$-rational G-invariant differential form on $G = \mathbf{G}_\infty = \mathbf{G}(\mathbb{R}) = \mathbb{S}U((2,1), \mathbb{C}) = \mathbb{S}U(\Phi, \mathbb{C})$. Let $\mathbf{G}$ be an arbitrary algebraic group defined over $\mathbb{Q}$. A $\mathbb{Q}$-rational differential form ω of higest degree on G is a differential form which can be written locally on open affine subschemes U of $\mathbf{G}$ as

$$\omega_U = g df_1 \wedge \ldots \wedge df_n,$$

where g, f_i belong to the coordinate ring $\mathbb{Q}[U]$. One can choose $f_1, \ldots, f_n$ as a system of local parameters at a point $x_o \in U$. For a G-invariant differential form of the above type it suffices to know ω at the unit point $e = 1_G \in \mathbf{G}(\mathbb{Q})$. There ω appears as linear map $\omega_e \in \Lambda^n T_e^\vee$, where T_e denotes the tangent space (with $\mathbb{Q}$-structure) in e and $T_e^\vee$ its dual (see e.g. [SC2], III, § 2.2). Up to a multiplicative rational constant there exists exactly one G-invariant non-trivial $\mathbb{Q}$-rational differential form of highest degree on G.

We turn back to our special unitary group. The LIE algebra $\mathfrak{g}(\mathbb{R})$ of G is defined by

$$^t\overline{X}\Phi + \Phi X = 0, \ \mathrm{Tr}(X) = 0.$$

We fix $K = \mathbb{Q}(\sqrt{-d}), \mathcal{O} = \mathcal{O}_K = \mathbb{Z}\cdot 1 + \mathbb{Z}\varepsilon$ as in section 5A.2, and consider

$$\mathfrak{g}(\mathbb{Q}) = \mathfrak{g}(\mathbb{R}) \cap \mathrm{Mat}_3(K), \ \mathfrak{g}(\mathbb{Z}) = \mathfrak{g}(\mathbb{R}) \cap \mathrm{Mat}_3(\mathcal{O}).$$

For the sake of simple notations we prefer to use $\mathrm{Mat}_3(\mathbb{C})$ as ambient space, but it is more precise to use the embedding (5A.2.1). We will keep it in mind. The $\mathbb{Z}$-lattice $\mathfrak{g}(\mathbb{Z})$ in $\mathfrak{g} = \mathfrak{g}(\mathbb{R})$ is generated by the elements

$$H_1 = \begin{pmatrix} \sqrt{-d} & 0 & 0 \\ 0 & -\sqrt{-d} & 0 \\ 0 & 0 & 0 \end{pmatrix}, \quad H_2 = \begin{pmatrix} 0 & 0 & 0 \\ 0 & \sqrt{-d} & 0 \\ 0 & 0 & -\sqrt{-d} \end{pmatrix},$$

$$H_3 = \begin{pmatrix} 0 & 1 & 0 \\ -1 & 0 & 0 \\ 0 & 0 & 0 \end{pmatrix}, \qquad H_4 = \begin{pmatrix} 0 & \varepsilon & 0 \\ -\overline{\varepsilon} & 0 & 0 \\ 0 & 0 & 0 \end{pmatrix}$$

$$M_1 = \begin{pmatrix} 0 & 0 & 1 \\ 0 & 0 & 0 \\ 1 & 0 & 0 \end{pmatrix}, \qquad M_2 = \begin{pmatrix} 0 & 0 & \varepsilon \\ 0 & 0 & 0 \\ \overline{\varepsilon} & 0 & 0 \end{pmatrix}, \tag{5A.3.1}$$

$$M_3 = \begin{pmatrix} 0 & 0 & 0 \\ 0 & 0 & 1 \\ 0 & 1 & 0 \end{pmatrix}, \qquad M_4 = \begin{pmatrix} 0 & 0 & 0 \\ 0 & 0 & \varepsilon \\ 0 & \overline{\varepsilon} & 0 \end{pmatrix}.$$

We consider the CARTAN decomposition

$$\mathfrak{g} = \mathfrak{k} + \mathfrak{m}, \ \mathfrak{k} = \sum_{i=1}^{4} \mathbb{R}H_i, \ \mathfrak{m} = \sum_{i=1}^{4} \mathbb{R}M_i.$$

Here $\mathfrak{k}$ is the LIE algebra of the maximal compact subgroup $\mathfrak{K} = \mathrm{Fix}_G(O) = S(\mathbb{U}(2) \times \mathbb{U}(1))$, and $\mathfrak{m}$ can be regarded as tangent space $T_O(\mathbb{B})$ at $O \in \mathbb{B}, \mathbb{B}$ the ball $\mathbb{B} = \mathfrak{K} \setminus G, O = \mathfrak{K} \setminus \mathfrak{K}$.

The $\mathbb{Q}$-rational differential form ω we look for is uniquely determined by the alternating multilinear map

$$\omega_e : \mathfrak{g}(\mathbb{Q}) \longrightarrow \mathbb{Q}, \ \omega_e(H_1, H_2, H_3, H_4, M_1, M_2, M_3, M_4) = 1. \tag{5A.3.2}$$

We will calculate local volumina with respect to this rational volume form ω. For this purpose we fix a prime p. For $A = (a_{ij}) \in \mathrm{Mat}_m(\mathbb{Z}_p)$ we define

$$v_p(A) = \min_{i,j} v_p(a_{ij}),$$

where $v_p : \mathbb{Q}_p \longrightarrow \mathbb{Z} \cup \{\infty\}, p \mapsto 1$, is the standard valuation. It is easy to check that

$$v_p(A \cdot B) \geq v_p(A) + v_p(B), A, B \in \mathrm{Mat}_m(\mathbb{Z}_p). \tag{5A.3.3}$$

Let

$$G_p(k) = \mathbf{G}(\mathbb{Z}_p)(k) = \left\{X \in \mathbf{G}(\mathbb{Z}_p); X \equiv E \bmod p^k\right\}$$

be the k-th local congruence subgroup and

$$\mathfrak{g}_p(k) = \mathfrak{g}(\mathbb{Z}_p)(k) = \{A \in \mathfrak{g}(\mathbb{Z}_p); v_p(A) \geq k\}.$$

Lemma 5A.3.1 *For $p \neq 2$, $k \geq 1$ or $p = 2$, $k \geq 2$ the exponential map*

$$\begin{array}{rcl} \exp : \mathfrak{g}_p(k) & \longrightarrow & G_p(k) \\ A & \mapsto & \sum\limits_{n=0}^{\infty} \frac{1}{n!} A^n \end{array}$$

is well-defined and bijective.

Proof The convergence of the series $\sum_{n=0} \frac{1}{n!} A^n$ follows from the elementary estimation

$$v_p(n!) \leq \frac{n-1}{p-1}, \quad n = 1, 2, 3, \ldots \tag{5A.3.4}$$

and from (5A.3.3) in the following manner:

$$v_p(A^n/n!) \geq v_p(A^n) - v_p(n!) \geq n v_p(A) - \frac{n-1}{p-1} > n\left(v_p(A) - \frac{1}{p-1}\right).$$

The inverse map is

$$\begin{aligned} X \mapsto lnX &= ln(E - (E - X)) \\ &= (E - X) + \frac{1}{2}(E - X)^2 + \frac{1}{3}(E - X)^3 + \ldots \end{aligned}$$

This series is convergent because

$$v_p((E - X)^n/n) \geq n v_p(E - X) - v_p(n) = n(v_p(E - X) - v_p(n)/n),$$

$$v_p(E - X) \geq k \geq 1 \text{ and } v_p(n)/n \leq \log_p(n)/n \underset{n\to\infty}{\longrightarrow} 0.$$

□

On the additive group of $\mathbb{Q}_p$ we consider the HAAR measure μ_p normalized by $\mu_p(\mathbb{Z}_p) = 1$. It holds that $\mu_p(p^k\mathbb{Z}_p) = 1/p^k$. The symbol $dt_1 dt_2 \cdots dt_n$ denotes the product measure $\mu_p \times \cdots \times \mu_p$ on $\mathbb{Q}_p^n$. It follows that

$$\int\limits_{p^{k_1}\mathbb{Z}_p\times\cdots\times p^{k_n}\mathbb{Z}_p} dt_1 dt_2 \cdots dt_n = \prod_{i=1}^{n} p^{-k_i}$$

with obvious notations. For $\mathfrak{g}(\mathbb{Q}_p)$ as above and the identification $\mathfrak{g}(\mathbb{Z}_p) = \mathbb{Z}_p^8$ it holds that

$$\int\limits_{\mathfrak{g}_p(k)} dt_1 \cdots dt_8 = p^{-8k} \tag{5A.3.5}$$

We localize now the G-invariant rational measure ω defined in (5A.3.2) to an invariant measure $\mathbf{G}(\mathbb{Q}_p)$ in the following manner. The exponential map *exp* in 5A.3.1 corresponds homeomorphically neighbourhoods V of $0 \in \mathbb{Q}_p^8$ to neighbourhoods U of the unit element $E \in G_p$:

$$\mathbb{Q}_p^8 \supset \mathfrak{g}_p(k) = V = V(k) \longrightarrow U = U(k) = G_p(k) \subset G_p.$$

The local measure ω_p, $\omega = f(x)dx_1 \cdots dx_n$, is defined on U by pull back:

$$\omega_p = |f(t)|_p dt_1 \cdots dt_n, (n = 8), \tag{5A.3.6}$$

where $|a|_p = p^{-v_p(a)}$for $a \in \mathbb{Q}_p$. We have to calculate

$$\omega_p(G_p(k)) = \int\limits_{G_p(k)} \omega_p = \int\limits_{G_p(k)} |f(t)|_p dt_1 \cdots dt_n.$$

It is necessary to determine $|f(t)|_p$ in (5A.3.6). We remember that ω has been uniquely determined by G-invariance and $\omega_E : \mathfrak{g} \longrightarrow \mathbb{R}$. For other points $A \in G$ one finds that

$$\omega_A = \left((dL_A)_E^{-1}\right)^* \omega_E$$

by shifting $L_A : G \xrightarrow{\sim} G,\ X \mapsto A \cdot X$. If $A = \exp T$ with the exponential map $\exp : \mathfrak{g} \longrightarrow G$, then

$$(d\exp)_T = (dL_A)_E \circ \frac{1-e^{-adT}}{adT}$$

(see HELGASON [HEL], II, § 1, Theorem 1.7). It follows that

$$\begin{aligned}(d\exp)_T^* \omega_A &= \left(\frac{1-e^{-adT}}{adT}\right)^* \omega_A \\ &= \left(\frac{1-e^{-adT}}{adT}\right)^* (dL_A)_E^* \left((dL_A)_E^{-1}\right)^* \omega_E \\ &= \left(\frac{1-e^{-adT}}{adT}\right)^* \omega_E\end{aligned}$$

This formula is valid also for our exponential function
$\exp : \mathfrak{g}_p(k) \longrightarrow G_p(k)$. Regarding T as variable we get

$$\int\limits_{G_p(k)} \omega_p = \int\limits_{T\in\mathfrak{g}_p(k)} \left|\det\left(\frac{1-e^{-adT}}{adT}\right)\right|_p dt_1\cdots dt_8,$$

where

$$\frac{1-e^{-adT}}{adT} = id - \frac{1}{2}(adT) + \frac{1}{3}(adT)^2 - + \cdots$$

and

$$(adT)^n(X) = T^nX - (n-1)T^{n-1}XT \ldots + (-1)^n XT^n, X \in \mathfrak{g}(\mathbb{Z}_p).$$

From $v_p(T) \geq k$ and (5A.3.3) follows $v_p((adT)^nX) \geq nk$, and (5A.3.4) implies

$$v_p\left(\frac{1}{n!}(adT)^{n-1}X\right) \geq (n-1)\left(k - \frac{1}{p-1}\right) > 0, \; n > 1.$$

Therefore it holds that

$$\left|\det\left(\frac{1-e^{-adT}}{adT}\right)\right|_p = |\det(id)|_p = 1$$

and via (5A.3.6)

$$\int\limits_{G_p(k)} \omega_p = \int\limits_{\mathfrak{g}_p(k)} dt_1\cdots dt_8.$$

Together with (5A.3.5) we get

$$\omega_p(G_p(k)) = \int\limits_{G_p(k)} \omega_p = p^{-8k} \tag{5A.3.7}$$

Proposition 5A.3.2 *Let ω be the $\mathbb{Q}$ -rational differential form on $G = \mathbf{G}(\mathbb{R}) = \mathbb{S}U(\Phi, \mathbb{C})$ defined in (5A.3.2) and ω_p the corresponding local measure on $\mathbf{G}(\mathbb{Q}_p)$ defined in (5A.3.6). Then it holds that*

$$\omega_p(\mathbf{G}(\mathbb{Z}_p)) = \delta \cdot \left(1 - \left(\frac{D}{p}\right)p^{-3}\right)(1 - p^{-2}).$$

where $D = D_{K/\mathbb{Q}}$ is the discriminant of $K = \mathbb{Q}(\sqrt{-d})$ and

$$\delta = \begin{cases} 1/4, & \text{if } p = 2 \text{ ramified in } K \\ 1, & \text{otherwise} \end{cases}$$

Proof The formula (5A.3.7) is valid for $k \geq 1$, $p \neq 2$ and also for $k \geq 2$, $p = 2$ (see Lemma 5A.3.1). The local measure ω_p is $\mathbf{G}(\mathbb{Q}_p)$-invariant by definition. Therefore it suffices to know the index $[\mathbf{G}(\mathbb{Z}_p) : G_p(k)]$ for the calculation of $\omega_p(\mathbf{G}(\mathbb{Z}_p))$. The formula follows now from Proposition 5A.2.9. □

5A.4 The Global Volume

We need a short introduction to the theory of TAMAGAWA measures. It can be found in KNESER'S article [KNE] in [C-F], Chap. X. Let $\mathbf{G}$ be a connected linear algebraic group defined over $\mathbb{Q}$, ω a $\mathbb{Q}$-rational G-invariant differential form of highest degree on G. A local HAAR measure on $\mathbf{G}(\mathbb{Q}_p)$ is defined by $\mathbb{Q}_p$ -analytic linearization as described in section 5A.3. Analogously, ω defines a HAAR measure ω_∞ on $G_\infty = \mathbf{G}(\mathbb{R}) = G$. Via linearization the measures ω_p, ω_∞ have been connected with the additive HAAR measures μ_p, μ_∞ on $\mathbb{Q}_p$ or $\mathbb{Q}_\infty = \mathbb{R}$ normalized by $\mu_p(\mathbb{Z}_p) = 1$ or $\mu_\infty([0,1]) = 1$, respectively. If $\mathbf{G}$ is semisimple, then the product $\prod \omega_p(\mathbf{G}(\mathbb{Z}_p))$ is absolutely convergent. Let $\mathbb{A} = \mathbb{A}_\mathbb{Q}$ be the adele ring of $\mathbb{Q}$ and $\mathbf{G}(\mathbb{A})$ the adele group of $\mathbf{G}$. The topological group $\mathbf{G}(\mathbb{A})$ is locally compact. The $\mathbb{Q}$-rational differential form ω defines in a unique manner a HAAR measure $\omega_\mathbb{A}$on $\mathbf{G}(\mathbb{A})$ satisfying

$$\omega_\mathbb{A}\left(\prod_{v \in S} U_v \times \prod_{v \in / S} \mathbf{G}(\mathbb{Z}_p)\right) = \prod_{v \in S} \omega_v(U_v) \cdot \prod_{v \notin S} \omega_v(\mathbf{G}(\mathbb{Z}_v)) \tag{5A.4.1}$$

for all finite subsets $S \subset \{\infty\} \cup \operatorname{Spec} \mathbb{Z}$ of valuations of $\mathbb{Q}$ and all open U_v in $\mathbf{G}(\mathbb{Q}_v)$ with compact closure in $\mathbf{G}(\mathbb{Q}_v)$.

Definition 5A.4.1 The HAAR measure $\omega_{\mathbb{A}}$ introduced above is called the TAMAGAWA *measure* on $\mathbf{G}(\mathbb{A})$.

The group $\mathbf{G}(\mathbb{Q})$ is considered as subgroup of $\mathbf{G}(\mathbb{A}) \subset \prod_v \mathbf{G}(\mathbb{Q}_v)$ via diagonal embedding. It is a discrete subgroup of $\mathbf{G}(\mathbb{A})$ with respect to the factor topology. If $\mathbf{G}$ is semisimple, then the volume $\omega_{\mathbb{A}}(\mathbf{G}(\mathbb{A})/\mathbf{G}(\mathbb{Q}))$ of a $\mathbf{G}(\mathbb{Q})$-fundamental domain is finite.

Definition 5A.4.2 If $\mathbf{G}$ is semisimple, then

$$\tau(\mathbf{G}) = \omega_{\mathbb{A}}(\mathbf{G}(\mathbb{A})/\mathbf{G}(\mathbb{Q}))$$

is called the TAMAGAWA *number* of $\mathbf{G}$.

The TAMAGAWA number $\tau(\mathbf{G})$ does not depend on the choice of ω.

Lemma 5A.4.3 *Let* $\mathbf{G}$ *be a semisimple linear algebraic group defined over* $\mathbb{Z}$ *with strong approximation (see 5A.2.10). Then*

$$\omega(\mathbf{G}(\mathbb{R})/\mathbf{G}(\mathbb{Z})) = \omega_\infty(\mathbf{G}(\mathbb{R})/\mathbf{G}(\mathbb{Z})) = \prod_p \omega_p(\mathbf{G}(\mathbb{Z}_p))^{-1} \cdot \tau(\mathbf{G}).$$

Proof Let $\tilde{\mathbb{A}} = \mathbb{R}\times\prod_p \mathbb{Z}_p \subset \mathbb{A}$ be the subring of integral adeles of $\mathbb{A} = \mathbb{A}_{\mathbb{Q}}$. We need the property (i) of Definition 5A.2.10: $G_\infty \cdot \mathbf{G}(\mathbb{Q})$ is dense in $\mathbf{G}(\mathbb{A})$. It follows that $\mathbf{G}\left(\tilde{\mathbb{A}}\right)\cdot\mathbf{G}(\mathbb{Q})$ is dense in $\mathbf{G}(\mathbb{A})$, hence $\mathbf{G}\left(\tilde{\mathbb{A}}\right)\mathbf{G}(\mathbb{Q})/\mathbf{G}(\mathbb{Q})$ is dense in $\mathbf{G}(\mathbb{A})/\mathbf{G}(\mathbb{Q})$. On the other hand $\mathbf{G}\left(\tilde{\mathbb{A}}\right)$ is a closed subset of $\mathbf{G}(\mathbb{A})$. Therefore $\mathbf{G}\left(\tilde{\mathbb{A}}\right)\cdot\mathbf{G}(\mathbb{Q})/\mathbf{G}(\mathbb{Q})$ is closed in $\mathbf{G}(\mathbb{A})/\mathbf{G}(\mathbb{Q})$. Altogether we get

$$\mathbf{G}\left(\tilde{\mathbb{A}}\right)\cdot\mathbf{G}(\mathbb{Q})/\mathbf{G}(\mathbb{Q}) = \mathbf{G}(\mathbb{A})/\mathbf{G}(\mathbb{Q}).$$

Furthermore, we have $\mathbf{G}\left(\tilde{\mathbb{A}}\right)\cap\mathbf{G}(\mathbb{Q}) = \mathbf{G}(\mathbb{Z})$. It follows that

$$\begin{aligned}\mathbf{G}\left(\tilde{\mathbb{A}}\right)/\mathbf{G}(\mathbb{Z}) &= \mathbf{G}\left(\tilde{\mathbb{A}}\right)/\mathbf{G}\left(\tilde{\mathbb{A}}\right)\cap\mathbf{G}(\mathbb{Q})\\ &= \mathbf{G}\left(\tilde{\mathbb{A}}\right)\cdot\mathbf{G}(\mathbb{Q})/\mathbf{G}(\mathbb{Q}) = \mathbf{G}(\mathbb{A})/\mathbf{G}(\mathbb{Q}).\end{aligned}$$

The multiplicative action of $\mathbf{G}(\mathbb{Z})$ on $\mathbf{G}\left(\tilde{\mathbb{A}}\right) = G_\infty\cdot\prod_p \mathbf{G}(\mathbb{Z}_p)$ is obviously effective. Therefore we can identify $(\mathbf{G}(\mathbb{R})/\mathbf{G}(\mathbb{Z}))\cdot\prod_p \mathbf{G}(\mathbb{Z}_p)$ with $\mathbf{G}\left(\tilde{\mathbb{A}}\right)/\mathbf{G}(\mathbb{Z})$ sending elements $\left(A \bmod^\times \mathbf{G}(\mathbb{Z})\right)\times\prod_p A_p$ to $\left(A\times\prod_p A_p\right)\bmod^\times\mathbf{G}(\mathbb{Z})$. Thus

$$\omega_{\mathbf{A}}\left((\mathbf{G}(\mathbb{R})/\mathbf{G}(\mathbb{Z}))\times\prod_p \mathbf{G}(\mathbb{Z}_p)\right)=\omega_{\mathbf{A}}\left(\mathbf{G}\left(\tilde{\mathbb{A}}\right)/\mathbf{G}(\mathbb{Z})\right)$$
$$=\omega_{\mathbf{A}}(\mathbf{G}(\mathbb{A})/\mathbf{G}(\mathbb{Q}))=\tau(\mathbf{G}).$$

We calculate the left-hand side componentwise. This is possible thanks to the approximative definition (5A.4.1). One has to understand $\omega_\infty(G/\mathbf{G}(\mathbb{Z}))$ as volume of a fundamental domain U in G with respect to the lattice subgroup $\mathbf{G}(\mathbb{Z})$ and U as union of a chain $U_1 \subset U_2 \subset U_3 \subset \cdots$ of subsets with compact closure in G. The decomposition

$$\tau(\mathbf{G})=\omega_{\mathbf{A}}\left((\mathbf{G}(\mathbb{R})/\mathbf{G}(\mathbb{Z}))\times\prod_p \mathbf{G}(\mathbb{Z}_p)\right)$$
$$=\omega_\infty(G/\mathbf{G}(\mathbb{Z}))\prod_p \omega_p(\mathbf{G}(\mathbb{Z}_p)) \tag{5A.4.2}$$

can be deduced now immediately from (5A.4.1). □

Theorem 5A.4.4 (Lai, [RAP], Theorem 6.6.1, see also [KOT]) *Let* $\mathbf{G}$ *be a*
semisimple, simply-connected, quasi-split, algebraic group over $\mathbb{Q}$. *Then the* TAMAGAWA *number* $\tau(\mathbf{G})$ *is equal to 1.*

□

Corollary 5A.4.5 *The* TAMAGAWA *number of the special unitary group* $\mathbf{G}$ *over* $\mathbb{Z}$ *defined in section 5A.2 corresponding to the imaginary quadratic number field* $K=\mathbb{Q}(\sqrt{-d})$ *is equal to 1.*

Proof The group $\mathbf{G}$(over $\mathbb{Q}$) is simple and simply-connected as already has been remarked after Corollary 5A.2.13. An algebraic group $\mathbf{G}$ over a field F is called *quasi-split* (over F), if there exists a BOREL subgroup $\mathbf{B}$ of $\mathbf{G}$ defined over F. We want to check that our groups $\mathbf{G}$ are quasi-split. For this purpose we let $G=\mathbb{S}U((2,1),\mathbb{C})$ act on the complex unit ball

$$\mathbb{B}=\left\{(z_1,z_2)\in\mathbb{C}^2;|z_1|^2+|z_2|^2<1\right\}\subset\mathbb{P}^2=\mathbb{P}^2(\mathbb{C}).$$

It is easy to see that the isotropy group of the boundary point (1,0) of $\mathbb{B}$ is a BOREL subgroup defined over $\mathbb{Q}$, namely $\mathbf{B}$ is a minimal parabolic subgroup of $\mathbf{G}$. Now we can apply Theorem 5A.4.4.

□

Theorem 5A.4.6 *Let $\mathbf{G}/\mathbb{Z}$ be the special unitary group of signature (2, 1) corresponding to $K = \mathbb{Q}(\sqrt{-d})$ with discriminant $D = D_{K/\mathbb{Q}}$ and ω the G-invariant $\mathbb{Q}$-rational differential form defined in (5A.3.2). Then the volume of a $\mathbf{G}(\mathbb{Z})$-fundamental domain in $G = \mathbf{G}(\mathbb{R})$ is equal to*

$$\omega(G/\mathbf{G}(\mathbb{Z})) = \int\limits_{G/\mathbf{G}(\mathbb{Z})} \omega = \delta^{-1} \cdot \zeta(2) \cdot L(3, \chi),$$

where

$$\delta = \begin{cases} 1/4, & \text{if 2 is ramified in } K \\ 1, & \text{otherwise,} \end{cases}$$

$\zeta(2)$ is the value of RIEMANN'S *Zeta function*

$$\zeta(s) = \sum_{n=1}^{\infty} 1/n^s$$

at $s = 2$ and $L(3, \chi)$ is the value at $s = 3$ of the L-series with DIRICHLET *character $\chi(n) = \left(\frac{D}{n}\right)$* (JACOBI *symbol):*

$$L(s, \chi) = \sum_{n=1}^{\infty} \chi(n)/n^s.$$

Proof Corollary 5A.4.5 and formula (5A.4.2) yield

$$1 = \tau(\mathbf{G}) = \omega_\infty(G/\mathbf{G}(\mathbb{Z})) \cdot \prod_p \omega_p(\mathbf{G}(\mathbb{Z}_p)).$$

The factors of the infinite product have been already calculated, see Proposition 5A.3.2. It follows that

$$\begin{aligned} \omega(G/\mathbf{G}(\mathbb{Z})) &= \delta^{-1} \cdot \prod_p \left(1 - p^2\right)^{-1} \prod_p \left(1 - \left(\frac{D}{p}\right) p^{-3}\right)^{-1} \\ &= \delta^{-1} \cdot \zeta(2) \cdot L(3, \chi). \end{aligned}$$

□

Main Theorem 5A.4.7 *Let $\Gamma = \mathbb{S}U((2,1), \mathcal{O}_K) = \mathbf{G}(\mathbb{Z})$ be the* PICARD *modular group of the field K acting on the complex 2-ball $\mathbb{B}$, $\mathbf{G}$ as in Theorem 5A.4.6. The volume of a fundamental domain with respect to Γ and the* EULER-CHERN *form c_2 of the G-invariant* BERGMANN *metric on $\mathbb{B}$ is equal to*

$$c_2(\mathbb{B}/\Gamma) = \varepsilon \cdot \frac{3|D|^{5/2}}{16\pi^5}\zeta(2) \cdot L(3,\chi) = \varepsilon \cdot \frac{3|D|^{5/2}}{32\pi^3}L(3,\chi) \tag{5A.4.3}$$

$$\textit{with } \varepsilon = \begin{cases} 3, & \textit{if } K = \mathbb{Q}(\sqrt{-3}) \\ 1, & \textit{otherwise} \end{cases}$$

Proof The verification of (5A.4.3) is subdivided in four steps.
Step 1. For different fields $K = \mathbb{Q}(\sqrt{-d}), d$ a squarefree natural number, we constructed different volume forms $\omega^{(d)}$ on $G = \mathbb{S}U((2,1),\mathbb{C})$ according to (5A.3.2). We set $\omega = \omega^{(1)}$ amd prove that

$$\omega^{(d)} = \frac{8}{\delta|D|^{5/2}}\omega^{(1)}, \ \delta = \begin{cases} 1/4, & 2 \text{ ramified in } K \\ 1, & \text{otherwise} \end{cases} \tag{5A.4.4}$$

Step. 2. Split $\omega = \omega^{(1)}$ with respect to a maximal compact subgroup $\mathfrak{K}$ of G into

$$\omega = \kappa \cdot \lambda, \tag{5A.4.5}$$

κ a $\mathfrak{K}$ -invariant form on $\mathfrak{K} = S(\mathbb{U}(2) \times \mathbb{U}(1)), \lambda$ a G-invariant form on the ball $\mathbb{B} = \mathfrak{K} \setminus G$.
Step 3. Express λ as multiple of the EULER-CHERN form on $\mathbb{B}$:

$$\lambda = \frac{\pi^2}{6}c_2 \tag{5A.4.6}$$

Step 4. Calculate the volume on $\mathfrak{K}$:

$$\int_{\mathfrak{K}} \kappa = 4\pi^3 \tag{5A.4.7}$$

Knowing the results of the four steps, which will be proved below, we obtain the formula (5A.4.3) of the Main Theorem in the following manner. Let Γ_o be a neat normal subgroup of Γ. Such subgroups exist by a theorem of BOREL [BO1]; more immediately one can take a suitable congruence subgroup. The group Γ_o acts freely on $\mathbb{B}$. If $\mathfrak{F}$ is a Γ_o-fundamental domain in $\mathbb{B} = \mathfrak{K} \setminus G$, then $\mathfrak{F}$ is understood as a set of $\mathfrak{K}$ -cosets in G. The union of these cosets is a Γ_o-fundamental domain in G. According to (5A.4.5) we split the integral calulation (see [HEL], X, § 1, Theorem 1.7). Together with (5A.4.4) one gets

$$\int\limits_{\mathfrak{K}} \kappa \cdot \int\limits_{\mathbb{B}/\Gamma_o} \lambda = \int\limits_{G/\Gamma_o} \kappa \cdot \lambda = \int\limits_{G/\Gamma_o} \omega = \omega(G/\Gamma_o) = \frac{\delta|D|^{5/2}}{8} \omega^{(d)}(G/\Gamma_o), \tag{5A.4.8}$$

where $\mathbb{B}/\Gamma_o$ and G/Γ_o stand for the corresponding fundamental domains. Now we apply the results of the steps 3 and 4 to the left-hand side of (5A.4.8).

$$\int\limits_{\mathfrak{K}} \kappa \cdot \int\limits_{\mathbb{B}/\Gamma_o} \lambda = (4\pi^3)\frac{\pi^2}{6} \cdot \int\limits_{\mathbb{B}/\Gamma_o} c_2 = \frac{2\pi^5}{3} c_2(\mathbb{B}/\Gamma_o).$$

Together with (5A.4.8) we obtain

$$c_2(\mathbb{B}/\Gamma_o) = \frac{3\delta|D|^{5/2}}{16\pi^5} \omega^{(d)}(G/\Gamma_o). \tag{5A.4.9}$$

The subgroup Γ_o acts effectively on $\mathbb{B}$. This is also true for the PICARD modular group Γ of the field K if $K \neq \mathbb{Q}(\sqrt{-3})$. In the case of EISENSTEIN numbers the element $\text{diag}(\rho, \rho, \rho)$, ρ a primitive cubic unit root, acts ineffectively on $\mathbb{B}$. This is the only exception. We multiply (5A.4.9) with $[\Gamma : \Gamma_o]$ and receive

$$c_2(\mathbb{B}/\Gamma) = \frac{3\varepsilon\delta|D|^{5/2}}{16\pi^5} \omega^{(d)}(G/\Gamma), \ \varepsilon = \begin{cases} 3, & \text{if } K = \mathbb{Q}(\sqrt{-3}) \\ 1, & \text{otherwise} \end{cases}$$

The right-hand side has been already calculated, see Theorem 5A.4.6. It follows that

$$c_2(\mathbb{B}/\Gamma) = \frac{3\varepsilon|D|^{5/2}}{16\pi^5} \zeta(2) L(3,\chi) = \varepsilon \frac{|D|^{5/2}}{32\pi^3} \cdot L(3,\chi)$$

because the value of $\zeta(s)$ at $s = 2$ is equal to $\frac{\pi^2}{6}$ (see e.g. [SER], II, § 4).

It remains to prove the results of the steps 1, 2, 3, 4.

Step 1. The volume form $\omega^{(d)}$ on G has been defined in (5A.3.2), (5A.3.1) by the correspondence

$$\omega_e^{(d)} : H_1 \wedge \cdots \wedge H_4 \wedge M_1 \wedge \cdots \wedge M_4 \mapsto 1, \left(H_i = H_i^{(d)}, M_j = M_j^{(d)}\right),$$

especially

$$\omega_e^{(1)} : h_1 \wedge \cdots \wedge h_4 \wedge m_1 \wedge \cdots \wedge m_4 \mapsto 1, \left(h_i = H_i^{(1)}, m_j = M_j^{(1)}\right).$$

Now it is easy to verify (5A.4.4).
Step 2. We remember the CARTAN decomposition

$$\mathfrak{g} = \text{Lie } G = \mathfrak{k} + \mathfrak{m} = \sum \mathbb{R}h_i + \sum \mathbb{R}m_i, \mathfrak{k} = \text{Lie } \mathfrak{K} = T_{e,\mathfrak{K}}, \mathfrak{m} = T_{O,\mathbb{B}}.$$

The components in (5A.4.5) are defined as invariant volume forms on $\mathfrak{K}$ or $\mathbb{B}$, respectively, corresponding

$$\kappa : h_1 \wedge \cdots \wedge h_4 \mapsto 1, \lambda : m_1 \wedge \cdots \wedge m_4 \mapsto 1.$$

Then we can write $\omega^{(1)} = \kappa \wedge \lambda = \kappa \cdot \lambda$.
Step 4. The maximal compact LIE subgroup $\mathfrak{K} = \mathbb{S}(\mathbb{U}(2) \times \mathbb{U}(1))$ has a double cover $\tilde{\mathfrak{K}} = S\mathbb{U}(2) \times \mathbb{U}(1)$:

$$\tilde{\mathfrak{K}} \longrightarrow \mathfrak{K}, \quad \left(\begin{array}{cc|c} \multicolumn{2}{c|}{\multirow{2}{*}{A}} & 0 \\ & & 0 \\ \hline 0 & 0 & \zeta \end{array}\right) \mapsto \left(\begin{array}{cc|c} \multicolumn{2}{c|}{\multirow{2}{*}{\zeta^{-1}A}} & 0 \\ & & 0 \\ \hline 0 & 0 & \zeta^2 \end{array}\right)$$

It follows that $\mathfrak{k} = \text{Lie } \mathfrak{K} = \text{Lie } \tilde{\mathfrak{K}}$, hence $\int_{\mathfrak{K}} \kappa = \frac{1}{2} \int_{\mathfrak{K}} \tilde{\kappa}$. The KILLING form $B = Tr(X \cdot Y)$ in $\mathfrak{k}$ defines a positive definite metric $-B$ on $\mathfrak{k}$. One checks that $h_1' = h_1 + 2h_2, h_1, h_4, h_3$ is an orthogonal basis of this metric. The GRAM matrix of this basis is equal to diag$(6, 2, 2, 2)$. The corresponding normalized volume form on $\mathfrak{K}$ is defined by

$$\beta : \frac{1}{\sqrt{6}} h_1' \wedge \frac{1}{\sqrt{2}} h_1 \wedge \frac{1}{\sqrt{2}} h_4 \wedge \frac{1}{\sqrt{2}} h_3 \mapsto 1.$$

Comparing with κ we have $\beta = 2\sqrt{3}\tilde{\kappa}$. We split β according to $\tilde{\mathfrak{K}} = \mathbb{S}\mathbb{U}(2) \times U(1)$ into $\beta = \beta_1 \beta_2$,

$$\begin{array}{rcl} \beta_1 : \mathbb{R}h_1' & \longrightarrow & \mathbb{R}, \\ h_1' & \mapsto & \sqrt{6} \end{array} \quad \begin{array}{rcl} \beta_2 : \Lambda^3(\mathbb{R}h_1 + \mathbb{R}h_4 + \mathbb{R}h_3) & \longrightarrow & \mathbb{R}; \\ h_1 \wedge h_4 \wedge h_3 & \mapsto & \sqrt{2^3} \end{array}$$

β_1, β_2 are considered as invariant volume forms on $\mathbb{U}(1)$ or $S\mathbb{U}(2)$, respectively. We compare them with well-known volume forms on spheres embedded in euclidean spaces. The plane sphere $S^1(\sqrt{6})$ with radius $\sqrt{6}$ is isometric to $(\mathbb{U}(1), \beta_1)$. By splitting of the coefficients $\begin{pmatrix} \alpha & -\overline{\beta} \\ \beta & \overline{\alpha} \end{pmatrix} \in \mathbb{S}U(2)$ in real and imaginary parts one obtains an isometry of $(\mathbb{S}U(2), \beta_2)$ onto the sphere $S^3(\sqrt{2})$ with radius $R = \sqrt{2}$ in $\mathbb{R}^4$. The multiplicative group actions change over to orthogonal group action in $\mathbb{R}^2$ or $\mathbb{R}^4$, respectively. Therefore, our invariant measures change over to the euclidean measures restricted to S^1 or S^3, respectively, denoted by ε_i, $i = 1, 2$. We receive

$$\int_{U(1)} \beta_1 = \int_{S^1(\sqrt{6})} \varepsilon_1 = 2 \cdot \sqrt{6}\pi,$$

$$\int_{SU(2)} \beta_2 = \int_{S^3(\sqrt{2})} \varepsilon_2 = 2\pi^2 R^3 = 2\sqrt{8}\pi^2$$

and finally

$$\int_{\mathfrak{K}} \kappa = \frac{1}{2} \int_{\tilde{\mathfrak{K}}} \tilde{\kappa} = \frac{1}{2} \cdot \frac{1}{2 \cdot \sqrt{3}} \int_{\tilde{\mathfrak{K}}} \beta = \frac{1}{4 \cdot \sqrt{3}} \int_{U(1)} \beta_1 \cdot \int_{SU(2)} \beta_2 = 4\pi^3.$$

Step 3. Here we use the spirit of HIRZEBRUCH'S proportionality principle connecting CHERN forms of non-compact symmetric spaces with CHERN forms of their compact symmetric duals; for instance, $\mathbb{P}^2$ is the symmetric dual of the 2–ball $\mathbb{B}$. For details we refer to HIRZEBRUCH'S paper [HI2]. CHERN forms on symmetric spaces are invariant forms by construction via the invariant metrics. So it suffices to compare the forms in one point, say in $O \in \mathbb{B} \subset \mathbb{P}^2$.

It turns out that the CHERN forms $c_2^{\mathbb{B}}$a nd $c_2^{\mathbb{P}^2}$ coincide in O:

$$c_{2,0}^{\mathbb{P}^2} = c_{2,0}^{\mathbb{B}} : \Lambda^4 T_{0,\mathbb{B}} = \Lambda^4 T_{0,\mathbb{P}^2} \longrightarrow \mathbb{R},$$

T_0 the real tangent space in O. This follows immediately from the construction (see [WEL]), using the local description of c_2 by means of curvature tensors, the CARTAN decomposition $\mathfrak{g} = \mathfrak{k} + \mathfrak{m}$ and $T_{0,\mathbb{B}} = \mathfrak{m}, T_{0,\mathbb{P}^2} = i\mathfrak{m}$. We have to describe $c_2 = c_2^{\mathbb{P}^2}$ in canonical coordinates (z_1, z_2) around $0 \in \mathbb{C}^2 \subset \mathbb{P}^2$ and to compare with λ constructed in Step 2. The projective plane $\mathbb{P}^2$ is considered as the GRASSMANN manifold $Gr(1,3)$ of complex lines in $\mathbb{C}^3$. We first determine the form $c_1(U)$ of the universal bundle $U = U_{1,3} \longrightarrow Gr(1,3)$ over $Gr(1,3)$ in the point O. The corresponding curvature form Θ can be found in [WEL], III, § 2, (2.11):

$$\Theta(0) = (dZ^* \wedge dZ)(0) = d\bar{z}_1 \wedge dz_1 + d\bar{z}_2 \wedge dz_2.$$

By definition we have

$$c(U) = \det \left(E + \frac{i}{2\pi} \Theta \right)$$

(see [WEL], III, § 3, def. 3.4), and $c_1(U)$ is the part of real degree 2 of $c(U)$. Hence

$$c_1(U) = \frac{i}{2\pi} (d\bar{z}_1 \wedge dz_1 + d\bar{z}_2 \wedge dz_2).$$

Let for a moment K be the canonical line bundle on $\mathbb{P}^2$. Its connection with the universal bundle is well-known (see [WEL], VI, Example 2.3):

$$K = U^{\otimes 3} = (H^{\vee})^{\otimes 3},$$

where H is the bundle of a projective line on $\mathbb{P}^2, H^{\vee}$ its dual. Knowing $c_1(U)$ we find

$$c_1(K) = \frac{3i}{2\pi}(d\bar{z}_1 \wedge dz_1 + d\bar{z}_2 \wedge dz_2).$$

The first CHERN form c_1 of the tangent bundle of $\mathbb{P}^2$ coincides with $-c_1(K)$ (see [WEL], I, § 4, formula IV), hence

$$c_1^2 = -\frac{9}{2\pi^2} dz_1 \wedge d\bar{z}_1 \wedge dz_2 \wedge d\bar{z}_2.$$

Since

$$c_1^2(\mathbb{P}^2) = (K^2) = \left((-3H)^2\right) = 9 = 3c_2(\mathbb{P}^2)$$

the volume form $3c_2$ on $\mathbb{P}^2$ is equal to the volume form c_1^2. Therefore,

$$\begin{aligned} c_2 = \frac{1}{3}c_1^2 &= \frac{3}{2\pi^2} dz_1 \wedge d\bar{z}_1 \wedge dz_2 \wedge d\bar{z}_2 \\ &= -\frac{3}{2\pi^2}(2/i)^2 dx_1 \wedge dy_1 \wedge dx_2 \wedge dy_2 \\ &= \frac{6}{\pi^2} dx_1 \wedge dy_1 \wedge dx_2 \wedge dy_2, \end{aligned}$$

where $z_k = x_k + iy_k$, $k = 1, 2$.

On the other hand the isomorphy $m \xrightarrow{\sim} T_O B$ corresponds to the vectors m_1, m_2, m_3, m_4 defined in Step 1 the tangent vectors $(1,0)$, $(i,0)$, $(0,1)$ or $(0,i)$, respectively, see [K-N], II, chap. XI, § 10, ex. 10.7. Therefore, the invariant differential form of Step 2 coincides locally around $O \in \mathbb{B}$ with $dx_1 \wedge dy_1 \wedge dx_2 \wedge dy_2 = \frac{\pi^2}{6} c_{2,0}$. The relation (5A.4.6) is the globalization of this local identity. This finishes the proof of the Main Theorem 5A.4.7.

□

6 $\mathbb{Q}$-Orbital Surfaces

6.1 Introduction

We want to extend the numerical theory of Galois coverings of algebraic surfaces to arbitrary surface coverings. For this purpose it is necessary to extend the notion of orbital surfaces. In the Galois theory orbital surfaces, orbital curves and points can be expressed by means of divisors and singularities. These are quite classical objects. The classical language does not work nicely in the general theory of surface coverings. Here we have to introduce sums of orbital curves and orbital points with rational coefficients. They burst open the old framework of divisor language. For a first motivation we remember to the idea of Riemann surfaces. Riemann constructed these surfaces by means of several exemplars of the complex affine line $\mathbb{A}^1 = \mathbb{A}^1(\mathbb{C}) = \mathbb{C}$ patched together along cuts. Then he transported geometry and complex analysis from $\mathbb{C}$ to Riemann surfaces. This fruitful work has been extended by many outstanding mathematicians and it is not finished until these days. For algebraic surfaces we converse in some sense the idea. Let X, Y be algebraic surfaces, say smooth, complex, compact, and $f : X \to Y$ a finite covering. The branch locus B_f of f is a (reduced) divisor on Y. For a simple illustration we refer to see Figure 6.1.1

The CHERN numbers $c_2(X), c_1^2(X)$ can be expressed as integrals

$$c_2(X) = \int\limits_X \gamma_2, \; c_1^2(X) = \int\limits_X \gamma_1^2, \tag{6.1.1}$$

γ_1, γ_2 (CHERN) forms representing the CHERN classes c_1 or c_2 of X, respectively. The central problem solved in this chapter is the following one: Is it possible to calculate the integrals (6.1.1) on the surface Y or, more precisely, which geometric and arithmetic informations coming from f are necessary to do that. Surely, the knowledge of the branch locus is not sufficient. Already in the GALOIS case one needs additional weights for the components which are equal to the corresponding ramification indices. In the GALOIS case we are in a rather happy situation. Namely, the ramification index does not depend on the choice of an irreducible preimage curve D of a component C of B_f. In the general case different preimages D of

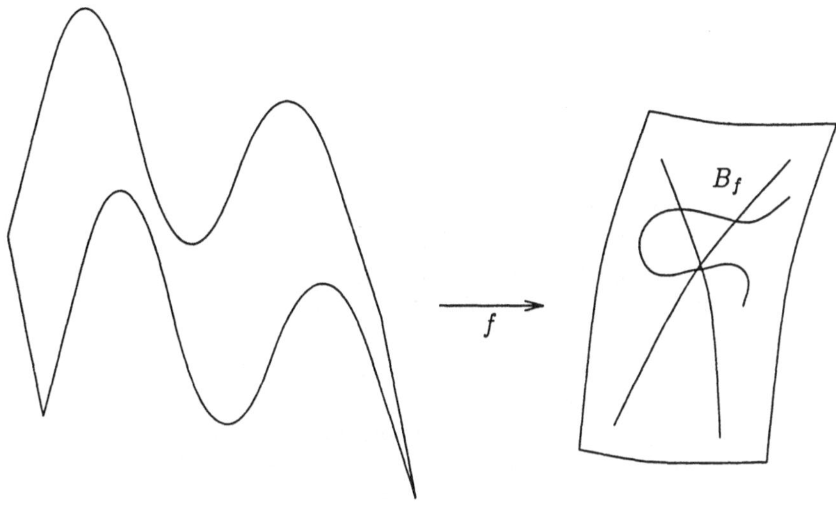

Figure 6.1.1

C have different ramification indices. Taking into consideration all these ramification indices we define the new finer objects $\mathbf{Y} = (Y, \mathbf{B}_f)$ called orbital surfaces again. The branch arrangement $\mathbf{B}_f$ will be a formal sum of orbital curves and orbital points with rational coefficients supported by B_f. We allow HIRZEBRUCH-JUNG singularities and crossing intersection of the components. All other cases can be birationally transformed to this central situation. In [HO6] the new objects have been called "rationally arranged surfaces" or "rationally arranged curves". Now we prefer to call them $\mathbb{Q}$-*orbital surfaces* . These surfaces form a category $\mathcal{C}$; the morphisms are called orbital morphisms. The old surface category C of algebraic surfaces appears as subcategory of $\mathcal{C}$. The old objects are considered as surfaces without arrangements or with "trivial" arrangements with weights equal to 1 (reduced orbital surfaces). We define orbital heights $\mathbf{c}(\mathbf{X}) \in \mathbb{Q}$ extending the definition of CHERN numbers $c(X) \in \mathbb{Z}$ of old objects for all rational CHERN symbols $c \in \mathbb{Q}c_1^2 + \mathbb{Q}c_2$, e.g. for $c = \tau, \chi, c_1^2, c_2 = e$. The definition will be extended to arranged morphisms. Formally we write $\mathbf{c}$ as a map $\mathbf{c} : \mathcal{C} \to \mathbb{Q}$.

Main Theorem. 6.1.1 *For any $c \in \mathbb{Z}\chi + \mathbb{Z}c_2$ the correspondence $X \mapsto c(X)$, $X \in Ob\ C$, can be extended orbitally to $\mathcal{C}$:*

$$\begin{array}{ccc} \mathbf{c}: \mathcal{C} & \longrightarrow & \mathbb{Q} \text{ (orbital heights)} \\ \updownarrow & \circlearrowright & \updownarrow \\ c: C & \longrightarrow & \mathbb{Z} \text{ (Chern numbers)} \end{array} \tag{6.1.2}$$

This means essentially that the degree formula $\mathbf{c}(\mathbf{X}) = d(\mathbf{f}) \cdot \mathbf{c}(\mathbf{Y})$ is valid for any orbital "finite" morphism $\mathbf{f}: \mathbf{X} \to \mathbf{Y}$. If $f: X \to Y$ is a finite morphism in the old sense, then it defines an orbital finite morphism $\mathbf{f}: X \to \mathbf{Y} = (Y, \mathbf{B}_f)$. The Main Theorem solves the problem of calculating $c(X)$ on Y. Namely, it follows immediately that $c(X) = \deg(f) \cdot \mathbf{c}(\mathbf{Y})$. So, in order to calculate $c(X)$ it is sufficient to know additionally to the geometric data of Y the branch arrangement $\mathbf{B}_f$ and the degree of f. With these data one can write down general and explicit formulas connecting the CHERN numbers of X with the CHERN numbers of Y (see [HUR]). We call them HURWITZ *type formulas* for CHERN numbers of algebraic surfaces. They are general because we allow (HIRZEBRUCH-JUNG) singularities. A short look at Galois coverings shows that this is necessary already for very simple cases.

Most consequently one applies the functoriality of the invariants $\mathbf{c}$ to finite coverings $f: X \to \mathbf{P}^2$ over the plane $\mathbf{P}^2$. If X is smooth and algebraic, then it is a projective surface. Therefore it can be finitely projected onto $\mathbf{P}^2$, e.g. by a general projection. A very important question is to prove inequalities for surface invariants of BOGOMOLOV-MIYAOKA-YAU type

$$c_1^2(X) \leq kc_2(X) \tag{6.1.3}$$

for surfaces in general and also for special surfaces with a view to arithmetical surfaces and the connection with effective solutions of diophantine equations (see PARSHIN [PAR], PARSHIN, ZARKHIN [P-Z]). The orbital heights light up the situation in a flexible manner. We change over to the orbital finite morphisms $\mathbf{f}: X \to \mathbf{P}_f^2 = (\mathbf{P}^2, \mathbf{B}_f)$ and set $\boldsymbol{\beta} = \mathbf{c}_1^2 - k\mathbf{c}_2$. Then we get the following subdivision of surfaces into two classes, which is not visible in the classical theory: In the first class, called the *general class*, we put all surfaces X admitting a finite covering $\mathbf{f}: X \to \mathbf{P}_f^2$ with $\boldsymbol{\beta}\left(\mathbf{P}_f^2\right) \leq 0$. All other surfaces of the considered type we put into the second class, called the *special class*. For surfaces of the general class the inequality (6.1.3) is obviously satisfied by the orbital property of $\boldsymbol{\beta}$. So the subdivision delegates the proof of the inequality to the special class. One checks then that the special class consists of surfaces of small KODAIRA dimension. Altogether the orbital heights allow to reduce the proof of inequalities of type (6.1.3) or similar types to the well-known theory of classification of surfaces with small KODAIRA dimension.

We illustrate this method for $k = 2$ and the class of surfaces admitting a regular finite covering onto $\mathbf{P}^2$, that means that the branch locus has normally crossing smooth components and no triple points (and also no higher multiple points). A little extension of the definition of orbital surfaces allowing some curve singularities and also triple points in the branch locus catch all smooth algebraic surfaces via general projections. For $k = 3$ we can then observe a subdivison described above. From this point of view the heights of orbital surfaces can also be considered as a new and finer geometric guide line for the numerical theory of arithmetical surfaces connected with diophantine equations.

6.2 Arrangements with Rational Coefficients

6.2.1 Let Y be a normal compact complex algebraic surface with (at most) HIRZEBRUCH—JUNG singularities and

$$B = B^1 + B^0, \ B^1 = \sum_i D_i, \ B^0 = \sum_i Q_j$$

a formal finite sum of smooth irreducible curves $D_i \subset Y$ and points $Q_j \in Y$ such that

(i) $\operatorname{supp} B^1 = \bigcup_i D_i$ has at most normal crossing branches.

(ii) If $D_i \neq D_j$, then each intersection point Q of D_i and D_j defines two opposite reduced arranged abelian points $(U_Q, Q; D_i, D_j)$ and $(U_Q, Q; D_j, D_i)$ denoted by Q or Q', respectively.

(iii) If $Q \in Y$ is a singularity lying on a component D_i of B^1, then there exists a curve D' on Y and an analytic neighbourhhod $U_Q \subset Y$ of Q such that $(U_Q, Q; D_i, D')$ is a reduced arranged abelian point.

(iv) Each singularity of Y and of supp B^1 belongs to $\operatorname{supp} B^0 = \bigcup_Q \{Q \in Y; Q \text{ is a component of } B^0\}$.

In contrast to the earlier definition of arrangements reflecting the situation of GALOIS coverings we allow that $D_i = D_j$ for different indices i, j and also $Q_k = Q_l$ for different indices k, l. It would be more precise to write for example B^0 as a tuple $(Q_1, Q_2, \ldots)$ with repetitions of some points.

Definition 6.2.2 A point arrangement $\mathbf{B^0}$ on Y (with rational coefficients) is a formal sum

$$\mathbf{B}^0 = \sum_j t_j \mathbf{Q}_j,\ t_j \in \mathbb{Q},$$

with $B^0 = \sum_j Q_j$ as above and $\mathbf{Q}_j$ an abelian point with reduction Q_j.

Definition 6.2.3 Let (U_D, D) be a reduced curve on Y denoted by D again, D smooth and compact. An *arrangement* on D (with rational weights) is a formal finite sum

$$\mathbf{B}^0 = \sum_j r_j \mathbf{Q}_j,\ r_j \in \mathbb{Q},$$

as above (on U_D instead of Y), $Q_j \in D$. The triple

$$\mathbf{D} = (U_D, vD; \mathbf{B}^0),\ v \in \mathbb{N}_+,$$

is called a $\mathbb{Q}$-*orbital curve.*

Definition 6.2.4 Let Y, $B = B^1 + B^0$ be as above satisfying properties (i), (ii), (iii), (iv). A (regular) $\mathbb{Q}$-*arrangement* (with rational coefficients) on Y is a formal sum

$$\mathbf{B} = \mathbf{B}^1 + \mathbf{B}^0 = \sum_i s_i \mathbf{D}_i + \sum_j t_j \mathbf{Q}_j, \tag{6.2.1}$$

where each $\mathbf{D}_i$ is a $\mathbb{Q}$-orbital curve in the sense of definition 6.2.3 and $\mathbf{B}^0$ a point arrangement on Y in the sense of definition 6.2.2. A regular $\mathbb{Q}$-*orbital surface* is a pair $\mathbf{Y} = (Y, \mathbf{B})$, $\mathbf{B}$ a $\mathbb{Q}$-arrangement on Y.

Convention 6.2.5 All arrangements on surfaces and curves considered in the previous chapters are now called *natural arrangements* because they have only natural weights and coefficients. The corresponding orbital surfaces are called sometimes *natural orbital surfaces* . In this chapter orbital surfaces are understood in the more general sense of Definition 6.2.4.

6.2.6 Now let $f : X \to Y$ be a finite covering of algebraic surfaces, Y as above and X of the same type. We call f regular, if

(i) for each point P there exist analytic neighbourhoods U_P and U_Q of P or $Q = f(P)$, respectively, such that the restriction $f_P : U_P \to U_Q$ is finite and the branch locus of f_P together with the ramification indices arranges Q to an abelian point $\mathbf{Q}_P$. That means that the branch locus of f_P has at most two (smooth) branches through Q arranging Q or only one with the same property, or $U_P^* \to U_Q^* = U_Q \setminus \{Q\}$ is unramified.

(ii) All irreducible one-dimensional components of the branch locus B_f of f are smooth curves.

Definition 6.2.7 Let $f : X \to Y$ be a regular finite covering. The *branch arrangement* of f on Y is the following $\mathbb{Q}$-arrangement on Y:

$$\mathbf{B}_f = \mathbf{B}_f^1 + \mathbf{B}_f^0 = \sum_{X \supset C \to D} \sum_{D \subset B_f^1} s_C \cdot \mathbf{D}_C + \sum_{X \ni P \to Q} \sum_{Q \in B_f^0} t_P \mathbf{Q}_P, \tag{6.2.2}$$

s_C and t_P are relative degrees

$$s_C = v_C \cdot \deg(f_C)/\deg(f), \ t_P = \deg(f_P)/\deg(f) \in \mathbb{Q}, \tag{6.2.3}$$

with restriction maps

$$f_C : C \to D, D \text{ a component of } B_f^1,$$
$$f_P : (U_P, P) \to (U_Q, Q), Q \in B_f^0;$$

v_C is the ramification index of f along C.

B_f^0 consists of all singularities of B_f and of Y, $\mathbf{D}_C$ is explained in the next

Definition 6.2.8 Let $f^C : (U_C, C) \to (U_D, D)$ be a finite covering of analytic surface germs along curves, which is regular in the sense of 6.2.6. The *branch arrangement* $\mathbf{B}^0(f^C)$ on (U_D, D) is the following point arrangement on D in the sense of definition 6.2.3:

$$\mathbf{B}^0(f^C) = \sum_{C \ni P \to Q} \sum_{Q \in D} r_P \cdot \mathbf{Q}_P, \ r_P = \deg(f_P)/\deg(f^C), \tag{6.2.4}$$

where $\mathbf{Q}_P$ is the abelian point coming from the restriction $f_P : (U_P, P) \to (U_Q, Q)$. The internal sum runs over all branch points on D of the restriction $f_C : C \to D$ of f^C and

$$\deg(f^C) = \deg(f_C) \cdot v_C,$$

v_C the ramification index of f^C along C. The arranged curve $\mathbf{D}_C = \mathbf{D}_{f^C}$ is defined as triple $\mathbf{D}_C = \mathbf{D}_{f^C} = (U_D, v_C D; \mathbf{B}^0(f^C))$.

We observe that not all arrangements on surfaces can be branch arrangements. The coefficients $s_C, t_p, r_P \in \mathbb{Q}$ are of special type. Counting preimages of points and curves they have to satisfy certain relations. In order to be more precise we give the following

Definition 6.2.9 Let Y be a surface as above and $\mathbf{B}$ an arrangement on Y. In addition to the notations of (6.2.1) we set

$$\mathbf{D}_i = \left(U_i, v_i D_i; \sum_k r_{ik} \mathbf{Q}_{ik} \right)$$

We call $\mathbf{B}$ an *honest arrangement* and $\mathbf{Y} = (Y, \mathbf{B})$ an *honest orbital surface,* if the following conditions are satisfied:

(i) $0 < r_{ik}, s_i, t_j \leq 1$

(ii) $\mathbf{D}_i \neq \mathbf{D}_m$ for $i \neq m, \mathbf{Q}_j \neq \mathbf{Q}_n$ for $j \neq n, \mathbf{Q}_{ik} \neq \mathbf{Q}_{i\lambda}$ for $\kappa \neq \lambda$,

(iii) $s(D) := \sum_{D_i = D} s_i = \begin{cases} 1 & \text{if } D_i = D \text{ exists} \\ 0 & \text{otherwise} \end{cases}$,

(iv) $t(Q) := \sum_{Q_j = Q} t_j = \begin{cases} 1 & \text{if } j \text{ with } Q_j = Q \text{ exists} \\ 0 & \text{otherwise} \end{cases}$,

(v) $r_i(Q) = \sum_{Q_{i\kappa} = Q} r_{i\kappa} = \begin{cases} 1 & \text{if } \kappa \text{ with } Q_{i\kappa} = Q \text{ exists} \\ 0 & \text{otherwise} \end{cases}$,

(vi) If $\mathbf{Q}_j = \mathbf{Q}_{i\kappa} \in \mathbf{D}_i \subset \mathbf{B}^1$, then $t_j = s_i \cdot r_{i\kappa}$.

By $D, D_i, Q, Q_j, Q_{i\kappa}$ we denote reduced arranged curves or points, respectively, $D_i, Q_j, Q_{i\kappa}$ being the reductions of $\mathbf{D}_i, \mathbf{Q}_j, \mathbf{Q}_{i\kappa}$. The properties in (iii) are called sometimes ***discrete decomposition of the unit*** (see [HO6] II; III.5).

Lemma 6.2.10 *Let $f : X \to Y$ be a regular finite covering. Then the branch arrangement $\mathbf{B}_f$ on f is an honest arrangement on Y and $\mathbf{Y}_f = (Y, \mathbf{B}_f)$ an honest orbital surface.*

Proof We work with the notations of (6.2.2) and (6.2.3). Let Q be a point of B_f^0, P a preimage, $f_P : U_P \to U_Q$ a finite restriction of f to small analytic neighbourhoods of P and Q. A general point $Q' \in U_Q$ has exactly $\deg(f_P)$ preimages in U_P and $\deg(f)$ preimages on X. Counting together we get

$$\deg(f) = \sum_{X \ni P \to Q} \deg(f_P) = \sum_{P \to Q} t_Q \cdot \deg(f). \tag{6.2.5}$$

Dividing by $\deg(f)$ we get the second relation of (iii).

Now let D be an irreducible component of the branch curve $\mathrm{supp}(B_f^1)$, $f^C : U_C \to U_D$ a finite restriction of f to small analytic neighbourhoods of C or D, respectively, C an irreducible component of $f^{-1}(D)$. We take again a general point $Q' \in U_D$. By a count of preimages we get the decomposition

$$\deg(f) = \sum_{C \to D} \deg(f^C) = \sum_{C \to D} v_C \cdot \deg(f_C) = \sum_{C \to D} s_C \cdot \deg(f). \tag{6.2.6}$$

The first relation of (iii) follows immediately. For the third relation we have to check that for $f^C : U_C \to U_D$

$$\sum_{C \ni P \to Q} r_P = 1 \tag{6.2.7}$$

with the notations of (6.2.4). We restrict f^C finitely to $f_P : U_P \to U_Q$, $U_P \subset U_C$, $U_Q \subset U_D$, in the usual manner. Counting the preimages of a general point $Q' \in U_Q$ near to D one gets

$$\deg(f^C) = \sum_{C \ni P \to Q} \deg(f_P) = \sum_{C \ni P \to Q} r_P \cdot \deg(f^C) \tag{6.2.8}$$

and finally (6.2.7). The properties (i), (ii) and (iv) are easy to verify looking at the definitions.

□

6.3 Finite Morphisms of $\mathbb{Q}$-Orbital Surfaces

6.3.1 As already mentioned in the introduction we consider the orbital surfaces as objects of a category $\mathcal{C}$, the category of orbital surfaces. A morphism (orbital morphism) $\mathbf{f} : \mathbf{X} \to \mathbf{Y}$ of $\mathcal{C}$ is nothing else but a morphism $f : X \to Y$ in the category of surfaces with the following properties:

(i) f is surjective (equivalently: f is generically finite);

(ii) $f(\mathrm{supp}(\mathbf{A})) \subseteq \mathrm{supp}(\mathbf{B})$,

where $\mathbf{A}$, $\mathbf{B}$ are the arrangements of $\mathbf{X}$ or $\mathbf{Y}$, respectively,

$$\mathrm{supp}(\mathbf{A}) = \mathrm{supp}(\mathbf{A}^1) \cup \mathrm{supp}(\mathbf{A}^0), \ \mathrm{supp}(\mathbf{B}) = \mathrm{supp}(\mathbf{B}^1) \cup \mathrm{supp}(\mathbf{B}^0),$$

see 6.2.82.1(i) and (iv).

We allow in Definition 6.3.1 also $\mathbf{X}$ (or $\mathbf{Y}$) to be trivially arranged, that means that the arrangement of $\mathbf{X}$ is a sum of different curves and points with weight 1. Such an arrangement is called a reduced arrangement and $\mathbf{X}$ a *reduced orbital surface.* From a numerical point of view we can identify two reduced arranged surfaces $\mathbf{X}$, $\mathbf{X}'$ if $X = X'$ and denote a reduced orbital surface simply by X. In this sense the category of compact complex algebraic surfaces with at most HIRZEBRUCH-JUNG singularities appears as subcategory $\mathcal{C}$ of $\boldsymbol{\mathcal{C}}$. So we dispose on "old" objects "old" morphisms (generically finite) belonging to $\mathcal{C}$ and on "new" objects and "new" morphisms belonging to $\boldsymbol{\mathcal{C}}$.

In general, an orbital surface $\mathbf{X} = (X, \mathbf{B})$ has no ingeniously defined reduction X because the rational coefficients of components of $\mathbf{B}$ with the same support cannot be summed up to 1. But each honest orbital surface $\mathbf{X}$ has a *reduction* $X = rcd(\mathbf{X})$ by Definition 6.2.9 (ii), (iii).

We want to define now finite morphisms in the category $\boldsymbol{\mathcal{C}}$. First we correspond to any finite (regular) morphism $f : X \to Y$ of $\mathcal{C}$ an orbital morphism $\mathbf{f} : X \to \mathbf{Y}$ in a natural manner. We take the same map f and have only to define the arranged objects. For X we can take the reduced orbital surface represented by (X, SgX), SgX the singular locus of X, and $\mathbf{Y} = \mathbf{Y}_f = (Y, \mathbf{B}_f)$ is defined by the branch arrangement $\mathbf{B}_f$ of f (see Definition 6.2.7). We call f and also $\mathbf{f}$ and *old* finite morphism. The old finite morphisms $\mathbf{f}$ are the first types of finite morphism in $\boldsymbol{\mathcal{C}}$. The other ones are closely related.

Definition 6.3.2 An orbital morphism $\mathbf{f} : \mathbf{X} \to \mathbf{Y}$ is called *finite* , if and only if there exists a commutative diagram:

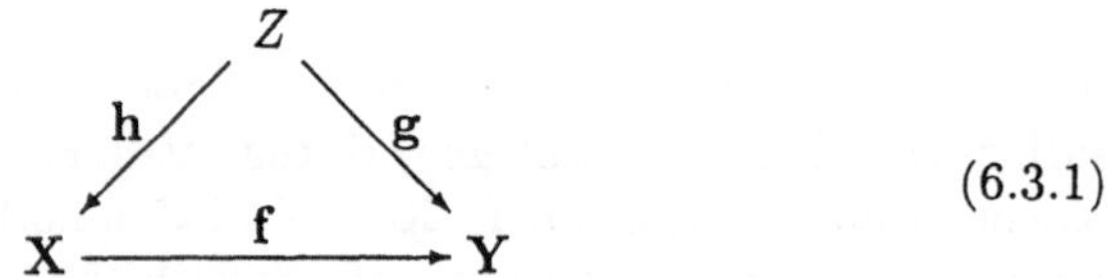

in $\boldsymbol{\mathcal{C}}$ such that $\mathbf{g}$ and $\mathbf{h}$ are old finite morphisms. We call $\mathbf{f}$ GALOIS-finite, if $\mathbf{g} : Z \to \mathbf{Y}$, $\mathbf{h} : Z \to \mathbf{X}$ in diagram (6.3.1) can be chosen as GALOIS coverings.

Each morphism $f' : X' \to Y$ of $\mathcal{C}$ can be decomposed into a birational and a finite morphism, see diagram:

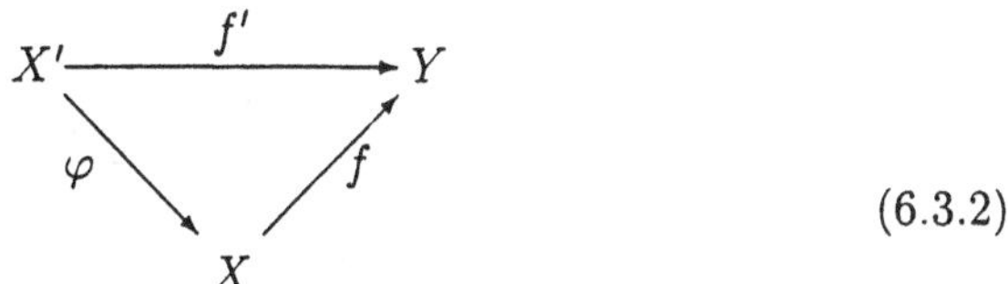

(6.3.2)

Namely, X, or more precisely $f : X \to Y$, is the normalization of Y in the function field $\mathbb{C}(X')$ of X' (see [MU6]). Since we assume that all our surfaces in $\mathcal{C}$ are normal, we get embeddings of local rings $\mathcal{O}_{X,x} \subset \mathcal{O}_{X',x'}$, defining a birational map $\varphi : X' \to X$ because the function fields are the same. We call an orbital morphism $\mathbf{f} : \mathbf{X} \to \mathbf{Y}$ *birational*, if the supporting morphism $f : X \to Y$ is birational in the old sense.

A very important role for the construction of orbital heights play base changes in $\boldsymbol{\mathcal{C}}$. In the old category $\mathcal{C}$ we call the diagram:

$$\begin{array}{ccc} X' & \xrightarrow{\varphi'} & X \\ {\scriptstyle f'}\downarrow & \otimes & \downarrow{\scriptstyle f} \\ Y' & \xrightarrow{\varphi} & Y \end{array} \qquad (6.3.3)$$

a base change diagram, if X' is the normalization of $X \times_Y Y'$, where the latter fibre product is taken in the category of all schemes over $\mathbb{C}$. The surface X has the usual universal property in the category $\mathcal{C}$ containing only normal surfaces (with HIRZEBRUCH-JUNG singularities). We restrict our attention to base change diagrams with birational φ and finite morphisms f. Then φ' is birational and f' is finite of the same degree as f. We extend the notion of base change from the old category to the category of orbital surfaces.

Definition 6.3.3 A commutative diagram (6.3.4) of orbital surfaces

$$\begin{array}{ccc} \mathbf{X}' & \xrightarrow{\varphi'} & \mathbf{X} \\ {\scriptstyle \mathbf{f}'}\downarrow & & \downarrow{\scriptstyle \mathbf{f}} \\ \mathbf{Y}' & \xrightarrow{\varphi} & \mathbf{Y} \end{array} \qquad (6.3.4)$$

is called a *base change diagram* , if it is an old base change diagram with old finite morphisms $\mathbf{f} : X \to \mathbf{Y}_f$, $\mathbf{f}' : X' \to \mathbf{Y}_{f'}$ or if there exists a commutative extension (6.3.5) of diagram (6.3.4)

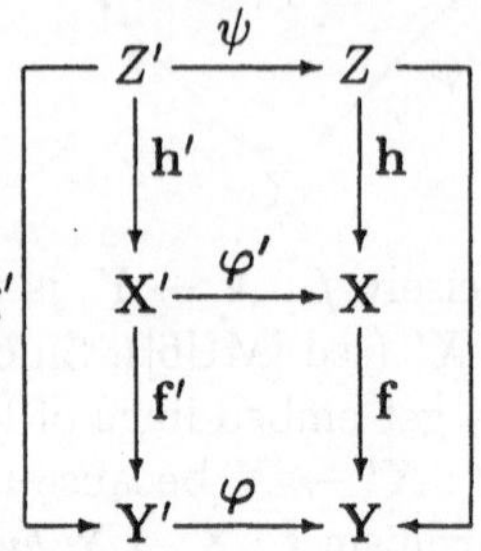

$$\begin{array}{ccc} \mathbf{Z}' & \xrightarrow{\psi} & \mathbf{Z} \\ \downarrow{\scriptstyle \mathbf{h}'} & & \downarrow{\scriptstyle \mathbf{h}} \\ \mathbf{X}' & \xrightarrow{\varphi'} & \mathbf{X} \\ \downarrow{\scriptstyle \mathbf{f}'} & & \downarrow{\scriptstyle \mathbf{f}} \\ \mathbf{Y}' & \xrightarrow{\varphi} & \mathbf{Y} \end{array} \qquad \mathbf{g}' : \mathbf{Z}' \to \mathbf{Y}', \quad \mathbf{g} : \mathbf{Z} \to \mathbf{Y} \tag{6.3.5}$$

such that $\mathbf{g}, \mathbf{h}, \mathbf{g}'$ and $\mathbf{h}'$ are old finite morphisms and the supporting old diagrams (6.3.6) in C

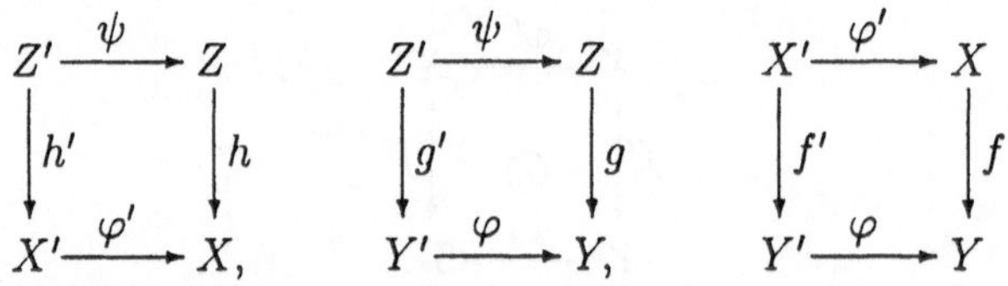

$$\begin{array}{ccc} Z' & \xrightarrow{\psi} & Z \\ \downarrow{\scriptstyle h'} & & \downarrow{\scriptstyle h} \\ X' & \xrightarrow{\varphi'} & X, \end{array} \qquad \begin{array}{ccc} Z' & \xrightarrow{\psi} & Z \\ \downarrow{\scriptstyle g'} & & \downarrow{\scriptstyle g} \\ Y' & \xrightarrow{\varphi} & Y, \end{array} \qquad \begin{array}{ccc} X' & \xrightarrow{\varphi'} & X \\ \downarrow{\scriptstyle f'} & & \downarrow{\scriptstyle f} \\ Y' & \xrightarrow{\varphi} & Y \end{array} \tag{6.3.6}$$

are base change diagrams.

It suffices to claim that two of the diagrams (6.3.6) have the base change property. From the definitions it follows that $\mathbf{f}$ and $\mathbf{f}'$ are finite orbital morphisms. We will only need base change diagrams (6.3.4) with birational φ, and φ', and we suppose this property at any place of this chapter.

A central place in the theory of orbital heights occupy relative orbital heights. They are defined on the category $\mathcal{C}/\mathcal{C}$ of relative objects $\mathbf{X}/\mathbf{Y}$, $\mathbf{X}, \mathbf{Y} \in \mathcal{C}$. The notation $\mathbf{X}/\mathbf{Y}$ is an abbreviation for a morphism $\mathbf{f} : \mathbf{X} \to \mathbf{Y}$ of $\mathcal{C}$. A morphism $\mathbf{X}'/\mathbf{X} \to \mathbf{Y}'/\mathbf{Y}$ in $\mathcal{C}/\mathcal{C}$ is a commutative diagram (6.3.4).

Definition 6.3.4 A morphism $\mathbf{F} : \mathbf{X}'/\mathbf{X} \to \mathbf{Y}'/\mathbf{Y}$ of relative orbital surfaces is called *finite* , if the corresponding diagram (6.3.4) is a base change diagram. The *degree* of a finite relative morphism $\mathbf{F}$ is defined by

$$\deg(\mathbf{F}) = \deg(\mathbf{f}) = \deg(f) = \deg(\mathbf{f}') = \deg(f').$$

We do not dispose in the category of orbital surfaces on kernels and cokernels as in the category of vector bundles over a manifold. We will fill this gap partly by means of relative objects in a tautological manner. Let $\varphi : \mathbf{X}' \to \mathbf{X}$ be a birational orbital morphism. The exceptional locus E' of φ can be thickened to an arrangement $\mathbf{E}'$ on $\mathbf{X}'$. Considering $\mathbf{E}'$ as subobject of $\mathbf{X}'$ we have a "short sequence"

$$\mathbf{E}' \longrightarrow \mathbf{X}' \longrightarrow \mathbf{X}$$

Instead of $\mathbf{E}'$ it will be convenient to consider the relative object $\mathbf{X}'/\mathbf{X}$ borrowed from the relative category $\mathcal{C}/\mathcal{C}$. We call by abuse of language

$$\mathbf{X}'/\mathbf{X} \cdots \mathbf{X}' \underset{\varphi}{\rightarrow} \mathbf{X} \tag{6.3.7}$$

the *tautological sequence* of φ. If $\mathbf{F} : \mathbf{X}'/\mathbf{X} \rightarrow \mathbf{Y}'/\mathbf{Y}$ is a finite relative orbital morphism, then we call

$$\begin{array}{ccccc} \mathbf{X}'/\mathbf{X} & \cdots & \mathbf{X}' & \rightarrow & \mathbf{X} \\ \mathbf{F}\downarrow & & \downarrow\mathbf{f}' & & \downarrow\mathbf{f} \\ \mathbf{Y}'/\mathbf{Y} & \cdots & \mathbf{Y}' & \rightarrow & \mathbf{Y} \end{array} \tag{6.3.8}$$

the corresponding tautological (base change) diagram. Remember that with $\mathbf{F}$ also $\mathbf{f}, \mathbf{f}'$ are finite and the right square of (6.3.8) is a base change diagram by Definition 6.3.4.

At the end of this section we introduce reductions of orbital finite morphisms. It makes no sense to define reductions of arbitrary orbital morphisms even if their support of arrangement is finite. The reduction of an orbital surface $\mathbf{X}$ is represented by (X, SgX). So it is in our notation the old object X itself endowed with a trivial kind of arranging.

Definition 6.3.5 Let $\mathbf{f} : \mathbf{X} \rightarrow \mathbf{Y}$ be a finite morphism in $\mathcal{C}$. the old finite morphism

$$\text{red }(\mathbf{f}) : X \longrightarrow \mathbf{Y} = \mathbf{Y}_f = (Y, \mathbf{B}_f)$$

is called the *reduction* of $\mathbf{f}, \mathbf{B}_f$ is the branch arrangement of $f : X \rightarrow Y$.

The definition is correct. One has to check that f is regular. This follows from the definitions 6.3.2 and 6.2.6. Namely, with notations of diagram (6.3.1), g is a regular finite covering and the branch locus B_f is less complicated then B_g because $B_f \subset B_g$.

The reduction will be used for GALOIS-finite coverings in $\mathcal{C}$. Assume that we have a commutative diagram (6.3.1) with regular GALOIS coverings $g : Z \rightarrow Z/G = Y$, $h : Z \rightarrow Z/H = X$. The arrangements of $\mathbf{X}, \mathbf{Y}$ have GALOIS weights. Therefore $\mathbf{X}, \mathbf{Y}$ are natural arrangements considered in the previous chapters. The reduction red$(\mathbf{f}) : X \rightarrow \mathbf{Y}$ changes the weights and produce rational coefficients at the components of $\mathbf{B}_f$. So the reductions are in general not simpler than the original orbital morphisms. They store information about the original arrangements of $\mathbf{X}$ and $\mathbf{Y}$.

6.4 Functorial Properties for Rational Invariants

Functorial surface invariants are functions

$$\begin{aligned} \mathbf{c}: \mathcal{C} \to \mathbb{Q},\ \mathcal{C}/\mathcal{C} \to \mathbb{Q} \\ \mathbf{X} \mapsto \mathbf{c}(\mathbf{X}),\ \mathbf{X}/\mathbf{Y} = \mathbf{f} \mapsto \mathbf{c}(\mathbf{f}) = \mathbf{c}(\mathbf{X}/\mathbf{Y}) \end{aligned}$$

on the class of objects of $\mathcal{C}$ and on the class of relative objects (morphisms) with rational values with some functorial properties, which we want to describe now.

If $\mathbf{f}$ is a finite morphism, then

$$\mathbf{c}(\mathcal{F}) \cdot \mathbf{c}(\mathbf{Y}) = \mathbf{c}(\mathbf{X}) \qquad \text{(Mult)}$$

where $\mathcal{F}$ denotes a general fibre of f and $\mathbf{c}(\mathcal{F}) = \deg(\mathbf{f}) = \deg(f)$.

If φ is a birational morphism, then

$$\mathbf{c}(\varphi) + \mathbf{c}(\mathbf{X}) = \mathbf{c}(\mathbf{X}'/\mathbf{X}) + \mathbf{c}(\mathbf{X}) = \mathbf{c}(\mathbf{X}') \qquad \text{(Add)}$$

where

$$\mathbf{X}'/\mathbf{X} \cdots \mathbf{X}' \underset{\varphi}{\to} \mathbf{X}$$

is the tautological sequence of φ.

For a base change diagram

$$\begin{array}{ccc} \mathbf{X}' & \underset{\varphi'}{\to} & \mathbf{X} \\ \mathbf{f}' \downarrow & \otimes & \downarrow \mathbf{f} \\ \mathbf{Y}' & \underset{\varphi}{\to} & \mathbf{Y} \end{array} \qquad (6.4.1)$$

it holds that

$$\mathbf{c}(\varphi') = \mathbf{c}(\mathbf{X}'/\mathbf{X}) = d \cdot \mathbf{c}(\mathbf{Y}'/\mathbf{Y}) = d \cdot \mathbf{c}(\varphi), \qquad \text{(B. Ch)}$$

where

$$d = \deg(\mathbf{f}) = \deg(\mathbf{f}') = \deg(f).$$

For a base change diagram (6.4) it holds that

$$\mathbf{c}(\mathbf{f}') = \mathbf{c}(\mathbf{f}) \qquad \text{(Shift)}$$

The first three postulates are called the multiplicative, additive or base change property of $\mathbf{c}$, respectively. The next two properties come near to the definition of a functor. For finite orbital morphisms $\mathbf{f}, \mathbf{h}$ it holds that

$$\mathbf{c}(\mathbf{f}) = 0 \tag{F.0}$$

$$\mathbf{c}(\mathbf{f} \circ \mathbf{h}) = \mathbf{c}(\mathbf{h}) + \deg(\mathbf{h})\mathbf{c}(\mathbf{f}) \tag{F.1}$$

Remember that a functor corresponds to each object of a category an object of a second category and to each morphism a morphism of the second category. The latter correspondence is multiplicative. Postulate (F.1) is a quasi-additive analogon of this property. A functor corresponds to a trivial (identical) morphism of an object to identical morphism of the image object. Postulate (F.0) distinguishes the finite morphisms as trivial ones from the numerical point of view corresponding them the additively neutral number 0. For reasons of proof we need also the weaker reduction property:

If $\mathbf{f} : \mathbf{X} \to \mathbf{Y}$ is finite and $f : X \to \mathbf{Y}_f$ its reduction, then the following implication is true:

$$\mathbf{c}(\mathbf{f}) = 0 \Longrightarrow \mathbf{c}(f) = 0. \tag{Red. 0}$$

The next postulate is satisfied in the most simplest way, namely by definition:

$$\mathbf{c}(\mathbf{g}) = \mathbf{c}(\mathbf{X}') - \deg(\mathbf{g}) \cdot \mathbf{c}(\mathbf{Y})5 \tag{Def}$$

for each orbital morphism $\mathbf{g} : \mathbf{X}' \to \mathbf{Y}$.

Before we establish the relations between the postulates we restrict some of them to special situations in order to find the weakest properties from which all other ones can be deduced. $(\mathrm{F.0})_{\mathbf{Gal}}$ is (F.0) restricted to GALOIS-finite coverings. $(\mathrm{F.0})_{\mathrm{old}}$ is (F.0) restricted to old finite morphisms $f : X \to \mathbf{Y}_f$. $(\mathrm{F.0})_{\mathrm{Gal}}$ is (F.0) restricted to reductions of GALOIS-finite coverings. $(\mathrm{Red.0})_{\mathrm{Gal}}$ is (Red.0) restricted to the same morphisms. Changing over to the category $\boldsymbol{C}'$ of relative objects we dispose also there on finite and birational morphisms. So we can transfer properties (...) for objects and morphisms in $\boldsymbol{C}$ to properties (...)' for relative objects and morphisms. For example (Mult)' and (F.0)' are clearly explained in the latter category. One identifies easily (Mult)' with (B. Ch) using the Definition 6.3.4.

6.4.1 The following scheme (6.4.2) summarizes the logical relations between several postulates

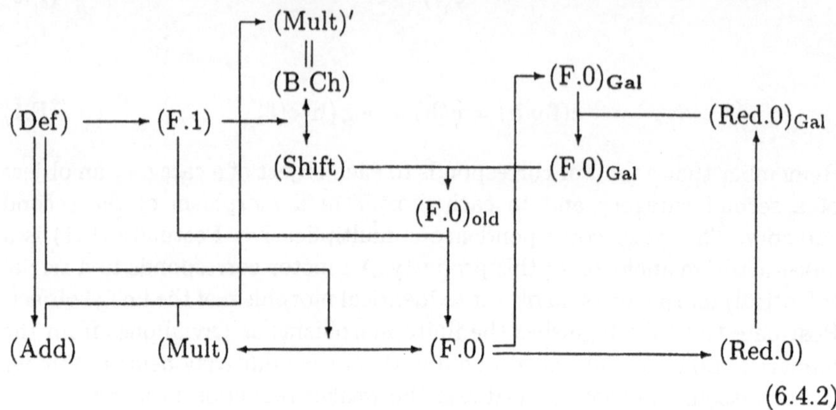

(6.4.2)

Proof of 6.4.1 (Def) $\Longrightarrow$ (Add) is a trivial implication.

(Def) $\Longrightarrow$ (F.1): Consider a composition $\mathbf{g}: \mathbf{Z} \underset{\mathbf{h}}{\rightarrow} \mathbf{X} \underset{\mathbf{f}}{\rightarrow} \mathbf{Y}$ in $\mathcal{C}$. Then

$$\begin{aligned}\mathbf{c}(\mathbf{f} \circ \mathbf{h}) &= \mathbf{c}(\mathbf{g}) = \mathbf{c}(\mathbf{Z}) - \deg(\mathbf{g})\mathbf{c}(\mathbf{Y}) \\ &= [\mathbf{c}(\mathbf{Z}) - \deg(\mathbf{h})\mathbf{c}(\mathbf{X})] + \deg(\mathbf{h})[\mathbf{c}(\mathbf{X}) - \deg(\mathbf{f})\mathbf{c}(\mathbf{Y})] \\ &= \mathbf{c}(\mathbf{h}) + \deg(\mathbf{h})\mathbf{c}(\mathbf{f})\end{aligned}$$

by (Def) and the multiplicative property of degrees.
With definition (Def) the equivalence of (F.0) and (Mult) is trivial.

(F.1), (B.Ch) $\Longrightarrow$ (Shift): With the notations of the base change diagram (6.4.1) we set $\mathbf{g} = \mathbf{f} \circ \varphi' = \varphi \circ \mathbf{f}'$. From (F.1) it follows that

$$\begin{aligned}&\mathbf{c}(\varphi') + \mathbf{c}(\mathbf{f}) = \mathbf{c}(\varphi') + \deg(\varphi')\mathbf{c}(\mathbf{f}) = \mathbf{c}(\mathbf{g}) \\ &= \mathbf{c}(\mathbf{f}') + \deg(\mathbf{f}')\mathbf{c}(\varphi).\end{aligned} \tag{6.4.3}$$

Now $\mathbf{c}(\mathbf{f}) = \mathbf{c}(\mathbf{f}')$ follows from the base change property. The reversed conclusion is also true.

(Mult), (Add) $\Longrightarrow$ (B. Ch): Look at the tautological base change diagram (6.3.8). Using (Add) and (Mult) one obtains

$$\begin{aligned}&\mathbf{c}(\mathbf{X}'/\mathbf{X}) = \mathbf{c}(\mathbf{X}') - \mathbf{c}(\mathbf{X}) = d\mathbf{c}(\mathbf{Y}') - d\mathbf{c}(\mathbf{Y}) = \\ &d[\mathbf{c}(\mathbf{Y}') - \mathbf{c}(\mathbf{Y})] = d\mathbf{c}(\mathbf{Y}'/\mathbf{Y}),\end{aligned}$$

where $d = \deg(\mathbf{f}') = \deg(\mathbf{f})$.

(F.1), $(\mathrm{F.0})_{\mathrm{old}}$ $\Longrightarrow$ (F.0): Let $\mathbf{f}: \mathbf{X} \rightarrow \mathbf{Y}$ be a finite orbital morphism and $\mathbf{g}: Z \underset{\mathbf{h}}{\rightarrow} \mathbf{X} \underset{\mathbf{f}}{\rightarrow} \mathbf{Y}$ a defining composition in the sense of definition 6.3.2. Applying $(\mathrm{F.0})_{\mathrm{old}}$ to $\mathbf{h}, \mathbf{g}$ we know that $\mathbf{c}(\mathbf{h}) = \mathbf{c}(\mathbf{g}) = 0$. The relation (F.1) yields $\mathbf{c}(\mathbf{f}) = 0$.

(Shift), $(\mathrm{F.0})_{\mathrm{Gal}} \Longrightarrow (\mathrm{F.0})_{\mathrm{old}}$: Let $f : X \to \mathbf{Y}_f$ be a finite branch morphism. We construct a base change diagram

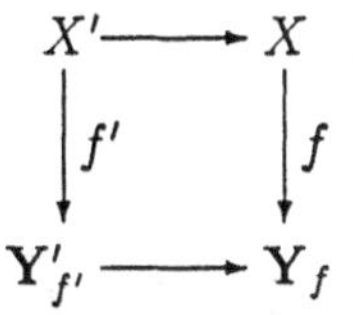

$$\begin{array}{ccc} X' & \longrightarrow & X \\ \downarrow{\scriptstyle f'} & & \downarrow{\scriptstyle f} \\ \mathbf{Y}'_{f'} & \longrightarrow & \mathbf{Y}_f \end{array} \tag{6.4.4}$$

with the reduction f' of a GALOIS-finite morphism $\mathbf{f}'$. Then $\mathbf{c}(f) = 0$ follows immediately from $\mathbf{c}(f') = 0$ by shifting. Let L be a finite GALOIS extension of the function field extension $\mathbb{C}(X)/\mathbb{C}(Y)$ and Z the normalization of X (or Y) in L. Then $X = Z/H$ for a subgroup H of $G = Gal(L/\mathbb{C}(Y))$ and $Y = Z/G$. The corresponding quotient morphisms are denoted by h or g, respectively. Unfortunately the branch locus B_g of g in Y does not look nicely. Removing singularities on Z and blowing up additionally some points we get a morphism $g' : Z' \longrightarrow Y' = Z'/G$ with a nice branch locus, more precisely, with a branch arrangement $\mathbf{B}_{g'}$ on Y' in our sense. On this way we obtain a base change diagram

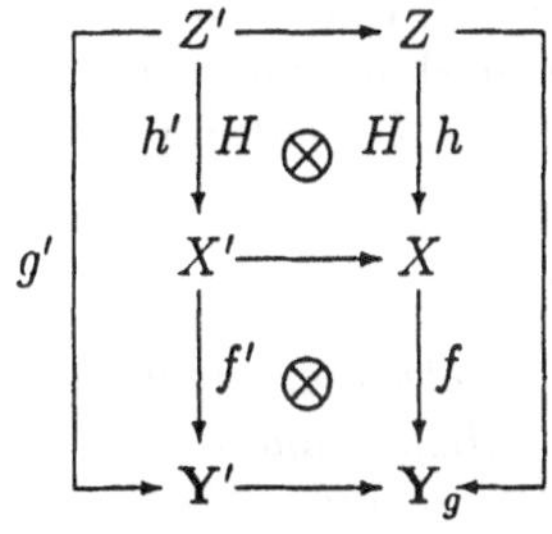

$$\begin{array}{ccccc} & Z' & \longrightarrow & Z & \\ & h' \downarrow H & \otimes & H \downarrow h & \\ g' & X' & \longrightarrow & X & g \\ & \downarrow f' & \otimes & \downarrow f & \\ & \mathbf{Y}' & \longrightarrow & \mathbf{Y}_g & \end{array} \tag{6.4.5}$$

with birational (finite) morphisms represented by the horizontal (vertical) morphisms. The coverings g' and h' are regular by construction, hence also f' is. Obviously, $f' : X' \to \mathbf{Y}'_{f'}$ is the reduction of the GALOIS-finite orbital morphism $\mathbf{f}' : \mathbf{X}'_{h'} \to \mathbf{Y}'_{g'}$. From $(\mathrm{F.0})_{\mathrm{Gal}}$ follows that $\mathbf{c}(f') = 0$. Since (6.4.4) is a base change diagram by definition and (6.4.5), the number $\mathbf{c}(f)$ vanishes by shifting.

We let $f : X \to \mathbf{Y}_f$ be the reduction of a GALOIS-finite covering $\mathbf{f} : \mathbf{X} \to \mathbf{Y}$. We know that $\mathbf{c}(\mathbf{f}) = 0$ by $(\mathrm{F.0})_{\mathbf{Gal}}$, and $(\mathrm{Red.0})_{\mathrm{Gal}}$ implies the vanishing of $\mathbf{c}(f)$.

The remaining implications $(\mathrm{F.0}) \Longrightarrow (\mathrm{F.0})_{\mathbf{Gal}}$ and $(\mathrm{F.0}) \Longrightarrow (\mathrm{Red.0}) \Longrightarrow (\mathrm{Red.0})_{\mathrm{Gal}}$ in scheme (6.4.2) are trivial.

Definition 6.4.2 A function $\mathbf{c} : \mathcal{C} \to \mathbb{Q}$ corresponding to each object $\mathbf{X}$ a rational number $\mathbf{c}(\mathbf{X})$ and to each morphism $\mathbf{f}$ a rational number $\mathbf{c}(\mathbf{f})$ is called *orbital* or *functorial*, if all properties appearing in the scheme (6.4.2) are satisfied. If $\mathbf{c}$ is functorial and not trivial, then we call the numbers $\mathbf{c}(\mathbf{X}), \mathbf{X} \in Ob\ (\mathcal{C})$, orbital heights.

The postulate (Def) is satisfied by definition knowing $\mathbf{c} : Ob(\mathcal{C}) \to \mathbb{Q}$. The dominating property is (Mult). From (Mult) follow all other orbital properties by our scheme (6.4.2). The definition of orbital heights could be restricted to this property. The main problem of this chapter is to prove the multiplicativity of given functions $\mathbf{c} : \mathcal{C} \to \mathbb{Q}$. The scheme (6.4.2) reflects the way of proof. One has to read (6.4.2) downwards forgetting the arrows with ascending directions. Looking at the first row of (6.4.2) we recognize the *key properties* of functoriality:

$$(\text{B.Ch }), (\text{Red.0 })_{\text{Gal}}, (\text{F.0 })_{\mathbf{Gal}}. \tag{6.4.6}$$

The latter property works only with natural arrangements coming from GALOIS coverings. They have been already investigated in the first three chapters. For the proof of $(\text{F.0})_{\mathbf{Gal}}$ we ran there already through a restricted version $(6.4.2_{\mathbf{Gal}}$ of the left part of diagram (6.4.2) and through a one-dimensional version of $(6.4.2)'_{\mathbf{Gal}}$ working only with natural arrangements, which appear in GALOIS-finite coverings. We restrict our attention to CHERN symbols $c \in \mathbb{Q}c_1^2 + \mathbb{Q}c_2$. For natural orbital surfaces $\mathbf{X}$ we already defined $\mathbf{c}(\mathbf{X}) \in \mathbb{Q}$, and we claim that $\mathbf{c} : \mathcal{C} \to \mathbb{Q}$ extends this definition to the category $\mathcal{C}$ of all orbital surfaces.

Proposition 6.4.3 *Under the above assumptions* $\mathbf{c} : \mathcal{C} \to \mathbb{Q}$ *is functorial if and only if* $(\text{Red.0})_{\text{Gal}}$ *is satisfied and* $\mathbf{c} : \mathcal{C}' \to \mathbb{Q}$ *defined by* (Def) *is functorial on the category* $\mathcal{C}'$ *of relative objects* $\mathcal{C}$.

Proof We extended functorially in the first three chapters the numerical results of equivariant K-Theory for algebraic surfaces. The propositions 3.3.9 and 3.4.5 yield $(\text{F.0})_{\mathbf{Gal}}$ for $\mathbf{c} = \mathbf{c}_2$ and τ. So $(\text{F.0})_{\mathbf{Gal}}$ is satisfied for all $\mathbf{c}$ by linear combination. Therefore, we can omit this property in (6.4.6). We dispose in $\mathcal{C}'$ on finite morphisms of old and new type defined by base change diagrams of old and new type (see definitons 6.3.3 and 6.3.4). The functoriality of $\mathbf{c} : \mathcal{C}' \to \mathbb{Q}$ is defined by the multiplicative property (Mult)$'$ for finite morphisms in $\mathcal{C}'$. Looking back to the key properties (6.4.6) we finish the proof of Proposition 6.4.3.

□

Proposition 6.4.3 delegates essentially the proof of the Main Theorem 6.1.1 to the corresponding theorem for the relative category $\mathcal{C}'$. There it will be proved by a geometric Local-Global Principle working with local functorial invariants. We need also the functorial invariants for orbital curves. These one-dimensional contributions have been also prepared earlier in the first two chapters by the investigation of natural orbital curves. The extension to all orbital curves will be managed in the same way as for orbital surfaces going through the scheme (6.4.2). All properties appearing there can be easily transferred to the one-dimensional orbital objects and morphisms. In order to avoid special purely analytical discussions we remark again that all our orbital curves and points live on algebraic surfaces. Moreover, all base change diagrams of orbital curves and points are assumed to live on base change diagrams of orbital surfaces:

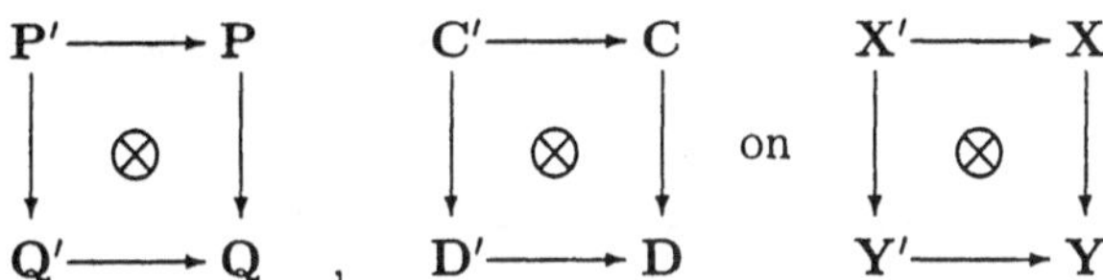

They are defined by this relative property. All properties appearing in the scheme (6.4.2) will be proved also for the category $\mathcal{C}_1$ of $\mathbb{Q}$-orbital curves. The relations between them are the same as for $\mathcal{C}$ drawn in (6.4.2). It is easy to write down an one-dimensional version $(6.4.2)_1$ of (6.4.2). The central property is $(\text{Mult})_1$ again. If we know that $(\text{Mult})_1$ is true, then we can work with all functorial properties. (Shift) is the most active property for the proof of (Mult), and (B. Ch) is a attackable by local methods involving local base change diagrams (see diagram (6.8.18) , below). The same is true in dimension 1.

6.5 Euler and Signature Heights

We start with a heuristical consideration. It is supported by the construction of orbital heights $\mathbf{e}$, $\boldsymbol{\tau}$ of natural orbital surfaces and their orbital decompositions, see Lemma 3.3.3 and Lemma 3.4.2. It is useful to keep in mind the connections between old invariants and the new functorial invariants in order to light up what we have to define in detail. Let $\mathbf{Y} = (Y, \mathbf{B})$ be an arbitrary orbital surface and $\mathbf{c} = \mathbf{e}$ or $\boldsymbol{\tau}$. The arrangement $\mathbf{B}$ is regarded as lower dimensional subobject of $\mathbf{Y}$. Setting all weights appearing in $\mathbf{B}$ equal to 1 we obtain a reduced subobject B of Y. If $\mathbf{Y}$ is an honest orbital surface, then B can be considered as cycle on Y. We draw an auxiliary diagram:

$$\begin{array}{ccccc} B & \longrightarrow & \mathbf{B} & \cdots & B/\mathbf{B} \\ \downarrow & & \downarrow & & \vdots \\ Y & \longrightarrow & \mathbf{Y} & \cdots & Y/\mathbf{Y} \end{array} \tag{6.5.1}$$

The objects on the right-hand side of (6.5.1) are understood as relative objects. The invariants in each row should be connected additively. The scheme language for divisors on algebraic varieties suggests to consider B as subobject of $\mathbf{B}$. We extend this suggestion to Y and $\mathbf{Y}$. Then we claim that the imaginary "factor object" $B/\mathbf{B}$ and $Y/\mathbf{Y}$ are numerically equivalent and work with "old" invariants $c(Y)$, $c(B)$ of Y and B and "new" invariants $\mathbf{c}(\mathbf{Y})$, $\mathbf{c}'(\mathbf{B})$. Using the additive connections in the rows we get the more precise

$$\mathbf{c}(\mathbf{Y}) = [c(Y) - c(B)] + \mathbf{c}'(\mathbf{B})5 \qquad \text{(Claim)}$$

In order to extend earlier definitions given for natural orbital surfaces we will work with $\mathbf{c}(\mathbf{B}) = -\mathbf{c}'(\mathbf{B})$. The arrangement $\mathbf{B}$ is a sum of orbital curves $\mathbf{D}$ and abelian points $\mathbf{Q}$ with rational coefficients. Knowing $\mathbf{c}(\mathbf{D})$ and $\mathbf{c}(\mathbf{Q})$ we will define $\mathbf{c}(\mathbf{B}) = \mathbf{c}(\mathbf{B}^1) + \mathbf{c}(\mathbf{B}^0)$ by (weighted) linear extension.

Definition 6.5.1 For $\mathbf{c} = \mathbf{e}$ or $\boldsymbol{\tau}$ and orbital surfaces $\mathbf{Y} = (Y, \mathbf{B})$, $\mathbf{B} = \mathbf{B}^1 + \mathbf{B}^0 = \sum_i s_i\mathbf{D}_i + \sum_j t_j\mathbf{Q}_j$, we set

$$\mathbf{c}(\mathbf{Y}) = c(Y) - c(B) - \mathbf{c}(\mathbf{B}), \tag{6.5.2.1}$$

$$\mathbf{c}(\mathbf{B}) = \mathbf{c}(\mathbf{B}^1) + \mathbf{c}(\mathbf{B}^0),\ c(B) = c(B^1) + c(B^0) = c(B^0) \tag{6.5.2.2}$$

$$\mathbf{c}(\mathbf{B}^0) = \sum_j t_j\mathbf{c}(\mathbf{Q}_j),\ \mathbf{c}(\mathbf{B}^1) = \sum_i s_i\lambda_c(\mathbf{D}_i)H_c(\mathbf{D}_i), \tag{6.5.2.3}$$

$$c(B^0) = \sum_j t_j c(Q_j),\ c(B^1) = \sum_i s_i\lambda_c(D_i)c(D_i) = 0, \tag{6.5.2.4}$$

with

$$\lambda_c(\mathbf{D}) = \lambda_c(v) = \begin{cases} \left(1 - \frac{1}{v}\right), & c = e \\ \left(v - \frac{1}{v}\right), & c = \tau \end{cases}, v = v_{\mathbf{D}} \tag{6.5.2.5}$$

$$\mathbf{c}(\mathbf{Q}) = \begin{cases} \frac{1}{d}\left(1 - \frac{1}{v}\right)\left(1 - \frac{1}{v'}\right), & c = e \\ 0, & c = \tau \end{cases}, \mathbf{Q} : \underset{v}{\circ} \xrightarrow{\langle d, e\rangle} \underset{v'}{\circ} \tag{6.5.2.6}$$

$$c(Q) = \begin{cases} \left(1 - \frac{1}{d}\right), & c = e \\ -4S(Q) = \frac{1}{3}\left(3l_Q + Tr(Q) - \frac{e}{d} - \frac{e'}{d}\right), & c = 3\tau \end{cases}, \tag{6.5.2.7}$$

where l_Q is the length of the minimal singularity resolution $E_Q \to Q$ of Q and $Tr(Q)$ the trace of the intersection matrix of E_Q. The definition of $H_c(\mathbf{D_i}) = \mathbf{c}(\mathbf{D}_i)$ is delegated to 6.5.2. For $c = ae + b\tau, a, b \in \mathbb{Q}$ we set

$$H_c(\mathbf{Y}) = \mathbf{c}(\mathbf{Y}) = a\mathbf{e}(\mathbf{Y}) + b\tau(\mathbf{Y})5 \tag{6.5.2.8}$$

and call it the (*orbital*) *c-height* of $\mathbf{Y}$, compare with and 4.10.1.

Definition 6.5.2 Let $\mathbf{D} = (U, vD; \sum_i r_i \mathbf{Q}_i)$ be an orbital curve in the sense of definition 6.2.3 and $\mathbf{c} = \tau$ or $\mathbf{e}$. We set

$$\begin{aligned} H_c(\mathbf{D}) = \mathbf{c}(\mathbf{D}) &= \left(c(\mathbf{D}) - \sum_i r_i c^1(Q_i)\right) - \sum_i r_i \mathbf{c}^1(\mathbf{Q}_i) \\ &= c(\mathbf{D}) - \sum_i r_i c^1(\mathbf{Q}_i), \end{aligned} \tag{6.5.3.1}$$

with

$$c(\mathbf{D}) = \begin{cases} e(D) = 2 - 2g(D), & c = e \\ (D^2)/3v, & c = \tau \end{cases} \tag{6.5.3.2}$$

$$c^1(\mathbf{Q}_i) = \begin{cases} 1 - \frac{1}{v_i d_i}, & c = e \\ -e_i/3vd_i, & c = \tau \end{cases}, \mathbf{Q}_i : \underset{v}{\circ} \xrightarrow{\langle d_i, e_i\rangle} \underset{v_i}{\circ} \tag{6.5.3.3}$$

$$\mathbf{c}^1(\mathbf{Q}_i) = c^1(\mathbf{Q}_i) - c^1(Q_i) = \begin{cases} \frac{1}{d_i}\left(1 - \frac{1}{v_i}\right), & c = e \\ \frac{e_i}{3d_i}\left(1 - \frac{1}{v}\right), & c = \tau \end{cases} \tag{6.5.3.4}$$

$H_e(\mathbf{D}) = \mathbf{e}(\mathbf{D})$ is called the EULER *height* and $H_\tau(\mathbf{D}) = \tau(\mathbf{D})$ the *signature height* of $\mathbf{D}$.

The definition 6.5.2 extends the earlier definitions (2.2.3) and (2.4.3) of heights of natural orbital curves. The definition 6.5.1 is compatible with the earlier definitions 3.3.1 and 3.4.1 because of the orbital decompositions 3.3.3 and 3.4.2. Indeed, we can write (5.2.1) as orbital decomposition

$$\mathbf{c}(\mathbf{Y}) = \mathbf{c}(Y) - \mathbf{c}(\mathbf{B}) = \mathbf{c}(Y) - \mathbf{c}(\mathbf{B}^1) - \mathbf{c}(\mathbf{B}^0), \tag{6.5.4}$$

where

$$\mathbf{c}(Y) = c(Y) - c(B) = c(Y) - c(B^0) = c(Y) - c(SgY) \tag{6.5.5}$$

in correspondence with the definition

$$\mathbf{c}(Y) = c(Y) - c(B) - \mathbf{c}(B),\ \mathbf{c}(B) = 0, c(B^1) = 0.$$

The orbital decomposition (6.5.4) contains only new invariants $\mathbf{c}$. Sometimes it is useful to express $\mathbf{c}(\mathbf{Y})$ by means of "old" invariants c, especially for applications. This is possible for old finite morphisms.

Proposition 6.5.3 *Let* $\mathbf{Y} = (Y, \mathbf{B})$ *be an honest orbital surface. Then, for* $c = e, \tau$ *it holds that*

$$\mathbf{c}(\mathbf{Y}) = c(Y) - \sum_{\mathbf{D}} s_{\mathbf{D}} \lambda_c(\mathbf{D}) c(\mathbf{D}) - \sum_{\mathbf{Q}} t_{\mathbf{Q}} c(\mathbf{Q}), \tag{6.5.6.1}$$

where $\mathbf{B} = \sum_{\mathbf{D}} s_{\mathbf{D}} \mathbf{D} + \sum_{\mathbf{Q}} t_{\mathbf{Q}} \mathbf{Q}$, *and for* $\mathbf{Q} : \underset{v}{\circ} \xrightarrow{\langle d, e\rangle} \underset{v'}{\circ}$

$$c(\mathbf{Q}) = \begin{cases} -1 + \frac{1}{v} + \frac{1}{v'} - \frac{1}{vv'd}, & c = e \\ \frac{1}{3}\left(3l_Q + T_r(Q) - \frac{e}{v^2 d} - \frac{e'}{v'^2 d}\right), & c = \tau \end{cases} \tag{6.5.6.2}$$

Proof We use orbital curves and points as indices. For the point arrangements $\mathbf{B}^0(\mathbf{D})$ of $\mathbf{D}$ we need double indices:

$$\mathbf{B}^0(\mathbf{D}) = \sum_{\mathbf{Q} \in \mathbf{D}} r_{\mathbf{Q}}^{\mathbf{D}} \cdot \mathbf{Q}. \tag{6.5.7}$$

The difference Δ of the right-hand sides of (6.5.4) and (6.5.6.1) is equal to

$$\Delta = (\mathbf{c}(Y) - c(Y)) - \sum_{\mathbf{D}} s_{\mathbf{D}} \lambda_c(\mathbf{D})(\mathbf{c}(\mathbf{D}) - c(\mathbf{D}))$$
$$- \sum_{\mathbf{Q}} t_{\mathbf{Q}}(\mathbf{c}(\mathbf{Q}) - c(\mathbf{Q}))$$

We have to show that Δ vanishes. From (6.5.5), (6.5.3.1) and (6.5.3.2) we obtain

$$\Delta = -c(SgY) + \sum_{\mathbf{D}} s_{\mathbf{D}}\lambda_c(\mathbf{D}) \sum_{\mathbf{Q}\in\mathbf{D}} r^{\mathbf{D}}_{\mathbf{Q}} c^1(\mathbf{Q})$$

$$-\sum_{\mathbf{Q}} t_{\mathbf{Q}}(\mathrm{c}(\mathbf{Q}) - c(\mathbf{Q})) \tag{6.5.8}$$

Now we need the following simple relation

$$s_{\mathbf{D}} \cdot r^{\mathbf{D}}_{\mathbf{Q}} = t_{\mathbf{Q}}. \tag{6.5.9}$$

It follows immediately from definition 6.2.9, especially (iv).

The definitions (6.5.2.5), (6.5.2.6), (6.5.2.7), (6.5.3.3) and (6.5.6.2) yield the following relation

$$c(\mathbf{Q}) - \mathrm{c}(\mathbf{Q}) - c(Q) = -\lambda_c(\mathbf{Q})c^1(\mathbf{Q}) - \lambda_c(\mathbf{Q}^\perp)c^1(\mathbf{Q}^\perp), \tag{6.5.10}$$

where $\mathbf{Q}^\perp : \underset{v'\ \langle d,e'\rangle\ v}{\circ\!\!\longrightarrow\!\!\circ}$ is the opposite point of $\mathbf{Q} : \underset{v\ \langle d,e\rangle\ v'}{\circ\!\!\longrightarrow\!\!\circ}$ and

$$\lambda_c(\mathbf{Q}) = \lambda_c(v). \tag{6.5.11}$$

Since $\mathbf{B}$ is an honest arrangement we can decompose $c(SgY)$ by definition 6.2.9, (iii) in the following manner:

$$c(SgY) = \sum_{\mathbf{Q}} t_{\mathbf{Q}} c(Q).$$

Together with (6.5.9) and (6.5.10) we can write the identity (6.5.8) as

$$\Delta = \sum_{\mathbf{D}} \lambda_c(\mathbf{D}) \sum_{\mathbf{Q}\in\mathbf{D}} t_{\mathbf{Q}} c^1(\mathbf{Q})$$

$$-\sum_{\mathbf{Q}} t_{\mathbf{Q}}[\lambda_c(\mathbf{Q})c^1(\mathbf{Q}) + \lambda_c(\mathbf{Q}^\perp)c^1(\mathbf{Q}^\perp)] \tag{6.5.12}$$

For $\mathbf{Q} \in \mathbf{D}$ we have $\lambda_c(\mathbf{D}) = \lambda_c(\mathbf{Q})$ by definition (6.5.11). Each point $\mathbf{Q}$ in the double sum of (6.5.12) appears together with its opposite point $\mathbf{Q}^\perp$. Therefore Δ vanishes. The proposition is proved.

□

6.6 Reduction of Galois-Finite Morphisms

This section is dedicated to the proof of the functional property $(\text{Red.0})_{\text{Gal}}$ appearing in the first row of the scheme (6.4.2)

Proposition 6.6.1 *Let* $\mathbf{f}: \mathbf{X} \to \mathbf{Y}$ *be a GALOIS-finite orbital morphism,* $f: X \to \mathbf{Y}_f$ *its reduction and* $c = e$ *or* τ. *If* $\mathrm{c}(\mathbf{X}/\mathbf{Y})$ *vanishes, then also* $\mathrm{c}(X/\mathbf{Y}_f) = 0$.

Proof By definition we have a commutative diagram

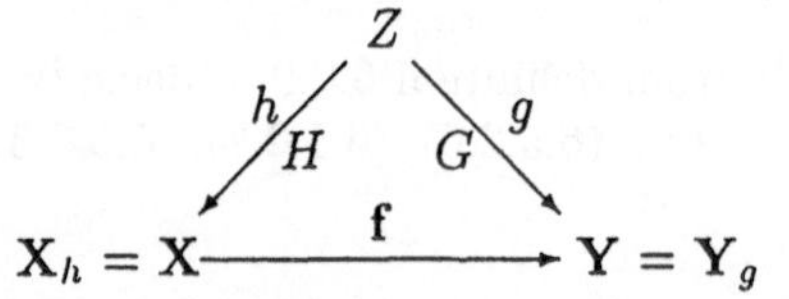

in $\mathcal{C}$ with GALOIS coverings h, g corresponding to GALOIS groups $H \subset G$ acting on Z. Fixing notations we set

$$\mathbf{X} = \mathbf{X}_h = (X, \mathbf{B}_h), \mathbf{Y} = \mathbf{Y}_g = (Y, \mathbf{B}_g), \mathbf{Y}_f = (Y, \mathbf{B}_f),$$

$$\mathbf{B}_h = \sum_c \mathbf{C}^h + \sum_P \mathbf{P}^h,\ \mathbf{B}_g = \sum_D \mathbf{D}^g + \sum_Q \mathbf{Q}^g$$

$$\mathbf{B}_f = \sum_C s_C \cdot \mathbf{D}_C + \sum_P t_P \mathbf{Q}_P,\ \mathbf{D}_C = (U_D, v_C \cdot D; \mathbf{B}^0(\mathbf{D}_C)),$$

$$\mathbf{B}^0(\mathbf{D}_C) = \sum_{P \in C} r_P^C \cdot \mathbf{Q}_P.$$

The electronic and atomic graphs fix further notations:

$\mathbf{P}^h$: 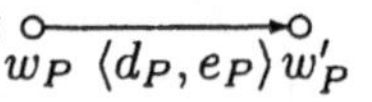

, $\mathbf{Q}^g$: 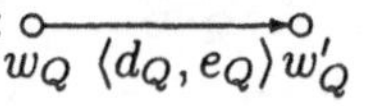

, $\mathbf{Q}_P$: 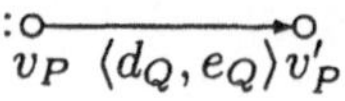

(6.6.1)

$\mathbf{C}^h$: 

, $\mathbf{D}^g$: 

(6.6.2)

So we use the letter w for GALOIS weights and the letter v for the branch weights of f. For

$$\begin{array}{ccccccc} \mathbf{P}^h & \in & \mathbf{C}^h & \qquad & P & \in & U_P \\ \downarrow & & \downarrow & & \downarrow & & \Big\downarrow {\scriptstyle A_P | f_P} \\ \mathbf{Q}^g & \in & \mathbf{D}^g & & Q & \in & U_Q \end{array}$$

A_P a finite abelian group (see Corollary 1.12.3), we get the following relations:

$$w_D/w_C = v_C = w_Q/w_P = v_P, \; w'_Q/w'_P = v'_P; 5 \tag{6.6.3}$$

$$\deg(f_P) = v_P \cdot v'_P d_Q/d_P. \tag{6.6.4}$$

The assumption of the proposition can be written as

$$\begin{aligned} &\mathbf{c}(X) - \mathbf{c}(\mathbf{B}^1_h) - \mathbf{c}(\mathbf{B}^0_h) = \\ &\deg(f)\big[\mathbf{c}(Y) - \mathbf{c}(\mathbf{B}^1_g) - \mathbf{c}(\mathbf{B}^0_g)\big]. \end{aligned}$$

We used the orbital decomposition (6.5.4) and the equivalence of (Mult) and (F.0) (see (6.4.2)). It follows that

$$\begin{aligned} &\mathbf{c}(X)/\deg(f) = \mathbf{c}(Y) - \sum_D \lambda_c(w_D)\mathbf{c}(\mathbf{D}^g) \\ &+ \frac{1}{\deg(f)}\sum_C \lambda_c(w_C)\mathbf{c}(\mathbf{C}^h) - \mathbf{c}(\mathbf{B}^0_g) \\ &+ \mathbf{c}(\mathbf{B}^0_h)/\deg(f) \end{aligned}$$

Since $\mathbf{B}_f$ is an honest arrangement we can use property (iii) of definition 6.2.9 to obtain

$$\begin{aligned} &\mathbf{c}(X)/\deg(f) = \mathbf{c}(Y) - \sum_D \sum_{C\to D} s_C\lambda_c(w_D)\mathbf{c}(\mathbf{D}^g) \\ &+ \sum_D \sum_{C\to D} \frac{1}{\deg(f)}\lambda_c(w_C)\mathbf{c}(\mathbf{C}^h) - \mathbf{c}(\mathbf{B}^0_g) \\ &+ \mathbf{c}(\mathbf{B}^0_h)/\deg(f) \end{aligned} \tag{6.6.5}$$

The coefficient s_C is by definition (6.2.3) the relative degree $v_C \cdot \deg(f_C)/\deg(f)$, hence

$$\frac{1}{\deg(f)}\mathbf{c}(\mathbf{C}^h) = \frac{\deg(f_C)}{\deg(f)}\mathbf{c}(\mathbf{D}^g) = \frac{1}{v_C}s_C \cdot \mathbf{c}(\mathbf{D}^g) \tag{6.6.6}$$

We used the functorial property (Mult) of $\mathbf{c}$ for the GALOIS-finite covering $\mathbf{C}^h \to \mathbf{D}^g$ of orbital curves (Theorems 2.2.4 and 2.4.2). The identity (6.6.6) allows to join the two double sums in (6.6.5):

$$\begin{aligned}
&\mathbf{c}(X)/\deg(f) = \mathbf{c}(Y) - \sum_D \sum_{C\to D} s_C[\lambda_c(w_D) - \lambda_c(w_C)/v_C]\mathbf{c}(\mathbf{D}^g)\\
&\quad - \mathbf{c}(\mathbf{B}_g^0) + \mathbf{c}(\mathbf{B}_h^0)/\deg(f),\\
&\mathbf{c}(X)/\deg(f) = \mathbf{c}(Y) - \sum_D \sum_{C\to D} s_C[\lambda_c(w_D) - \lambda_c(w_C)/v_C]c(\mathbf{D}^g)\\
&\quad + \sum_D \sum_{C\to D} s_C[\lambda_c(w_D) - \lambda_c(w_C)/v_C]c^1(\mathbf{D}^g)\\
&\quad - \mathbf{c}(\mathbf{B}_g^0) + \mathbf{c}(\mathbf{B}_h^0)/\deg(f)
\end{aligned} \tag{6.6.7}$$

The coefficients in the double sums are

$$\lambda_c(w_D) - \lambda_c(w_C)/v_C = \begin{cases} 1 - 1/v_C = \lambda_e(v_C), & c = e \\ \left(v_C - \frac{1}{v_C}\right)w_C = \lambda_\tau(v_C)w_C, & c = \tau \end{cases} \tag{6.6.8}$$

We used the first identity of (6.6.3) and the definitions of λ_c in (6.5.2.5). Now we change over in the first double sum of (6.6.7) from $c(\mathbf{D}^g)$ to $c(\mathbf{D}_C)$. For $c = e$ there is no difference, but for $c = \tau$ we have to substitute w_D by v_C according to (6.5.3.2). The coefficients (6.6.8) compensate the situation

$$[\lambda_c(w_D) - \lambda_c(w_C)/v_C]c(\mathbf{D}^g) = \lambda_c(v_C)c(\mathbf{D}_C). \tag{6.6.9}$$

The second double sum in (6.6.7) is denoted by T or E for $c = \tau$ or e, respectively. We transform them separately.

$$\begin{aligned}
T &= \sum_D \sum_{C\to D} s_C\left(1 - \frac{1}{v_C^2}\right)w_D \cdot \tau^1(\mathbf{D}^g)\\
&= -\sum_D \sum_{C\to D} s_C\left(1 - \frac{1}{v_C^2}\right)\frac{1}{3}\sum_{Q\in D} e_Q/d_Q\\
&= -\sum_D \sum_{C\to D} \frac{1}{3}\left(1 - \frac{1}{v_C^2}\right)s_C \sum_{P\in C} r_P^C e_Q/d_Q\\
&= -\sum_D \sum_{C\to D} \frac{1}{3}\left(1 - \frac{1}{v_C^2}\right) \sum_{\substack{P\in C\\ P\to Q}} t_P e_Q/d_Q
\end{aligned} \tag{6.6.10}$$

$$= -\sum_{\mathbf{Q}_P} \frac{1}{3} t_P \left[\left(1 - \frac{1}{v_P^2}\right) e_Q / d_Q + \left(1 - \frac{1}{v_P'^2}\right) e_Q' / d_Q \right]$$

We made use of honest arrangement properties 6.2.9, (iii), (iv), see also (6.5.9). By definition (6.5.6.2) we can write

$$T = - \sum_{\mathbf{Q}_P \in \mathbf{B}_f^0} t_P [\tau(\mathbf{Q}_P) - \tau(Q)] = - \sum_{\mathbf{Q}_P \in \mathbf{B}_f^0} t_P \tau(\mathbf{Q}_P) + \sum_{Q} \tau(Q)$$

We substitute T in (6.6.7) and remember that $\tau\left(\mathbf{B}_g^0\right) = \tau(\mathbf{B}_h^0) = \mathbf{0}$ by definition (6.5.2.6).

$$\begin{aligned} \tau(X)/\deg(f) = \tau(Y) &- \sum_{D} \sum_{C \to D} s_C \cdot \lambda_\tau(v_C) \tau(\mathbf{D}_C) \\ &- \sum t_P \tau(\mathbf{Q}_P) + \tau(SgY) \\ &= \tau(Y) - \sum s_\mathbf{D} \lambda_\tau(\mathbf{D}) \tau(\mathbf{D}) - \sum t_\mathbf{Q} \tau(\mathbf{Q}), \end{aligned}$$

hence

$$\boldsymbol{\tau(X) = \deg(f) \cdot \tau(Y)}$$

by Proposition 6.5.3. The multiplicative property for τ at f is verified.

A similar procedure works for $c = e$. We transform

$$\begin{aligned} E &= \sum_{D} \sum_{C \to D} s_C \left(1 - \frac{1}{v_C}\right) e^1(\mathbf{D}^g) \\ &= \sum_{D} \left(\sum_{C \to D} s_C \right) e^1(\mathbf{D}^g) - \sum_{D} \sum_{C \to D} \frac{s_C}{v_C} \sum_{Q \in D} \left(1 - \frac{1}{w_Q' d_Q}\right) \\ &= \sum_{D} e^1(\mathbf{D}^g) - \sum_{D} \sum_{C \to D} (s_C / v_C) \sum_{Q \in D} \sum_{C \ni P \to Q} r_P^C \left(1 - \frac{1}{w_Q' d_Q}\right) \\ &= \sum_{D} e^1(\mathbf{D}^g) - \sum_{D} \sum_{C \to D} \sum_{Q \in D} \sum_{C \ni P \to Q} \frac{1}{v_P} t_P \left(1 - \frac{1}{w_Q' d_Q}\right) \end{aligned}$$

In the last step we used relations (6.6.3) and 6.2.9, (iv). We change over to point sums taking into account that together with $\mathbf{Q}_P$ the twin brother $\mathbf{Q}_P^{\perp}$ has to appear. One obtains

$$
\begin{aligned}
E &= \sum_Q \sum_{P\to Q} t_P \left[\left(1-\frac{1}{w_Q d_Q}\right)+\left(1-\frac{1}{w'_Q d_Q}\right)\right] \\
&- \sum_Q \sum_{P\to Q} t_P \left[\frac{1}{v_P}\left(1-\frac{1}{w'_Q d_Q}\right)+\frac{1}{v'_P}\left(1-\frac{1}{w_Q d_Q}\right)\right] \\
&= \sum_Q \sum_{P\to Q} t_P \left[1-\frac{1}{w_Q d_Q}+1-\frac{1}{w'_Q d_Q}-\frac{1}{v_P}\right] \\
&+ \sum_Q \sum_{P\to Q} t_P \left[\frac{1}{v_P w'_Q d_Q}-\frac{1}{v'_P}+\frac{1}{v'_P w_Q d_Q}\right]
\end{aligned}
\tag{6.6.11}
$$

The last two summands in (6.6.7) can be written in the following manner:

$$
\begin{aligned}
\mathbf{e}(\mathbf{B}^0_g) &= \sum_Q \frac{1}{d_Q}\left(1-\frac{1}{w_Q}\right)\left(1-\frac{1}{w'_Q}\right) \\
&= \sum_Q \sum_{P\to Q} t_P \left[\frac{1}{d_Q}-\frac{1}{w_Q d_Q}-\frac{1}{w'_Q d_Q}+\frac{1}{w_Q w'_Q d_Q}\right]
\end{aligned}
\tag{6.6.12}
$$

$$
\begin{aligned}
\mathbf{e}(\mathbf{B}^0_h)/\deg(f) &= \sum_P \mathbf{e}(\mathbf{P}^h)/\deg(f) = \sum_P t_P \mathbf{e}(\mathbf{P^h})/\deg(f_P) \\
&= \sum_Q \sum_{P\to Q} \frac{d_P}{v_P v'_P d_Q} t_P \frac{1}{d_P}\left(1-\frac{1}{w_P}\right)\left(1-\frac{1}{w'_P}\right) \\
&= \sum_Q \sum_{P\to Q} t_P \left[\frac{1}{v_P v'_P d_Q}-\frac{1}{w_Q v'_P d_Q}-\frac{1}{v_P w'_Q d_Q}+\frac{1}{w_Q w'_Q d_Q}\right]
\end{aligned}
\tag{6.6.13}
$$

Summing up one receives

$$
\begin{aligned}
&E - \mathbf{e}(\mathbf{B}^0_g) + \mathbf{e}(\mathbf{B}^0_h)/\deg(f) \\
&= \sum_Q \sum_{P\to Q} t_P \left[\left(1-\frac{1}{d_Q}\right)+\left(1-\frac{1}{v_P}-\frac{1}{v'_P}+\frac{1}{v_P v'_P d_Q}\right)\right] \\
&= \sum_Q \left(\sum_{P\to Q} t_P\right)\left(1-\frac{1}{d_Q}\right) - \sum_P t_P e(\mathbf{Q}_P) \\
&= \sum_Q e(Q) - \sum_{\mathbf{Q}_P} t_P e(\mathbf{Q}_P)
\end{aligned}
$$

Now look back to (6.6.7) again. Then we get together with (6.6.9) and (6.5.6.1)

$$\begin{aligned}
&\mathbf{e}(X)/\deg(f)\\
&= \mathbf{e}(Y) - \sum_{\mathbf{D}_C} s_C \lambda_e(\mathbf{D}_C) e(\mathbf{D}_C) - \sum_{\mathbf{Q}_P} t_P e(\mathbf{Q}_P) + c(SgY)\\
&= e(Y) - \sum_{\mathbf{D}_C} s_C \lambda_e(\mathbf{D}_C) e(\mathbf{D}_C) - \sum_{\mathbf{Q}_P} t_P e(\mathbf{Q}_P) = \mathbf{e}(\mathbf{Y}_f).
\end{aligned}$$

The proof of Proposition 6.6.1 is finished.

□

We must dispose on the same functorial properties appearing in (6.4.2) for the invariants **c** of orbital curves. The corresponding properties are denoted by $(\cdots)_1$ or $(\cdots)^1$ instead of $(\cdots)$ We want to prove $(\text{Red.0})^1_{\text{Gal}}$.

Proposition 6.6.2 *Let* $\mathbf{f} : \mathbf{C} \to \mathbf{D}$ *be a GALOIS-finite morphism of orbital curves,* $f : C \to \mathbf{D}_f$ *its reduction and* $c = e$ *or* τ*. If* $\mathbf{c}(\mathbf{C}/\mathbf{D})$ *vanishes, then also* $\mathbf{c}(C/\mathbf{D}_f) = 0$.

Proof By definition we have a commutative diagram:

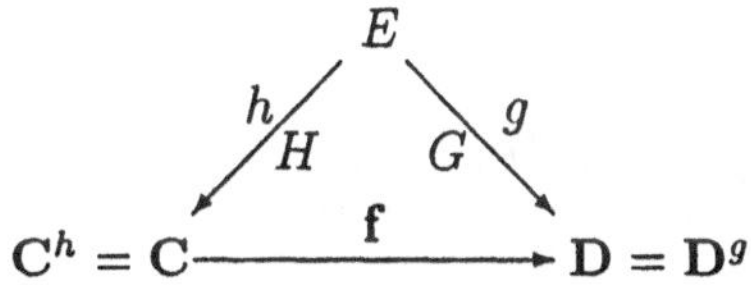

$$\begin{array}{ccccc}
 & & E & & \\
 & {}^{h}\swarrow_{H} & & {}_{G}\searrow^{g} & \\
\mathbf{C}^h = \mathbf{C} & & \xrightarrow{\quad \mathbf{f} \quad} & & \mathbf{D} = \mathbf{D}^g
\end{array} \tag{6.6.14}$$

in category $\mathcal{C}^1$ of orbital curves with GALOIS coverings h, g corresponding to GALOIS groups $H \subset G$ acting on (U_E, E). Fixing notations for the point arrangements we set

$$\mathbf{B}^0(\mathbf{C}) = \sum_P \mathbf{P}^h, \ \mathbf{B}^0(\mathbf{D}) = \sum_Q \mathbf{Q}^g,$$
$$\mathbf{B}^0(\mathbf{D}_f) = \mathbf{B}^0(\mathbf{D}_C) = \sum_P r_P \mathbf{Q}_P$$

Furthermore, we use the same notations as described in the electronic and atomic graphs (6.6.1) and (6.6.2). Then the relations (6.6.3) and (6.6.4) are satisfied. The degree of $\mathbf{f} : \mathbf{C} \to \mathbf{D}$ and of $f : C \to \mathbf{D}_C$ is equal to the degree $\deg(f_C)$ of the restriction $f_C : C \to D$. The assumption of the proposition is

$$\mathbf{c}(\mathbf{C}^h) = \deg(f_C) \cdot \mathbf{c}(\mathbf{D}^g). \tag{6.6.15}$$

We have to prove that

$$\mathbf{c}(C) = \deg(f_C) \cdot \mathbf{c}(\mathbf{D}_C). \tag{6.6.16}$$

We prove (6.6.16) separately for $\mathbf{c} = \mathbf{e}$ and $\mathbf{c} = \tau$. For $c = e$ we write (6.6.15) explicitly by means of definition 6.5.2:

$$e(C) - \left(1 - \frac{1}{w'_P d_P}\right) = \deg(f_C) \cdot \left(e(D) - \sum_Q \left(1 - \frac{1}{w'_Q d_Q}\right)\right).$$

On both sides we add $\sum_P \left(\frac{1}{d_P} - \frac{1}{w'_P d_P}\right)$. Then

$$\mathbf{e}(C) = \deg(f_C) e(D) - \sum_Q \deg(f_C)\left(1 - \frac{1}{w'_Q d_Q}\right) + \sum_P \left(\frac{1}{d_P} - \frac{1}{w'_P d_P}\right). \tag{6.6.17}$$

Now we use the third relation of (iii) in definition 6.2.9, or better its version (6.2.7), and (6.6.3), (6.6.4):

$$v_C - \deg(f_C) = \sum_{P \to Q} \deg(f_P) = \sum_{P \to Q} v_P v'_P d_Q / d_P = v_C \sum_{P \to Q} \frac{w'_Q d_Q}{w'_P d_P}$$

A corresponding substitution of $\deg(f_C)$ in (6.6.17) yields

$$\begin{aligned} \mathbf{e}(C) &= \deg(f_C) e(D) - \sum_Q \sum_{P \to Q} \frac{w'_Q d_Q}{w'_P d_P}\left(1 - \frac{1}{w'_Q d_Q}\right) \\ &+ \sum_Q \sum_{P \to Q} \left(\frac{1}{d_P} - \frac{1}{w'_P d_P}\right) \\ &= \deg(f_C) e(D) - \sum_Q \sum_{P \to Q} v'_P \frac{d_Q}{d_P}\left(1 - \frac{1}{v'_P d_Q}\right) \end{aligned}$$

Remember that r_P is defined as relative degree

$$r_P = \frac{\deg(f_P)}{\deg(f^C)} = \frac{\deg(f_P)}{v_C \cdot \deg(f_C)}.$$

Together with (6.6.3) and (6.6.4) it follows that

$$\deg(f_C)\cdot r_P = v'_P d_Q/d_P,$$

hence

$$\mathbf{e}(C) = \deg(f_C)\left[e(D) - \sum_{\mathbf{Q}_P} r_P\left(1 - \frac{1}{v'_P d_Q}\right)\right] = \deg(f_C)\cdot \mathbf{e}(\mathbf{D}_f).$$

Now we prove (6.6.16) for $c=\tau$. The assumption (6.6.15) can be written explicitly as

$$\left((C^2) + \sum_P e_P/d_P\right)/3w_C = \deg(f_C)\cdot\left[(D^2) + \sum_Q e_Q/d_Q\right]/3w_D,$$

or

$$\tau(C) = \deg(f_C)\left[(D^2) + \sum_Q e_Q/d_Q\right]/3v_C.$$

Since $\sum_{P\to Q} r_P = 1$ by 6.2.9, (iii) it follows that

$$\tau(C) = \deg(f_C)\left[(D^2) + \sum_Q \sum_{P\to Q} r_P e_Q/d_Q\right]/3v_C$$

$$= \deg(f_C)\left[\tau(\mathbf{D}_C) - \sum_{\mathbf{Q}_P} r_P \tau^1(\mathbf{Q}_P)\right] = \deg(f_C)\cdot\tau(\mathbf{D}_f).$$

The proposition is proved.

□

Remark 6.6.3 Up to (B. Ch) all functorial properties in the first row of the logical scheme (6.4.2) have been verified. The base change property must be first studied locally. This will be done in the next section.

6.7 Local Base Changes

We call (6.7.1) a *local base change diagram*

$$\begin{array}{ccc} \mathbf{P}_0 & \xrightarrow{\psi^0_P} & \mathbf{P} \\ \downarrow \mathbf{f}'_{\mathbf{P}} & & \downarrow \mathbf{f}_{\mathbf{P}} \\ \mathbf{Q}_0 & \xrightarrow{\varphi^0_Q} & \mathbf{Q} \end{array} \tag{6.7.1}$$

if $\mathbf{P}_0$, $\mathbf{P}$, $\mathbf{Q}_0$, $\mathbf{Q}$ are arranged abelian points and if there exists a globalization

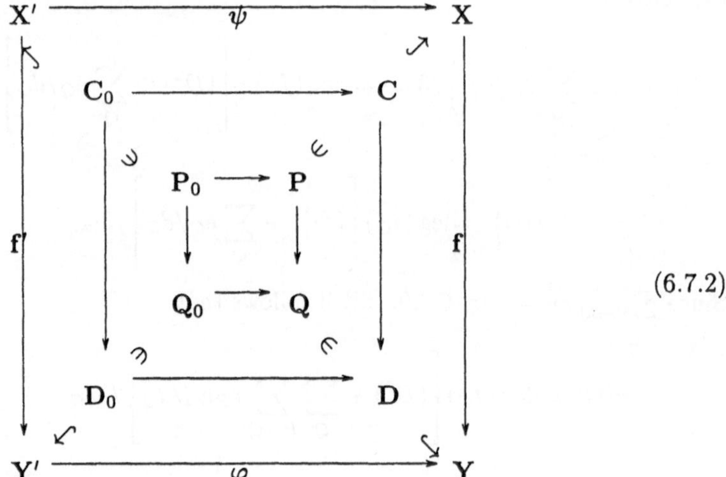

(6.7.2)

such that the external square is a base change diagram in $\mathcal{C}$ with the following properties:

(i) The exceptional loci of ψ and φ are disjoint sums of linear trees of projective lines whose intersection points are cyclic singularities.

(ii) The restrictions of ψ and φ to C_0 or D_0, respectively, induce isomorphisms of smooth algebraic curves.

The inclusions in diagram (6.7.2) mean that the weights of curves and points have been chosen correctly. For example the first weight of $\mathbf{Q}$ is the weight of $\mathbf{D}$, and the weight of $\mathbf{D}$ is equal to 1 or the weight of $\mathbf{D}$ as component of the arrangement of $\mathbf{Y}$. More precisely, if $\mathbf{f} : \mathbf{X} \to \mathbf{Y}$ is an old finite morphism, then $\mathbf{D} = \mathbf{D}_C$ with the corresponding ramification weight v_C. For a new finite morphism we have $Z \to \mathbf{X} \to \mathbf{Y}$ with old finite morphisms $Z \to \mathbf{X}$ and $Z \to \mathbf{Y}$. Then we suppose that $\mathbf{C} = \mathbf{C}_E$, $\mathbf{D} = \mathbf{D}_E$ for a preimage curve $E \subset Z$.

The external base change square of (6.7.2) induces also mixed local base change diagrams. For $\mathbf{Q} \in \mathbf{Y}$ we first define the *mixed base change diagram*

$$\begin{array}{ccc} \mathbf{E_Q}(\varphi) & \longrightarrow & \mathbf{Q} \\ \downarrow & & \downarrow \\ \mathbf{Y'} & \xrightarrow{\varphi} & \mathbf{Y} \end{array} \tag{6.7.3}$$

where $E_Q(\varphi)$ is the reduced preimage of Q on Y' along φ and $\mathbf{E_Q}(\varphi)$ the natural linear arrangement supported by $E_Q(\varphi)$ with weights borrowed from $\mathbf{Y'}$ in the same manner as for $\mathbf{D} \subset \mathbf{Y}$. For $\mathbf{P} \in \mathbf{X}$ we get a similar mixed base change diagram. On this way the external square of (6.7.2) can be restricted to the following mixed local base change diagram.

$$\begin{array}{ccc} \mathbf{E_P}(\psi) & \xrightarrow{\psi_P} & \mathbf{P} \\ \downarrow & & \downarrow \\ \mathbf{E_Q}(\varphi) & \xrightarrow{\varphi_Q} & \mathbf{Q} \end{array} \tag{6.7.4}$$

The horizontal morphisms are supported by analytic contractions, e.g. $U_P(\psi) \to U_P$, $U_P(\psi)$ an analytic neighbourhood of $E_P(\psi)$ on X' and U_P the image neighbourhood of P on X.

The most important property of local base change diagrams is the following:

6.7.1 The vertical arrows in diagrams (D.7.1) and (6.7.4) are quotient morphisms by a finite abelian group. The group acts on each component of $E_P(\psi)$ fixing intersection points.

This follows immediately from Corollary 1.12.3 and from STEIN'S extension theorem 1.12.2 combined with HARTOG'S theorem applied to automorphisms restricted to $U_P^* = U_P \setminus \{P\} \cong U_P(\psi) \setminus E_P(\psi) \subset U_P(\psi)$. They can be extended to $U_P(\psi)$. For an old finite morphism $\mathbf{f} = f : X \to \mathbf{Y}$ the weights of the components of $\mathbf{E_Q}(\varphi)$ are the quotient weights of the local group action along $E_P(\psi)$. For new finite morphisms we have a locally GALOIS-finite situation because of a composition $Z \to \mathbf{X} \to \mathbf{Y}$ with old finite morphisms $Z \to \mathbf{X}$ and $Z \to \mathbf{Y}$ which are locally abelian.

The GALOIS-finite local morphisms and diagrams have been investigated carefully from a numerical point of view in the first two chapters. There we already introduced local base changes as morphisms of the relative category by means of group actions. We only supposed that there exists a globalization (6.7.2). From a local numerical point of view one can forget the special choice of globalization. We found already local invariants with a functorial behaviour. With the notations of (6.7.1), (6.7.2) and the graphs

$$\mathbf{Q}^0: \quad \underset{w=w_0}{\circ} \xrightarrow{\langle d_0, e_0\rangle} \underset{w_0'}{\circ}, \qquad \mathbf{Q}: \quad \underset{w}{\circ} \xrightarrow{\langle d, e\rangle} \underset{w'}{\circ}$$

we define in analogy with (2.3.7), (2.4.9) and (2.4.7)

$$\mathrm{e}(\mathbf{Q}^0/\mathbf{Q}) = \mathbf{e_P}(\mathbf{D}^0/\mathbf{D}) = -e^1(\mathbf{Q}^0) + e^1(\mathbf{Q}) = -\frac{1}{w'd} + \frac{1}{w_0'd_0} \tag{6.7.5}$$

$$\begin{aligned}\tau(\mathbf{Q}^0/\mathbf{Q}) = \tau_{\mathbf{P}}(\mathbf{D}^0/\mathbf{D}) &= \frac{1}{3w}(-a^-(Q^0/Q) + e_0/d_0 - e/d), \\ &= -\frac{1}{3w}a^-(Q^0/Q) - \tau^1(\mathbf{Q}_0) + \tau^1(\mathbf{Q})\end{aligned} \tag{6.7.6}$$

where $a^-(Q^0/Q)$ is the left absorptive number of $\varphi^0_{\mathbf{Q}}$ determined by comparing minimal resolutions of Q^0 or Q, respectively (Definition 1.18.3). The base change properties for these local invariants are already known.

Proposition 6.7.2 *With the notations of the diagrams (6.7.1) and (6.7.2) the following base change property holds for $c = e, \tau$:*

$$\mathrm{c}(\mathbf{P}^0/\mathbf{P}) = d(\mathbf{f_P}) \cdot \mathrm{c}(\mathbf{Q}^0/\mathbf{Q}), \tag{B.Ch^0}$$

where $d(\mathbf{f_P}) = w'_Q d_Q / w'_P d_P$ is the reduced degree of $\mathbf{f_P}$ (see (1.16.16)). □

For $c = e$ the result follows from the definition (6.7.5), or (2.3.7) and Lemma 1.16.4. For $c = \tau$ we refer to (2.4.10).

We turn our attention now to the mixed local base change diagram (6.7.4). Write $\mathbf{E_P}$, $\mathbf{E_Q}$ instead of $\mathbf{E_P}(\psi)$ or $\mathbf{E_P}(\varphi)$, respectively. We fix further notations with the atomic graph of

$$\mathbf{E_Q} = \mathbf{E}_1 + \ldots + \mathbf{E}_l + \mathbf{Q}^0 + \mathbf{Q}^1 + \ldots + \mathbf{Q}_l,$$

$$\mathbf{E_Q}: \quad \underset{w=w_0}{\circ} \xrightarrow{\langle d_0, e_0\rangle} \underset{w_1}{\bullet} \xrightarrow{\langle d_1, e_1\rangle} \underset{w_2}{\bullet} \cdots \to \underset{w_l}{\bullet} \xrightarrow{\langle d_l, e_l\rangle} \underset{w_{l+1}=w'}{\circ}$$

We already defined in (1.17.9) and (3.4.15)

$$\mathbf{e}(\mathbf{E_Q}/\mathbf{Q}) = \mathbf{e}(\varphi_\mathbf{Q}) = \sum_{i=0}^{l} \frac{1}{w_i w_{i+1} d_i} - ww'd \tag{6.7.7}$$

$$\tau(\mathbf{E_Q}/\mathbf{Q}) = \tau(\varphi_\mathbf{Q}) = \sum_{i=1}^{l} \frac{1}{w_i}\tau(\mathbf{E}_i) + \frac{1}{w}\tau(\mathbf{Q}^0/\mathbf{Q}) + \frac{1}{w'}\tau(\mathbf{Q}_e^\perp/\mathbf{Q}^\perp) \tag{6.7.8}$$

As usual $\mathbf{Q}_e^\perp$, $\mathbf{Q}^\perp$ denote the opposite arranged abelian points of $\mathbf{Q}_e$ or $\mathbf{Q}$, respectively. We already proved the following

Proposition 6.7.3 *With our notations the following base change property for mixed local base change diagrams and for $c = e$ or τ holds:*

$$\mathbf{c}(\mathbf{E_P}/\mathbf{P}) = \deg(\mathbf{f_P}) \cdot \mathbf{c}(\mathbf{E_Q}/\mathbf{Q}),$$

where $\deg(\mathbf{f_P}) = \deg(f_P) = w_Q w'_Q d_Q / w_P w'_P d_P = |A_P|$, *$A_P$ the group of the local quotient morphism $f_P : P \to \mathbf{Q} = P/A_P$.*

For $c = e$ this property is nothing else than the degree formula I,(1.17.14). For $c = \tau$ and $\mathbf{f_P} = f_P$ we refer to (3.4.22). The general GALOIS-finite case can be easily deduced from this reduced case. □

6.8 Global Base Changes

The aim of this section is to prove the base change property (B.Ch) for $\mathbf{c} = \mathbf{e}, \boldsymbol{\tau}$. The other functorial properties in the first row of (6.4.2) have been already verified. We explain the connection with the local versions of (B.Ch). Roughly spoken the connection can be established by the geometric Local-Global-Principle. For GALOIS-finite arranged morphisms this has already been done in the proof of 3.3.8. It is convenient to understand what happens in terms of diagrams. We use the notations of the diagrams (6.7.1), (6.7.2) and (6.7.3). Furthermore, we take also in consideration tautological sequences defined in (6.3.7). Together we receive diagrams of type:

$$\begin{array}{ccccc} \mathbf{E_Q}/\mathbf{Q} \cdots & \mathbf{E_Q} & \longrightarrow & \mathbf{Q} \\ \downarrow & \downarrow & \otimes & \downarrow \\ \mathbf{Y}'/\mathbf{Y} \cdots & \mathbf{Y}' & \xrightarrow{\varphi} & \mathbf{Y} \end{array} \tag{6.8.1}$$

The local decomposition 3.3.6 allows to understand $\mathbf{Y}'/\mathbf{Y}$ numerically as sum of "relative fibres" $\mathbf{E_Q}/\mathbf{Q}$:

$$\mathbf{Y}'/\mathbf{Y} \approxeq \sum_{\mathbf{Q}} \mathbf{E_Q}/\mathbf{Q} \tag{6.8.2}$$

which means that

$$c\left(\mathbf{Y}'/\mathbf{Y}\right) = \sum_{\mathbf{Q}} c(\mathbf{E_Q}/\mathbf{Q}) \tag{L.D}$$

The geometric Local-Global Principle in the GALOIS situation is nothing else but the combination of the local decomposition (L.D) with $(\mathrm{B.Ch})^{1,0}$ together with the conclusion (B.Ch)

$$(L.D),\ (B.Ch)^{1,0} \Longrightarrow (B.Ch) \tag{L-G.P}$$

The same principle works for arranged curves:

$$(L.D)^1,\ (B.Ch)^0 \Longrightarrow (B.Ch)^1 \tag{(L - G.P)^1}$$

Everything is proved for GALOIS-finite morphisms. We have to transfer the proofs to old finite morphisms $f : X \to \mathbf{Y}_f$ and $f_C : C \to \mathbf{D}_C$ and their birational modifications. Here we have to work with sums of relative fibres with rational coefficients. Let us start with orbital curves.

The analogon of (6.8.1) is the tautological base change diagram:

$$\begin{array}{ccccc} \mathbf{Q}_0/\mathbf{Q} & \cdots & \mathbf{Q}_0 & \longrightarrow & \mathbf{Q} \\ \downarrow & & \downarrow & & \downarrow \\ \mathbf{D}_0/\mathbf{D} & \cdots & \mathbf{D}_0 & \longrightarrow & \mathbf{D} \end{array} \tag{6.8.3}$$

the right part sitting in (6.7.2). The arrangements of $\mathbf{D}$, $\mathbf{D}^0$ are denoted by

$$\mathbf{B}(\mathbf{D}) = \sum r_{\mathbf{Q}}\mathbf{Q},\ \mathbf{B}(\mathbf{D}_0) = \sum r^0_{\mathbf{Q}}\mathbf{Q}_0.$$

Since $f_P : P \to \mathbf{Q} = \mathbf{Q}_P$, $f'_{P_0} : P_0 \to \mathbf{Q}_0$ are local quotient morphisms by the same finite abelian group A, we get

$$\begin{aligned} r_{\mathbf{Q}} &= \deg(f_P)/\deg\left(f^C\right) = |A|/\deg(f_C)\cdot v_C \\ &= |A|/\deg\left(f'_C\right)\cdot v_{C'} = r^0_{\mathbf{Q}}. \end{aligned}$$

It is quite natural to regard

$$\mathbf{D}^0/\mathbf{D} \approx \sum \mathbf{r_Q} \cdot (\mathbf{Q}_0/\mathbf{Q})$$

as analogon of (6.8.2). Really we can prove that

$$\mathbf{c}(\mathbf{D}^0/\mathbf{D}) = \sum r_{\mathbf{Q}} \mathbf{c}(\mathbf{Q}_0/\mathbf{Q})$$

for $\mathbf{c} = \mathbf{e}$ or $\boldsymbol{\tau}$.

Proof By definition 6.5.2 we have

$$\mathbf{c}(\mathbf{D}^0/\mathbf{D}) = [c(\mathbf{D}^0) - c(\mathbf{D})] - \left[\sum_{\mathbf{Q}_0} r^0_{\mathbf{Q}} c^1(\mathbf{Q}_0) - \sum_{\mathbf{Q}} r_{\mathbf{Q}} c^1(\mathbf{Q})\right] \tag{6.8.4}$$

The curves D^0, D are isomorphic and $\mathbf{D}_0$, $\mathbf{D}$ have the same weight v. Therefore, the first summand vanishes if $c = e$, hence

$$\begin{aligned}\mathbf{e}(\mathbf{D}^0/\mathbf{D}) &= -\sum_{\mathbf{Q}} r_{\mathbf{Q}} \left[e^1(\mathbf{Q}_0) - e^1(\mathbf{Q})\right] \\ &= \sum_{\mathbf{Q}} r_{\mathbf{Q}} \mathbf{c}(\mathbf{Q}_0/\mathbf{Q})\end{aligned}$$

by definition (6.7.5).

For $c = \tau$ we have

$$\tau(\mathbf{D}^0) - \tau(\mathbf{D}) = \frac{1}{3v}((D^{02}) - (D^2)) = -\frac{1}{3v} \sum_{\mathbf{Q}} a^-(Q^0/Q)$$

by definition 6.5.2 and (1.18.7) (see also (2.4.8) and (2.4.7)). The orbital covering $f_C : C \to \mathbf{D}$ defines an honest arrangement on $\mathbf{D} = \mathbf{D}_C$ in the sense of the last relation in 6.2.9,(iii). With our notations it can be written as $\sum_{P \to Q} r_P = 1$. We distribute the left absorptive number $a^-(Q^0/Q)$ to the several arranged points $\mathbf{Q} = \mathbf{Q}_P$ supported by Q with coefficients $r_P = r_{\mathbf{Q}}$:

$$\tau(\mathbf{D}^0) - \tau(\mathbf{D}) = -\frac{1}{3v} \sum_{\mathbf{Q}} r_{\mathbf{Q}} a^-(Q^0/Q)$$

Together with (6.8.4) and (6.7.6) we obtain

$$\begin{aligned}\boldsymbol{\tau}(\mathbf{D}^0/\mathbf{D}) &= \sum_{\mathbf{Q}} r_{\mathbf{Q}} \left[-\frac{1}{3v} a^-(Q^0/Q) - \tau^1(\mathbf{Q}_0) + \tau^1(\mathbf{Q})\right] \\ &= \sum_{\mathbf{Q}} \boldsymbol{\tau}(\mathbf{Q}^0/\mathbf{Q}).\end{aligned}$$

The relation $(L.D)^1$ of local decomposition is proved.

Following the geometric Local-Global Principle $(L - G.P)^1$ we come to the end of the verification of all functorial properties of heights of orbital curves described in (6.4.2)

Proposition 6.8.1 *Let*

$$\begin{array}{ccc} \mathbf{C}_0 & \xrightarrow{\psi_C} & \mathbf{C} \\ \downarrow {\mathbf{f}'^0_{\mathbf{C}}} & & \downarrow \mathbf{f}_{\mathbf{C}} \\ \mathbf{D}_0 & \xrightarrow{\varphi_D} & \mathbf{D} \end{array}$$

be a base change diagram of orbital curves. Then for $\mathbf{c} = \tau$ *or* $\mathbf{e}$ *it holds that*

$$\mathbf{c}(\mathbf{C}_0/\mathbf{C}) = \deg(f_C) \cdot \mathbf{c}(\mathbf{D}_0/\mathbf{D}) \qquad (\mathrm{B.Ch}^1)$$

Proof It suffices to prove $(\mathrm{B.\ Ch})^1$ for old finite morphisms $C \to \mathbf{D} = \mathbf{D}_C$ because the other ones are connected with old finite morphisms by a tower $E \to \mathbf{C} \to \mathbf{D}$ with two old finite morphisms starting from E, and a base change is defined in the same manner by old base changes. With obvious notations using $(L.D)^1$ and $(B.Ch)^0$ we find

$$\begin{aligned} \mathbf{c}(C_0/C) &= \sum_P \mathbf{c}(P_0/P) = \sum_P d(f_P) \cdot \mathbf{c}(\mathbf{Q}_P^0/\mathbf{Q}_P) \\ &= \sum_Q \sum_{P \to Q} d(f_P) \cdot \mathbf{c}(\mathbf{Q}_P^0/\mathbf{Q}_P) \\ &= \sum_Q \sum_{P \to Q} \deg(f_C) \cdot r_P \mathbf{c}(\mathbf{Q}_P^0/\mathbf{Q}_P) \\ &= \deg(f_C) \cdot \sum_{\mathbf{Q}} r_{\mathbf{Q}} \mathbf{c}(\mathbf{Q}^0/\mathbf{Q}) \\ &= \deg(f_C) \cdot \mathbf{c}(\mathbf{D}^0/\mathbf{D}). \end{aligned}$$

We used

$$\deg(f_C) \cdot r_P = \deg(f_C) \cdot \frac{\deg(f_P)}{v \cdot \deg(f_C)} = \frac{\deg(f_P)}{v} = \frac{v \cdot v' \cdot d_Q}{v \cdot d_P}$$

$$= \frac{v' d_Q}{d_P} = d(f_P).$$

Proposition 6.8.1 is proved.

□

We change over again to the second dimension.

Proposition 6.8.2 *Let*

$$\begin{array}{ccc} \mathbf{X}' & \xrightarrow{\psi} & \mathbf{X} \\ \Big\downarrow{\mathbf{f}'} & & \Big\downarrow{\mathbf{f}} \\ \mathbf{Y}' & \xrightarrow{\varphi} & \mathbf{Y} \end{array}$$

be a base change diagram of orbital surfaces. Then, for $\mathbf{c} = \tau$ *or* $\mathbf{e}$ *it holds that*

$$\mathbf{c}(\mathbf{X}'/\mathbf{X}) = \deg(\mathbf{f}) \cdot \mathbf{c}(\mathbf{Y}'/\mathbf{Y}).$$

Proof We can assume that $\mathbf{f} = f : X \to \mathbf{Y} = \mathbf{Y}_f$ is an old finite morphism and $f' = \mathbf{f}'$, too. Furthermore, we suppose without loss of generality that φ and $\psi = \psi$ are locally linear in the sense of (6.7.2), (i) because each finite base change diagram can be decomposed into finitely many base change diagrams with locally linear orbital birational morphisms. In this case

6.8.3 *the finite morphisms* $\mathbf{f}, \mathbf{f}'$ *are locally abelian,*

that means that $f_P : P \to \mathbf{Q}_P$ and $f_{P_0} : P_0 \to \mathbf{Q}_{P_0}$ are quotient morphisms by finite abelian groups A_P acting around P and P_0.

Following the geometric Local-Global-Principle we prove

$$\mathbf{c}(\mathbf{Y}'/\mathbf{Y}) = \sum_{\mathbf{Q}} t_{\mathbf{Q}} \mathbf{c}(\mathbf{E}_{\mathbf{Q}}/\mathbf{Q}) \tag{L.D}$$

with $t_{\mathbf{Q}} = t_P = \deg(f_P)/\deg(f)$ for $\mathbf{Q} = \mathbf{Q}_P$.

Definition 6.8.4 Let $\varphi : \mathbf{Y}' \to \mathbf{Y}$ be an orbital birational morphism and $\mathbf{Q}$ an abelian point on $\mathbf{Y}$ with linear preimage arrangement

$$\mathbf{E}_{\mathbf{Q}} = \sum_{i=1}^{l_Q} \mathbf{E}_i + \sum_{i=0}^{l_Q} \mathbf{Q}_i.$$

For $c = e$ or τ the local invariant $\mathbf{c}_{\mathbf{Q}}(\mathbf{Y}'/\mathbf{Y})$ is defined by

$$\mathbf{c}_{\mathbf{Q}}(\mathbf{Y}'/\mathbf{Y}) = c_Q(Y'/Y) - \lambda_c(v)\mathbf{c}(\mathbf{Q}_0/\mathbf{Q}) - \lambda_c(v')\mathbf{c}\left(\mathbf{Q}_0'/\mathbf{Q}'\right)$$
$$- \sum_{i=1}^{l_Q} \lambda_c(v_i)\mathbf{c}(\mathbf{E}_i) - \left[\sum_{i=0}^{l_Q} \mathbf{c}(\mathbf{Q}_i) - \mathbf{c}(\mathbf{Q})\right] \tag{6.8.5}$$

where

$$c_Q(Y'/Y) = \pm l_Q + c(Q) - \sum_{i=0}^{l_Q} c(Q_i), \begin{cases} +\text{for} & c = e \\ -\text{for} & c = \tau \end{cases} \tag{6.8.6}$$

$\mathbf{Q} = (Q; vC, v'C')$, v_i is the weight of $\mathbf{E}_i$ and $\mathbf{D}_0$, $\mathbf{D'}_0$ are the arranged curves on $\mathbf{Y}'$ lying over $\mathbf{D}$ or $\mathbf{D}'$, respectively.

Lemma 6.8.5 *Let*

$$\begin{array}{ccc} X' & \xrightarrow{\psi} & X \\ \downarrow{\mathbf{f}'} & & \downarrow{\mathbf{f}} \\ \mathbf{Y}' & \xrightarrow{\varphi} & \mathbf{Y} \end{array}$$

be a base change diagram in $\mathcal{C}$ *with old finite orbital morphisms* $\mathbf{f}, \mathbf{f}'$ *and locally linear morphisms* ψ *and* φ. *Then for* $c = e, \tau$ *it holds that*

$$\mathbf{c}(\mathbf{Y}'/\mathbf{Y}) = \sum_{\mathbf{Q}} t_{\mathbf{Q}} \mathbf{c}_{\mathbf{Q}}(\mathbf{Y}'/\mathbf{Y}). \tag{6.8.7}$$

Proof We set

$$\Delta = \mathbf{c}(\mathbf{Y}'/\mathbf{Y}) = \mathbf{c}(\mathbf{Y}') - \mathbf{c}(\mathbf{Y}). \tag{6.8.8}$$

Then Δ has a decomposition

$$\Delta = \Delta^2 - \Delta^1 - \Delta^1_{\mathbf{exc}} - \Delta^0, \tag{6.8.9}$$

where

$$\begin{aligned} \Delta^2 &= \mathbf{c}(Y') - \mathbf{c}(Y), \\ \Delta^1 &= \sum_{\mathbf{D}} \lambda_c(\mathbf{D}) s_{\mathbf{D}}(\mathbf{c}(\mathbf{D}_0) - \mathbf{c}(\mathbf{D})), \\ \Delta^1_{\mathbf{exc}} &= \sum_{\mathbf{Q}} \sum_{i=1}^{l_Q} \lambda_c(\mathbf{E}_i(\mathbf{Q})) s_{\mathbf{E}_i(\mathbf{Q})} \mathbf{c}(\mathbf{E}_i(\mathbf{Q})) \\ \Delta^0 &= \sum_{\mathbf{Q}} \left[\sum_{i=0}^{l_Q} t_{\mathbf{Q}_i} \mathbf{c}(\mathbf{Q}_i) - t_{\mathbf{Q}} \mathbf{c}(\mathbf{Q}) \right] \end{aligned} \tag{6.8.10}$$

This follows immediately from the orbital decompositions of $\mathbf{c}(\mathbf{Y}')$ and $\mathbf{c}(\mathbf{Y})$ defined in (6.5.4), 6.5.1. We remark that $l_Q = 0$ is allowed. In this case $\Delta^1_{\mathbf{exc}}(\mathbf{Q}) = 0$ and $\mathbf{Q}_0 \xrightarrow{\sim} \mathbf{Q}$, hence also the $\mathbf{Q}$-summand of Δ^0 vanishes. We want to prove that there are local decompositions

$$\Delta^i = \sum_{\mathbf{Q}} t_{\mathbf{Q}} \Delta^i(\mathbf{Q}), i = 0, 1, 2,$$

$$\Delta^1_{\text{exc}} = \sum_{\mathbf{Q}} t_{\mathbf{Q}} \cdot \Delta^1_{\text{exc}}(\mathbf{Q}) \tag{6.8.11}$$

such that

$$\mathbf{c}_{\mathbf{Q}}(\mathbf{Y}'/\mathbf{Y}) = \Delta^2(\mathbf{Q}) - \Delta^1(\mathbf{Q}) - \Delta^1_{\text{exc}}(\mathbf{Q}) - \Delta^0(\mathbf{Q}) \tag{6.8.12}$$

If this is done, then (6.8.7) follows from, (6.8.8), (6.8.9), (6.8.11) and (6.8.12):

$$\begin{aligned} \mathbf{c}(\mathbf{Y}'/\mathbf{Y}) &= \Delta = \Delta^2 - \Delta^1 - \Delta^1_{\text{exc}} - \Delta^0 \\ &= \sum_{\mathbf{Q}} t_{\mathbf{Q}} (\Delta^2(\mathbf{Q}) - \Delta^1(\mathbf{Q}) - \Delta^1_{\text{exc}}(\mathbf{Q}) - \Delta^0(\mathbf{Q})) \\ &= \sum_{\mathbf{Q}} t_{\mathbf{Q}} \mathbf{c}_{\mathbf{Q}}(\mathbf{Y}'/\mathbf{Y}). \end{aligned}$$

First we decompose Δ^2 using $\mathbf{Q} = \mathbf{Q}_P$ for a point $P \in X$ and discrete decompositions of the unit in 6.2.9, (iii).

$$\begin{aligned} \Delta^2 &= \mathbf{c}(Y') - \mathbf{c}(Y) = c(Y') - c(Y) - (c(Sg(Y')) - c(Sg(Y))) \\ &= \sum_{Q} \left[\pm l_Q + c(Q) - \sum_{i=0}^{l_Q} c(Q_i) \right] = \sum_{Q} \left(\sum_{P \to Q} t_P \right) c_Q(Y'/Y) \\ &= \sum_{\mathbf{Q}} t_Q c_Q(Y'/Y) \end{aligned}$$

With

$$\Delta^2(\mathbf{Q}) = c_Q(Y'/Y) = \pm l_Q + c(Q) - \sum_{i=0}^{l_Q} c(Q_i) \tag{6.8.13}$$

we proved the Δ^2-part of (6.8.11). The Δ^0-part is trivially satisfied setting

$$\Delta^0(\mathbf{Q}) = \sum_{i=0}^{l_Q} \mathbf{c}(\mathbf{Q}_i) - \mathbf{c}(\mathbf{Q}) \tag{6.8.14}$$

because $t_{\mathbf{Q}_i} = t_{\mathbf{Q}}$ in the last identity of (6.8.1). Namely, $\mathbf{Q} = \mathbf{Q}_P$ is a quotient point P/A_P (see 6.8.3). The action of A_P around P extends to an action around the resolution curve $E_P = \sum_{i=1}^{l_Q} L_i \subset X'$ of P. It can be restricted to an action around each component L_i of E_P. Especially, $\mathbf{Q}_i$ is a quotient point P_i/A_P, $P_i \in L_i$. By definition we have

$$t_{\mathbf{Q}} = t_P = \deg(f_P)/\deg(f) = |A_P|/\deg(f) = t_{P_i} = t_{\mathbf{Q}_i}.$$

Now we look for local decomposition of Δ^1_{exc} with coefficient $t_{\mathbf{Q}}$ at the $\mathbf{Q}$-contribution. We consider the coefficient $s_i = s_{\mathbf{E}_i}$ of $\mathbf{E}_i = \mathbf{E}_i(\mathbf{Q})$ in (6.8.11). With the above notations we have $\mathbf{E}_i = L_i/A_P$. Therefore

$$s_i = v_i \cdot \deg(f_{L_i})/\deg(f) = |A_P|/\deg(f) = t_P = t_{\mathbf{Q}},$$

and the exceptional part of (6.8.11) is satisfied if we set

$$\Delta^1_{\text{exc}}(\mathbf{Q}) = \sum_{i=1}^{l_{\mathbf{Q}}} \lambda_c(\mathbf{E}_i(\mathbf{Q}))\mathbf{c}(\mathbf{E}_i(\mathbf{Q})). \tag{6.8.15}$$

With obvious notations we decompose Δ^1 using the local decomposition $(\text{L.D})^1$ and $s_{\mathbf{D}} \cdot r_{\mathbf{Q}} = t_{\mathbf{Q}}$ for $\mathbf{Q} \in \mathbf{D}$ (see Lemma 6.2.10 and 6.2.9, (iv)).

$$\begin{aligned}\Delta^1 &= \sum_{\mathbf{D}} \lambda_c(\mathbf{D}) s_{\mathbf{D}} \mathbf{c}(\mathbf{D}_0/\mathbf{D}) = \sum_{\mathbf{D}} \lambda_c(\mathbf{D}) \sum_{\mathbf{Q}\in\mathbf{D}} s_{\mathbf{D}} r_{\mathbf{Q}} \mathbf{c}(\mathbf{Q}_0/\mathbf{Q}) \\ &= \sum_{\mathbf{D}} \sum_{\mathbf{Q}\in\mathbf{D}} \lambda_c(\mathbf{Q}) t_{\mathbf{Q}} \mathbf{c}(\mathbf{Q}_0/\mathbf{Q}) \\ &= \sum_{\mathbf{Q}} t_{\mathbf{Q}} \left\{ \lambda_c(v_{\mathbf{Q}})\mathbf{c}(\mathbf{Q}_0/\mathbf{Q}) + \lambda_c\left(v'_{\mathbf{Q}}\right)\mathbf{c}\left(\mathbf{Q}'_0/\mathbf{Q}'\right) \right\}\end{aligned}$$

The Δ^1-decomposition in (6.8.11) holds for

$$\Delta^1(\mathbf{Q}) = \lambda_c(v_{\mathbf{Q}})\mathbf{c}(\mathbf{Q}_0/\mathbf{Q}) + \lambda_c\left(v'_{\mathbf{Q}}\right)\mathbf{c}\left(\mathbf{Q}'_0/\mathbf{Q}\right). \tag{6.8.16}$$

The definitions of local contributions $\Delta^i(\mathbf{Q})$, $\Delta^1_{\text{exc}}(\mathbf{Q})$ confirm the relation (6.8.12). Lemma 6.8.5 is proved.

□

Lemma 6.8.6 *If the assumption of Lemma 6.8.5 is satisfied, then for each orbital point* $\mathbf{Q} \in \mathbf{Y}$ *it holds that*

$$\mathbf{c}_{\mathbf{Q}}(\mathbf{Y}'/\mathbf{Y}) = \mathbf{c}(\mathbf{E}_{\mathbf{Q}}/\mathbf{Q})$$

Proof We transform the summands of $\mathbf{c}_{\mathbf{Q}}(\mathbf{Y}'/\mathbf{Y})$ in (6.8.5) separately. The first summand has been decomposed in (6.8.6). We introduce

$$\delta_c(v)=\begin{cases}1, & c=e\\ v, & c=\tau\end{cases};\ \text{then } \lambda_c(v)=\left(\delta_c(v)-\frac{1}{v}\right), v=v_{\mathbf{Q}},$$

$$c(\mathbf{Q}_0/\mathbf{Q})=\begin{cases}0, & c=e\\ -a^-(Q_0/Q)/3v, & c=\tau\end{cases}$$

From (6.7.5), (6.7.6) we obtain

$$\lambda_c(v)\mathbf{c}(\mathbf{Q}_0/\mathbf{Q})=\left(\delta_c(v)-\frac{1}{v}\right)\left[c(\mathbf{Q}_0/\mathbf{Q})-c^1(\mathbf{Q}_0)+c^1(\mathbf{Q})\right]$$

$$=\delta_c(v)c(\mathbf{Q}_0/\mathbf{Q})-\delta_c(v)\left[c^1(\mathbf{Q}_0)-c^1(\mathbf{Q})\right]-\frac{1}{v}\mathbf{c}(\mathbf{Q}_0/\mathbf{Q}).$$

$$\sum_{i=1}^{l}\lambda_c(v_i)\mathbf{c}(\mathbf{E}_i)=\sum_{i=1}^{l}\delta_c(v_i)\mathbf{c}(\mathbf{E}_i)-\sum_{i=1}^{l}\mathbf{c}(\mathbf{E}_i)/v_i$$

$$=-\sum_{i=1}^{l}\delta_c(v_i)\left[c^1\left(\mathbf{Q}_{i-1}'\right)+c^1(\mathbf{Q}_i)\right]+\sum_{i=1}^{l}\delta_c(v_i)c(\mathbf{E}_i)-\sum_{i=1}^{l}\frac{1}{v_i}\mathbf{c}(\mathbf{E}_i).$$

Altogether we find that

$$\mathbf{c}_{\mathbf{Q}}(\mathbf{Y}'/\mathbf{Y})=\frac{1}{v}\mathbf{c}(\mathbf{Q}_0/\mathbf{Q})+\frac{1}{v'}\mathbf{c}\left(\mathbf{Q}_0'/\mathbf{Q}'\right)+\sum_{i=1}^{l}\mathbf{c}(\mathbf{E}_i)/v_i$$

$$-\delta_c(v)c(\mathbf{Q}_0/\mathbf{Q})-\delta_c(v')c\left(\mathbf{Q}_0'/\mathbf{Q}'\right)-\sum_{i=1}^{l}\delta_c(v_i)c(\mathbf{E}_i)$$

$$+\sum_{i=0}^{l}[\delta_c(v_i)c^1(\mathbf{Q}_i)+\delta_c\left(v_i'\right)c^1\left(\mathbf{Q}_i'\right)-\mathbf{c}(\mathbf{Q}_i)-c(Q_i)]$$

$$-\delta_c(v)c^1(\mathbf{Q})-\delta_c(v')c^1(\mathbf{Q}')+\mathbf{c}(\mathbf{Q})+c(Q)\pm l. \tag{6.8.17}$$

Now, we calculate separately for $c=e$ and $c=\tau$ the summands of (6.8.17):

$$\sum_{i=1}^{l}\delta_c(v_i)c(\mathbf{E}_i)=\begin{cases}2l, & c=e\\ \frac{1}{3}\sum\limits_{i=1}^{l}(E_i^2), & c=\tau\end{cases}$$

$$\delta_c(v)c(\mathbf{Q}_0/\mathbf{Q})+\delta_c(v')c\left(\mathbf{Q}_0'/\mathbf{Q}'\right)$$

$$=\begin{cases}0, & c=e\\ -\frac{1}{3}\left[a^-(Q_c/Q)+a^-\left(Q_0'/Q'\right)\right]=-\frac{1}{3}(a^-+a^+), & c=\tau\end{cases}$$

$$\mathrm{c}(\mathbf{Q}) + c(Q) - \delta_c(v)c^1(\mathbf{Q}) - \delta_c(v')c^1(\mathbf{Q}')$$
$$= \begin{cases} \left(1 - \frac{1}{vd} - \frac{1}{v'd} + \frac{1}{vv'd}\right) - \left(1 - \frac{1}{v'd}\right) - \left(1 - \frac{1}{vd}\right) = -\left(1 - \frac{1}{vv'd}\right), & c = e \\ \frac{1}{3}[3l + Tr(Q)], & c = \tau \end{cases}$$

$$\mathrm{c}(\mathbf{Q}_i) + c(Q_i) - \delta_c(v_i)c^1(\mathbf{Q}_i) - \delta_c\left(v_i^{'}\right)c^1\left(\mathbf{Q}_i^{'}\right)$$
$$= \begin{cases} -\left(1 - \frac{1}{v_i v_i^{'} d_i}\right) = -\left(1 - \frac{1}{v_i v_{i+1} d_i}\right), & c = e \\ \frac{1}{3}[3l_i + Tr(Q_i)], & c = \tau, l_i = l(Q_i) \end{cases}$$

Summing up we obtain for $c = e$, $v = v_0$, $v' = v_{l+1}$, $\mathbf{Q}_0' = \mathbf{Q}_l'$, ...

$$\mathrm{e}_{\mathbf{Q}}(\mathbf{Y}'/\mathbf{Y}) = \frac{1}{v}\mathrm{e}(\mathbf{Q}_0/\mathbf{Q}) + \frac{1}{v'}\mathrm{e}\left(\mathbf{Q}_0^{'}/\mathbf{Q}^{'}\right)$$
$$+ \sum_{i=1}^{l} \frac{1}{v_i}\mathrm{e}(\mathbf{E}_i) - 2l - \left(1 - \frac{1}{vv'd}\right) + \sum_{i=0}^{l}\left(1 - \frac{1}{v_i v_i^{'} d_i}\right) + l$$
$$= \frac{1}{v_0}\left(\frac{1}{v_1 d_0} - \frac{1}{v'd}\right) + \frac{1}{v'}\left(\frac{1}{v_l d_l} - \frac{1}{vd}\right)$$
$$+ \sum_{i=1}^{l} \frac{1}{v_i}\left(\frac{1}{v_{i+1}} + \frac{1}{v_{i-1}d_{i-1}}\right) + \frac{1}{vv'd} - \sum_{i=0}^{l} \frac{1}{v_i v_{i+1} d_i}$$
$$= -\frac{1}{vv'd} + \sum_{i=0}^{l} \frac{1}{v_i v_{i+1} d_i} = \mathrm{e}(\mathbf{E}_{\mathbf{Q}}/\mathbf{Q})$$

(see (6.7.7)).

In the same manner we obtain for $c = \tau$

$$\tau_{\mathbf{Q}}(\mathbf{Y}'/\mathbf{Y}) = \frac{1}{v}\tau(\mathbf{Q}_0/\mathbf{Q}) + \frac{1}{v'}\tau\left(\mathbf{Q}_0^{'}/\mathbf{Q}^{'}\right) + \sum_{i=1}^{l} \tau(\mathbf{E}_i)/v_i$$
$$- \frac{1}{3}\left[-(a^- + a^+) + \sum_{i=1}^{l}(E_i^2) + \sum_{i=0}^{l}(3l_i + Tr(Q_i)) + 3l\right]$$
$$+ \frac{1}{3}[3l + Tr(Q)]$$

The sum on the right-hand side appeared already in (3.4.14). According to this identity it is equal to $\tau(\varphi_{\mathbf{Q}})$ in the sense of the definitions (3.4.12) and (6.7.8). This number coincides with $\tau(\mathbf{E}_{\mathbf{Q}}/\mathbf{Q})$ by the notation given in (6.7.8). The proof of Lemma 6.8.6 is finished.

□

We continue the proof of Proposition 6.8.2. The local decomposition (L.D) is an immediate consequence of our two lemmas. Namely, from (6.8.8) and Lemma 6.8.6 we get

$$\mathbf{c}(\mathbf{Y}'/\mathbf{Y}) = \sum_{\mathbf{Q}} t_{\mathbf{Q}} \mathbf{c}_{\mathbf{Q}}(\mathbf{Y}'/\mathbf{Y}) = \sum_{\mathbf{Q}} t_{\mathbf{Q}} \mathbf{c}(\mathbf{E}_{\mathbf{Q}}/\mathbf{Q}).$$

Finally, we have only to apply the base change property $(\text{B.Ch})^{1,0}$ (see Proposition 6.7.3). With the notations of the diagram (6.8.7) we receive

$$\begin{aligned}
\mathbf{c}(X'/X) &= \sum_{P\in X} \mathbf{c}_P(X'/X) = \sum_P \mathbf{c}(E_P/P) \\
&= \sum_P \deg(f_P) \cdot \mathbf{c}(\mathbf{E}_{\mathbf{Q}_P}/\mathbf{Q}_P) \\
&= \deg(f) \sum_P t_P \mathbf{c}(\mathbf{E}_{\mathbf{Q}_P}/\mathbf{Q}_P) \\
&= \deg(f) \sum_{\mathbf{Q}} t_{\mathbf{Q}} \mathbf{c}(\mathbf{E}_{\mathbf{Q}}/\mathbf{Q}) = \deg(f) \cdot \mathbf{c}(\mathbf{Y}'/\mathbf{Y}).
\end{aligned}$$

Proposition 6.8.2 is proved for "old" base change diagrams with locally linear birational morphisms ψ, φ. At the begin of the proof we observed already that the base change property extends via locally linear decompositions to arbitrary old base changes.

The general case can be reduced to the old case by definition of a general base change diagram by means of old base changes. With the notations of a defining diagram (6.3.5) we get

$$\mathbf{c}(\mathbf{X}'/\mathbf{X}) = \frac{1}{\deg(h)} \mathbf{c}(Z'/Z) = \frac{\deg(g)}{\deg(h)} \mathbf{c}(\mathbf{Y}'/\mathbf{Y}) = \deg(f) \mathbf{c}(\mathbf{Y}'/\mathbf{Y})$$

This completes the proof of Proposition 6.8.2.

□

At the end of this section we collect the connections between local decompositions and base change properties in the following logical scheme (6.8.18) in addition to the scheme (6.4.2):

In order to understand the dottedd arrows in (6.8.18) we refer back to the proof for GALOIS-finite base changes of orbital curves, orbital surfaces, and $c = \tau$, see Theorem 2.4.2 and Proposition 3.4.5. We only used special cases of $(\text{B.Ch})^0$ or $(\text{B.Ch})^{1,0}$ for the inductive proof of $(\text{B.Ch})^1$ or (B.Ch),

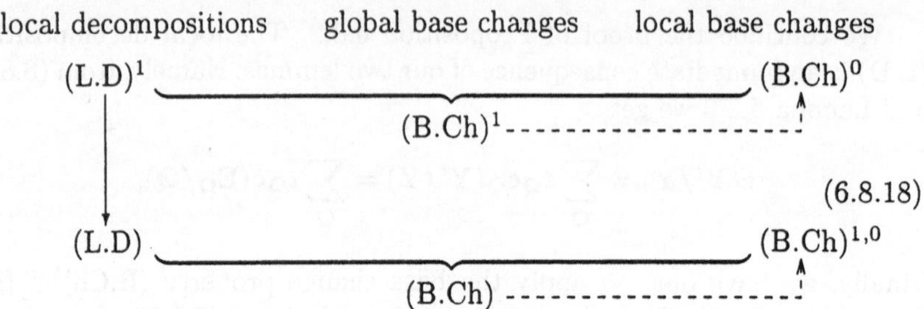

(6.8.18)

respectively, and deduced later the local base changes from the globall ones by shifting.

Now the Main Theorem 6.1.1 is proved. We have only to put together the two schemes (6.4.2) and (6.8.18). By Remark 6.6.3 all functorial properties of **c** appearing in the first row of (6.4.2) have been verified. All other functorial properties follow from these, especially the strongest and most applicable multiplicative property (Mult).
The base change property has been proved by a geometric Local-Global-Principle. It can be understood as a special inductive procedure from points over curves to surfaces. Maybe, it can be continued to higher dimensions. But already in the case of 3–folds one has to expect complicated constructions.
Using the 1–dimensional version of the scheme (6.4.2) we also proved the

Proposition 6.8.7 *For a finite (regular) covering* $\mathbf{f}: \mathbf{C} \to \mathbf{D}$ *of orbital curves the functorial property*

$$H_c(\mathbf{C}) = \mathbf{c}(\mathbf{C}) = \deg(f_C) \cdot \mathbf{c}(\mathbf{D}) = \deg(f_C) H_c(\mathbf{D})$$

holds for all $c \in \mathbb{Q}e + \mathbb{Q}\tau$.

□

6.9 Explicit Hurwitz Formulas for Finite Surface Coverings

We present in this section as an immediate application of the theory of functorial invariants explicit and universal formulas connecting CHERN invariants of surfaces X, Y joined by a finite regular morphism $f : X \to Y$. Remember that f is called regular if $\mathbf{f} : X \to \mathbf{Y}_f$ belongs to our category $\mathcal{C}$, which means that $\mathbf{B}_f$ is a regular arrangement in the sense of Definition 6.2.4. The corresponding formulas are called of HURWITZ type because of HURWITZ' genus formula connecting the EULER numbers of compact RIEMANN surfaces C, D joined by a finite covering $C \to D$. In HURWITZ' genus formula the curves C, D are assumed to be smooth. All other cases of finite coverings $C_1 \to D_1$ can be reduced to the smooth case by resolution of singularities as described in diagram:

$$\begin{array}{ccc} C & \xrightarrow{\psi_1} & C_1 \\ \downarrow{\scriptstyle f_1'} & & \downarrow{\scriptstyle f_1} \\ D & \xrightarrow{\varphi_1} & D_1 \end{array} \tag{6.9.1}$$

One has to resolve the singularities of D_1 by means of a birational morphism φ_1. then C, f_1', ψ_1 are defined by base changes. The curve C is the normalization of C_1 in the function field $\mathbb{C}(C_1)$ of C_1. On this way each finite covering f_1 can be birationally transformed to a finite covering of smooth curves. In this sense the HURWITZ genus formula solves generally the problem of calculation of $e(C_1)$ by $e(D_1)$, or vice versa, for finite coverings $C_1 \to D_1$.

In the case of algebraic surfaces it is generally not possible to find for a given finite covering $f_1 : X_1 \to Y_1$ a commutative diagram:

$$\begin{array}{ccc} X & \xrightarrow{\psi} & X_1 \\ \downarrow{\scriptstyle f'} & & \downarrow{\scriptstyle f_1} \\ Y & \xrightarrow{\varphi} & Y_1 \end{array} \tag{6.9.2}$$

with birational morphisms ψ, φ and smooth surfaces X, Y. Counter examples of simple GALOIS coverings can be easily found (e.g. the 3-sheeted cover of $\mathbb{P}^2$ branched along the quadrihedral Figure,(3.7.1)). But it is possible to modify f_1 to f' such that X, Y have only HIRZEBRUCH-JUNG singularities. Namely, by the theory of HIRONAKA (or ZARISKI) we can simultaneously resolve the singularities of Y_1 and singularities of the branch divisor such that on Y the branch curve has only normal crossings. Defining X by normal base change we find a diagram such that f' is finite, X normal. The preimages of intersection points of two components of the branch locus are HIRZEBRUCH-JUNG singularities by Proposition 3.5.12 (in the case of smooth Y we can also refer to [BPV], III, Theorem 5.2). In this manner each finite covering f_1 can be modified to a regular finite covering f' connected with f_1 by a base change diagram (6.9.2). In this sense the formulas of HURWITZ type relating CHERN numbers of X and Y along regular finite coverings are general formulas. The relations of CHERN numbers of X_1 and Y_1 can be found along ψ, f', φ in diagram (6.9.2) for any finite covering $f_1 : X_1 \to Y_1$.

Before we establish the formulas in dimension two we look back for a moment to the case of curves.

Proposition 6.9.1 *Let $f : (X, C) \to (Y, D)$ be a regular finite covering of embedded curves $D \subset X$, $C \subset Y$ in the sense of diagram (6.7.2). Then it holds that*

$$e(C) = \deg(f_C) e(D) + \sum_{P \in C} \left(1 - \frac{v'_P d_{f(P)}}{d_P}\right) \tag{6.9.3}$$

$$(C^2) + \sum_{P \in C} \frac{e_P}{d_P} = \frac{\deg(f_C)}{v} \left[(D^2) + \sum_{Q \in D} e_Q/d_Q \right] \tag{6.9.4}$$

Proof As usual we consider the orbital curve $\mathbf{D} = \mathbf{D}_C$ with branch arrangement

$$\mathbf{B}(\mathbf{D}) = \sum_{\mathbf{Q} \in \mathbf{D}} r_{\mathbf{Q}} \mathbf{Q} = \sum_{Q} \sum_{C \ni P \to Q} r_P \mathbf{Q}_P,$$

$$\mathbf{Q}_P : \underset{v_P}{\circ} \xrightarrow{\langle d_Q, e_Q \rangle} \underset{v'_P}{\circ} \quad , \qquad P : \underset{1}{\circ} \xrightarrow{\langle d_P, e_P \rangle} \underset{1}{\circ}$$

$v = v_P = v_{\mathbf{Q}} = v_{\mathbf{C}} = v_D$ the ramification index of f along D, $r_P = r_{\mathbf{Q}} = \deg(f_P)/\deg(f^C) = v'_P \cdot d_Q/d_P \cdot \deg(f_C)$.

The formulas (6.9.3) and (6.9.4) follow immediately from the orbital property

$$\mathbf{c}(C) = \deg(f_C) \cdot \mathbf{c}(\mathbf{D}) \tag{Mult^1}$$

of $\mathbf{c} = \mathbf{e}$ and τ applied to $f_C : C \to \mathbf{D}$. Namely, by definition 6.5.2 the relations $(\mathrm{Mult})^1$ has to be read as

$$e(C) - \sum_{P\in C}\left(1 - \frac{1}{d_P}\right)$$

$$= \deg(f_C)\left[e(D) - \sum_{Q}\sum_{P\to Q}\frac{v'_P d_Q}{d_P\cdot\deg(f_C)}\left(1 - \frac{1}{v'_P d_Q}\right)\right]$$

$$(C^2) + \sum_{P\in C}\frac{e_P}{d_P} = \frac{\deg(f_C)}{v}\left[(D^2) + \sum_{Q}\sum_{P\to Q} r_P\cdot\frac{e_Q}{d_Q}\right]$$

We remember that $\sum_{P\to Q}\cdot r_P = 1$ by 6.2.9, (iii). This proves (6.9.4). The other relation follows also immediately.

□

For algebraic surfaces we have two kinds of HURWITZ type formulas corresponding to the going-up problem and the going-down problem. The first problem supposes for a regular finite covering $f: X \to Y$ the knowledge of the CHERN numbers of Y, the data of the curves and points of the branch locus, the global degree and local degrees. It asks for the CHERN numbers of X. The going-down problem supposes the knowledge of the CHERN numbers of X, the data of the ramification locus $R^f \subset X$, local degrees and the global degree. It asks for the CHERN numbers of Y.

Proposition 6.9.1 teaches us that it is equivalent to know selfintersection numbers and EULER numbers of the branch curves on Y and of the ramification curves on X. It allows some variations in the suppositions of our two problems. But we will solve the two problems in the pure form given above. We begin with the going-up problem $c = e, \tau$. The qualitative version of its solution is the functorial property (Mult) combined with the orbital decomposition (6.5.4).

Theorem 6.9.2 *(Going-up formulas) For a regular finite covering $f: X \to Y$ of algebraic surfaces it holds that*

$$e(X) = \deg(f)e(Y) - \sum_{D\subset Y}\left[\deg(f) - \sum_{C\to D}\deg(f_C)\right]e(D)$$
$$+ \sum_{Q\in Y}\sum_{P\to Q}\frac{d_Q}{d_P}\left(v_P v'_P - v_P - v'_P + \frac{d_P}{d_Q}\right) \qquad (6.9.5)$$

$$\tau(X) = \deg(f)\tau(Y) - \frac{1}{3}\sum_{D\subset Y}\left[\deg(f) - \sum_{C\to D}\deg(f_C)/v_C\right](D^2)$$
$$- \deg(f)\sum_{Q\in Y}\left(l_Q + \frac{1}{3}Tr(Q)\right) + \sum_{P\in X}\left(l_P + \frac{1}{3}Tr(P)\right)$$
$$+ \frac{1}{3}\sum_{P\in X}\frac{1}{d_P}\left[e_{f(P)}\frac{v'_P}{v_P} + e'_{f(P)}\frac{v_P}{v'_P} - e_P - e'_P\right] \qquad (6.9.6)$$

Proof We already expressed in Proposition 6.5.3 the orbital decomposition (6.5.4) in terms of old invariants. With the notations of 6.5.3 the multiplicative law

$$\mathbf{c}(X) = \deg(f)\cdot\mathbf{c}(\mathbf{Y}_f)$$

changes over to

$$\mathbf{c}(X) = \deg(f)\left[c(Y) - \sum_{\mathbf{D}} s_{\mathbf{D}}\lambda_c(\mathbf{D})c(\mathbf{D}) - \sum_{\mathbf{Q}} t_{\mathbf{Q}}c(\mathbf{Q})\right]$$
$$\mathbf{c}(X) = c(X) - \sum_{P\in X} c(P) = c(X) - \sum_{Q}\sum_{P\to Q} c(P),$$

hence

$$c(X) = \deg(f)\cdot c(Y) - \sum_{D}\sum_{C\to D}\lambda_c(v_C)\deg(f_C)v_C c(\mathbf{D}_C)$$
$$+ \sum_{Q}\sum_{P\to Q}[c(P) - \deg(f_P)c(\mathbf{Q}_P)] \qquad (6.9.7)$$

Now, we use definitions (6.5.6.2) and (6.5.2.7)

$$e(P) - \deg(f_P)e(\mathbf{Q}_P)$$
$$= \left(1 - \frac{1}{d_P}\right) - \frac{v_P v'_P d_Q}{d_P}\left[-1 + \frac{1}{v_P} + \frac{1}{v'_P} - \frac{1}{v_P v'_P d_Q}\right]$$
$$= \frac{d_Q}{d_P}\left(v_P v'_P - v_P - v'_P + \frac{d_P}{d_Q}\right)$$

$$\tau(P) - \deg(f_P)\tau(\mathbf{Q}_P) = \frac{1}{3}\left[3l_P + Tr(P) - \frac{e_P}{d_P} - \frac{e'_P}{d_P}\right]$$
$$- \frac{v_P v'_P d_Q}{3d_P}\left[3l_Q + Tr(Q) - \frac{e_Q}{v_P^2 d_Q} - \frac{e'_Q}{v'^2_P d_Q}\right]$$
$$= \frac{1}{3}[3l_P + Tr(P)] - \frac{1}{3}\deg(f_P)[3l_Q + Tr(Q)]$$
$$+ \frac{1}{3d_P}\left[e_Q \frac{v'_P}{v_P} + e'_Q \frac{v_P}{v'_P} - e_P - e'_P\right]$$

Remember that

$$\sum_{P \to Q} \deg(f_P) = \deg(f)$$

by (6.2.5). Now the point parts of (6.9.5) and (6.9.6) are clear. We have to transform the curve part in (6.9.7)

$$\lambda_e(v_C) \cdot \deg(f_C) v_C e(\mathbf{D}_C) = \left(1 - \frac{1}{v_C}\right) v_C \deg(f_C) e(C)$$
$$= [-\deg(f_C) + s_C \cdot \deg(f)] e(C),$$
$$\lambda_\tau(v_C) \deg(f_C) v_C \tau(\mathbf{D}_C)$$
$$= \frac{1}{3}\left[s_C \cdot \deg(f) - \frac{1}{v_C}\deg(f_C)\right](D^2)$$

We remember that

$$\sum_{C \to D} s_C \cdot \deg(f) = \deg(f)$$

by (6.2.5). Substitutions in (6.9.7) yield the relations (6.9.5) and (6.9.6). Theorem 6.9.2 is proved.

□

The going-down theorem below has also a qualitative functorial version. For its formulation we extend only a little bit the notion of natural orbital surfaces admitting not only natural numbers as weights of curves and points, but also their reciprocals. We do not work with rational coefficients. A *reciprocal arrangement* on a surface X with at most HIRZEBRUCH-JUNG singularities is a finite formal sum

$$\mathbf{R} = \mathbf{R}_1 + \mathbf{R}_0 = \sum_i \mathbf{C}^i + \sum_j \mathbf{P}^j \tag{6.9.8}$$

such that

$$\mathbf{R}' = \sum_i \mathbf{C}_i + \sum_j \mathbf{P}_j$$

is a natural arrangement on X and if

$$\mathbf{C}_i = \left(U_i, v_i C; \sum_{\mathbf{P}_j \in \mathbf{C}_i} \mathbf{P}_j \right)$$

then

$$\mathbf{C}^i = \left(U_i, \frac{1}{v_i} C; \sum_{\mathbf{P}_j \in \mathbf{C}_i} \mathbf{P}^j \right)$$

and if

$$\mathbf{P}_j = \left(U_{P_j}, P_j; v_{P_j} \mathbf{C}, v'_{P_j} \mathbf{C}' \right)$$

then

$$\mathbf{P}^j = \left(U_{P_j}, P_j; \frac{1}{v_{P_j}} \mathbf{C}, \frac{1}{v'_{P_j}} \mathbf{C}' \right).$$

It is easy to define more generally arrangements with rational weights but we only need the reciprocal type introduced above. In opposition to the branch arrangement of a regular finite covering $f : X \to Y$ of surfaces we define the ramification arrangement as a reciprocal arrangement

$$\mathbf{R}^f = \mathbf{R}_1^f + \mathbf{R}_0^f = \sum_C \mathbf{C}^f + \sum_P \mathbf{P}^f,$$

where $\mathbf{P}^f = (X, P; v_P^{-1} C_P, {v'}_P^{-1} C'_P)$ are reciprocal abelian points, C_P, C'_P the curve germs through P with ramification indices v_P, v'_P with respect to f and

$$\mathbf{C}^f = \left(Y, v_C^{-1} C; \sum_{P \in R_0^f \cap C} \mathbf{P}_C^f \right),$$

where $\mathbf{P}_C^f$ is the arranged reciprocal abelian point $\mathbf{P}^f$ with first weighted curve germ $v_C^{-1} C$, v_C the ramification index of f along C. We always assume that $R^f = \operatorname{supp}\left(\mathbf{R}^f\right)$ is regular in the sense of 6.2.1 (i), (ii), (iii), (iv).

Definition 6.9.3 $\mathbf{R}^f = \sum_{C \subset R_f^1} \mathbf{C}^f + \sum_{P \in R_f^2} \mathbf{P}^f$ is called the *inverse* (or *reciprocal*) arrangement of f (or of $\mathbf{B}_f$), and $\mathbf{X}^f = (X, \mathbf{R}^f)$ the *reciprocal orbital surface* of f. If $\mathbf{X}$ is a reciprocal orbital surface with reciprocal arrangement (6.9.8), $c = e$ or τ, then we call

$$\mathbf{c}(\mathbf{X}) = \mathbf{c}(X) - \sum_i \lambda_c\left(\frac{1}{v_i}\right)\mathbf{c}(\mathbf{C}^i) - \sum_j \mathbf{c}(\mathbf{P}_j) \tag{6.9.9}$$

the *Euler height* or *signature height* of $\mathbf{X}$, respectively.

In analogy with (6.5.2.6) the number $\mathbf{c}(\mathbf{P})$ is defined as

$$\mathbf{c}(\mathbf{P}) = \begin{cases} \frac{1}{d}\left(1 - \frac{1}{v^{-1}}\right)\left(1 - \frac{1}{v'^{-1}}\right) = \frac{1}{d}(1-v)(1-v'), & c = e \\ 0, & c = \tau \end{cases}$$

$$\mathbf{P}: \quad \underset{v^{-1}}{\circ} \xrightarrow{\langle d, e\rangle} \underset{v'^{-1}}{\circ}$$

In analogy with 6.5.2 we define also

$$\mathbf{c}(\mathbf{C}) = c(\mathbf{C}) - \sum_j c^1(\mathbf{P}^j),$$

$$c(\mathbf{C}) = \begin{cases} e(C), & c = e \\ (C^2)/3v^{-1} = \frac{1}{3}v(C^2), & c = \tau \end{cases}$$

$$c^1(\mathbf{P}) = \begin{cases} 1 - \frac{1}{v'^{-1}d} = 1 - \frac{v'}{d}, & c = e \\ -e/3v^{-1}d = -ve/3d, & c = \tau \end{cases}$$

for a reciprocal orbital curve $\mathbf{C} = (U_C, v^{-1}C; \sum \mathbf{P}^j)$.

Theorem 6.9.4 *(Inverse version of (Mult)). Let $f: X \to Y$ be a regular finite surface covering, $\mathbf{X}^f = ((X, \mathbf{R}^f)$ the corresponding reciprocal orbital surface with inverse arrangement $\mathbf{R}^f$. Then for $c = e, \tau$ or any rational linear combination of e, τ it holds that*

$$\mathbf{c}(\mathbf{X}^f) = \deg(f) \cdot \mathbf{c}(Y) \tag{6.9.10}$$

Proof It suffices to work with $c = e, \tau$. From the Main Theorem we know that $\mathbf{c}$ is multiplicative along $\mathbf{f}: X \to \mathbf{Y}_f$, which means that

$$\begin{aligned}\mathbf{c}(X) &= \deg(f) \cdot \mathbf{c}(\mathbf{Y}_f) \\ &= \deg(f)\left[\mathbf{c}(Y) - \sum_C \lambda_c(v_C)s_C\mathbf{c}(\mathbf{D}_C) - \sum_P t_P\mathbf{c}(\mathbf{Q}_P)\right]\end{aligned}$$

with the usual notations. We have to compare this expression with

$$\mathbf{c}(\mathbf{X}^f) = \mathbf{c}(X) - \sum_C \lambda_c(\nu_C)\mathbf{c}(\mathbf{C}^f) - \sum_P \mathbf{c}(\mathbf{P}^f), \; \nu_C = v_C^{-1}. \tag{6.9.11}$$

It suffices to check that

$$\begin{aligned}\mathbf{c}(\mathbf{R}^f) &= \sum_C \lambda_c(\nu_C)\mathbf{c}(\mathbf{C}^f) + \sum_P \mathbf{c}(\mathbf{P}^f)\\ &= -\deg(f)\left[\sum_C \lambda_c(v_C)s_C\mathbf{c}(\mathbf{D}_C) + \sum_P t_P\mathbf{c}(\mathbf{Q}_P)\right]\\ &= -\deg(f)\cdot\mathbf{c}(\mathbf{B}_f)\end{aligned} \tag{6.9.12}$$

We set

$$\nu_C' = v_C'^{-1}, \nu_P = v_P^{-1}, \nu_P' = v_P'^{-1}.$$

For $c = \tau$ we transform

$$\begin{aligned}&-\deg(f)\cdot\tau(\mathbf{B}_f) = -\deg(f)\left[\tau(\mathbf{B}_f^1) + \tau(\mathbf{B}_f^0)\right] = -\deg(f)\cdot\tau(\mathbf{B}_f^1)\\ &= -\sum_C \lambda_\tau(v_C)\cdot v_C\cdot\deg(f_C)\mathbf{c}(\mathbf{D}_C)\\ &= -\sum_C\left(v_C - \frac{1}{v_C}\right)v_C\tau(C)\\ &= \sum_C\left(\nu_C - \frac{1}{\nu_C}\right)\frac{1}{3\nu_C}\left[(C^2) + \sum_{P\in C} c_P/d_P\right]\\ &= \sum_C \lambda_\tau(\nu_C)\tau(\mathbf{C}^f) = \tau\left(\mathbf{R}_1^f\right) = \tau(\mathbf{R}^f).\end{aligned}$$

For $c = e$ we obtain in the same manner

$$\begin{aligned}&-\deg(f)\mathbf{e}(\mathbf{B}_f)\\ &= -\deg(f)\left[\sum_C\left(1 - \frac{1}{v_C}\right)s_C\mathbf{e}(\mathbf{D}_C) + \sum_P t_P\mathbf{e}(\mathbf{Q}_P)\right]\\ &= -\sum_C\left(1 - \frac{1}{v_C}\right)v_C\deg(f_c)\mathbf{e}(\mathbf{D}_C) - \sum_P\deg(f_P)\mathbf{e}(\mathbf{Q}_P)\\ &= \left(1 - \frac{1}{\nu_C}\right)\mathbf{e}(C) - \sum_Q\sum_{P\to Q}\frac{v_P v_P' d_Q}{d_P}\left[\frac{1}{d_Q}\left(1 - \frac{1}{v_P}\right)\left(1 - \frac{1}{v_P'}\right)\right]\\ &= \sum_C\left(1 - \frac{1}{\nu_C}\right)\left[e(C) - \sum_{P\in C}\left(1 - \frac{1}{d_P}\right)\right]\\ &- \sum_P\frac{1}{d_P}(v_P - 1)\left(v_P' - 1\right)\end{aligned}$$

$$
\begin{aligned}
&-\sum_C \left(1-\frac{1}{\nu_C}\right)\left[e(C)-\sum_{P\in C}\left(1-\frac{1}{\nu'_C d_P}\right)\right] \\
&+2\sum_P \left(1-\frac{1}{\nu_P}\right)\left(\frac{1}{d_P}-\frac{1}{\nu'_P d_P}\right)-\sum_P \frac{1}{d_P}\left(1-\frac{1}{\nu_P}\right)\left(1-\frac{1}{\nu'_P}\right) \\
&=\sum_C \lambda_e(\nu_C)\mathbf{e}(\mathbf{C}^f)+\sum_P \frac{1}{d_P}\left(1-\frac{1}{\nu_P}\right)\left(1-\frac{1}{\nu'_P}\right) \\
&=\sum_C \lambda_e(\nu_C)\mathbf{e}(\mathbf{C}^f)+\sum_P \mathbf{e}(\mathbf{P}^f)=\mathbf{e}(\mathbf{R}^f).
\end{aligned}
$$

The relation (6.9.10) is verified and the Theorem 6.9.4 is proved.

□

Corollary 6.9.5 *(Going-down formulas of HURWITZ type) Assume that the conditions of Theorem 6.9.4 are satisfied. Then for $c = e, \tau$ it holds that*

$$
\begin{aligned}
&\deg(f)\cdot e(Y) \\
&= e(X)+\sum_C (v_C-1)e(C)+\sum_P\left(1-v_P-v'_P+\frac{v_P v'_P d_{f(P)}}{d_P}\right)
\end{aligned}
\tag{6.9.13}
$$

$$
\begin{aligned}
&\deg(f)\cdot\tau(Y) \\
&=\tau(X)+\frac{1}{3}\sum_C (v_C^2-1)(C^2)+\deg(f)\sum_{Q\in Y}\left(l_Q+\frac{1}{3}Tr(Q)\right) \\
&-\sum_{P\in X}\left(l_P+\frac{1}{3}Tr(P)\right) \\
&+\frac{1}{3}\sum_P \frac{1}{d_P}\left[e_P v_P^2-e_{f(P)}v_P v'_P-e'_{f(P)}v_P v'_P+e'_P v'^2_P\right]
\end{aligned}
\tag{6.9.14}
$$

Proof The reciprocal arrangement $\mathbf{R}^f$ can be considered as an honest arrangement in the spirit of Definition 6.2.9 because we have only trivial coefficients r_P, s_C, t_P equals to 1 (or 0). It does not matter to change even from natural weights to positive rational ones. Then we can apply Proposition 6.5.3 in order to express $\mathbf{c}(\mathbf{X}^f)$ in terms of "old" invariants:

$$\mathbf{e}(\mathbf{X}^f) = e(X) - \sum_C \left(1 - \frac{1}{\nu_C}\right) e(\mathbf{C}^f) - \sum_P e(\mathbf{P}^f),$$

$$\tau(\mathbf{X}^f) = \tau(X) - \sum_C \left(\nu_C - \frac{1}{\nu_C}\right) \tau(\mathbf{C}^f) - \sum_P \tau(\mathbf{P}^f),$$

$$\mathbf{e}(\mathbf{X}^f) = e(X) + \sum_C (v_C - 1) e(C) + \sum_P \left(1 - v_P - v'_P + v_P v'_P / d_P\right),$$

$$\tau(\mathbf{X}^f) = \tau(X) + \sum_C \left(v_C - \frac{1}{v_C}\right) v_C (C^2)$$

$$- \sum_P \frac{1}{3} \left(3 l_P + Tr(P) - \frac{e_P}{d_P} v_P^2 - \frac{e'_P}{d_P} v'^2_P\right)$$

On the other hand we have

$$\mathbf{e}(Y) = e(Y) - e(SgY) = e(Y) - \sum_Q \left(1 - \frac{1}{d_Q}\right)$$

$$\tau(Y) = \tau(Y) - \frac{1}{3} \sum_Q \left(3 l_Q + Tr(Q) - \frac{e_Q}{d_Q} - \frac{e'_Q}{d_Q}\right)$$

and

$$\deg(f) \cdot \sum_Q \left(1 - \frac{1}{d_Q}\right) = \sum_Q \sum_{P \to Q} \deg(f_P) \left(1 - \frac{1}{d_Q}\right)$$

$$= \sum_P \frac{v_P v'_P d_Q}{d_P} \left(1 - \frac{1}{d_Q}\right),$$

$$\deg(f) \cdot \sum_Q (3 l_Q + Tr(Q)) - \frac{e_Q}{d_Q} - \frac{e'_Q}{d_Q}$$

$$= \deg(f) \sum_Q (3 l_Q + Tr(Q)) - \sum_P \left(\frac{e_Q}{d_P} + \frac{e'_Q}{d_P}\right) v_P v'_P,$$

where $Q = f(P)$. Now we substitute the expressions into (6.9.12), and we receive (6.9.13) and (6.9.14).

□

The going-down formula for the signature can be simplified a littlebit if we change over to the minimal resolutions $\tilde{X}, \tilde{Y}$ of X or Y, respectively. By definition (3.2.6) and the definition of l_P as length of the minimal resolution of the cyclic singularity P, we have

$$\tau(X)-\sum_P l_P=\tau\left(\tilde{Y}\right),\ \tau(Y)-\sum_Q l_Q=\tau\left(\tilde{Y}\right).$$

So we get the following

Corollary 6.9.6 *Under the assumption of Theorem 6.9.4 the signatures of the minimal resolutions $\tilde{X},\tilde{Y}$ of singularities of X or Y, respectively, are related by*

$$\deg(f)\left[\tau\left(\tilde{Y}\right)-\frac{1}{3}\sum_{Q\in Y}Tr(Q)\right]$$
$$=\tau\left(\tilde{X}\right)+\frac{1}{3}\sum_{C\subset X}\left(v_C^2-1\right)\left(C^2\right)-\frac{1}{3}\sum_{P\in X}Tr(P)$$
$$+\frac{1}{3}\sum_P\left[e_Pv_P^2-\left(e_{f(P)}+e'_{f(P)}\right)v_Pv'_P+e'^2_Pv'^2_P\right]/d_P.\Box$$

$\Box$

For the sake of applications we also present the formulas for the special cases when both surfaces X and Y are smooth.

Corollary 6.9.7 *Let $f:X\to Y$ be a regular finite covering of smooth (compact complex algebraic) surfaces. Then*

$$\deg(f)e(Y)=e(X)+\sum_{C\subset X}(v_C-1)e(C)+\sum_{P\in X}(v_P-1)\left(v'_P-1\right)$$

(e-down)

$$e(X)=\deg(f)e(X)-\sum_{D\subset Y}\left[\deg(f)-\sum_{C\to D}\deg(f_C)\right]e(C)$$
$$+\sum_{P\in X}(v_P-1)\left(v'_P-1\right) \qquad \text{(e-up)}$$

$$\deg(f)\tau(Y)=\tau(X)+\frac{1}{3}\sum_{C\subset X}\left(v_C^2-1\right)\left(C^2\right) \qquad (\tau\text{-down})$$

$$\tau(X) = \deg(f)\tau(Y) - \frac{1}{3}\sum_{D\subset Y}\left[\deg(f) - \sum_{C\to D}\deg(f_C)/v_C\right](D^2) \qquad (\tau\text{-up})$$

□

The going-up formula (e-up) can be found in IVERSEN'S paper [IVE]. IVERSEN found also a going-up formula for the arithmetic genus χ involving intersections of the branch curve components with a canonical divisor. His formulas are more general than (e-up) and (χ-up) because he only assumed that X, Y are smooth and f does not need to be regular. Since the branch locus may be complicated IVERSEN'S local contributions are not easy to describe. Our formulas have the advantage that the knowledge of a canonical divisor is superfluous. Moreover, our extension to surfaces with singularities give the general formulas (up to modifications of finite coverings).

Our formulas have a nice coordinate interpretation, which will also accentuate the special role of e, τ among all rational CHERN symbols $c \in \mathbb{Q}c + \mathbb{Q}\tau$. In surface geography one corresponds to a smooth surface X two CHERN numbers, say $c_2(X), c_1^2(X)$. They are considered as coordinates of a point in the real plane $\mathbb{R}^2$. The surfaces with fixed coordinate pair form an algebraic family (see GIESEKER [GIE]). The surface geography asks for all points in $\mathbb{Z}^2 \subset \mathbb{R}^2$ which can be realized as CHERN coordinates of an algebraic surface. For more details we refer to [F-H] and the related literature given there. We correspond to normal surfaces X with (at most) HIRZEBRUCH-JUNG singularities, pairs of rational numbers $\mathbf{c}(X)$, say $\mathbf{e}(X)$ and $\tau(X)$, and call it an ***orbital coordinate pair*** of X. Obviously it extends the old correspondence for smooth surfaces.

Consider now a regular finite map $f : X \to Y$. Working with pairs of invariants we can write the going-down version (6.9.10) of the functorial property (Mult) as multiplication with a simple diagonal matrix:

$$\begin{pmatrix}\mathbf{e}(Y)\\ \tau(Y)\end{pmatrix} = \begin{pmatrix}1/\deg(f) & 0\\ 0 & 1/\deg(f)\end{pmatrix}\begin{pmatrix}\mathbf{e}(\mathbf{X}^f)\\ \tau(\mathbf{X}^f)\end{pmatrix}$$

Analogously, we have for

$$\begin{pmatrix}\chi\\ \mathbf{c}_1^2\end{pmatrix} = \begin{pmatrix}1/4 & 1/4\\ 2 & 3\end{pmatrix}\begin{pmatrix}\mathbf{e}\\ \tau\end{pmatrix}$$

$$\begin{pmatrix}\chi(Y)\\ \mathbf{c}_1^2(Y)\end{pmatrix} = \begin{pmatrix}1/\deg(f) & 0\\ 0 & 1/\deg(f)\end{pmatrix}\begin{pmatrix}\chi(\mathbf{X}^f)\\ \mathbf{c}_1^2(\mathbf{X}^f)\end{pmatrix} \qquad (6.9.15)$$

The orbital decompositions (6.9.9) of $\mathbf{c}(\mathbf{X}^f)$ suggest to define also coordinate pairs for embedded curves $C \subset X$ and embedded points $P \in C \subset X$. We correspond to C the (e,τ)-coordinates $e(C)$ and (C^2). The selfintersection number (C^2) has to be taken as usual on the minimal resolution $\tilde{X}$ of singularities of X. Furthermore, we correspond to the reciprocal abelian point

$$\mathbf{C} = \mathbf{C}^f \ni \mathbf{P} = \mathbf{P}^f : \underset{v_P}{\circ} \xrightarrow{\langle d_P, e_P \rangle} \underset{v'_P}{\circ}, \quad v_P = v_C$$

and to its twin brother $\mathbf{P}^\perp \in \mathbf{C}'$ the e,τ-coordinates

$$\mathbf{P} \mapsto \begin{pmatrix} \left(v'_P + 1\right)/2d_P - 1 \\ (v_P + 1)e_P/3d_P \end{pmatrix}, \quad \mathbf{P}^\perp \mapsto \begin{pmatrix} (v_P + 1)/2d_P - 1 \\ \left(v'_P + 1\right)e'_P/3d_P \end{pmatrix}$$

Then the going-down formulas can be written by means of matrix multiplications in the following manner:

$$\begin{pmatrix} \mathrm{e}(Y) \\ \tau(Y) \end{pmatrix} = \Delta \cdot \begin{pmatrix} \mathrm{e}(X) \\ \tau(X) \end{pmatrix} + \sum_C \Delta_{\mathbf{C}} \cdot \begin{pmatrix} e(C) \\ (C^2) \end{pmatrix}$$
$$+ \sum_{\mathbf{P}} \Delta_{\mathbf{P}} \begin{pmatrix} \left(v'_P + 1\right)/2d_P - 1 \\ (v_P + 1)\, e_P/3d_P \end{pmatrix} \tag{6.9.16}$$

where Δ, $\Delta_{\mathbf{C}}$, $\Delta_{\mathbf{P}}$ are universal diagonal matrices depending only on ramification indices and the degree of f:

$$\Delta = \begin{pmatrix} 1/\deg(f) & 0 \\ 0 & 1/\deg(f) \end{pmatrix},$$
$$\Delta_{\mathbf{C}} = \frac{1}{\deg(f)} \begin{pmatrix} v_C - 1 & 0 \\ 0 & \frac{1}{3}(v_C^2 - 1) \end{pmatrix},$$
$$\Delta_{\mathbf{P}} = \frac{1}{\deg(f)} \begin{pmatrix} v_P - 1 & 0 \\ 0 & v_P - 1 \end{pmatrix}$$

Note that the point P appears twice in the last sum of (6.9.16), namely as arranged point $\mathbf{P} \in \mathbf{C}$ and as opposite arranged point $\mathbf{P}^\perp \in \mathbf{C}'$. In the coordinates and the multiplication matrix one has to interchange v_P by v'_P and e_P by e'_P.

Proof of (6.9.16). We use (6.9.10) and (6.9.11) for $\mathbf{c} = \mathbf{e}, \boldsymbol{\tau}$:

$$\deg(f)\mathbf{c}(Y) = \mathbf{c}(\mathbf{X}^f) = \mathbf{c}(X) - \mathbf{c}(\mathbf{R}^f). \tag{6.9.17}$$

Suitable expressions for $\mathbf{c}(\mathbf{R}^f)$ have been found in the proof of (6.9.12):

$$\mathbf{e}(\mathbf{R}^f) = \sum_C (1 - v_C)\left[e(C) - \sum_{P \in C}\left(1 - \frac{1}{d_P}\right)\right] - \sum_P \frac{1}{d_P}(v_P - 1)\left(v'_P - 1\right)$$

$$\tau(\mathbf{R}^f) = \frac{1}{3}\sum_C \left(\frac{1}{v_C} - v_C\right) v_C \left[(C^2) + \sum_{P \in C} e_P/d_P\right]$$

It follows that

$$\begin{aligned}
-\mathbf{e}(\mathbf{R}^f) &= \sum_C (v_C - 1)e(C) \\
&\quad - \sum_P \left[(v_P - 1)\left(1 - \frac{1}{d_P}\right) + \left(v'_P - 1\right)\left(1 - \frac{1}{d_P}\right)\right] \\
&\quad - \sum_P \left[(v_P - 1)\left(v'_P - 1\right)/d_P\right] \\
&= \sum_C (v_C - 1)e(C) + \sum_P (v_P - 1)\left(1 + v'_P - 2d_P\right)/2d_P \\
-\tau(\mathbf{R}^f) &= \frac{1}{3}\sum_C (v_C^2 - 1)(C^2) + \frac{1}{3}\sum_{\mathbf{P}} (v_P^2 - 1)e_P/d_P
\end{aligned}$$

Now (6.9.16) follows from (6.9.17).

□

From the numerical point of view each regular finite covering $f : X \to Y$ of surfaces has a universal matrix presentation with respect to any pair of rational CHERN symbols. We can find easily the matrix presentation with respect to the pair χ, $\mathbf{c}_1^2$ joining $(\chi(Y), \mathbf{c}_1^2(Y))$ with $(\chi(X), \mathbf{c}_1^2(X))$ in a universal linear manner.

$$\begin{pmatrix} \chi(Y) \\ \mathbf{c}_1^2(Y) \end{pmatrix} = A \begin{pmatrix} \chi(X) \\ \mathbf{c}_1^2(X) \end{pmatrix} + \sum_C A_{\mathbf{C}} \begin{pmatrix} e(C) \\ (C^2) \end{pmatrix}$$

$$+ \sum_{\mathbf{P}} A_{\mathbf{P}} \begin{pmatrix} \left(v'_P + 1\right)/2d_P - 1 \\ (v_P - 1)\, e_P/3d_P \end{pmatrix}$$

$$A = \frac{1}{\deg(f)}\begin{pmatrix} 1 & 0 \\ 0 & 1 \end{pmatrix}, \quad A_{\mathbf{C}} = \frac{1}{\deg(f)}\begin{pmatrix} (v_C - 1)/4 & (v_C^2 - 1)/12 \\ 2(v_C - 1) & (v_C^2 - 1) \end{pmatrix},$$

$$A_{\mathbf{P}} = \begin{pmatrix} (v_P - 1)/4 & (v_P - 1)/4 \\ 2(v_P - 1) & 3(v_P - 1) \end{pmatrix} \qquad (6.9.18)$$

Proof We have only to multiply (6.9.16) from the left side with the matrix $M = \begin{pmatrix} 1/4 & 1/4 \\ 2 & 3 \end{pmatrix}$ according to the transformation (6.9.15). Then

$$A = M\Delta M^{-1} = \Delta, \; A_{\mathbf{C}} = M \cdot \Delta_{\mathbf{C}}, \; A_{\mathbf{P}} = M \cdot \Delta_{\mathbf{P}}.$$

□

The matrix presentations of f allow to understand a finite covering as a discrete motion, as an immediate jump. Comparing with RIEMANN surfaces one can imagine that $\deg(f)$ sheets of X jump together to the surface Y, curves C on X jump together to curves D on Y, and also points jump together. Working with CHERN coordinates of surfaces, curves and points we found matrix presentations of these jumps in a universal manner. The functorial invariants $\mathbf{c}$ of the corresponding orbital surfaces $\mathbf{X}^f$ or $\mathbf{Y}$ appear as invariants of our jumps up to degree multiplication. So the orbital theory can be understood as invariant theory of finite surface coverings imagined as jumping motions. Now look back to the matrix presentations and compare (6.9.18) with (6.9.16). Then the role of e, τ among all CHERN symbols becomes clear. Up to multiples these are the only CHERN coordinates with universal jumping matrices in diagonal form in all dimensions 2, 1, 0, if we use the most natural invariants $e(C)$ and (C^2) as 1-coordinates of our surfaces. By linear transformation it is not difficult to find for a given pair $b, c \in \mathbb{Q}e \oplus \mathbb{Q}\tau$ other but more complicated coordinates for $\mathbf{C}$ and $\mathbf{P}$ such that the corresponding universal matrix presentations have diagonal form in any dimension. Namely, instead of $(e(C), (C^2))$ we could work with 1–coordinates $\left(e(C), \frac{1}{3}(v_C + 1)(C^2)\right)$ of $\mathbf{C}$. Then all multiplication matrices for (e, τ) lie in the center of $\mathbb{G}l_2(\mathbb{C})$. We can change over to arbitrary linearly independent pairs (b, c) of CHERN symbols, say $\binom{b}{c} = M\binom{e}{\tau}$, and to new curve and point coordinates multiplying the old ones with M. It turns out that the multiplication matrices are the same as before because conjugation with M does not change elements of the center of $\mathbb{G}l_2(\mathbb{C})$. For $(b, c) = (\chi, c_1^2)$ one gets in this way the following version of (6.9.18):

$$\deg(f)\begin{pmatrix} \chi(Y) \\ c_1^2(Y) \end{pmatrix} = \begin{pmatrix} \chi(X) \\ c_1^2(X) \end{pmatrix} + \sum_{\mathbf{C}} \begin{pmatrix} v_C - 1 & 0 \\ 0 & v_C - 1 \end{pmatrix} \begin{pmatrix} \chi(\mathbf{C}) \\ c_1^2(\mathbf{C}) \end{pmatrix}$$

$$+ \sum_{\mathbf{P}} \begin{pmatrix} v_{\mathbf{P}} - 1 & 0 \\ 0 & v_{\mathbf{P}} - 1 \end{pmatrix} \begin{pmatrix} \chi(\mathbf{P}) \\ c_1^2(\mathbf{P}) \end{pmatrix}, \tag{6.9.19}$$

where for this moment the corresponding coordinates of $\mathbf{C}$ and $\mathbf{P}$ are

$$\begin{pmatrix} \chi(\mathbf{C}) \\ c_1^2(\mathbf{C}) \end{pmatrix} = \begin{pmatrix} \frac{1}{4}e(C) + \frac{1}{12}(v_C+1)(C^2) \\ 2e(C) + (v_C+1)(C^2) \end{pmatrix},$$

$$\begin{pmatrix} \chi(\mathbf{P}) \\ c_1^2(\mathbf{P}) \end{pmatrix} = \begin{pmatrix} \left(3 + e_P - 6d_P + 3v_P^{'} + 2e_Pv_P\right)/24d_P \\ \left(1 + e_P - 2d_P + v_P^{'} + e_Pv_P\right)/d_P \end{pmatrix} \qquad (6.9.20)$$

We see that for any version the formulas for χ, c_1^2 are more complicated than those for e and τ. The construction of functorial invariants is a rather long procedure. It would be possible but not convenient to work for instance with χ instead of τ. The general and explicit HURWITZ formula for χ could not be found before in this direct manner because of these complications. Taking up besides the EULER number the diagonalizing signature we could light up the situation.

6.10 Finite Coverings of Ruled Surfaces and the Inequality $c_1^2 \leq 2c_2$

The watershed in surface geography is the line $c_1^2 = 2c_2$ in the real (c_1^2, c_2)-plane. It divides the class of surfaces into two subclasses: the surfaces X with $c_1^2 \leq 2c_2$ and the surfaces with positive signature because of $\tau = \frac{1}{3}(c_1^2 - 2c_2)$. There was a conjecture about the non-existence of simply-connected surfaces with positive signature of general type which has been disproved (see e.g. [MI2]). MOISHEZON/TEICHER [M-T] and also CHEN/XIAO [CHN] found counter-examples. On the other hand PERSSON'S investigations in [PE1] of genus-two fibrations coming from double coverings of ruled surfaces along sextics belong to the class of surfaces X with $c_1^2(X) \leq 2c_2(X)$. The CHERN coordinates (c_1^2, c_2) of these surfaces fill almost the set of all possible points of the (c_1^2, c_2)-plane under the watershed. The watershed is furthermore a serious line in surface geography which is not completely unterstood. There are also other interesting lines in surface geography, for example the border line $c_1^2 = 3c_2$. MIYAOKA proved in [MI1] after preparatory results of VAN DE VEN (see [VEN]) that $c_1^2(X) \leq 3c_2(X)$ holds for all (smooth, algebraic, compact, complex) surfaces of general type. The inequality is sharp because of the existence of ball quotient surfaces corresponding to neat cocompact ball lattices (HIRZEBRUCH [HI2], BOREL [BO1]).

The surface geography suggests to consider more generally for any $c \in \mathbb{Q}c_2 + \mathbb{Q}c_1^2$ the c-line $c = 0$ in the real (c_1^2, c_2)-plane. For example, the τ-line is nothing else than the watershed. For $\beta = c_1^2 - 3c_2$ the β-line is the border line considered above. Let $\mathfrak{S}$ be a class of (smooth, compact complex algebraic) surfaces. We want to divide $\mathfrak{S}$ for given c into two classes, the c-general class and the c-special class. For this purpose we look at projections (finite morphisms) of a surface $X \in \mathfrak{S}$ onto the projective plane. Such projections exist because each smooth compact complex algebraic surface is projective, hence a subvariety of a projective space $\mathbb{P}^N, N \geq 3$ (see e.g. [BPV], IV.(5.5)). Assume that for certain projections $f : X \to \mathbb{P}^2$ we can explain the corresponding orbital surface $\mathbf{P}^2_f$ and the c-height $\mathbf{c}\left(\mathbf{P}^2_f\right)$. Then we can give the following definition:

The *c-general subclass* $\mathfrak{S}^-(c) = \mathfrak{S}^-\left(\mathbb{P}^2, c\right)$ of $\mathfrak{S}$ consists of all surfaces $X \in \mathfrak{S}$ such that there exists a projection $f : X \to \mathbb{P}^2$ with $\mathbf{c}\left(\mathbf{P}^2_f\right) \leq 0$.

The *c-special subclass* $\mathfrak{S}^+(c) = \mathfrak{S}^+\left(\mathbb{P}^2, c\right)$ of $\mathfrak{S}$ consists of all surfaces of $\mathfrak{S}$ which are not c-general.

We receive a disjoint union

$$\mathfrak{S} = \mathfrak{S}^-(c) \sqcup \mathfrak{S}^+(c),$$

As first example we consider the class $\mathfrak{S}$ of all surfaces for which there exists a regular projection onto $\mathbb{P}^2$. We remember that a regular projection $f : X \to \mathbb{P}^2$ has by definition a branch locus with smooth components and f is locally abelian, that means that the worst local branch loci are two smooth curve germs intersecting each other transversally.

Theorem 6.10.1 *For each regular finite cover X of $\mathbb{P}^2, X \neq \mathbb{P}^2$; it holds that*

$$c_1^2(X) \leq 2c_2(X).$$

Proof. There are two possibilities:

(i) X belongs to the τ-general class $\mathfrak{S}^-(\tau)$

(ii) X belongs to the τ-special class $\mathfrak{S}^+(\tau)$.

In the first case the conclusion of the theorem follows immediately from the functorial property of $\tau = \mathbf{c}_1^2 - 2\mathbf{c}_2$. Namely, using $f : X \to \mathbb{P}^2$ with $\tau\left(\mathbf{P}^2_f\right) \leq 0$ we get

$$\frac{1}{3}\left(c_1^2(X) - 2c_2(X)\right) = \tau(X) = \deg(f) \cdot \tau\left(\mathbf{P}^2_f\right) \leq 0.$$

So the multiplicative property of τ delegates immediately the proof of the inequality for surfaces belonging to the τ-special class $\mathfrak{S}^+(\tau)$. We show that this class consists really of special surfaces in the sense of small KODAIRA-dimension.

Lemma 6.10.2 *If X is a τ-special surface of $\mathfrak{S}$, then its KODAIRA-dimension is equal to $-\infty$. In other words: X is equal to $\mathbb{P}^2$ or a blown up (minimal) ruled surface.*

The lemma delegates our problem to the knowledge of CHERN number relations for surfaces which are birationally equivalent to ruled surfaces. If $F \to B$ is a ruled surface, B the base curve, then

$$c_1^2(F) = 8(1-g),\ c_2(F) = 4(1-g), \tag{6.10.1}$$

where $g = g(B)$ is the genus of B (see e.g. [BPV], VI, Table 10). Therefore the signature $\tau(F)$ vanishes. Since X is not the projective plane, by assumption, there exists a birational morphism $X \to F$ onto a ruled surface because the ruled surfaces are the minimal models in the class of surfaces $X \neq \mathbb{P}^2$ with KODAIRA dimension $-\infty$. Each birational morphism can be decomposed into a finite number of σ-processes. If $\sigma : Z \to Y$ is a σ-process, then $\tau(Z) = \tau(Y) - 1$. Therefore,

$$\tau(X) \leq \tau(F) = 0.$$

Thus, the inequality of the theorem is true also for the surfaces $X \neq \mathbb{P}^2$ of the τ-special class $\mathfrak{S}^+(\tau)$.

Proof of Lemma 6.10.2 The class of surfaces which are birational equivalent to a ruled surface and the class of surfaces with KODAIRA-dimension equal to $-\infty$ coincide (see e.g. [BPV], VI, Theorem (6.1.1)). By definition, a surface X belongs to the latter class if and only if all plurigonera $P_m(X)$ vanish for $m \geq 1$. So we have to show that

$$P_m(X) = \dim H^0(X, \mathcal{O}(K_X)) = 0, m = 1, 2, 3, \ldots,$$

where K_X denotes a canonical divisor of X and $\mathcal{O}(K_X)$ the corresponding invertible sheaf. Equivalently, we have to prove that the complete linear system $|mK_X|$ is void, that means that there does not exist an effective divisor D such that $D \equiv mK_X$ (linear equivalence).

This statement follows from

6.10.3 For $X \in \mathfrak{S}^+(\tau), f : X \to \mathbb{P}^2$ a regular projection, it holds that $(K_X \cdot f^*(L)) < 0$ for all projective lines $L \subset \mathbb{P}^2$.

Namely, assume that $D \geq 0$ is an effective divisor in $|mK_X|$. Then we can find a projective line $L \subset \mathbb{P}^2$ such that $f^*(L)$ and D have no common components. Hence

$$(K_X \cdot f^*(L)) = \frac{1}{m}(mK_X \cdot f^*(L)) = \frac{1}{m}(D \cdot f^*(L)) \geq 0.$$

This contradicts with 6.10.3.

Now, we prove 6.10.3. Let

$$\mathbf{B}_f = \sum_{\mathbf{D}} s_{\mathbf{D}}\mathbf{D} + \sum_{\mathbf{Q}} t_{\mathbf{Q}}\mathbf{Q} = \sum_{D}\sum_{C\to D} s_C\mathbf{D}_C + \sum_{Q}\sum_{P\to Q} t_P\mathbf{Q}$$

be the branch arrangement of f on $\mathbb{P}^2, \mathbf{P}_f^2 = (\mathbb{P}^2, \mathbf{B}_f)$. Then

$$\begin{aligned}
\tau(\mathbf{P}_f^2) &= \tau(\mathbb{P}^2) - \sum_{D}\sum_{C\to D} s_C\tau(\mathbf{D}_C) - \sum_{P} t_P\tau(\mathbf{Q}_P) \\
&= 1 - \frac{1}{3}\sum_{D}\left[\sum_{C\to D} s_C\left(1 - \frac{1}{v_C^2}\right)(D^2)\right] \\
&= 1 - \frac{1}{3}\sum_{D}\left[\left(\sum_{C\to D} s_C\right) - \sum_{C\to D}\frac{\deg(f_C)}{v_C\deg(f)}\right](D^2) \\
&= 1 - \frac{1}{3}\sum_{D}\left[1 - \sum_{C\to D}\frac{\deg(f_C)}{v_C\deg(f)}\right](D^2)
\end{aligned}$$

by 6.2.9 (iii) because $\mathbf{B}_f$ is an honest arrangement (Lemma 6.2.10). By the assumption of 6.10.3 the signature height $\tau\left(\mathbf{P}_f^2\right)$ is positive. Therefore we get the inequality

$$-3 + \sum_{D}\left[\frac{1 - \sum_{C\to D}\deg(f_C)}{v_C\deg(f)}\right](D^2) < 0 \tag{6.10.2}$$

By [BPV], I.16

$$K_X = f^*(K_{\mathbb{P}^2}) + R_f,\ R_f = \sum_{C\subset X}(v_C - 1)C$$

is a canonical divisor on X. We intersect K_X with the inverse image of a projective line $L \subset \mathbb{P}^2$.

$$\begin{aligned}
&(K_X \cdot f^*(L)) = (f^*(K_{\mathbb{P}^2}) \cdot f^*(L)) + (R_f \cdot f^*(L)) \\
&= \deg(f) \cdot (K_{\mathbb{P}^2} \cdot L) + (f_*(R_f) \cdot L),
\end{aligned}$$

where

$$f_*(C) = \deg(f_C) \cdot f(C) = \deg(f_C) \cdot D \tag{6.10.3}$$

We used the degree formula and the projection formula for intersections of divisors of surfaces (see e.g. [KUR], 5.1(9)). Since $K_{\mathbb{P}^2} \equiv -3L$ we obtain

$$(K_X \cdot f^*(L)) =$$

$$= \deg(f)\left(-3 + \sum_D \left[\left(\sum_{C\to D} \frac{\deg(f_C)\cdot v_C}{\deg(f)}\right) - \frac{\deg(f_C)}{\deg(f)}\right](D\cdot L)\right)$$

$$= \deg(f)\left(-3 + \sum_D \left[1 - \sum_{C\to D} \frac{\deg(f_C)}{\deg(f)}\right](D\cdot L)\right) \qquad (6.10.4)$$

For a comparison with (6.10.3) we use the simple estimation

$$0 = 1 - \sum_{C\to D} s_C = 1 - \sum_{C\to D} \frac{\deg(f_C)\cdot v_C}{\deg(f)} \leq 1 - \sum_{C\to D} \frac{\deg(f_C)}{\deg(f)}$$

$$\leq 1 - \sum_{C\to D} \frac{\deg(f_C)}{v_C \cdot \deg(f)}$$

Together with (6.10.3) and (6.10.4) one obtains

$$(K_X \cdot f^*(L)) \leq \deg(f)\left(-3 + \sum_D \left[1 - \sum_{C\to D} \frac{\deg(f_C)}{\deg(f)}\right](D^2)\right)$$

$$\leq \deg(f)\left(-3 + \sum_D \left[1 - \sum_{C\to D} \frac{\deg(f_C)}{v_C\cdot\deg(f)}\right](D^2)\right) < 0.$$

This proves 6.10.3, Lemma 6.10.2 and Theorem 6.10.1.

□

The following statement is the contraposition of the theorem.

Corollary 6.10.4 *Each (smooth, complex, compact, algebraic) surface $X \neq \mathbb{P}^2$ with positive signature cannot be projected regularly onto $\mathbb{P}^2$.*

□

We want to extend the theorem to regular covers of all minimal models of surfaces with negative KODAIRA dimension. Originally the author was inspired by a conjecture of PERSSON [PE2]: If X is a genus-two fibration, then $c_1^2(X) \leq 2c_2(X)$. Genus-two fibrations can be constructed as double covers of ruled surfaces branched along a sextic. The conjecture has been proved in the meantime by XIAO [XIA]. The conjecture of PERSSON has been generalized by himself in [PE2]: For any hyperelliptic fibration X (smooth, of dimension two) the inequality $c_1^2(X) \leq 2c_2(X)$ holds. As far as the author knows the conjecture has not been proved until now. Hyperelliptic curves are plane curves C of equation type $Y^2 = p(X)$, $p(X)$ a polynomial in X. The involution $(x, y) \mapsto (x, -y)$ gives rise to a GALOIS action on models of the hyperelliptic fibration X. The GALOIS quotient is a model of a ruled surface. The branch locus may have complicated singularities. The quotient morphism must be modified in order to get a regular covering. PERSSON'S hyperelliptic conjecture could be checked in this way. We concentrate our attention to arbitrary regular coverings of ruled surfaces.

6.10.5 Theorem, *Let $F \neq \mathbb{P}^2$ be a (minimal) ruled surface, X a smooth surface and $f : X \to F$ a regular covering. Then the inequality*

$$c_1^2(X) \leq 2c_2(X)$$

holds.

it Proof. We follow the general line of the proof of Theorem 6.10.1. Let $\mathcal{F}$ be the class of all ruled surfaces and $\mathfrak{S}(\mathcal{F})$ the class of all surfaces allowing a regular projection (finite covering) onto a ruled surface. Then we get a (disjoint) decomposition

$$\mathfrak{S}(\mathcal{F}) = \mathfrak{S}^-(\mathcal{F}, \tau) \sqcup \mathfrak{S}^+(\mathcal{F}, \tau)$$

into the subclasses of τ-general surfaces and τ-special surfaces, more precisely: $\mathfrak{S}^-(\mathcal{F}, \tau)$ consists of all surfaces X which can be regularly projected onto a ruled surface F, say $f : X \to F$, such that $\tau(\mathbf{F}_f) \leq 0$. $\mathfrak{S}^+(\mathcal{F}, \tau)$ consists of all surfaces which can be regularly projected onto a ruled surface F, but for any such projection $f : X \to F$ it holds that $\tau(\mathbf{F}_f) > 0$.

As in the proof of Theorem 6.10.5 we have no troubles with τ-general surfaces. The functorial property (Mult) lifts $\tau(\mathbf{F}_f) \leq 0$ to $0 \geq \tau(X) = \tau(X) = \frac{1}{3}\left(c_1^2 - 2c_2\right)$. The proof is delegated to τ-special surfaces again. As before the theorem follows now from the following

Lemma 6.10.6 *The τ-special class $\mathfrak{S}^-(\mathcal{F}, \tau)$ consists only of surfaces with negative KODAIRA dimension.*

Proof. We follow the strategy of the proof of Lemma 6.10.2. Let X be a τ-special surface of our class. In order to prove that the plurigenera vanish, it suffices to show that

6.10.7 *for any regular projection $f : X \to F$ onto a ruled surface F/B, B the base curve, and all fibres L of F/B it holds that*

$$(K_X \cdot f^*(L)) < 0.$$

For the proof we need basic information about curve theory on a ruled surface F/B. They can be found in HARTSHORNE'S book [HAR], Chapter V. The group Num(F) of numerical equivalence classes of divisors on F with respect to the intersection product is a free abelian group of rank 2. More precisely, there is a section S of $F \to B$ such that

$$(S^2) = -e, \text{ Num } (X) \cong \mathbb{Z}S \oplus \mathbb{Z}f,$$

where e is a well-defined invariant of F/B and f a fibre (see [HAR], V.2.3), $(f^2) = 0$. We distinguish three cases

a) $e < 0$, b) $e = 0$, c) $e > 0$.

Lemma 6.10.8 *If $D \neq S$ is an irreducible curve on F which is not a fibre of F/B. Then $(D^2) \geq 0$. Moreover, in the cases a), b) all irreducible curves have non-negative selfintersection numbers, and in the case c) it holds that*

$$(D^2) > a^2 e \textit{ for } D \approxeq aS + bf \, (\textit{numerical equivalence}). \tag{6.10.5}$$

Proof. Obviously we have $(D^2) = -a^2e + 2ab$.

Case a): Since D is not a fibre there are only the following two possibilities: $a = 1$, $b \geq 0$ or $a \geq 2$, $2b \geq ae$ (see [HAR], V.2.21). The estimation $(D^2) \geq 0$ is an immediate consequence.

Case b): If $D \neq S$ is not a fibre, then $a > 0$, $b > ae$ ([HAR], V.2.20). It follows that $ab > a^2e$, hence

$$(D^2) = -a^2e + ab + ab > -a^2e + a^2e + ab = ab > 0.$$

In the cases a), b) also $(S^2) = -e$ is non-negative, and $(f^2) = 0$ for fibres f is always satisfied. Therefore all irreducible curves have non-negative selfintersection.

Case c): Here we have the same inequalities $a > 0$, $b > ae$ as in b) by the same reference, hence $(D^2) > ab > a^2e$. This finishes the proof of Lemma 6.10.8.

We have to prove 6.10.7 now with the assumption of Lemma 6.10.6. Setting, with obvious notions,

$$\begin{aligned} \mathbf{B}_f &= \sum_{\mathbf{D}} s_{\mathbf{D}}\mathbf{D} + \sum_{\mathbf{S}} s_{\mathbf{S}}\mathbf{S} + \sum_{\mathbf{Q}} t_{\mathbf{Q}}\mathbf{Q} \\ &= \sum_{D \neq S} \sum_{C \to D} s_C \mathbf{D}_C + \sum_{T \to S} s_T \mathbf{S}_T + \sum_{Q} \sum_{P \to Q} t_P \mathbf{Q}_P \end{aligned}$$

we can calculate the signature height of $\mathbf{F}_f = (F, \mathbf{B}_f)$.

$$\begin{aligned}\tau(\mathbf{F}_f) &= \tau(F) - \sum_{D \neq S} \sum_{C \to D} s_C \tau(\mathbf{D}_C) - \sum_{T \to S} s_T \tau(\mathbf{S}_T) - \sum_P t_P \tau(\mathbf{Q}_P) \\ &= \tau(F) - \frac{1}{3} \sum_{D \neq S} \left[\left(\sum_{C \to D} s_C \right) - \sum_{C \to D} s_C/v^2 \right] (D^2) \\ &\quad - \frac{1}{3} \left[\left(\sum_{T \to S} s_T \right) - \sum_{T \to S} s_T/v^2 \right]\end{aligned}$$

Since $\tau(F) = 0$ (see (6.10.1)), $\tau(\mathbf{F}_f) > 0$ by the assumption $F \in \mathfrak{S}^+(\mathcal{F}, \tau)$,
$\sum_{C \to D} s_C = \sum_{T \to S} s_T = 1$ (honest arrangement) we receive the following estimation:

$$\sum_{D \neq S} \left[1 - \sum_{C \to D} s_C/v_C^2 \right] (D^2) + \left[1 - \sum_{T \to S} s_T/v_T^2 \right] (S^2) < 0 \tag{6.10.6}$$

This inequality can only be satisfied if $(S^2) < 0$ by Lemma 6.10.8. This is only possible in case c). So, we can assume that $-(S^2) = +e > 0$. With

$$D \approx a_D S + b_D f, a_D = (D \cdot f) = (D \cdot L)$$

(6.10.6) can be written as

$$\sum_{D \neq S} \left[1 - \sum_{C \to D} s_C/v_C^2 \right] (D^2) < \left[1 - \sum_{T \to S} s_T/v_T^2 \right] e,$$

and from (6.10.5) it follows that

$$\sum_D \left[1 - \sum_{C \to D} s_C/v_C^2 \right] a_D^2 e < \left[1 - \sum_T s_T/v_T^2 \right] e,$$

$$0 \leq \sum_D \left[1 - \sum_{C \to D} s_C/v_C^2 \right] a_D < \left[1 - \sum_T s_T/v_T^2 \right] \leq 1,$$

$$\sum_{D \neq S} \left[1 - \sum_{C \to D} \frac{\deg(f_C)}{v_C \cdot \deg(f_C)} \right] (D \cdot L) + \left[1 - \sum_{T \to S} \frac{\deg(f_T)}{v_T \cdot \deg(F)} \right] < 2, \tag{6.10.7}$$

where $L \simeq \mathbb{P}^1$ is an arbitrary fibre of F/B.

Now we estimate $(K_X \cdot f^*L)$. We know that

$$K_X = f^*(K_F) + R_f, \; K_F \approxeq -2S + kf \approxeq -2S + kL,$$

for a suitable integer k by [HAR], V.2.10. As in the proof of 10.5 we apply the degree formula for curve intersections, the projection formula and (6.10.3).

$$\begin{aligned}
&(K_X \cdot f^*(L)) = ((f^*(K_F) + R_f) \cdot f^*(L)) \\
&= \deg(f) \cdot (K_F \cdot L) + (f_*(R_f) \cdot L) \\
&= \deg(f) \cdot (-2) + \sum_C (v_C - 1) \deg(f_C) \cdot (f(C) \cdot L) \\
&= \deg(f) \left\{ -2 + \sum_{D \neq S} \left[\left(\sum_{C \to D} s_C \right) - \sum_{C \to D} \frac{\deg(f_C)}{\deg(f)} \right] (D \cdot L) \right. \\
&\qquad \left. + \left[\left(\sum_{T \to S} s_T \right) - \sum_{T \to S} \frac{\deg(f_T)}{\deg(f)} \right] (S \cdot L) \right\} \\
&= \deg(f) \left\{ -2 + \sum_D \left[1 - \sum_{C \to D} \frac{\deg(f_C)}{\deg(f)} \right] (D \cdot L) \right. \\
&\qquad \left. + \left[1 - \sum_T P \frac{\deg(f_T)}{\deg(f)} \right] \right\}
\end{aligned}$$

The last expression has been already estimated in (6.10.7). This proves 6.10.7, Lemma 6.10.6 and Theorem 6.10.5.

□

Corollary 6.10.9 *If $X \neq \mathbb{P}^2$ is a smooth compact complex algebraic surface with positive signature, then it cannot be regularly projected neither onto $\mathbb{P}^2$ not onto any ruled surface.*

Discussion 6.10.10 The attentive reader could ask for a purely classical proof of the inequality of the Theorems 6.10.1 and 6.10.6 without using branch arranged orbital surfaces and the functorial property (Mult) of τ. Indeed, after the proofs have been found one can translate them back because (Mult) is nothing else than a translation of the formulas of HURWITZ type for CHERN numbers. This section dealt only with finite coverings $X \to Y$ of smooth surfaces. The functorial invariant theory was originally made for general finite coverings admitting singularities. In the smooth cases one disposes already on IVERSEN'S formulas [IVE]. In contrast to our general formulas IVERSEN needs the knowledge of a canonical divisor on Y. It appears like an "unknown variable" in his formulas depending on Y. With our formulas we removed this "unknown variable". We only need the knowledge of the branch locus. This is a progress even in the smooth cases. But for the applications in this section the occurrence of the canonical divisor in IVERSEN'S formulas is not a handicap because the basic surfaces of our coverings are $\mathbb{P}^2$ and ruled surfaces, where the canonical divisors are (numerically) known. So, one can work really with these formulas.

But the question is more delicate. The multiplicative property of functorial invariants gives birth to the proof method dividing the considered surfacees into two classes $\mathfrak{S}^+$ and $\mathfrak{S}^-$. This watershed method is not visible without the knowledge of orbital invariants. One had to write $\tau\left(\mathbf{P}_f^2\right) \leq 0$ in explicit terms without any multiplicative motivation. For other invariants $\mathbf{c}$ the subdivision is even more complicated and the translated explicit version is surely not visible and not motivated. For the author it seems to be important to dispose on a new *flexible* method for finding CHERN number inequalities for classes of surfaces.

Discussion 6.10.11 *The general inequality of* MIYAOKA *for non-special (smooth compact complex algebraic) surfaces settles down in our watershed strategy.*

At first one has to study carefully general projections of surfaces onto $\mathbb{P}^2$. It turns out that the worst singularities of the branch locus are cusps and intersection points of two smooth curve germs (see [MI2], Lemma 3.3). A curve cusp can be resolved by three σ-processes or by one projective line supporting two cyclic singularities, one of type $\langle 2.1 \rangle$ and the other of type $\langle 3,1 \rangle$. We refer to pictures V.(P.2.9) and V,(P.2.3). The branch arrangement of the PICARD modular surface $\mathbb{B}/\Gamma$ of the full GAUSS lattice contains the curve cusp P_2 (see V,(P.2.9)). This curve singularity is removed in V,(P.2.3) by means of the three lines L_2, L_{21}, L_{23}. If we contract L_{22} and L_{23}, then we have a one-line resolution of the curve cusp described above. In this manner we get a commutative diagram:

$$\begin{array}{ccc} X' & \longrightarrow & X \\ \downarrow f' & & \downarrow f \\ Y' & \xrightarrow{\varphi} & \mathbb{P}^2 \end{array} \qquad (6.10.8)$$

where f is the general projection we started with, φ the resolution of cusp points described above, and (6.10.8) is a base change diagram in the category of normal surfaces, that means that X' is the normalization of X in the function field $\mathbb{C}(X)$. Obviously, f' is a regular covering, hence $\mathbf{Y}'_{\mathbf{f}'}$ and $\mathbf{c}(\mathbf{Y}'_{\mathbf{f}'})$ are explained. Especially f' is locally abelian. Moreover, f itself is locally GALOIS, that means that we can find for each point $P \in X$ analytic neighbourhoods U_P, U_Q, where $Q = f(P)$, and a finite group G_P acting on U_P such that $U_Q = U_P/G_P$. For curve cusp we refer to the example of $P_2 \in \mathbb{B}/\Gamma_1$ in picture V,(P.2.9), which has obviously a local uniformization. The general situation $f_P : U_P \to U_Q$ is isomorphic to this special one.Triple points of curves and their uniformizations have been considered in 3.6, see also Figure (3.8.1).

Similarly, as in the case of global GALOIS coverings, we can extend the notions of arrangements, orbital surfaces and functorial invariants $\mathbf{c}$ to the situation of general projections, using the local GALOIS property described above, and base change diagrams:

$$\begin{array}{ccc} X' & \xrightarrow{\psi} & X \\ \downarrow \mathbf{f}' & & \downarrow \mathbf{f} \\ \mathbf{Y}'_f & \xrightarrow{\varphi} & \mathbf{P}^2_f \end{array} \tag{6.10.9}$$

derived from (6.10.8).

Now, let $\mathfrak{S}$ be the class of all (smooth, compact, complex, algebraic) surfaces and $\beta = c_1^2 - 3c_2$. The β-line in the (c_1^2, c_2)-plane splits $\mathfrak{S}$ into

$$\mathfrak{S} = \mathfrak{S}^-(\beta) \sqcup \mathfrak{S}^+(\beta),$$

where $\mathfrak{S}^-(\beta)$ consists of all surfaces X possessing a general projection $f : X \to \mathbb{P}^2$ with inequality $\beta\left(\mathbf{P}_f^2\right) \leq 0$. These are the *$\beta$-general surfaces.* The complement $\mathfrak{S}^+(\beta)$ of $\mathfrak{S}^-(\beta)$ in $\mathfrak{S}$ consists of the *β-special surfaces.* Since the inequality $\beta\left(\mathbf{P}_f^2\right) \leq 0$ lifts to $c_1^2(X) \leq 3c_2(X)$ via functoriality of β for $X \in \mathfrak{S}^-(\beta)$ MIYAOKA'S result has the following translation:

Proposition 6.10.12 *Any β-special surface has negative KODAIRA dimension.*

□

Such surfaces X exist in the birational equivalence classes of ruled surfaces. They have negative CHERN numbers $c_1^2(X), c_2(X)$. An intrinsic proof in the style of the proofs of 6.10.3 and 6.10.8 via inverse images of $K_{\mathbb{P}^2}$ and general lines L in $\mathbb{P}^2$ should be an interesting theme for a special article. At least one finds a useful translation (simplification ?) of MIYAOKA'S proof.

Index

Bibliography

[AP1] P. Appell. Sur les fonctions hypergéométriques de deux variables et sur des équations linéaires aux dérivées partielles. *C. R. Acad. Sci.*, 90:296, 731, 977, 1880.

[AP2] P. Appel. *Fonctions hypergéométriques et hypersphériques, polynoms d'Hermite.* Gauthier-Villars, Paris, 1926.

[ART] M. Artin. *Lipman's proof of resolutions of singularities for surfaces*, pages Ch. XI in [C-S].

[AS1] M.F. Atiyah and I.M. Singer. The index of elliptic operators I. *Ann. of Math.*, 87:484–530, 1968.

[AS2] M.F. Atiyah and I.M. Singer. The index of elliptic operators III. *Ann. of Math.*, 87:546–604, 1968.

[B-B] W.L. Baily and A. Borel. Compactification of arithmetic quotients of bounded symmetric domains. *Ann. Math.*, 84:442–528, 1966.

[BAS] C. Banica and O. Stanasila. *Algebraic methods in the global theory of complex spaces.* John Wiley & Sons, London, New York, Sydney, Toronto, 1974.

[BHH] G. Barthel, T. Höfer, F. Hirzebruch. *Geradenkonfigurationen und algebraische Flächen.* Asp. Math. D4, Vieweg, Braunschweig, Wiesbaden, 1987.

[BPV] W. Barth, C. Peters, and A. van de Ven. *Compact complex surfaces.* Erg. Math. u. Grenzgeb. (3. Folge) 4, Springer, Berlin, Heidelberg. New York, 1985.

[B-K] G. Barthel and L. Kaup. Homotopieklassifikation einfach zusammenhängender normaler kompakter komplexer Flächen. *Math. Ann.*, 212:113–144, 1974.

[BOEG] S. Böge. Schiefhermitesche Formen über Zahlkörpern und Quaternionenschiefkörpern. *Journ. f. reine u. angew. Math.*, 221:85–113, 1966.

[BOEL] R. Bölling. Bemerkungen über Klassenzahlen und Summen von Jacobi-Symbolen. *Math. Nachr.*, 90:159–172, 1979.

[BO1] A. Borel. Compact Clifford-Klein forms of symmetric spaces. *Topology*, 2:111–122, 1963.

[BO2] A. Borel. *Introduction aux groupes arithmétiques.* Herman, Paris, 1969.

[BO3] A. Borel. *Linear algebraic groups.* Benjamin, New York, Amsterdam, 1969.

[BO4] A. Borel. Some metric properties of arithmetic quotients of symmetric spaces and an extension theorem. *Journ. Diff. Geom.*, 6:543–560, 1972.

[BOS] S.I. Borevich and I.R. Shafarevich. *Number theory (Russian).* Nauka, Moskva, 1985.

[BOU] Bourbaki. *Algèbre, Livre II.* Herman, Paris, 1960.

[BRA] H. Braun. Zur Theorie der hermiteschen Formen. *Abh. Sem. d. Hansischen Univ. Hamburg*, 14:61–150, 1941.

[BRI] E. Brieskorn. Rationale Singularitäten komplexer Flächen. *Inv. Math.*, 4:336–358, 1967.

[CAR] H. Cartan. Quotient d'un espace analytique par un group d'automorphismes. *Algebraic Geometry and Topology, Princeton Univ. Press*, pages 90–102, 1957.

[C-F] J.W.S. Cassels and A. Fröhlich. *Algebraic number theory.* Acad. Press, London, 1967.

[CHN] Z. Chen. On the geography of surfaces - simply connected minimal surfaces with positive index. *Math. Ann.*, 277:141–164, 1987.

[CHV] C. Chevalley. Invariants of finite groups generated by reflections. *Am. Journ. Math.*, 77:778–782, 1955.

[C-S] G. Cornell and J.-H. Silverman, editors. *Arithmetic geometry.* Springer, New York, 1986.

[DED] R. Dedekind. *Erläuterungen zu den vorstehenden Fragmenten.* Number 436-447. in: Riemanns Werke, 1876.

[D-M] P. Deligne and G. Mostow. Monodromy of hypergeometric functions and non-lattice integral monodromy. *Publ. Math. I.H.E.S.*, 63:5–90, 1986.

[DWI] H.B. Dwight. *Tables of integrals and other mathematical data.* Macmillan Comp., New York, 1961.

[EIS] G. Eisenstein. Théorèmes arithmétiques. *Crelle Journ. f. reine u. angew. Math.*, 27:36–37, 1844.

[FE1] J.-M. Feustel. *Über die Spitzen von Modulflächen zur zweidimensionalen komplexen Einheitskugel.* Preprint Series. Akad. Wiss. DDR, ZIMM, Berlin, 1977.

[FE2] J.-M. Feustel. *Volumina der Fundamentalgebiete der Spiegelscheiben Picardscher Modulgruppen.* Preprint Series 13. Akad. Wiss. DDR, ZIMM, Berlin, 1979.

[FE3] J.-M. Feustel. Spicgelungs- und Spitzenkontributionen zum arithmetischen Geschlecht Picardscher Modulflächen. Master's thesis, Akad. Wiss. DDR, ZIMM, 1980.

[FE4] J.-M. Feustel. *Klassifikation der elliptischen Fixpunkte bezüglich der Wirkung der Picardschen Modulgruppe auf die komplexe Einheitskugel.* Preprint Series 30. Akad. Wiss. DDR, Inst. Math., Berlin, 1981.

[FE5] J.-M. Feustel. Arithmetik und Geometrie Picardscher Modulflächen. Habil thesis, Akad. Wiss. DDR, Inst. Math, Berlin, 1987.

[FE6] J.-M. Feustel. *Kurvenarrangements auf Picardschen Modulflächen.* Preprint Series 8. Akad. Wiss. DDR, Inst. Math,, Berlin, 1989.

[FE7] J.-M. Feustel. Zur groben Klassifikation der Picardschen Modulflächen. *Math. Nachr.*, 118:215–251, 1984.

[F-H] J.-M. Feustel, R.-P. Holzapfel. Symmetry points and Chern invariants of Picard modular surfaces. *Math. Nachr.*, 111:7–40, 1983.

[FRE] E. Freitag. über die Struktur der Funktionenkörper zu hyperabelschen Gruppen I. *Journ. f. reine u. angew. Math.*, 247:97–117, 1971.

[GIE] D. Gieseker. Global moduli for surfaces of general type. *Inv. Math.*, 43:233–282, 1977.

[G-R] H. Grauert and R. Remmert. Komplexe Räume. *Math. Ann.*, 136:245–318, 1958.

[GRO] A. Grothendieck and J. Dieudonné. Eléments de géométrie algébrique III. *Publ. Math. I.H.E.S.*, 11, 17, 1961, 1963.

[HAR] F. Hartshorne. *Algebraic geometry*, Volume 52, *Grad. Texts in Math.* Springer, Berlin-Heidelberg-New York, 1977.

[HEL] S. Helgason. *Differential geometry and symmetric spaces.* Academic Press, New York-London, 1962.

[HEM] J.C. Hemperly. The parabolic contribution to the number of linearly independent automorphic forms on a certain bounded domain. *Amer. Journ. Math*, 94:1078–1100, 1972.

[HER] M. Hervé. *Several complex variables (Local theory).* Tata Lectures. Oxford Univ. Press, Bombay, 1963.

[HI1] F. Hirzebruch. Über vierdimensionale Riemannsche Flächen mehrdeutiger analytischer Funktionen von zwei komplexen Veränderlichen. *Math. Ann.*, 126:1–22, 1953.

[HI2] F. Hirzebruch. Automorphe Formen und der Satz von Riemann-Roch. *Symp. Intern. Top. Alg., Univ. de Mexico*, pages 129–144, 1958.

[HI3] F. Hirzebruch. *Topological methods in algebraic geometry*, Volume 131, *Grundl. d. math. Wiss., 3rd ed.* Springer, New York, 1966.

[HI4] F. Hirzebruch. Chern numbers of algebraic surfaces - an example. *Math. Ann.*, 266:351–356, 1984.

[H-Z1] F. Hirzebruch and D. Zagier. *The Atiyah-Singer theorem and elementary number theory.* Publish or Perish, Boston, 1974.

[H-Z2] F. Hirzebruch and D. Zagier. Intersection numbers of curves on Hilbert modular surfaces and modular forms of Nebentypus. *Inv. Math.*, 36:57–113, 1976.

[HOEF] T. Höfer. *Ballquotienten als verzweigte Überlagerungen der projektiven Ebene.* PhD thesis, Univ. Bonn, 1985.

[HOI] R.-P. Holzapfel. Arithmetische Kugelquotientenflächen I/II. *Sem. Ber. Math. Humb. Univ. Berlin*, 14, 1978.

[HOV] R.-P. Holzapfel. Arithmetische Kugelquotientenflächen V/VI. *Sem. Ber. Math. Humb. Univ. Berlin*, 21, 1979.

[HO1] R.-P. Holzapfel. A class of minimal surfaces in the unknown region of surface geography. *Math. Nachr.*, 98:211–233, 1980.

[HO2] R.-P. Holzapfel. Arithmetic curves on ball quotient surfaces. *Ann. Glob. Analysis and Geom.*, 1(2):21–90, 1983.

[HO3] R.-P. Holzapfel. Basic two-dimensional versions of Hurwitz' genus formula. *Ann. Glob. Anal. and Geom.*, 4(1):1–70, 1986.

[HO4] R.-P. Holzapfel. *Geometry and Arithmetic around Euler partial differential equations.* Dt. Verlag d. Wiss./ Reidel Publ. Comp., Berlin/ Dordrecht, 1986.

[HO5] R.-P. Holzapfel. Chern numbers of algebraic surfaces - Hirzebruch's examples are Picard modular surfaces. *Math. Nachr.*, 126:255–273, 1986.

[HO6] R.-P. Holzapfel. Discrete analysis of surfaces coverings I, II. *Rev. Roum. Math.*, 33:(3) 197–232, (4) 305–348, 1988.

[HO7] R.-P. Holzapfel. *Hierarchies of endomorphism algebras of abelian varieties corresponding to Picard modular surfaces*, Volume 190, Schriftenr. "Komplexe Mannigfaltigkeiten", Erlangen, 1994.

[HO8] R.-P. Holzapfel. *The ball and some Hilbert problems.* Lect. in Math. ETH Zürich. Birkhäuser, Basel-Boston-Berlin, 1995.

[HUR] Hurwitz. Über Riemannsche Flächen mit gegebenen Verzweigungspunkten. *Math. Ann.*, 39:1–61, 1891.

[I-S] H. Inose and T. Shioda. On singular K3 surfaces. *in: Complex analysis and algebraic geometry, coll. papers ded. to K. Kodaira, Iwanami Shoten, Cambridge Univ. Press*, pages 119–136, 1977.

[IVE] B. Iversen. Numerical invariants and multiple planes. *Amer. Journ. Math.*, 62:968–996, 1970.

[JAC] R. Jacobowitz. Hermitian forms over local fields. *Am. Journ. Math.*, 84:441–465, 1962.

[K-N] S. Kobayashi, K. Nomizu. Foundation of differential geometry. *Intersci. Tracts Pure and Appl. Math.* 15, vol. II, J. Wiley and Sons, 1969.

[KNE] M. Kneser. *Semisimple algebraic groups*, Ch. X in [C-F].

[KOB] R. Kobayashi. Einstein-Kähler metrics on open algebraic surfaces of general type. *Tohoku Math. Journ.*, 37:43–77, 1985.

[KOT] R.E. Kottwitz Tamagawa numbers. *Ann. Math.* 127:629–646, (1988).

[KNS] R. Kobayashi and F. Sakai S. Nakamura. A numerical characterization of ball quotients for normal surfaces with branch loci. *Proc. Jap. Acad.*, 65(Ser. A., No. 7):238–241, 1989.

[KUD] S. Kudla. Intersection numbers for quotients of the complex ball and Hilbert modular forms. *Invent. Math.*, 47:189–208, 1978.

[KUR] H. Kurke. *Vorlesungen über algebraische Flächen.* Teubner, Leipzig, 1982.

[LA1] S. Lang. *Introduction to algebraic and abelian functions.* Addison-Wesley Publ. Comp., Massachusetts, 1972.

[LA2] S. Lang. *Introduction to modular forms*, Volume 222, *Grundl. d. Math. Wiss.* Springer, Berlin-Heidelberg-New York, 1976.

[LA3] S. Lang. *Fundamentals of diophantine geometry.* Springer, New York-Berlin-Heidelberg, 1983.

[L-R] R.P. Langlands and D. Ramakrishnan, editors. *The Zeta Functions of Picard modular surfaces. Lectures delivered at a CRM Workshop in the spring of 1988.* Les Publications CRM, Montreal, 1992.

[LEH] J. Lehner. Discontinuous groups and automorphic functions. *Math. Surveys*, 18(Amer. Math. Soc., Providence, Rhode Island), 1964.

[LEV] R. Le Vavasseur. Sur le système d'équation aux dérivées partielles simultanées auxquelles satisfait la séries hypergéométriques à deux vairables. *Journ. Fac. Toulouse*, 7:1–205, 1896.

[MAL] S. Mac Lane. *Homology*, Volume 114, *Grundl. d. Math. Wiss.* Springer, Berlin-Göttingen-Heidelberg, 1963.

[MI1] Y. Miyaoka. On the Chern numbers of surfaces of general type. *Invent. Math.*, 42:225–237, 1977.

[MI2] Y. Miyaoka. On algebraic surfaces with positive index. Technical report, Preprint, Tokyo, 1985.

[M-T] B.G. Moishezon and M. Teicher. Simply-connected algebraic surfaces of positive index. *Invent. Math.*, 89:601–643, 1987.

[MU1] D. Mumford. The topology of normal singularities of an algebraic surface and a criterion for simplicity. *Publ. Math. I.H.E.S.*, 9:229–246, 1961.

[MU2] D. Mumford. *Geometric invariant theory.* Erg. Math. u. Grenzgeb. Springer, Berlin-Heidelberg-New York, 1965.

[MU3] D. Mumford. *Abelian varieties.* Tata lectures. Oxford Univ. Press, 1970.

[MU4] D. Mumford. Hirzebruch's proportionality theorem in the non-compact case. *Inv. Math.*, 42:239–272, 1977.

[MU5] D. Mumford. *Tata lectures on theta II.* Number 43 in Progr. in Math. Birkhäuser, 1984.

[MU6] D. Mumford. *The red book of varieties and schemes.* Number 1358 in Lecture Notes. Springer, 1994.

[MRT] D. Mumford and Y. Tai, M. Rapoport. *Smooth compactification of locally symmetric varieties.* Math. Sci. Press, Brooklin, Mass., 1975.

[NER] A. Neron. *Modèles minimaux des variétés abéliennes sur les corps locaux et globaux*, Volume 24, Publ. Math. I.H.E.S., 1965.

[OTR] G. Otremba. Zur Theorie der hermiteschen Formen in imaginärquadratischen Zahlkörpern. *Journ. f. reine u. angew. Math.*, 249:1–19, 1969.

[PAR] A.N. Parshin. On the use of branched coverings in the theory of diophantine equations (russian). *Math. Sbornik*, 180(2):244–259, 1989.

[P-Z] A.N. Parshin and Yu.G. Zarhin. *Finiteness problems in diophantine geometry (Russian)*, Appendix in [LA3]. Mir, Moscow, 1986.

[PE1] U. Persson. Chern invariants of surfaces of general type. *Comp. Math.*, 43:3–58, 1981.

[PE2] U. Persson. An introduction to the geography of surfaces of general type. *Algebraic Geometry I, Proc. of 1985 AMS Summer Inst. Bowdoin*, pages 195–218, 1987.

[PI1] E. Picard. Sur une extension aux fonctions de deux variables indépendentes analogues aux fonctions modulaires. *Acta mat.*, 2:114–135, 1883.

[PI2] E. Picard. Sur les fonctions hyperfuchsiennes provenant des séries hypergéométriques de duex variables. *Ann. École Norm. sup.*, 62(3-e ser.):357–384, 1885.

[PIN] H. Pinkham. Normal surface singularities with $\mathbb{C}^*$-action. *Math. Ann.*, 227:183–193, 1977.

[P-S] I.I. Pjatezkij-Shapiro. *Automorphic functions and the geometry of classical domaians.* Gordon & Breach, Science Publishers, New York-London-Paris, 1969.

[PLA] V.P. Platonov. The problem of strong approximation and the Kneser-Tits conjecture for algebraic groups. *Izv. Akad. Nauk SSSR*, Ser. Mat. 33(6), 1969.

[POC] L. Pochhammer. über hypergeometrische Funktionen höherer Ordnung. *Journ. Math*, 71:316–362, 1870.

[PRI] I.I. Priwalow. *Einführung in die Funktionentheorie II*, Volume 22, *Math. Nat. wiss. Bibl.* Teubner, Leipzig, 1959.

[RAG] M.S. Raghunatan. *Discrete subgroups of Lie groups*, Volume 68, *Erg. Math. u. Grenzgeb.* Springer, Berlin-Heidelberg-New York, 1972.

[RAP] M. Rapoport. Détermination du nombre de Tamagawa des groupes semi-simples quasi-déployés sur des corps de numbres. *Preprint*, 1975.

[SER] J.-P. Serre. *Course d'arithmétique.* Presses Univ. de France, Paris, 1970.

[SC1] I.R. Shafarevich et al. *Algebraic surfaces (Russian).* Number 75 of Trudi Mat. Inst. Steklov. Nauka, Moskva, 1965.

[SC2] I.R. Shafarevich. *Foundations of algebraic geometry (Russian).* Nauka, Moskva, 1972.

[SHA] P. Shanahan. *The Atiyah-Singer index theorem.* LNM 638. Springer, New York, 1978.

[SH1] G. Shimura. Arithmetic of unitary groups. *Ann. Math.*, 79(2):369–409, 1964.

[SH2] G. Shimura. On purely transcendental fields of automorphic functions of several variables. *Osaka Journ. Math.*, 1(1):1–14, 1964.

[SH3] G. Shimura. Arithmetic of unitary groups. *Ann. Math.*, 79:369–409, 1964.

[SH4] G. Shimura. *Introduction to the arithmetic theory of automorphic functions.* Iwanami Shoten Publ. & Princeton Univ. Press, 1971.

[SV1] O.W. Shvartsman. *On the factor space of an arithmetic group acting on the complex ball (Russian).* PhD thesis, Moscow, 1974.

[SV2] O.W. Shvartsman. On the factor space of one of the simplest arithmetic groups in the complex ball (russian). *Usp. Math. Nauk*, 28(4):202–203, 1974.

[SOU] C. Soulé. *Hermitian vector bundles on arithmetic varieties.* Lecture. AMS Summer Research Institute, Santa Cruz, 1995.

[VEN] A. van de Ven. On the Chern numbers of surfaces of general type. *Inv. Math.*, 36:285–293, 1976.

[VIN] E. Vinberg. The two most algebraic K3 surfaces. *Math. Ann.*, 265:1–21, 1983.

[WEL] R.O. Wells. *Differential analysis on complex manifolds.* Prentice-Hall, Englewood Cliffs, New Jersey, 1973.

[XIA] G. Xiao. *Surfaces fibrées en courbes de genre deux*, Volume 1137, *LNM.* Springer, New York-Heidelberg-Tokyo, 1985.

[YAU] S.-T. Yau. Calabi's conjecture and some new results in algebraic geometry. *Proc. Nat. Acad. Sci. USA*, 74:1798–1799, 1977.

[YO1] M. Yoshida. Volume formula for certain discrete reflection groups in $\mathbb{PU}(2,1)$. *Mem. Fac. Sc., Kyushu Univ.*, 36(Ser. A, No. 1):1–11, 1982.

[YO2] M. Yoshida. *Fuchsian differential equations*, Volume E11, *Asp. Math.* Vieweg, Braunschweig-Wiesbaden, 1987.

[YO3] M. Yoshida. *Hypergeometric Functions, My Love*, Vieweg, Braunschweig-Wiesbaden, 1997.

[ZIN] Th. Zink. Über die Spitzen einiger arithmetischer Untergruppen unitärer Gruppen. *Math. Nachr.*, 79:315–320, 1979.